International Conference of Solid State Science and Technology (ICSSST 2017) and Workshop on Advanced Materials Technology: Growth and Characterization

Enhancing Dimensional Network of Science and Technology

Journal of Physics: Conference Series
Volume 1083

AA006149

Bayan Lepas, Malaysia
13 - 16 November 2017

Editors:

Mohd. Zubir Mat Jafri
Shahrom Mahmud
Ahmad Fairuz Omar
Suhana Arshad

ISBN: 978-1-5108-7090-1
ISSN: 1742-6588

Printed from e-media with permission by:

Curran Associates, Inc.
57 Morehouse Lane
Red Hook, NY 12571

Some format issues inherent in the e-media version may also appear in this print version.

Copyright© (2017) by the Institute of Physics
All rights reserved. The material featured in this book is subject to
IOP copyright protection, unless otherwise indicated.

Printed by Curran Associates, Inc. (2018)

For permission requests, please contact the Institute of Physics
at the address below.

Institute of Physics
Dirac House, Temple Back
Bristol BS1 6BE UK

Phone: 44 1 17 929 7481
Fax: 44 1 17 920 0979

techtracking@iop.org

Additional copies of this publication are available from:

Curran Associates, Inc.
57 Morehouse Lane
Red Hook, NY 12571 USA
Phone: 845-758-0400
Fax: 845-758-2633
Email: curran@proceedings.com
Web: www.proceedings.com

Table of contents

Volume 1083

The International Conference of Solid State Science and Technology (ICSSST 2017)

13–16 November 2017, Olive Tree Hotel, Penang, Malaysia

Accepted papers received: 9 August 2018
Published online: 10 September 2018

Preface

Papers

Ceramics and Glasses

Electrochemical and Photonic Devices

Nanoscience and Nanotechnology

Other Applied Sciences

IOP Conf. Series: Journal of Physics: Conf. Series **1083** (2018) 011001 doi:10.1088/1742-6596/1083/1/011001

PREFACE

The International Conference of Solid State Science and Technology (ICSSST) is an annual event in Malaysia. In addition to this conference event, a workshop on Advanced Materials Technology: Growth and Characterization (AMTGC) has also been conducted. This 6th ICSSST and AMTGC workshop are jointly organized by the School of Physics and Institute of Nano Optoelectronics Research and Technology (INOR), Universiti Sains Malaysia as host, in cooperation with the Malaysian Solid State Science and Technology Society (MASS). The aim of this conference is to provide a platform for discussions on solid state science and technology in all aspects of physical and chemical properties, preparation and characterization techniques, optical properties; macro-, micro- and nano-structures. Furthermore, ICSSST 2017 which focuses on three key aspects of advanced technologies, namely nanoscience, nanotechnology and materials science, will provide the best platform for ideas and intellectual exchanges to catapult trans-disciplinary research, collaborations and networking among expert in the many areas of solid state studies. The four days conference and workshop were well attended by over 200 participants, comprising mostly of scholars and researchers of solid-state research field from countries such as the USA, UK, Singapore, Japan, Kazakhstan, Iran, Nigeria, Indonesia and Malaysia.

The ICSSST 2017 has covered recent developments in the following main topics:

- Advanced Material Synthesis and Crystal Growth Technology
- Biomaterial
- Carbon and Related Materials
- Ceramics and Glasses
- Devices and Materials for Biology and Medicine
- Electrochemical & Solid State
- Lasers and Optoelectronics
- Magnetic Materials
- Materials and Energy
- Metals and Alloys
- Nanoscience and Nanotechnology

Content from this work may be used under the terms of the Creative Commons Attribution 3.0 licence. Any further distribution of this work must maintain attribution to the author(s) and the title of the work, journal citation and DOI.
Published under licence by IOP Publishing Ltd

- Optical and Dielectric Materials
- Organic Materials and Applications
- Photonic Devices and Optoelectronic Integration
- Polymers and Composites
- Semiconductors and Devices
- Solid State Theory, Simulation and Computation
- Superconductors
- Thin Films and Nanostructures
- Crystallography
- Other Solid State Research

This volume of **Journal of Physics: Conference Series includes 64 papers** (Oral presenters and posters) devoted to various branches of solid state physics. We would like to thank the members of both the Organizing Committee and Reviewers for their patience and considerable effort to improve the manuscripts. Additionally, we deeply thank the authors for their high-level contributions and we are convinced that this volume will be of great interest for the international community of people working in solid state physics.

Thank you.

The Editors,
Prof Dr. Mohd. Zubir Mat Jafri
(*Universiti Sains Malaysia, USM*)
Assoc. Prof. Dr. Shahrom Mahmud
(*Universiti Sains Malaysia, USM*)
Dr. Ahmad Fairuz Omar
(*Universiti Sains Malaysia, USM*)
Dr. Suhana Arshad
(*Universiti Sains Malaysia, USM*)

IOP Conf. Series: Journal of Physics: Conf. Series **1083** (2018) 011001 doi:10.1088/1742-6596/1083/1/011001

SST2017 ORGANIZING COMMITTEE

Patron
Honourable Prof. Datuk Dr. Asma Ismail
Vice Chancellor
Universiti Sains Malaysia

Advisors
Prof. Dr. Azlan Abdul Aziz
Prof. Dr. Haslan Abu Hassan
Prof. Dr. Abdul Halim Shaari

Chairman
Prof. Dr. Abdul Razak Ibrahim

Deputy Chairman I
Prof. Dr. Mohd. Zubir Mat Jafri

Deputy Chairman II
Assoc. Prof. Dr. Ng Sha Shiong

Secretary I
Dr. Iskandar Shahrim Mustafa

Secretary II
Dr. Suhana Arshad

Treasurer I
Dr. Nik Noor Asyikin Nik Abd Razak

Treasurer II
Mrs. Siti Khadijah Mohd. Bakhori

Registration
Assoc. Prof. Dr. Yam Fong Kwong
Dr. Saiful Najmee Mohamad
Dr. Beh Khi Poay
Mr. Faris Hidayat Ahmad Suhaimi
Mr. Ng Yu Zhang

Multimedia, Publicity and Promotion
Dr. Nik Noor Asyikin Nik Abd Razak
Mrs. Ainul Huda Ayu Anuar Aziz
Ms. Nur Awanis Sirat
Ms. Nor Izzati Hashim

Publication and Scientific Materials
Prof. Dr. Mohd. Zubir Mat Jafri
Assoc. Prof. Dr. Shahrom Mahmud
Dr. Ahmad Fairuz Omar
Mr. Tan Chun Ho

Protocols and Management on Conference Tentative, Program & Abstract Book
Dr. Mohd Mahadi Halim
Dr. Mohd Marzaini Bin Mohd Rashid
Dr. Ng Sha Shiong
Dr. Iskandar Shahrim Mustafa
Dr. Nur Shafawati Rosli
Dr. Nuridayanti Che Khalib
Mrs. Junaidah Abu bakar
Mrs. Nordiyana Marzaki
Mrs. Norazlina Abd Malik
Mrs. Dian Alwani Zainuri
Ms. Siti Nur Shafiqa Abdul Rahim
Ms. Nurizati Rosli
Mr. Muhamad Akram Zaki Roslan

Workshop on nano-optoelectronics
Assoc. Prof. Dr. Ng Sha Shiong
Mr. Mohd Anas Ahmad

Exhibitions and sponsorship
Assoc. Prof. Dr. Wong Chow Jeng
Dr. Wan Maryam Binti Wan Ahmad Kamil
Ms. Nur 'Izazi Maad Tazri
Ms. Nur Fadzliana Ramli

Beverages, Banquet, Banquet Tentative
Dr. Azhar Abdul Rahman
Dr. Tan Kok Chooi
Dr. Suhana Arshad
Mr. Noor Aswafi Ahmad Zaini
Mr. Mohd. Rizal Mohamad Rodin

IOP Conf. Series: Journal of Physics: Conf. Series **1083** (2018) 011001 doi:10.1088/1742-6596/1083/1/011001

Mr. Hazhar Hassan
Mrs. Haslinda Musa

Technical and Logistic
Dr. Mohd Zamir Pakhuruddin
Mr. Norazam Ariffin
Mr. Mohd Noor Fakarruddin Abdullah
Mr. Mohamed Mustaqim Abu Bakar
Mr. Jasnisuhardi Abdul Manab
Mr. Muhammad Anis Ibnu Hajar

Mr. Shahil Ahmad Khosaini

Certificates, Gifts, Conference Kit
Dr. Nurul Zahirah Noor Azman
Dr. Ramzun Maizan Ramli
Ms. Munirah Jamil
Mr. Muhammad Hazritz Haslan
Ms. Fatin Izzati Hamarudin
Ms. Siti Zulaikha Mohd Sharifuddin Al-Fajri

IOP REVIEWERS

Prof. Dr. Mohd. Zubir Mat Jafri
Assoc. Prof. Dr. Shahrom Mahmud
Dr. Ahmad Fairuz Omar
Dr. Suhana Arshad
Prof. Dr. Abdul Razak Ibrahim
Assoc. Prof. Dr. Ng Sha Shiong
Dr. Iskandar Shahrim Mustafa
Dr. Nik Noor Asyikin Nik Abd Razak
Mrs. Siti Khadijah Mohd. Bakhori
Assoc. Prof. Yam Fong Kwong
Dr. Saiful Najmee Mohamad
Dr. Mohd Mahadi Halim
Dr. Mohd Marzaini Bin Mohd Rashid
Dr. Wan Maryam Binti Wan Ahmad Kamil
Dr. Tan Kok Chooi
Dr. Mohd Zamir Pakhuruddin
Dr. Nurul Zahirah Noor Azman
Dr. Ramzun Maizan Ramli
Assoc. Prof. Dr. Norzaini Zainal
Prof. Dato' Dr. Ahmad Shukri Mustapa Kamal
Dr. Nursakinah Suardi
Dr. Nordiana Mohd Muztaza
Dr. Naser Mahmoud Ahmed
Assoc. Prof. Dr. Quah Ching Kheng
Assoc. Prof. Dr. Yoon Tiem Leong
Dr. Beh Khi Poay

Photographs

Figure 1 : Registration

Figure 2 : The Registration Committee

IOP Conf. Series: Journal of Physics: Conf. Series **1083** (2018) 011001 doi:10.1088/1742-6596/1083/1/011001

Figure 3 : The Exhibition

Figure 4 : Keynote Session

IOP Conf. Series: Journal of Physics: Conf. Series **1083** (2018) 011001 doi:10.1088/1742-6596/1083/1/011001

Figure 5 : The Participants

Figure 6 : Tea-Break Session

IOP Conf. Series: Journal of Physics: Conf. Series **1083** (2018) 011001 doi:10.1088/1742-6596/1083/1/011001

Figure 7 : Group Photo

IOP Conf. Series: Journal of Physics: Conf. Series **1083** (2018) 011002 doi:10.1088/1742-6596/1083/1/011002

Peer review statement

All papers published in this volume of Journal of Physics: Conference Series have been peer reviewed through processes administered by the proceedings Editors. Reviews were conducted by expert referees to the professional and scientific standards expected of a proceedings journal published by IOP Publishing.

Content from this work may be used under the terms of the Creative Commons Attribution 3.0 licence. Any further distribution of this work must maintain attribution to the author(s) and the title of the work, journal citation and DOI.
Published under licence by IOP Publishing Ltd

IOP Conf. Series: Journal of Physics: Conf. Series **1083** (2018) 012001 doi:10.1088/1742-6596/1083/1/012001

Effect of Heat Treatment on Structural Properties of Silica Borotellurite Glasses Containing MnO

Zaitizila I. [1,a]**, Halimah M.K.** [1,b*] **, Muhammad F.D.** [1,c] **, Nurisya M.S.** [1,2,d]

Glass and Dielectric Lab, Department of Physics, Faculty of Science, University Putra Malaysia
[2]Lab. of Comp. Sciences & Mathematical Physics, Institute for Mathematical Research (INSPEM), University Putra Malaysia

[a]zaitizila@yahoo.com, [b]hmk6360@gmail.com, [c]farahdiana@upm.edu.my, [d]risya@upm.edu.my

Abstract. $\{[(TeO_2)_{0.7}(B_2O_3)_{0.3}]_{0.8}[SiO_2]_{0.2}\}_{1-x}\{MnO_2\}_x$ where x = 0.00, 0.01, 0.02, 0.03, 0.04 and 0.05 molar fraction were prepared by using melt quenching technique. Heat treatment of the glass sample was performed at 600 °C for 3 hours. The structure of the glass samples was characterized by using Fourier Transform Infrared (FTIR) and X-ray diffraction (XRD). Crystalline phase was recorded after parent glass went through heat treatment process. The FTIR spectra for parent glass and heat treated glass samples validate the presence of TeO_4, TeO_3, BO_4, BO_3 and Si-O(Si) in the glass network. After the samples were heat treated, vibration from manganese group were detected in FTIR spectra.

1. Introduction
Tellurite based glasses have attracted much attention from researches due to their high refractive index, low melting point, good infrared transmission, high dielectric constant, low phonon maxima and excellent transparency in the far infrared region [1-2]. Borate glasses have been known to have the nature of amorphous materials according to the specific structure and physical properties [3]. As reported earlier, the borate glasses can be used as inorganic hosts for transition metal and increase the glass forming ability [4-5]. The purpose of this research is to study the crystallization and structural changes in the glass network. For this purpose, XRD and FTIR were used as combined technique to investigate the changes that occur in the glass system before and after the heat treatment process. This research attempts to study the effect of heat treatment on FTIR and XRD of the glass system. There is also less research done on glass system doped with transition metal oxide to be applied as UV absorbance glass.

2. Experimental methods
Glasses of different compositions in the system $\{[(TeO_2)_{0.7}(B_2O_3)_{0.3}]_{0.8}[SiO_2]_{0.2}\}_{1-x}\{MnO_2\}_x$ where x = 0.00, 0.01, 0.02, 0.03, 0.04 and 0.05 molar fraction were fabricated via the melt quenching technique. The mixed chemicals were pre-heated in alumina crucible at 400°C for 1 hour and then melted at 1100°C for 3 hours in an electric furnace (Carbolite). The glass sample was quench into a stainless steel cylindrical shape mould before being transferred for annealing process at 400°C for 2 hours. The fabricated glass samples were heat-treated at 600 °C for 3 hours. The infrared spectra were measured using FTIR spectrometer (Perkin Elmer) in the region of between 280-4000 cm^{-1}. X-ray diffraction

Content from this work may be used under the terms of the Creative Commons Attribution 3.0 licence. Any further distribution of this work must maintain attribution to the author(s) and the title of the work, journal citation and DOI.
Published under licence by IOP Publishing Ltd

(XRD) studies were performed at room temperature on the as-quenched and heat-treated samples to confirm their amorphous and crystalline states.

3. Result and discussion

3.1 FTIR

The IR spectra for unheated and heat-treated glass samples are shown in Fig. 1 and Fig. 2 respectively. Based on Fig. 1, it can be seen that there are three obvious absorption bands for all glass samples. The bands are assigned to tellurite, borate and silica glass network. Two conventional regions for tellurite group can be observed, whereby the first region from 600 to 650 cm^{-1} exists due to vibrations of TeO_4 tetragonal pyramids [6]. On the other hand, the second region which is observed from 671 to 714 cm^{-1} occurs due to the stretching vibration of equatorial and axial Te-bonds in the TeO_4 and TeO_3 unit [7]. As for the borate groups, three bands are observed at 820 cm^{-1}, 1234cm^{-1} and 1380 cm^{-1} which are assigned to the B-O bond stretching of BO_4 units [8], BO_3 unit with non-bridging oxygen existing in boron-oxygen network and the stretching of B-O bonds of various borate arrangements [9-10]. The band at 1070 cm^{-1} is attributed to Si-O(Si) anti-symmetric stretching [11]. The bands around 3090-3120 cm^{-1} correspond to the hydroxyl groups and hydrogen bonds [12].

The heat-treated glasses show remarkable changes in FTIR spectra. There is addition of the bands in heat-treated glass. The new absorption bands that appear at 310-327 cm^{-1} and 514-540 cm^{-1} are assigned to vibration of Mn^{2+} and Mn^{3+} as well as Mn-O stretching mode respectively [13-14]. Moreover, the presence of small kinks at 470 cm^{-1} can be explained by the symmetric stretching mode of Te-O-Te chain unit [12].

In the FTIR spectra, it can be seen that the presence of additional bands and sharp peaks in heat-treated glasses when compared to room temperature glass. This might be due to strong vibrational frequency of the various groups present in glass network as the samples were heat-treated. Moreover, increasing the concentration of the dopant might have resulted in glass structure that is more open as the MnO breaks the network structure of the glass former and consequently produces NBO. As the amount of MnO in the glass system increases, the absorption bands of the FTIR spectra become deeper as can be clearly seen at 1070 cm^{-1} band. According to Central Force Model, the increasing intensity of Si-O(Si) absorption band may be caused by the structural changes and structural rearrangement that occur in the glass system [15-16]. In this regards, the detection of Mn and Te-O-Te functional group in the heat-treated samples might be the result of the rearrangement of the glass structure after the heat treatment process.

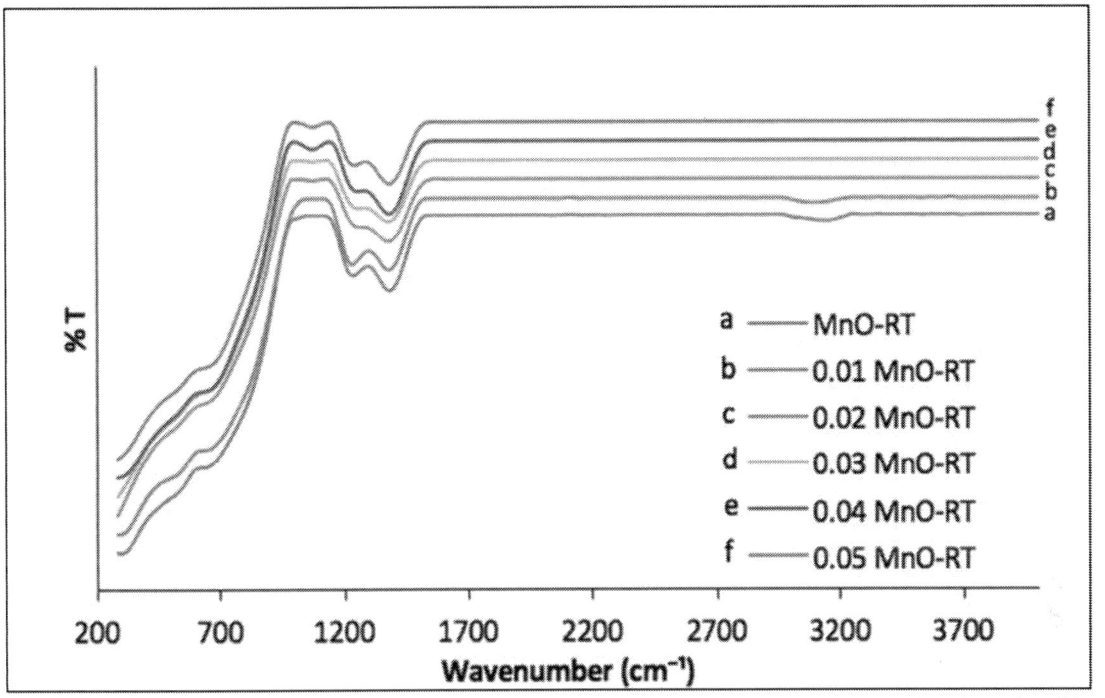

FIG. 1. FTIR spectra for glass samples at room temperature corresponding to: (a) undoped MnO_2, (b) 0.01 MnO_2, (c) 0.02 MnO_2, (d) 0.03 MnO_2, (e) 0.04 MnO_2, (f) 0.05 MnO_2.

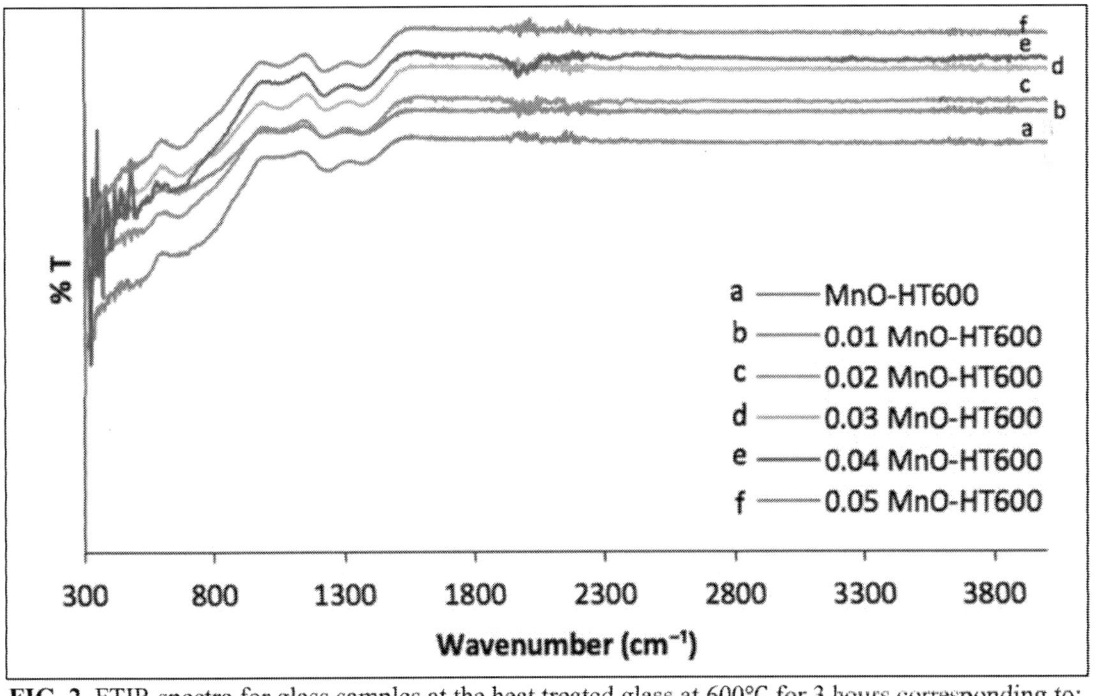

FIG. 2. FTIR spectra for glass samples at the heat treated glass at 600°C for 3 hours corresponding to: (a) undoped MnO_2, (b) 0.01 MnO_2, (c) 0.02 MnO_2, (d) 0.03 MnO_2, (e) 0.04 MnO_2, (f) 0.05 MnO_2.

3.2 XRD

Fig. 3 and Fig.4 show the XRD patterns for as-quenched and heat treated glass samples respectively. No discrete or continuous sharp peaks can be observed from the XRD spectra, as shown in Fig 3. Broad hump with the absence of sharp line spectra in the XRD pattern indicates the presence of long-range structural disorder and amorphous characteristic of the glass samples.

The glass samples tend to show crystalline peaks after being heated at 600 °C. However, only undoped glass sample and glass samples with 0.01 and 0.02 molar fraction show crystalline peaks. These crystalline peaks show a decreasing intensity as concentration of the dopant increases. Glass samples with undoped manganese oxide have three phases (**β**-Boron, SiO_2 and B_2O_3) which appear for 2**θ**. The crystalline size of the undoped sample is 88 nm. On the other hand, the heated-treated glass with 0.01 molar fraction of manganese only has one phase which is tellurium silicon ($Te_{16}Si_{38}$). The crystallization peaks appear at around 20.87°, 32.00°, 34.33° and 53,90°. The crystalize size for this sample has decreased to 37 nm.

From Fig. 4, it can be observed that the heated glass samples with 0.02 molar fraction of manganese is still in glassy nature with small crystalline traces of SiO_2. The crystalline size of this sample decreases to 24 nm with the addition of manganese oxide. Glass samples with 0.03 – 0.05 molar fraction of manganese oxide are also still in the glassy state. Nucleation temperature of the glass samples that lies within and near to the heat treatment temperature range as well as the addition of manganese that will reduce the ionic potential and decrease the amount of crystallization might have led to the glassy nature of the samples [3].

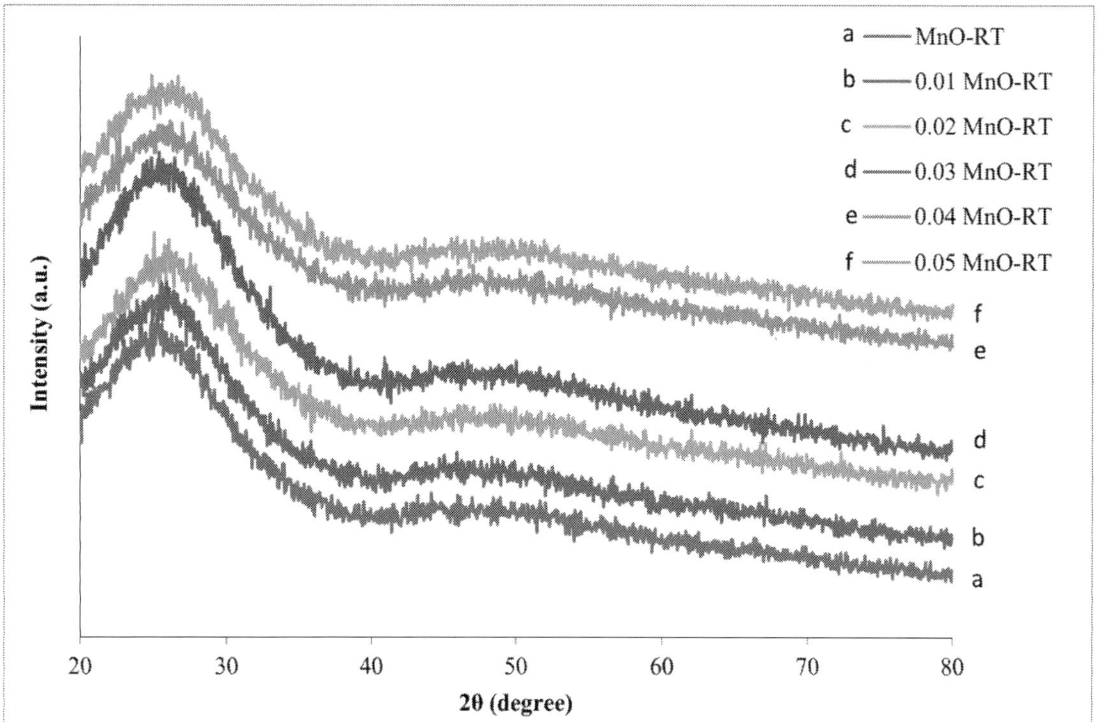

FIG. 3. The XRD for glass samples at room temperature corresponding to: (a) undoped MnO_2, (b) 0.01 MnO_2, (c) 0.02 MnO_2, (d) 0.03 MnO_2, (e) 0.04 MnO_2, (f) 0.05 MnO_2.

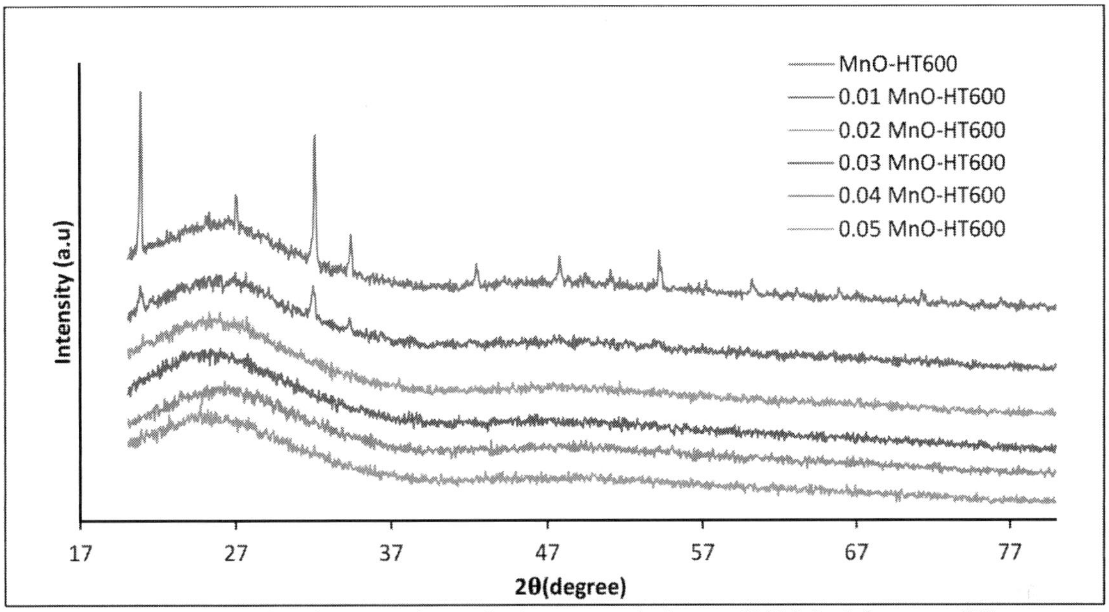

FIG. 4. The XRD for glass samples at the heat treated glass at 600°C for 3 hours at different concentration of MnO_2.

4. Conclusion

Silica boro tellurite glass doped with manganese has been successfully fabricated by using melt quenching method. The results from FTIR spectra show that the pristine glasses consist of BO_4, BO_3, TeO_4, TeO_3, Si-O(Si) functional group. After the heat treatment process, Mn^{2+}, Mn^{3+}, Mn-O as well as Te-O-Te functional groups were detected. The XRD spectra analysis reveals a crystallization peaks after the heat treatment process only for undoped sample and sample with 0.01 and 0.02 molar fraction of manganese oxide. On the other hand, the glass samples that did not undergo heat treatment process present an amorphous characteristic. The potential application of this material system is as a UV absorbance glass.

Acknowlegement

The authors appreciate the financial support for the work from University Putra Malaysia (UPM) through GP-IPS (grant number: 9504400).

References

[1] A.K. Yakhkind (1966). *Tellurite glasses.* Journal of the American Ceramic society, 49(12), 670-675.

[2] G.V. Prakash, D.N. Rao, A.K. Bhatnagar. *Linear optical properties of niobium-based tellurite glass.* Solid State Comm. 2001, 119, 39–44.

[3] I. Kashif, A.A. Soliman, E.M. Sakr, A. Ratep. (2013). *XRD and FTIR studies the effect of heat treatment and doping the transition metal oxide on $LiNbO_3$ and $LiNb_3O_8$ nano-crystallite phases in lithium borate glass system*, Spectrochimica Acta Part A: Molecular and Biomolecular Spectroscopy, 113, 15-21.

[4] A. Terczyńska-Madej, K. Cholewa-Kowalska, M. Łączka. (2011). *Coordination and valence state of transition metal ions in alkali-borate glasses.* Optical Materials, 33(12), 1984-1988.

[5] L. Hasnimulyati, M.K. Halimah, A. Zakaria, S.A. Halim, M. Ishak, *A comparative study of the experimental and the theoretical elastic data of Tm^{3+} doped zinc borotellurite glass*, Mater. Chem. Phys. 192 (2017) 228–234.

[6] H.A.A. Sidek, S. Rosmawati, Z.A. Talib, M.K. Halimah, W.M. Daud, *Synthesis and optical*

properties of ZnO-TeO_2 glass system, Am. J. Appl. Sci. 6 (2009) 1489.
[7] El-Mallawany, R.A., 2002. Tellurite Glasses Handbook, *Physical Properties and Data*. CRC Press, Baca Raton, FL., ISBN: 0849303680, pp: 540.
[8] B. Karthikeyan, C.S. Suchand Sandeep, J. Cha, H. Takebe, R. Philip, S. Mohan, *Optical properties and ultrafast optical nonlinearity of Yb^{3+} doped sodium borate and bismuthate glasses*, J. Appl. Phys. 103 (2008).
[9] G. Padmaja, P. Kistaiah, *Infrared and Raman Spectroscopic Studies on Alkali Borate Glasses* : Evidence of Mixed Alkali Effect, (2009) 2397–2404.
[10] V. Naresh, S. Buddhudu. *Structural, thermal, dielectric and ac conductivity properties of lithium fluoro-borate optical glasses*, Ceram. Int. 38 (2012) 2325–2332.
[11] M. Sitarz. *Influence of modifying cations on the structure and texture of silicate – phosphate glasses*, 887 (2008) 237–248.
[12] Z.A. Said Mahraz, M.R. Sahar, S.K. Ghoshal, M. Reza Dousti. *Concentration dependent luminescence quenching of Er^{3+}-doped zinc boro-tellurite glass*, J. Lumin. 144 (2013) 139–145.
[13] D. Moncke, E.I. Kamitsos, D. Palles, R. Limbach, A. Winterstein-Beckmann, T. Honma, Z. Yao, T. Rouxel, L. Wondraczek. *Transition and post-transition metal ions in borate glasses: Borate ligand speciation, cluster formation, and their effect on glass transition and mechanical properties*, J. Chem. Phys. 145 (2016).
[14] K.S. Pugazhvadivu, K. Ramachandran, K. Tamilarasan, *Synthesis and characterization of cobalt doped manganese oxide nanoparticles by chemical route*, Phys. Procedia. 49 (2013) 205–216.
[15] E.G. Parada, P. González, J. Pou, J. Serra, D. Fernández, B. León, M. Pérezamor, J. Pou, J. Serra, D. Ferna, *Aging of photochemical vapor deposited silicon oxide thin films Aging of photochemical vapor deposited silicon oxide thin films*, 436 (1996) 1–6.
[16] G. Kaur, M. Kumar, A. Arora, O.P. Pandey, K. Singh. *Influence of Y_2O_3 on structural and optical properties of SiO_2–BaO–ZnO–xB_2O_3–$(10-x)$ Y_2O_3 glasses and glass ceramics*, 357 (2011) 858–863.

IOP Conf. Series: Journal of Physics: Conf. Series **1083** (2018) 012002 doi:10.1088/1742-6596/1083/1/012002

Synthesis and Characterisation of AWO_4 (A = Mg, Zn) Tungstate Ceramics

Eza Nuraiena Sota[1, a *], Fadhlina Che Ros [1,b] and Jumiah Hassan[2,c]

[1]Department of Physics, Centre for Defence Foundation Studies,
Universiti Pertahanan Nasional Malaysia, Kem Perdana Sungai Besi, 57000 Kuala Lumpur, Malaysia
[2]Department of Physics, Faculty of Science, Universiti Putra Malaysia, 43400 Serdang Selangor, Malaysia

[a]ezanuraiena@gmail.com, [b]fadhlina@upnm.edu.my, [c]jumiah@upm.edu.my

Abstract. The AWO_4 (A = Mg, Zn) tungstate ceramics have been prepared using conventional solid-state route and characterised. Samples were analysed X-ray diffraction (XRD), scanning electron microscopy (SEM) and impedance spectroscopy (IS). Phase-pure $ZnWO_4$ and $MgWO_4$ were successfully obtained at temperatures 1000°C for 24 and 12 hours, respectively, where both compounds had a wolframite-like structure monoclinic, space group *P2*/c (13). Preliminary results of electrical properties of $ZnWO_4$ indicated that $ZnWO_4$ is a dielectric insulator with permittivity, $\varepsilon_r \sim 20$ at room temperature. The inhomogeneous electrical properties of $ZnWO_4$ is observed where the grain boundary and bulk contributions effects were seen at temperature excess of 100°C.

1. Introduction

In the past decade, microwave dielectric resonators and antennas have been developed for applications in communication systems such as cellular phone, broadcasting satellite and global positioning systems. Today, dielectric ceramic components need to be miniaturized, multilayered devices and packaged with conducting layers of minimal dielectric losses, in order to be reliable and competitive to meet the telecommunication and broadband industries' demand [1]. Therefore, three important properties such as high permittivity, ε_r, high quality factor, Q_f and near zero temperature coefficient of resonant frequency, TC_f are essential and should be precisely controlled in order for a material to be potentially used as dielectric resonator.

A relatively high dielectric constant enables the dielectric resonator to be smaller due to the physical size of the microwave device whereas the TC_f value should be less than 10 ppm/°C and temperature dependent, so that the signal would not drift in and out of resonance in any weather conditions. Q_f value which also associate with *tan* δ, should be high enough i.e. $Q > 30{,}000$ at 1 GHz so as to reduce the risk of cross-talk within a certain range of frequency [1,2].

Alkaline earth metal tungstates, AWO_4 compounds of scheelite and wolframite structures have drawn much interest in microwave applications due to their low permittivities (<12) but high Q_f values (> 50 000 GHz). For example, mineral scheelite, $CaWO_4$ is used for its optical properties (scintillation counters, lasers, optical fibres) and wolframite (Fe/Mn) WO_4 are commonly used as raw material for tungsten metal production [2]. Synthesis and the preparation of the materials are important and may

Content from this work may be used under the terms of the Creative Commons Attribution 3.0 licence. Any further distribution of this work must maintain attribution to the author(s) and the title of the work, journal citation and DOI.
Published under licence by IOP Publishing Ltd

contribute to the factors in accomplishing the desired properties of the samples. *Pullar* et al. reported that there was a small amount of unidentified second phase in $MgWO_4$ whilst for $ZnWO_4$, phase-pure sample has been achieved at 900°C. *Santucci* et al. [3] also reported that high purity of $ZnWO_4$ was achieved at 1000°C, 24 hours. The synthesis, phase analysis and structure of AWO4 (A = Mg, Zn) ceramics was reported in this paper, and the sample was compared with literatures. However, limited studies on the electrical properties of both compounds were reported; in this paper only $ZnWO_4$ sample is available to be reported in the electrical properties section. In this work, all samples were prepared using conventional solid-state technique using high purity oxides.

2. Methodology

WO_3 (dried at 100°C), ZnO and MgO (dried at 800° C) with 99.99, 99.9 and 99.95% purity respectively, were used as reagents. Appropriate amount of each reagent was mixed in an agate mortar and pestle with acetone to form a slurry and grind until dry. Pellets were prepared by uniaxially pressing powders using stainless steel die, placed on a bed of sacrificial powder of the same composition in an alumina crucible and reacted at temperatures in the range 850 – 1100°C for 3 hours in an ambient air with heating rate of 5°C/min. The processes of regrinding and reacting were repeated until samples were phase pure by XRD.

The crystalline phases of the sintered ceramics powders were analysed using X-ray powder diffraction, XRD (D8 Advanced) with Cu $K\alpha_1$ radiation at CRIM, UKM and the patterns were analysed using DIFFRAC EVA Software. The micrographs of the samples were obtained using ultra high-resolution scanning electron microscope, FESEM (FEI NOVA NANOSEM 230) with an energy dispersive X-Ray, EDX, at ITMA, UPM.

For electrical properties measurements, sintered pellets were polished and silver electrode was pasted onto the opposite surfaces before dried them for overnight. Impedance measurements were performed using Agilent 4294A Impedance Analyzer, UPM at various temperatures and the data were corrected for overall geometry.

3. Results and Analysis

3.1. Phase Analysis

Powder diffraction data for $ZnWO_4$ and $MgWO_4$ are shown in Figures 1 and 2, respectively. Peak positions of $ZnWO_4$ and $MgWO_4$ samples were indexed on a monoclinic unit cell, space group P2/c (13). Lattice parameters for $ZnWO_4$: a = 4.6935 Å, b = 5.7211 Å and c = 4.9300 Å and for $MgWO_4$: a = 4.6906 Å, b = 5.6782 Å and c = 4.9315 Å. These lattice parameters are comparable with the ICDD cards 01-078-4464 and 01-073-0562, respectively.

Both compounds, $ZnWO_4$ and $MgWO_4$ were initially calcined at 850°C and several impurities were identified. For $ZnWO_4$, un-reacted WO_3 was seen at ~ 26°, 27° and 28° whilst the un-reacted ZnO was detected at 31°, 47° and 48°. As the temperature gradually increased to 1000°C, the impurities were steadily disappeared therefore, single-phase $ZnWO_4$ is attained after firing at 1000°C, 24 h. As for $MgWO_4$, at temperature of 850°C, un-reacted peaks of WO_3 were identified at ~ 27°, 33° and 35°, but as the temperature gradually increased to 1000°C, the un-reacted peaks gradually decreased and a phase-pure $MgWO_4$ was achieved at 1000°C, 12 hours. However, according to literature [2], *Pullar* et. al reported a small trace of unidentified second phase at ~ 37.5° and therefore failed to achieve single-phase $MgWO_4$. Peak intensities for both compositions corresponded well with the ICDD cards, except for 010 and 020. Both peaks were relatively higher than that of the standard cards' intensity, which may be due to higher degree of crystallinity and increase in the crystallite size [5].

Energy dispersive X-ray analysis, EDX has been carried out for all samples. Figure 3 shows the presence of Zn, W and O elements in $ZnWO_4$ whilst in $MgWO_4$, the existence of Mg, W and O elements were observed, Figure 4 indicating the purity of the compounds. The analysis shows that the calculated ratio of Zn:W and Mg:W are both 50:50 (within the experimental accuracy) parallel in comparing to 50% and 50% of Zn, Mg and W in stoichiometric $ZnWO_4$ and $MgWO_4$. Subsection

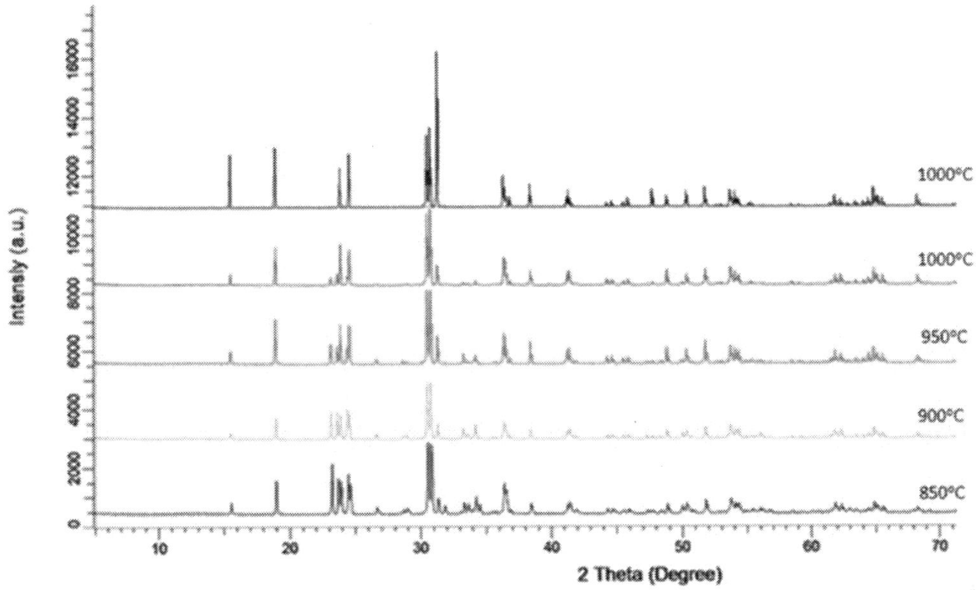

Figure 1: X-ray diffraction patterns for $ZnWO_4$ ceramic at 1000°C for 24 hours.

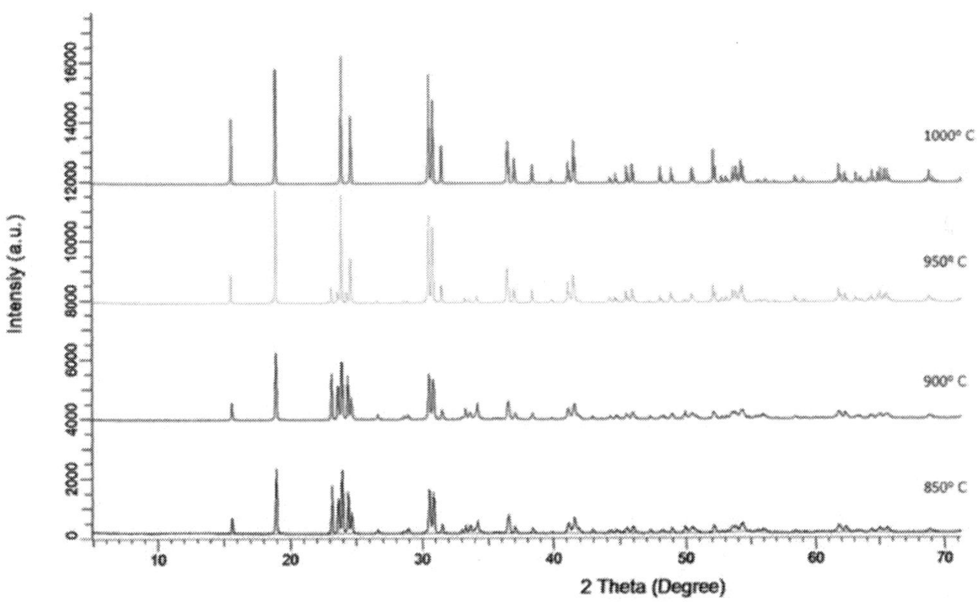

Figure 2: X-ray diffraction patterns of $MgWO_4$ ceramic at 1000°C for 12 hours.

IOP Conf. Series: Journal of Physics: Conf. Series **1083** (2018) 012002 doi:10.1088/1742-6596/1083/1/012002

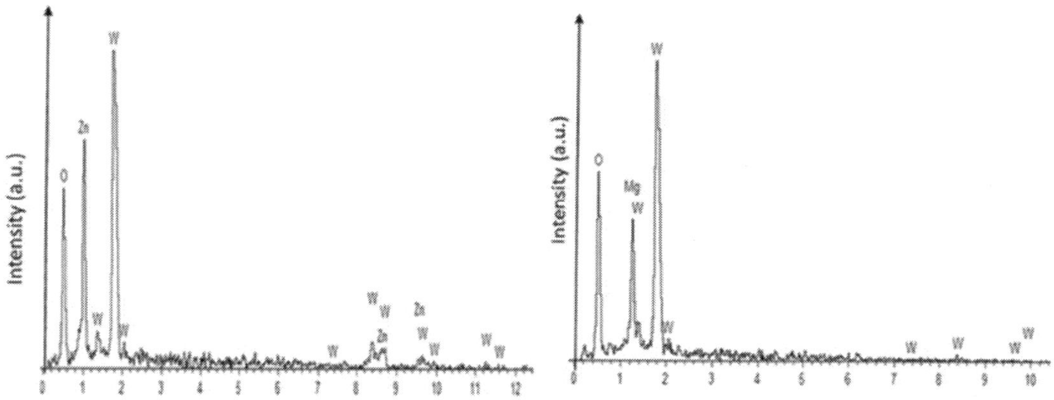

(a) (b)

Figure 3: (a) EDX spectrum for (a) $ZnWO_4$ sintered at 1000°C/24h and (b) $MgWO_4$ sintered at 1000°C/12h.

3.2. Scanning Electron Microscopy Analysis

Figure 4 shows the scanning electron micrographs of phase-pure $ZnWO_4$ and $MgWO_4$ samples, respectively. Sintering temperature effect of 1000°C, 24 hours on crystallite size of $ZnWO_4$ contributed to well crystallization of uniform rock-like structure depicted in Figure 4(a). For $MgWO_4$ sample, it shows well sintered at 1000°C, 12 hours despite its low density which can be observed from the microstructure shown in Figure 4(b). Literature reported that with further increment in firing time and temperature, a non-uniform of grain growth will be observed resulting to the existence of extremely large grain due to the discontinuous grain growth (DGG) [2]. The DGG phenomenon will resulted to a very poor Q-value where it may decline sharply even if the density and the permittivity, ε_r continue to rise. Therefore, synthesis of microwave materials should be carried out in care so that the materials' quality i.e. Q_f and TC_f will not be diminished by any parameters during the synthesis.

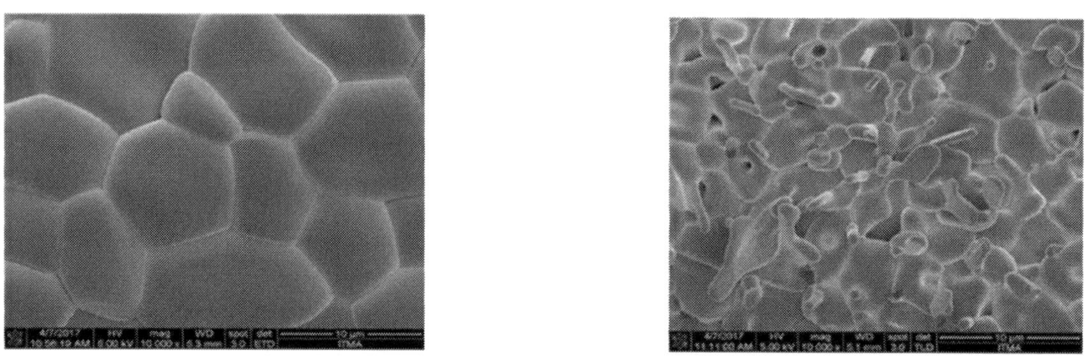

(a) (b)

Figure 4: FESEM micrograph of (a) $ZnWO_4$ sintered at 1000°C for 24 hours and (b) $MgWO_4$ sintered at 1000°C for 12 hours.

3.3. Electrical properties of $ZnWO_4$

Preliminary results of electrical properties of $ZnWO_4$ sample were carried out using impedance spectroscopy (IS), at different temperatures, in air. The electrical data were recorded over the frequency range 10 Hz to 1 MHz and analysed in conventional impedance format as a function of frequency at fixed temperature. The pellet density measured using Archimedes technique was ~ 91% and data were corrected for overall pellet geometry.

Figure 5(a) shows temperature-dependence of real part of capacitance, *C'*, vs frequency, *f* for $ZnWO_4$ at temperature range room temperature to 400°C. At 30°C, the capacitance data is independent of frequency and temperature that the capacitance value of plateau ~ 1.8 pF/m. As the temperature increased to 100°C, the data showed *C'* drop effects where two distinct capacitance plateaus were observed. The low-frequency dielectric response is perhaps dominated by grain boundary and the high-frequency response by the bulk contributions. As the temperatures raised, the time constant of the low-frequency plateaus shift to higher frequencies, and it is apparent that these plateaus values are independent of temperatures and frequencies.

The conductivity, σ' data as a function of frequency for selection of higher temperatures in the range of 250°C to 400°C for $ZnWO_4$ is also presented in Figure 5(b). At temperatures below 300°C, frequency-dependence of *ac* conductivities at high frequency are observed where the data sets do not show a frequency-independent conductivity plateau at lower frequencies. The *dc* conductivity could be observed and measured only at temperatures ≥ 300°C. It shows evidence of lower frequency *dc* plateau which changes to temperature- and frequency-dependent behaviour of power law response at higher frequencies.

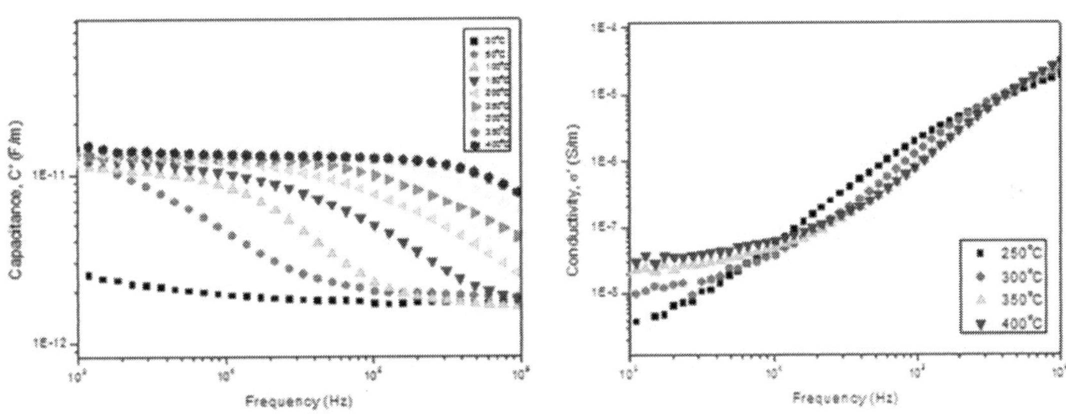

(a) (b)

Figure 5: (a) The frequency dependence of the capacitance, C' data of $ZnWO_4$ covering range of temperatures 30°C to 400°C and (b) conductivity, σ' vs frequency, *f* for 250° to 400°C.

4. Conclusion

Single-phase $MgWO_4$ and $ZnWO_4$ ceramics have been successfully prepared using solid state conventional technique. Both compounds were phase-pure at temperatures 1000°C for 24 hours and 1000°C for 12 hours, respectively; all samples show monoclinic unit cell, space group *P*2/c (13) structure, which are comparable to literatures. The purity of these compositions has been confirmed by X-ray diffraction and EDX analysis indicating the absence of impurity. Scanning electron microscopy pictures of $ZnWO_4$ composition sintered at 1000°C for 24 hours reveals the uniform rock-like structures, meanwhile for $MgWO_4$ composition sintered at 1000°C, 12 hours, a low densification of

$MgWO_4$ may be seen in the micrograph. IS response shows that $ZnWO_4$ exhibits low permittivity, $\varepsilon_r \sim$ 20 at room temperature; two distinct capacitance plateaus at low-frequency and high-frequency are observed, which may indicate grain boundary and bulk response. At temperatures below 300°C, frequency-dependence *ac* conductivities are only observed at high frequency and with further increment in temperature, *dc* conductivities appeared. Further investigations in electrical as well as microwave properties of $MgWO_4$, $ZnWO_4$-doped and $MgWO_4$-doped compositions will be carried out soon. Substitution of other element at A-site of AWO_4 is expected to enhance properties such as the permittivity, ε_r, quality factor, Q_f and the temperature coefficient of frequency, TC_f of the composition.

Acknowledgements
This research is supported by the Fundamental Research Grant Scheme (FRGS) grant no. FRGS/1/2017/STG07/UPNM/02/2 and Research Acculturation Collaborative Effort (RACE) grant no. RACE/F3/SG7/UPNM/8.

References
[1] Ian M. Reaney, Microwave dielectric ceramics for resonators and filters in mobile phone networks, *J. Am. Ceram. Soc.* 89 (7) 2006 2063 – 2072.
[2] R.C. Pullar, S. Farrah, N. McN Alford, $MgWO_4$, $ZnWO_4$, $NiWO_4$ and $CoWO_4$ microwave dielectric ceramics, *Journal of the European Ceramic Society* 27 2007 1059 – 1063.
[3] A.R. Phani, M. Passacantando, L. Lozzi, S. Santucci, Structural characterization of bulk ZnWO4 prepared by solid state method, *Journal of Material Science* 35 2000 4879-4883.
[4] Li X., Song Z. and Qu B., Shape-controlled electrochemical synthesis of $SrWO_4$ crystallites and their optical properties, *Short communication Ceramics Int.* 40 2014 1205-1208.
[5] Nidhi K., Sinha E., Rout S. K. And Kar M., Structural optical and microwave dielectric properties of $Sr_{1-x}Ca_xWO_4$ ceramics prepared by the solid state reaction route, *Ceramics Int.* 39 2013 9627-9635.
[6] Rainer Schmidt and Derek C. Sinclair, $CaCu_3Ti_4O_{12}$ (CCTO) ceramics for capacitor applications, Capacitors: *Theory of Operation, Behaviour and Safety Regulations*, Nova Science Publishers, Inc., 2013. pages
[7] E. S. Kim, S. H. Kim, B. I. Lee, Low-temperature sintering and microwave dielectric properties of $CaWO_4$ ceramics for LTCC applications, *Journal of the European Ceramics Society* 26 2006 2101-2104.
[8] L. Cheng, P. Liu, S.Qu, H. Zhang, Microwave dielectric properties of AWO_4 (A = Ca, Ba, Sr) ceramics synthesized via high energy ball milling method, *Journal of Alloys and Compounds* 581 2013 553-557.
[9] L. Li, H. Cai, X. Yu, Q. Liao, Z. Gao, Structure analysis and microwave dielectric properties of $Ca_xZn_{1-x}Sn_{0.08}Ti_{1.92}Nb_2O_{10}$ ceramics, *Journal of Alloys and Compounds* 584 2014 315-321.
[10] S. W. Xia, Y. G. Zhang, L. W. Shi, M. M. Zhang, Enhanced microwave dielectric properties of $ZnTa_2O_6$ ceramics with Sb^{5+} ion substitution, *Material Letters* 124 2014 64-66.

IOP Conf. Series: Journal of Physics: Conf. Series **1083** (2018) 012003 doi:10.1088/1742-6596/1083/1/012003

Structural and Magnetic Properties of Eu^{3+} Doped Borotellurite Glass Containing Manganese Nanoparticles

Siti Maisarah Aziz[1,a], M.R.Sahar[2, b*] E.S.Sazali[3,]

[1,2]Advanced Optical Material Research Group, Department of Physics, Faculty Science, Universiti Teknologi Malaysia,81310, Skudai, Johor Bahru, Malaysia.

[a]sitimaisarahaziz@yahoo.com, *[b]mrahim057@gmail.com_ezzashuhada@utm.my

Abstract. We report the structural and magnetic properties of $30B_2O_3$-(59-x)TeO_2-10MgO-1Eu_2O_3-$x$$Mn_3O_4$ glass prepared via melt quenching method. The glass samples are characterized by X-ray Diffraction (XRD), Raman Spectroscopy and Vibrating Sample Magnetometer (VSM). The existence of broad hump in X-ray diffraction (XRD) pattern verifies the amorphous nature of glasses. Raman spectra displays a shift in the vibrational modes of TeO_4 and TeO_3 units and thereby indicate an alteration in the glass network structure due to the incorporation of Mn_3O_4 NPs. The influence of varying Mn_3O_4 nanoparticles (NPs) concentrations on the magnetic were determined. The glass saturation magnetization and susceptibility were founded to be in the range of $(5.45\times10^{-2} \pm 0.27 - 8.49\times10^{-2} \pm 0.42)$ $emug^{-1}$ and $(4.54 \times10^{-6} \pm 0.23 - 7.08 \times10^{-6} \pm 0.35)$ $emuOe^{-1}g^{-1}$ respectively. The obtained hysteresis curve indicates that the glass samples are paramagnetic materials which may be useful for developing new magneto-optical devices.

1. Introduction

Spectroscopic properties of borotellurite (BT) glasses have been widely investigated for their scientific interest and practical applications [1, 2]. BT glasses have promising optical materials due to their high refractive index, low phonon energy and higher transparency in the infrared spectrum [3, 4]. Luminescence properties of BT glass is one of the important characteristic that can be used as a strong indicator to hunt for the new functional material. Incorporation of nanoparticles in BT glass shows remarkable changes in optical properties of lanthanides [5]. Among nanoparticles, magnetic nanoparticles are remarkable which is useful for magneto-optic devices. Several researchers have widely reported about manganese nanoparticles (Mn_3O_4 NPs) due to their extraordinary properties such as large specific surface area, low dimensionality and quantum confinement effect compared with bulk materials [6, 7]. Other than that, it exhibits high saturation field, high field irreversibility and extra anisotropy contributions [8]. However, not many efforts were dedicated towards the incorporation of Mn_3O_4 NPs in this glass system. In this paper, a new series of Mn_3O_4 NPs embedded BT glass doped trivalent europium have been prepared and their structural and magnetic studies are performed and reported.

Content from this work may be used under the terms of the Creative Commons Attribution 3.0 licence. Any further distribution of this work must maintain attribution to the author(s) and the title of the work, journal citation and DOI.
Published under licence by IOP Publishing Ltd

2. Experimental

Raw materials of Tellurium dioxide (TeO_2) (99.9%), Boric acid (H_3BO_3) (99.0%), Magnesium oxide (MgO) (99.0%), Europium oxide (Eu_2O_3)(99.0%) and Manganese oxide (Mn_3O_4)(99.0%) are weighed and mixed according to the nominal composition of $(59\text{-}x)TeO_2\text{-}30B_2O_3\text{-}10MgO\text{-}1Eu_2O_3\text{-}xMn_3O_4$ (where x = 0.5, 1.0, 1.5 mol %). The batch is placed in an platinum crucible for 1 h in the furnace of 900 °C before it is cooled down to room temperature. The amorphousness of the obtained sample is identified by X-ray Diffraction pattern using the Siemens Diffractometer D5000. Diffraction pattern was collected in the range of diffraction angle (2θ) from 10° to 90°. The bonding vibrational modes were obtained using Raman Spectroscopy in the range of 200 to 1600 cm^{-1}, performed on a Horiba Jobin Yvon (Model HR800 UV). Magnetic properties such saturation magnetization (M_s), coercive force (H_c), remanence (M_r) and squareness (M_r/M_s) of the glasses were determined using a Vibrating Sample Magnetometer (VSM) (Model: Lake Shore's new 7400 series) for low moment measurement capabilities with field up to 12 kOe. The magnetic susceptibility of the glass system were calculated by [9],

$$M = \chi_m H \tag{1}$$

where χ_m is magnetic susceptibility, M is a magnetization and H is a magnetic field. All measurements were carried out at room temperature. The batch compositions of the prepared glasses are as follows:

BTMEMn0.5 : $58.5TeO_2\text{-}30B_2O_3\text{-}10MgO\text{-}1Eu_2O_3\text{-}0.5Mn_3O_4$
BTMEMn1.0 : $58TeO_2\text{-}30B_2O_3\text{-}10MgO\text{-}1Eu_2O_3\text{-}1.0Mn_3O_4$
BTMEMn1.5 : $57.5TeO_2\text{-}30B_2O_3\text{-}10MgO\text{-}1Eu_2O_3\text{-}1.5Mn_3O_4$

3. Results and discussions

Fig. 1 shows the XRD patterns of the synthesized glass sample. The absence of any sharp peak in the XRD patterns verifies their amorphous nature. A broad hump around 25- 35° and 50° is related to diffuse scattering that indicates the presence of short-range atomic arrangement in the matrix [10].

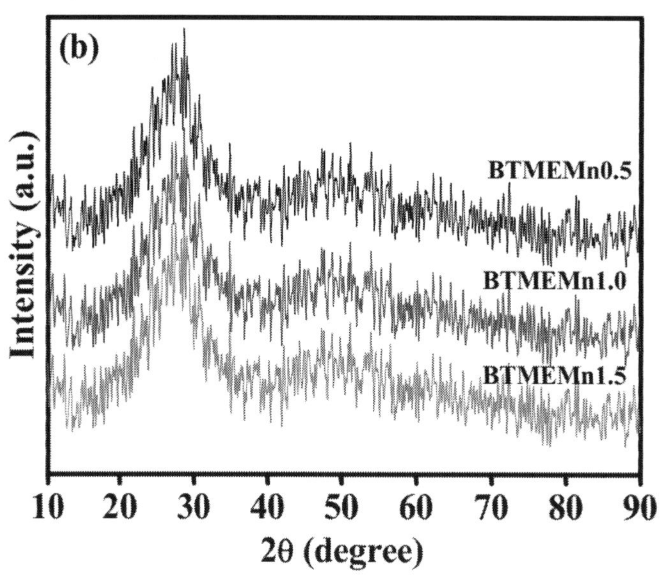

Fig. 1: XRD patterns of glass system

Fig. 2 shows the occurrence of prominent bands in the Raman spectra. It can be seen that there is a presence of various structural units in the glass system. The Raman band near 450-485 cm^{-1} indicates the characteristic of the bending and stretching vibrations mode of the Te-O-Te or O-Te-O linkages [11]. The appearance of the broadband around 665-682 cm^{-1} and 750-778 cm^{-1} is corresponding to the presence of stretching vibration mode between tellurium atom and bridging oxygen in the TeO_4 tbp units [3] and vibration of the TeO_3 tp units with non-bridging oxygen (NBO), respectively [12].

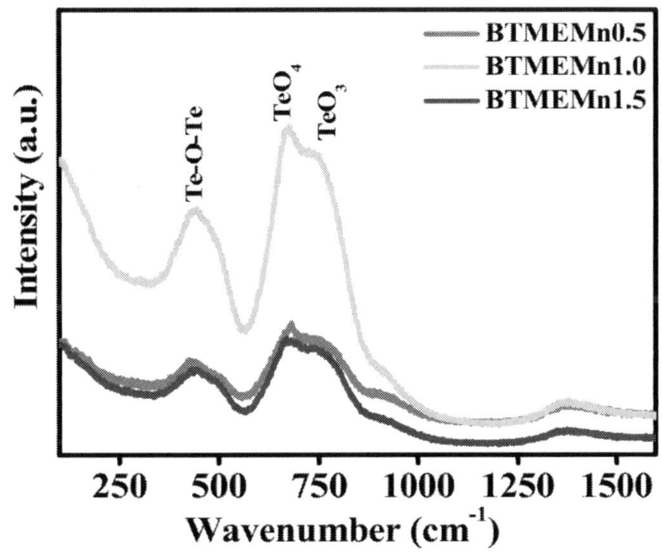

Fig.2. Raman spectra of the prepared glass system

From Fig.2, it is noticed that the peak at 665-682 cm^{-1} is referred to the TeO_4 tbp group in the present glass. It is observed that, as the amount of Mn_3O_4 NPs is increased, the peak of TeO_4 tbp is displaced from 665 cm^{-1} toward a higher wavenumber and reaches 682 cm^{-1} at 1.0 mol % of Mn_3O_4 NPs. This is attributed to the formation of more TeO_3 units at the expense of TeO_4 units [48]. Formation of large number of Te-O bonds in TeO_3 units has strengthened the glass network. However, as the amount of Mn_3O_4 NPs is beyond 1.0 mol%, the vibration peaks slightly shifted toward a lower wavenumber. This shift might indicate structural alteration. The magnetization versus external magnetic field (M-H) curves was recorded at room temperature for variation concentration of Mn_3O_4 NPs is shown in Fig. 3.

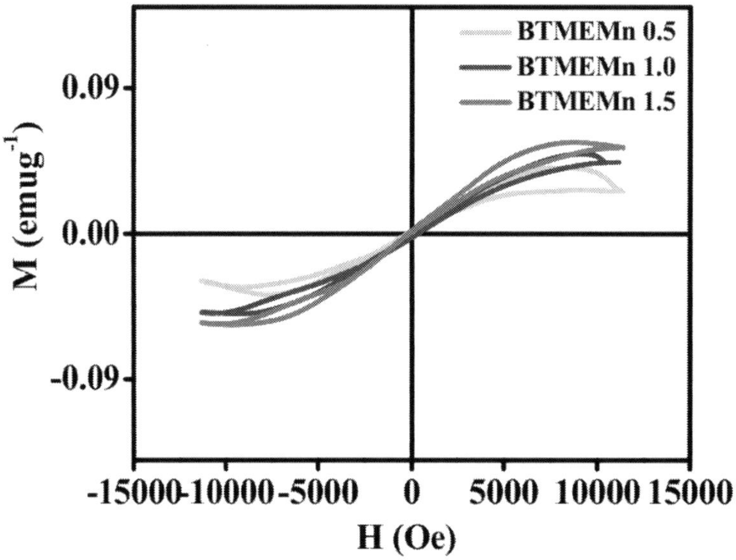

Fig.3. The magnetization against external magnetic field of prepared glass samples

Fig.3 shows the magnetization of glass sample increases with increasing of Mn_3O_4 NPs concentration. This is as a result of more nanoparticles moment are able to align toward applied field [9]. The increment of magnetization is also associated to the effect of interaction among the magnetic ions of manganese in the glass matrix. This is ascribed to the difference in the surrounding of magnetic atoms which is contributed by the spin configuration and surface area of magnetic particles that has been reported elsewhere [13, 14]. The value of saturation magnetization (M_s), remanent magnetization (M_r), coercivity (H_c) and susceptibility (χ_m) of the prepared NPs are enlisted in Table 1.

Table 1: Values of M_s, M_r, H_c, and χ_m of prepared BT glasses

Sample	Magnetic parameters			
	M_s (emug^{-1}) ×10^{-2}	M_r (emug^{-1}) ×10^{-3}	H_c (Oe)	χ_m (emuOe^{-1}g^{-1}) ×10^{-6}
BTMEMn0.5	5.45 ± 0.27	0.97 ± 0.04	162.01 ± 8.10	4.54 ± 0.23
BTMEMn1.0	8.11 ± 0.40	4.06 ± 0.20	1424.42 ± 71.22	6.76 ± 0.34
BTMEMn1.5	8.49 ± 0.42	2.30 ± 0.11	81.61 ± 4.08	7.08 ± 0.35

From Table 1, it is observed that the remanent magnetization increases by addition of Mn_3O_4 NPs. This presumably is due to the single domain possessing a grain structure with a domain specific atomic field direction. This atomic direction remains unaffected by the influence of external magnetic field which contributes the rotation domain and produce large energy which enhance the magnetization [15]. The calculated value of H_c is ranged between 81.61 to 1424.42 Oe which is slightly increased as the Mn_3O_4 NPs is increased up to 1.0 mol %. Furthermore, the H_c start to decrease as the Mn_3O_4 NPs increase. This variation is due to the increment in the surface anisotropy and spin glass-like state [9]. It is also observed that the susceptibility increased as the Mn_3O_4 NPs increased. This is due to the single ion anisotropy of magnetic ions of Eu^{3+} is affected the Mn_3O_4 NPs local field environment where the main contribution is from the changes of electron spin. Hence, the presence of the Mn_3O_4 NPs in BTMEMn glass with positive susceptibility extremely indicates that all the glass samples behave in paramagnetic nature.

4. Conclusion

The structural and magnetic properties of Eu^{3+}:Mn doped BT glass has successfully been studied and has been prepared by melt quenching technique. The amorphous nature of glasses are confirmed by XRD. The Raman spectra were strongly influenced by the variation of Mn_3O_4 NPs concentration. These NPs revealed paramagnetic hysteresis behaviour.

Acknowledgement

The authors gratefully acknowledge the financial support from RMC, UTM through the research grant (VOTE: 4F752, 4L657, 16H41 and 13J81) and thank Faculty of Science UTM for providing the measurement facilities.

Reference

[1] M. N. Azlan, M. K. Halimah, S. Z. Shafinas, W. M. Daud, Electronic polarizability of zinc borotellurite glass system containing erbium nanoparticles, Mater Express 5 (2015) 211-218.

[2] S. P. Hashim, H. A. Sidek, M. K. Halimah, K. A. Matori, W. M. Yusof, M. H. Zaid, The effect of remelting on the physical properties of borotellurite glass doped with manganese, Int J Mol Sci 14 (2013) 1022-30.

[3] A. Muhammad Noorazlan, H. Mohamed Kamari, S. S. Zulkefly, D. W. Mohamad, Effect of Erbium Nanoparticles on Optical Properties of Zinc Borotellurite Glass System, J Nanomater 2013 (2013) 1-8.

[4] M. Halimah, W. Daud, H. Sidek, A. Zainal, A. Zainul, H. Jumiah, Structural analysis of borotellurite glass, Am. J. Appl. Sci. 5 (2007) 323-327.

[5] Z. A. S. Mahraz, M. Sahar, S. Ghoshal, Reduction of non-radiative decay rates in boro-tellurite glass via silver nanoparticles assisted surface plasmon impingement: Judd Ofelt analysis, J Lumin (2017)

[6] A. M. Abdel Hakeem, Room-temperature ferromagnetism in $Zn1{-}xMnxO$, J Magn Magn Mater 322 (2010) 709-714.

[7] I. Ardelean, N. Mureşan, P. Păşcuţă, EPR and magnetic susceptibility studies of manganese ions in $70TeO_2{\cdot}25B_2O_3{\cdot}5SrO$ glass matrix, Mater Chem and Phys 101 (2007) 177-181.

[8] M. Augustin, D. Fenske, I. Bardenhagen, A. Westphal, M. Knipper, T. Plaggenborg, Manganese oxide phases and morphologies: A study on calcination temperature and atmospheric dependence, Beilstein J Nanotechnol 6 (2015) 47-59.

[9] S. A. M. Azmi, M. R. Sahar, Optical response and magnetic characteristic of samarium doped zinc phosphate glasses containing nickel nanoparticles, J Magn Magn Mater 393 (2015) 341-346.

[10] Z. A. Said Mahraz, M. R. Sahar, S. K. Ghoshal, M. Reza Dousti, Concentration dependent luminescence quenching of Er^{3+}-doped zinc boro-tellurite glass, J Lumin 144 (2013) 139-145.

[11] K. Maheshvaran, P. K. Veeran, K. Marimuthu, Structural and optical studies on Eu^{3+} doped boro-tellurite glasses, Solid State Sci. 17 (2013) 54-62.

[12] N. Kaur, A. Khanna, Structural characterization of borotellurite and alumino-borotellurite glasses, J Non-Cryst. Solids 404 (2014) 116-123.

[13] A. Dehelean, E. Culea, Magnetic behaviour of europium ions in some tellurite glasses obtained by the sol-gel method, Journal of Physics: Conference Series 182 (2009) 012064.

[14] I. Edelman, O. Ivanova, I. Ivantsov, D. Velikanov, E. Petrakovskaja, A. Artemenko, Magnetic properties and morphology of manganese ferrite nanoparticles in glasses, IOP Conference Series: Materials Science and Engineering 25 (2011) 012017.

[15] P. Anigrahawati, M. R. Sahar, S. K. Ghoshal, Influence of Fe3O4 nanoparticles on structural, optical and magnetic properties of erbium doped zinc phosphate glass, Mater Chem Phys 155 (2015) 155-161.

IOP Conf. Series: Journal of Physics: Conf. Series **1083** (2018) 012004 doi:10.1088/1742-6596/1083/1/012004

The Physical and Optical Studies of Erbium Doped Borosilicate Glass

Nur Ain Nabilah Razali[1,a], Iskandar Shahrim Mustafa[1,2,b], Nurul Zahirah Noor Azman[1,2], Halimah Mohamed Kamari[3], Amalina Abd Rahman[1], Khairiah Rosli[1], Noor Syamimi Taib[1], Nur Adilah Tajuddin[1]

[1]School of Physics, Universiti Sains Malaysia Main Campus, 11800 USM, Penang, Malaysia.
[2]Institute of Nano-optoelectronic, Universiti Sains Malaysia, Bukit Jambul, 11900 Bayan Lepas, Penang, Malaysia.
[3]Department of Physics, Faculty of Science, Universiti Putra Malaysia, 43400 Serdang, Selangor, Malaysia.

[a]nurainnabilah_razali@yahoo.com, [b]iskandarshah@usm.my

Abstract. A glass series of erbium doped zinc borosilicate glass system was prepared by using the melt-quenching method. The absorption spectra revealed several bands at visible range which correspond to the following transitions (from the ground state); $^4F_{5/2}$ + $^4F_{7/2}$ + $^2H_{11/2}$ + $^4S_{3/2}$ + $^4F_{9/2}$ + $^4I_{9/2}$. The density and molar volume obtained increased and decreased respectively as the concentration of erbium oxide varied from 0 mol % to 1.0 mol %. Meanwhile, oxygen packing density (OPD) increased as the concentration of erbium oxide increased up to 3.0 mol % due to the build-up of bridging oxygen. The indirect band gap energy was achieved from 2.56 eV to 2.80 eV and the Urbach energy values lies between 0.15 eV and 0.61 eV.

1. Introduction

Borosilicate glass is a glass which the main components are silicon dioxide and boron oxide. Borosilicate glass has been produced in a wide range of compositions. The batch materials can be divided into three categories based on their role in the process: glass former, glass intermediate and glass modifier. The most important of any glass batch is always the glass former. Each glass has one or more components which act as the primary source of the structure. These components are commonly designated as glass formers and also called as network formers. These components are the basis for the generic name used for the glass. In this project, both boron oxide and silicon dioxide (SiO_2) act as network former and namely the glass samples are known as borosilicate glasses [1]. Zinc oxide operates as a stabilizer substance in the glass network of ternary and quaternary glass. With the addition of zinc oxide, it will raise the surface tension of the molten and recover the crystallization stability within the substance of the existing section. The rare-earth elements act as glass intermediate. The rare-earth elements (REE) consist of La, Ce, Pr, Nd, Pm, Sm, Eu, Gd, Tb, Dy, Ho, Er, Tm and Yb. The rare earths (RE) are highly valued for their unique properties especially as optically active elements in their ionized state for lasers. When incorporated in crystalline or amorphous hosts, the rare earths exist as ion^{+3} [2]. The main objectives of the present study are to identify the effect of erbium concentration on physical and linear optical properties of borosilicate glass system, and to explore the possibility the use of Er^{3+} doped

Content from this work may be used under the terms of the Creative Commons Attribution 3.0 licence. Any further distribution of this work must maintain attribution to the author(s) and the title of the work, journal citation and DOI.
Published under licence by IOP Publishing Ltd

borosilicate glass towards nano-optoelectronic devices. The schematic study on the physical density and absorption erbium doped borosilicate glass had been done. The nonlinear refractive index was analyzed to determine the possibility of the glass materials for nano-optoelectronic field.

2. Materials and Method

2.1 Sample Preparation

Melt quenching technique was used to produce multi-composition glass samples [(x % Er_2O_3) - (50 - x % ZnO) - (20 % B_2O_3) - (30 % SiO_2)] where x= 0, 0.5, 1.0, 2.0, 3.0 and 4.0 mol %. The high purity of chemical powder (99.9% purity grade) of erbium (III) oxide, Er_2O_3 (REacton, Alfa Aesar), zinc oxide, ZnO (Assay, Alfa Aesar) and boron oxide, B_2O_3 (Assay, Alfa Aesar) were used for sample preparation. The network former silicon dioxide SiO_2 99.37 %, which was obtained from rice husk ash (RHA) is used during the glass fabrication [3]. The raw materials raw materials (7 g) were thoroughly mixed in an alumina crucible and melted at 1300 °C for 2 h in an electric furnace. The melts were quenched on a 250 °C preheated stainless steel mould at in an electric furnace. The stainless-steel mould was left to cool down towards the room temperature. The obtained samples were then polished carefully in order to achieve parallel surface for physical and optical properties measurement.

2.2 Physical Properties Measurement

The density of the glass sample was measured based on Archimedes principle by using distilled water as immersion liquid. Density, molar volume and oxygen packing density were calculated by using equation 1, 2 and 3, respectively.

$$\rho_s = \rho_{dw} \left(\frac{w_{air}}{w_{air} - w_{dw}}\right) \tag{1}$$

Where *M*=Molecular weight, ρ=density and n= number of oxygen atoms per each composition [4].

$$V_m = M/\rho \tag{2}$$

$$\text{Oxygen packing density} = (1000 \times \rho \times n)/M \tag{3}$$

2.3 Optical Properties Measurement

UV-Visible 1800 Shimadzu Spectrophotometer was used to perform optical absorbance measurement of glass structure at room temperature. The spectrometer was set up at absorption mode. The range of wavelength used was between 200 nm to 1000 nm and the absorbance range was from 0 A to 4 A.

3. Results and Discussions

3.1 Physical Properties

The range of thickness for the glass samples is between 4.51 mm to 5.55 mm. The variation of density, molar volume and oxygen packing density (OPD) values as function of Er_2O_3 contents are listed in Table 1. The density is found to be increased from 3.46g/cm³ to 3.95 g/cm³ with the increment percentage mole of Er_2O_3. The density depends on the multi-composition of the glass network. Eventually, higher structure compactness will produce higher density of the glass. The increment in density indicates that the number of oxygen packing density (OPD) in the glass network increase and decrease in non-bridging oxygen. The increment in molar volume is due to the augmentation in the structure bonding length or the increment in inter-atomic spacing between atoms [5].

3.2 Optical Properties

The absorption spectra of the prepared glass samples consist of 7 fundamental absorption edge, which located at 409, 452, 490, 527, 656, 808 and 980 nm wavelengths as shown in Figure 1. The absorption spectra consist of 7 transitions originating from the ground state $^4I_{15/2}$ to the excited states $^4F_{5/2}$ + $^4F_{7/2}$ +

$^4H_{11/2}$ + $^4S_{3/2}$ + $^4F_{9/2}$ + $^4I_{9/2}$. The fundamental absorption edges of the spectra were slightly shifted towards the higher wavelength with the addition of erbium oxide. The shifted of the absorption edge were attributed to structural rearrangements of the glass [6].

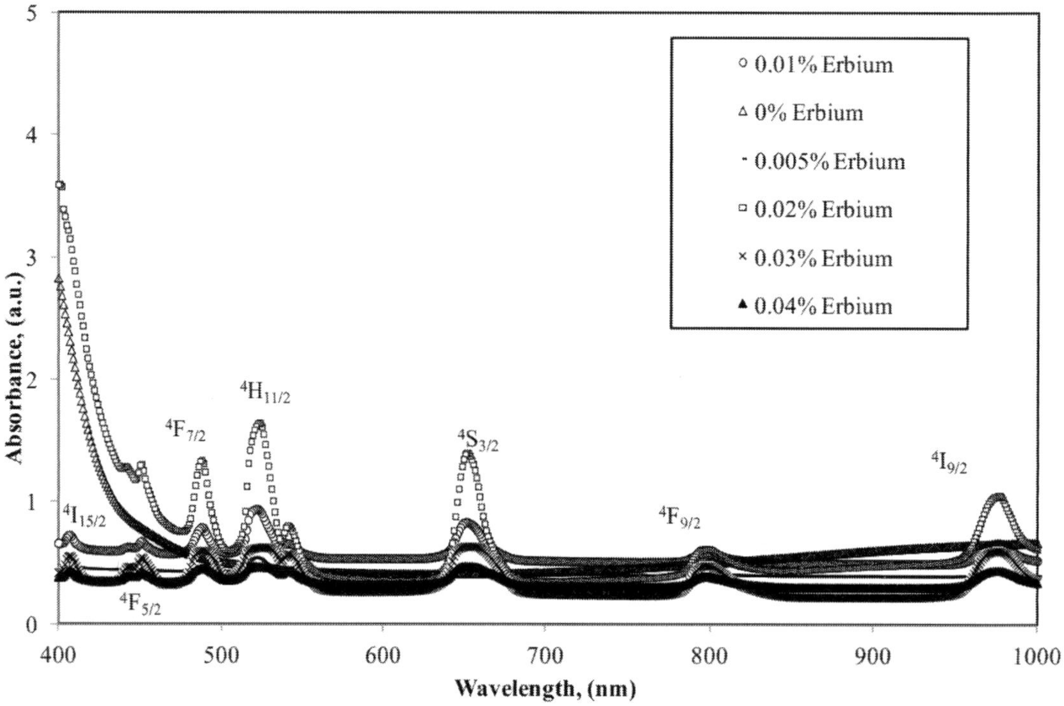

Figure 1: The absorbance of [(x % Er_2O_3) - (50 - x % ZnO) - (20 % B_2O_3) - (30 % SiO_2)] glass structure.

The absorption coefficient $\propto(\omega)$ can be shown as a function of photon energy (for the direct and indirect band gap energy:

$$\propto(\omega) = B\frac{(\hbar\omega - E_{opt})^n}{\hbar\omega} \quad (4)$$

Where n is the index number, B is constant, E_{opt} is optical band gap energy and is incident photon energy [7]. The Tauc's plot can be strained between $(\propto\hbar\omega)^{1/n}$ and ($\hbar\omega$) by substituting n= ½ for direct and n=2 for indirect as illustrated in Figure 2. Whereas, the Urbach energy was calculated using the gradient obtained from Figure 3 based on the given expression:

$$\alpha(\omega) = B\exp(\hbar\omega / \Delta E) \quad (5)$$

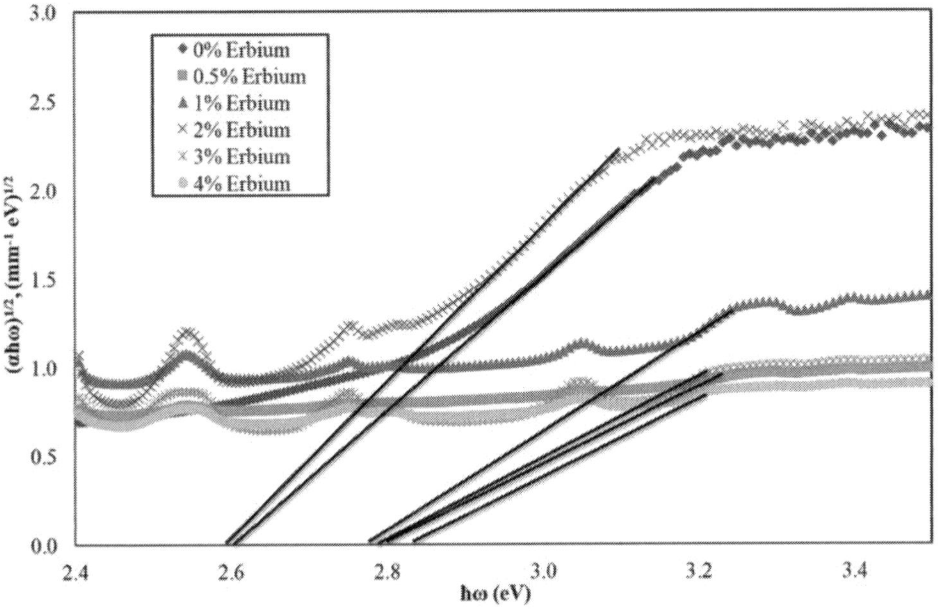

Figure 2: The extrapolation of indirect band gap energy.

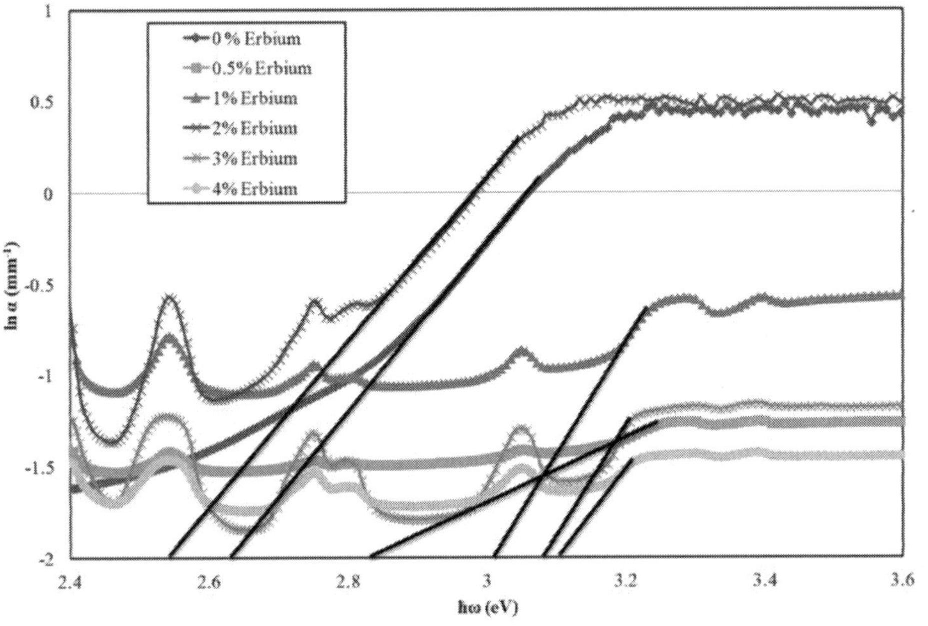

Figure 3: Extrapolating the straight line of Urbach energy

The increment of indirect band gap energy due to the decrement in non-bridging oxygen (NBO) is illustrated in Table 1. Higher Urbach energy indicates greater tendency to convert weak bonds into defects [8].

Table 1: Physical properties and Indirect band gap energy, Urbach energy, refractive index, molar refraction and polarizability of [(x % Er_2O_3) - (50 - x % ZnO) - (20 % B_2O_3) - (30 % SiO_2)] glass structure

Mole percentage of Er_2O_3 (%)	Indirect band gap energy (eV)	Urbach energy (eV)	Refractive index,n	Molar refraction, Rm (cm^3)	Polarizability, α_m (10^{-24})	Density (g/cm^3)	Molar Volume (cm^3)	OPD (g-atom/l)
0	2.56	0.22	2.53	7.09	2.81	3.46 $\pm$0.005	21.00	80.97
0.5	2.72	0.61	2.48	6.87	2.72	3.57 $\pm$0.008	20.80	82.23
1.0	2.74	0.21	2.47	6.77	2.68	3.67 $\pm$0.017	20.59	83.52
2.0	2.56	0.23	2.53	7.03	2.79	3.78 $\pm$0.003	20.82	83.57
3.0	2.76	0.17	2.46	6.87	2.72	3.89 $\pm$0.009	20.99	83.85
4.0	2.80	0.35	2.45	6.89	2.76	3.95 $\pm$0.016	21.44	83.03

Refractive index parameter depends on density and polarizability of the glass materials [9]. Refractive index values were calculated by using the following equation.

$$\frac{n^2-1}{n^2+2} = 1 - \sqrt{\frac{E_{opt}}{20}} \tag{6}$$

Where n= refractive index, E_{opt} = optical band gap energy [10].

The increment in refractive index at 2.0 mol % is because of the addition rare earth ions to the network which leads to the formation of dense packing of rare earth modifiers into the host materials. The higher amounts of dense packing network signify the higher value of refractive index of the glass. The electronic polarizability, α_m is proportional to the molar refraction, by introducing the Avogadro's number through the following equation:

$$R_m = (4\pi\alpha_m N_A)/3 \tag{7}$$

$$R_m = \frac{n-1}{n+2}\mathbf{Vm} \tag{8}$$

Where R_m= molar refraction, V_m= molar volume. This relationship gives the average value for molar refraction for isotropic substance such as glass [11].The polarizability decreased with an increased concentration of erbium oxide as illustrated in Table 1. The increment in polarizability at 2.0 mol % and 4.0 mol % are because of high polarity possession which tends to form NBO. Each of Er^{3+} will break the bridging oxygen (BO) and lead to formation of NBO atoms. Nevertheless, the lower formation of NBO in the glass network, the lower polarizability of the glass obtained.

4. Conclusion

The melt-quenching method has been used to fabricate erbium doped zinc borosilicate glass system. Several absorption bands are observed in UV–vis absorption spectra for all the glass samples. The absorption bands of erbium and erbium nanoparticles doped zinc borotellurite glasses are located at 409, 452, 490, 527, 656, 808, and 980 nm of wavelengths. The observed absorption bands are due to the 4f–4f transition in erbium and zinc ions in the glass network. The obtained values of physical and linear optical properties of erbium doped zinc borosilicate glass system may contribute towards the development of solid state devices, optical displays and optical memory devices.

Acknowledgment
The author acknowledges Univerisiti Sains Malaysia for the financial support for this research under two USM (Short Term Grant) which are 304 / PFIZIK / 6313152 and 304 / PFIZIK / 6313249. Other financial supports grant is from postgraduate USM that is 308/AIFIZIK/415403.

References

[1] Shelby J.E. 2005 Introduction to Glass Science and Technology. 2nd ed. The Royal Chemistry.
[2] Walsh B.M. 2006 Advances in Spectroscopy for Lasers and Sensing, Springer.
[3] Mustafa I.S, Razali N.A.N , Ibrahim A.R and Kamari H.M 2015 From Rice Husk to Transparent Radiation Protection Material, *Jurnal Intelek,* 9, pp. 1–6.
[4] Althaf M., and Chaudhr A.M 2010 Physical Properties of Lithium Containing Cadmium Phosphate Glasses. *Journal Modern Physics*, 1, pp.201-05.
[5] Halimah M.K, Daud W., Sidek H., Zaidan H., and Zainal A. 2010 Optical properties of ternary tellurite glasses. *Materials Science-Poland* 28(1), pp. 173-180.
[6] Said Mahraz Z.A, Sahar M.R, and Ghoshal S.K, 2014 Band gap and polarizability of boro-tellurite glass: Influence of erbium ions. *Journal of Molecular Structure*, 1072(1), pp. 238–241.
[7] Sushama D. and Predeep P. 2014 Thermal and Optical Studies of Rare Earth Doped Tungston-Tellurite Glass. *International Journal of Applied Physics and Mathematics*, 4(2), pp.139-43.
[8] Ashur Z., Mahraz S., Sahar M.R, Ghoshal S.K, and Dousti M.R 2013 Concentration dependent luminescence quenching of Er^{3+}-doped zinc. *Journal of Luminescence*, 144, pp. 139–145.
[9] Azlan M.N, Halimah M.K, Siti S.Z and Daud W.M 2013 Effect of Erbium Nanoparticles on Optical Properties of Zinc Borotellurite Glass System. *Journal of Nanomaterials*, pp.168.
[10] Said Mahraz Z.A, Sahar M.R, Ghoshal S.K, and Reza Dousti M., M 2013 Concentration Dependent Luminescence Quenching of Er^{3+}- Doped Zinc Boro Tellurite Glass. *Journal of Luminescence*, 144, pp.139-45.
[11] Dimitrov V. and T. Komatsu 1999 Electronic Polarizability, Optical Basicity and Non-Linear Optical Properties of Oxide Glasses. *Journal of Non Crystalline Solids*, 249, pp.160-79.

Physical and Shielding Protection Parameterization of PbO-ZnO-B_2O_3-SiO_2 Glass Network

Nur Ain Nabilah Razali[1, a], Iskandar Shahrim Mustafa[1,2,b], Nurul Zahirah Noor Azman[1,2], Halimah Mohamed Kamari[3], Intan Sofea Mat Hashim[1], Nurulshuhada Ahmad[1], Nur Afini Norazam[1], Che Amira Shahira Che Asudin[1]

[1]School of Physics, Universiti Sains Malaysia Main Campus, 11800 USM, Penang, Malaysia.
[2]Institute of Nano-optoelectronic, Universiti Sains Malaysia, Bukit Jambul, 11900 Bayan Lepas, Penang, Malaysia.
[3]Department of Physics, Faculty of Science, Universiti Putra Malaysia, 43400 Serdang, Selangor, Malaysia.

[a]nurainnabilah_razali@yahoo.com, [b]iskandarshah@usm.my

Abstract. Glass series of quaternary lead-zinc-borosilicate glass was fabricated using melt-quenching method. It was successfully prepared using silicon dioxide (SiO_2 99.37 %) as the network former, which was obtains from rice husk ash (RHA). The thickness of the prepared glass samples used during the measurement was 0.3, 0.4 and 0.5 cm. As the percentage of PbO increased from 5% to 40%, the density and molar volume of glass sample increased from 3.719 g/cm^3 to 5.010 g/cm^3 and 21.44 cm^3 to 25.82 cm^3, respectively. The significant modification was due to augmentation in bond length and increment in inter-atomic spacing between atoms with the presence of Pb, Zn and B atoms. Nevertheless, the oxygen packing density decreased from 79.29 g-atom/l to 65.84 g-atom/l. The mass attenuation coefficient was found to be increased for 0.3 cm, 0.4 cm and 0.5 cm. The escalating mass attenuation coefficient was due to the increment of PbO percentage mole, which related to the emergent photoelectric absorption of the glass sample. Thus, it also affects the decreasing trend of Half Value Layer (HVL) and the increasing of Effective atomic number (Z_{eff}).

1. Introduction

The extensive usage of ionizing radiation in various applications motivated nuclear engineers, radiologists and radiation physicists to edge on appropriate designs for radiation shielding. In designing radiation shields for X-rays and gamma rays, some relevant parameters such as mass attenuation coefficient, buildup factor, mean free path and atomic effective number Z_{eff} are calculated to estimate the radiation shielding properties [1]. Furthermore, mass attenuation coefficient and Z_{eff} is a very functional parameter especially involving multi-element materials such as compounds, mixtures and composites. Different concretes are commonly used as shielding materials towards nuclear radiation exposure due to cost effectiveness. However, the variation of moisture in concrete due to temperature gradient adds uncertainty to the calculations of attenuation coefficient. There are a great deal of literature studying towards X-ray and gamma photon shielding in concretes and glass systems [2]. Previous researchers investigated different types of glasses as a new shielding material and reported valuable

Content from this work may be used under the terms of the Creative Commons Attribution 3.0 licence. Any further distribution of this work must maintain attribution to the author(s) and the title of the work, journal citation and DOI.
Published under licence by IOP Publishing Ltd

data. Recently, there are numerous literatures reported the study on X-ray and gamma-ray shielding properties for altered glasses such as aluminosilicate glass with PbO and Bi_2O_3 [3]. Lead (Pb)-incorporated glasses are commercially available for nuclear radiation shielding applications. Lead glasses have many extraordinary properties such as excellent infrared transmission, large density, high nonlinear optical susceptibility and high refractive index. Boron oxide (B_2O_3) is one of the commonly used glass formers or intermediate with a lower price (cost-effectiveness) with fine optical transparency and demonstrates hefty glass forming tendency, good thermal stability and moderate rare-earth (RE) ion solubility [4]. Meanwhile, zinc oxide (ZnO) plays a twofold function both as network former (at higher concentration) and glass modifier (at a lower concentration) and imparts extended UV optical transparency, lower melting temperatures and enhanced glass forming region in oxide glasses. In the present work, we have investigated physical properties and radiation shielding properties such as mass attenuation coefficient and Z_{eff} for several multi-composition silicate glasses.

2. Materials and Method

2.1 Sample Preparation

Melt quenching technique was used to produce multi-composition glass samples [x% PbO - (50-x) % ZnO - 20% B_2O_3 - 30% SiO_2] where x= 0, 5, 10, 20, 30 and 40 mole %. The high purity of chemical powder (99.9% purity grade) of lead (II) oxide, PbO (REacton, Alfa Aesar), ZnO (Assay, Alfa Aesar) and B_2O_3 (Assay, Alfa Aesar) were used for sample preparation. The network former silicon dioxide SiO_2 99.37 %, which was obtained from rice husk ash (RHA) was used during the glass fabrication [3]. The raw materials raw materials (7 g) were thoroughly mixed in an alumina crucible and melted at 1100 °C for 2 h in an electric furnace. The melts were quenched on a 250 °C preheated stainless steel mould at in an electric furnace. The stainless-steel mould was left to cool down to room temperature. The obtained samples were then polished carefully to achieve parallel surface for physical and radiation shielding properties measurement. The thickness used for the mass attenuation measurement was 0.3 cm, 0.4 cm and 0.5 cm.

2.2 Physical Properties Measurement

The density of the glass sample was measured based on Archimedes principle by using distilled water as immersion liquid. Density, molar volume and oxygen packing density were calculated by using equation 1, 2 and 3, respectively.

$$\rho_s = \rho_{dw}\left(\frac{w_{air}}{w_{air} - w_{dw}}\right) \quad (1)$$

Where *M*=Molecular weight, ρ=density and n= number of oxygen atoms per each composition [5].

$$V_m = M/\rho \quad (2)$$

$$\text{Oxygen packing density} = (1000 \times \rho \times n)/M \quad (3)$$

2.3 Gamma Attenuation Measurement

Gamma-ray spectrometry system with NaI(Tl) detector high efficiency was used with present study. The linear and mass attenuation coefficients were measured using gamma-ray spectrometry based on NaI (Tl) scintillation detector with energies of 59.54 keV from americium-241 radiation source. The gamma rays attenuation were measured on a scintillator detector using the 3″×3″ NaI(Tl) detector having an energy resolution of 7.5% at 662 keV (ORTEC 905-4 series). The sample was positioned in front of the source at distance of 13.5 cm as illustrated in Figure 1. The optimum distances between the source and the detector were 20 cm. The spectra were recorded using a MAESTRO®-32 PC-based multi-channel analyzer (MCA).

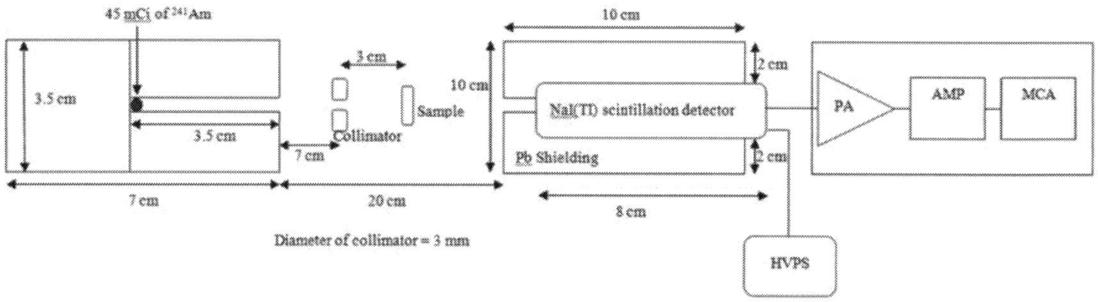

Figure 1: Radiation attenuation parameterization measurement setup using ^{241}Am as source of gamma ray interactions at energy 59.54 keV.

3. Results and Discussions

3.1 Physical Properties

The glass produced was yellowish in colour due to the properties of PbO. The density and molar volume was calculated by using Equation 1 and 2. The increase in density and molar volume as the percentage of PbO increase was shown in Table 1 due to the molecular weight of PbO was higher than ZnO-SiO_2-B2O_3. The increase in molar volume was due to the augmentation in bond length or an increase in interatomic spacing between atoms [6]. The decrement of oxygen packing density by the increment of the percentage of PbO was illustrated in Table 1 due to the increase in molecular weight of the composition of PbO in the glass sample. But, the oxygen packing density of 0% of PbO at 80.8129128 g-atom/l was less than the oxygen packing density of 5 % of PbO at 81.9620271 g-atom/l because of the PbO act as the network intermediate, where Pb^{2+} has strongly interacts with O2^-.

Table 1 : The physical properties of (x) PbO - (0.5-x) ZnO – (0.2) B_2O_3 - (0.3) SiO_2 glass structure

% mole of PbO	Density (g/cm^3)	Molar Volume (cm^3)	Oxygen Packing Density (g-atom/l)
0	3.45 ± 0.007	21.04	80.81
5	3.84 ± 0.010	20.74	81.96
10	4.07± 0.003	21.33	79.71
20	4.56± 0.015	22.17	76.68
30	4.89± 0.009	23.54	72.22
40	5.18± 0.018	24.99	68.02

3.2 Gamma Attenuation Analysis

The linear attenuation coefficient (μ) was calculated by using Eq.4 below:

$$N = N_o e^{-\mu x} \qquad (4)$$

where N refer to the number of transmitted photon, N_0 is the number without an absorber and x is the thickness of the glass sample. In this research, the thicknesses of the glass samples were from 3mm to 5mm. The mass attenuation coefficients can be calculated this Eq.5:

$$\mu_m = {}^{\mu}/_{\rho} \qquad (5)$$

Where the μ_m refers to mass attenuation coefficients, the μ refer to linear attenuation and ρ refer to density of glass sample [7]. Half-value layer (HVL) is the thickness of material that decreases radiation intensity to 50% of the initial value. The HVL is computed accordingly using expression [8]:

$$HVL = {0.693}/{\mu} \quad (6)$$

The linear and mass attenuation coefficients (μ) as tabulated in Table 2 were increased as the PbO mole percentage increased for thickness 0.3, 0.4 and 5 cm due to the elevated value of Pb atom that has higher atomic number as compared to other atom. The linear and mass attenuation coefficients were increased with the emergent photoelectric absorption of the glass sample [9]. The mass attenuation increased when the PbO mole percentage and thickness increased. The photoelectric absorption increased with the increment in atomic number of the glass system. The Compton scattering and photoelectric absorption was found as the main interaction process that could observe small difference in the PbO concentration in the glass matrix. The Compton scattering is found to be decreased as the percentage of PbO increased. The values of photoelectric absorptions were found to be affectd to the photon at the lower photon energy [10]. At lower photon energy, the photoelectric absorptions were clearly increased with the increment of the PbO concentration, in a similar manner as that of the total mass attenuation coefficients. Thus, the photoelectric absorption interactions increase with the increase in atomic number of glass system. The higher the mass attenuation coefficient will leads to high shielding protection of the glass sample. This decreased in HVL values as illustrated in Table 2 can be conclude that the glass sample can be act as shielding protection from gamma rays [11]. It is due to the increase in photoelectric absorption that had been discussed above.

Table 2 : Linear attenuation , mass attenuation coefficient and half value layer of The physical properties of (x) PbO - (0.5-x) ZnO – (0.2) B_2O_3 - (0.3) SiO_2 glass structure

PbO % mole	**Linear attenuation coefficient , μ (cm^{-1})**			**Mass Attenuation Coefficient , μ_m (cm^{-2})**			**Half Value layer, HVL (cm)**		
	Thickness (cm)								
	0.3	0.4	0.5	0.3	0.4	0.5	0.3	0.4	0.5
0	1.79	3.91	5.20	0.52	1.13	1.50	0.60	0.18	0.13
5	5.34	8.60	8.63	1.39	2.24	2.25	0.13	0.08	0.08
10	11.52	9.66	10.83	2.83	2.37	2.66	0.06	0.07	0.06
20	12.26	12.15	13.64	2.69	2.67	2.99	0.05	0.06	0.05
30	17.50	13.91	18.35	3.58	2.84	3.75	0.04	0.05	0.04
40	18.42	19.90	19.08	3.56	3.84	3.69	0.03	0.03	0.03

3.3 Effective Atomic Number Analysis

Effective atomic number (Z_{eff}) is the attenuation of photons in complex media consist several elements which depended on energy of photon with interaction process. Moreover, Z_{eff} performed an important role in the attenuation of photons in a complex medium as well as it is also useful in the calculations of dose in radiation therapy. Based on previous researched, it is reported that it can be considered to be a single fictitious element having an effective atomic number Z_{eff} when gamma rays interact with a heterogeneous material consisted of a number of elements in varying proportions. Effective atomic number, Z_{eff} for all types of materials, compounds and mixtures can be calculated by using the formula below:

$$Z_{eff} = \frac{\sigma_a}{\sigma_e} \quad (7)$$

given $\sigma_{a=\frac{\sigma_t}{\sum_i n_i}}$ is the effective atomic cross section where $\sigma_{t=\frac{M\mu_m}{N_A}}$.

σt is the total photon interaction cross-section of the glasses which is determined with the help of the mass attenuation coefficient of glass. M=$\sum_i A_i n_i$ is the molecular weight of the sample, *M* is the atomic weight of the earth element, ni the number of formula units of a molecule whereas *NA* is the

Avogadro's number. Besides, $\sigma_e = \frac{1}{N_A}\sum_i \frac{f_i A_i}{Z_i}(\mu_m)$. σ_e is the total electronic cross section where f_i denotes the fractional abundance of the element i and Z_i is the atomic number of constituent element. Hence, the effective atomic number (Z_{eff}) is calculated as it is related to σ_a and σ_e [12].

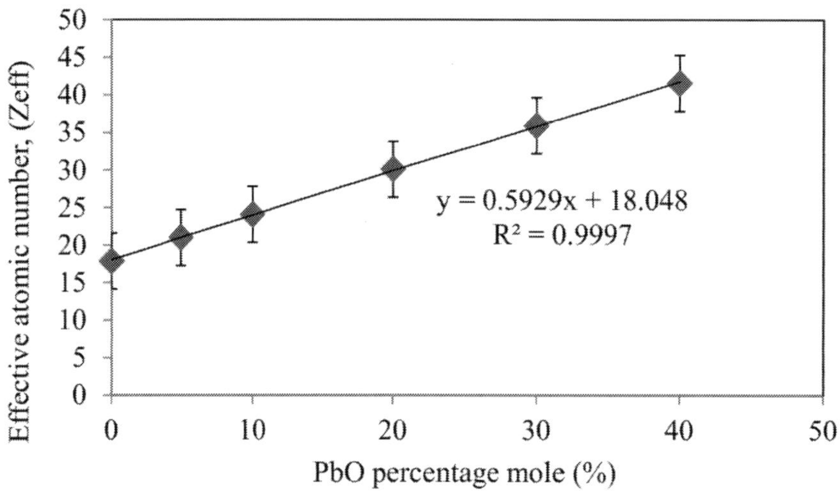

Figure 2: Effective atomic numbers, Z_{eff} at different percentages of PbO.

As illustrated in Figure 2, it is observed that Z_{eff} is increased by the increment of PbO percentage mole in the glass systems and also in some photon energies. It also investigated that Z_{eff} is a convenient parameter for representing the attenuation of X-rays and gamma-rays in a complex medium and particularly for the calculation of the dose in radiation therapy [13]. In this study, it is obtained that the Z_{eff} values of Boron ores are significantly lower than PbO which is in between 80.1 and 1250 keV. Thus, energy dependence of the atomic number is clearly showed that composite materials having high Z_{eff} values absorbed effectively the incoming photons. So, the values of mass attenuation coefficient and Z_{eff} decreased as photon energy increased. Whereas, mean free path values increased by the increment of photon energy. As a result, lead-zinc-borosilicate glass with percentage of lead more than 30 percent is good shielding material.

4. Conclusion

The melt-quenching method has been used to fabricate lead-zinc-borosilicate glass system. The multi-composition glasses were successfully fabricated based on the network former silicon dioxide, SiO_2 99.37 %, which was obtained from rice husk ash (RHA). The glass structure was found to be suitable for shielding protection against radiation due to its physical properties, mass attenuation coefficient values and high values of atomic effective number Z_{eff}. Moreover, the shielding property was essentially due to the high atomic number of PbO which lend itself as a robust radiation protection material.

Acknowledgment

The author acknowledges University Science Malaysia for the financial support for this research under two USM (Short Term Grant) which are 304 / PFIZIK / 6313152 and 304 / PFIZIK / 6313249. Other financial supports grant is from postgraduate USM that is 308/AIFIZIK/415403.

References

[1] Murat Kurudirek 2017 Heavy metal borate glasses: Potential use for radiation shielding. *Journal of Alloys and Compounds*, 727, pp. 1227-1236.

[2] Manohara S.R., Hanagodimath S.M., Gerward L. 2009 Photon interaction and energy absorption in glass: a transparent gamma ray shield. *Journal of Nuclear Materials*, 393, pp. 465-472.

[3] Singh K.J., Kaur S., Kaundal R.S. 2014 Comparative study of gamma ray shielding and some properties of $PbO–SiO_2–Al_2O_3$ and $Bi_2O_3–SiO_2–Al_2O_3$ glass systems. *Radiation Physics and Chemistry* 96, pp. 153-157.

[4] G.Lakshminarayana, S.O.Baki, Kawa M.Kaky, M.I.Sayyed, H.O.Tekin, A.Lira, I.V.Kityk, M.A.Mahdi 2017 Investigation of structural, thermal properties and shielding parameters for multicomponent borate glasses for gamma and neutron radiation shielding applications, *Journal of Non-Crystalline Solids*, 471, pp. 222-237.

[5] Altaf M. and Chaudhry M.A 2010 Physical properties of lithium containing cadmium phosphate glasses. *Journal of Modern Physics*, 1(04), pp. 201.

[6] Halimah M.K, Daud W.M, Sidek H.A.A, Zaidan A.W and Zainal A.S 2010 Optical properties of ternary tellurite glasses. *Material Science Poland*, 28.1, pp. 173-180.

[7] Gupta S. and Sidhu G. S. 2013 Measurement of total and partial mass attenuation coefficients of oxide glasses: A radiation field. *Int. J. Mod. Eng.* Res 3.6, pp. 3830-3835.

[8] El-Mallawany R., Sayyed M.I., and Dong M.G. 2017 Comparative shielding properties of some tellurite glasses: Part 2. Journal of Non-Crystalline Solids, 474, pp. 16-23.

[9] Singh N., Singh R., Singh K.J and Singh K. 2005 γ-ray shielding properties of lead and bismuth borosilicate glasses. *Glass technology* 46.4, pp. 311-314.

[10] Chanthima N., Kaewkhao J., Limkitjaroenporn P., Tuscharoen S., Kothan S., Tungjai M., Kaewjaeng S., Sarachai S., Limsuwan P. 2017 Development of $BaO–ZnO–B_2O_3$ glasses as a radiation shielding material. Radiation Physics and Chemistry, 137, pp. 72-77.

[11] Sharma G., Singh K., Mohan S., Singh H. and Bindra S. 2006 Effects of gamma irradiation on optical and structural properties of $PbO–Bi_2O_3–B_2O_3$ glasses. *Radiation Physics and Chemistry* 75.9, pp. 959-966.

[12] Shams, Issa A.M. 2016 Effective atomic number and mass attenuation coefficient of $PbO–BaO–B_2O_3$ glass system. *Radiation Physics and Chemistry*, 120, 33-37.

[13] Harvinder, Singh, Singh Kulwant, Gerward Leif, Singh Kanwarjit, Singh Sahota Hari, and Nathuram Rohila 2003 $ZnO-PbO-B_2O_3$ glasses as gamma-ray shielding materials. *Nuclear Instruments and Methods in Physics Research Research B: Beam interactions with materials and atoms*, 207, no. 3, 257-262.

Effects of Oxygen Annealing on Kaolin-Doped Zinc Oxide Discs

F Azhari[1], S Mahmud[1], S K Mohd Bakhori[1], M F Che Pa[1], N S Mohd Latiff[1] and A F Omar[1]

[1] School of Physics, Universiti Sains Malaysia, 11800, USM, Pulau Pinang, Malaysia

fahimy_azhari@hotmail.com

Abstract. ZnO nanostructures were used to prepare sintered ZnO dics that were pure ZnO discs (ZnO-100) and kaolin-doped ZnO discs (ZnO-97) containing 3 weight % of kaolin. The ZnO powder was mixed via ball milling process, pressed into discs and sintered at 1000 °C followed by annealing at 700 °C under oxygen ambient. The annealed discs were labeled as ZnO-100/O_2 and ZnO-97/O_2. Optical, structural and electrical properties were characterized using Raman scattering, XRD and current-voltage measurement. The crystallite size of ZnO was calculated using XRD data in that pure ZnO-100 had crystallite size of 33.034 nm and ZnO-100/O_2 oxygen annealed had 41.972 nm. The XRD data also showed oxygen annealed kaolin-doped ZnO-97/O_2 had a smaller crystallite size of 13.054 nm while unannealed kaolin-doped ZnO-97 had 14.285 nm. Kaolin doping and oxygen annealing on ZnO discs appeared to affect crystallite size. The resistance measured on the ZnO disc surface revealed that doped ZnO-97 had lower resistance (2 MΩ) compared to ZnO-100 (6 MΩ). On the other hand, annealed ZnO-97 O_2 had a lower resistance (0.8 MΩ) compared to annealed ZnO-100 O_2 (30 MΩ). The aluminum element in kaolin could have served as a dopant that lowered the potential barrier for electrical conduction to take place.

1. Introduction

Over the years, ZnO has been used widely in optoelectronics for many applications. Such as LED's, sensors, solar cells and electronic devices. ZnO exhibit versatile characterictics, such as chemical sensitivity to numerous gases, high chemical stability, low cost, radiation hardness, high temperature resistance and flexibility for fabrication [1,2]. ZnO has been recognized to have many advantages. Firstly, it is a semiconductor that has direct and wide bandgap of 3.37 eV and it has large excitation binding energy (60 meV). These properties make ZnO an important and useful metal oxide which exhibits near-ultraviolet emission and photonics [3]. Secondly, ZnO is piezoelectric and can be used as electromechanically coupled sensors and transducers. Lastly, ZnO is biosafe, biocompatible and low in toxicity, which can be used for various biomedical purposes [4]. ZnO is zinc compounds which are generally known as safe product by U.S Food and Drug Administration [5].

Al and Si doped ZnO show reasonable sensor responses through lowering resistance values than undoped ZnO. Lowering the resistance make this device suitable for sensory and metal oxide varistor application. Structural, optical and electrical properties that had been studied were discussed in order to obtain the relationship how these properties are related to doping, sintering and annealing. In this study, 100 % ZnO disc and 97% ZnO disc doped with 3% kaolin were fabricated. The samples were annealed with oxygen for 700 °C for an hour.

Content from this work may be used under the terms of the Creative Commons Attribution 3.0 licence. Any further distribution of this work must maintain attribution to the author(s) and the title of the work, journal citation and DOI.
Published under licence by IOP Publishing Ltd

2. Methodology

ZnO powder was synthesis by using French process as described by Mahmud [6]. The powder was divided into four group and pressed into disc form. Subsequently the discs were sintered at 1000 °C for 1 hour soaking time. Then, the crystalline structure of the discs was confirmed by using XRD PanAnalytical X'Pert Pro at 20-80 degree of 2Θ. The optical properties of ZnO was studied using Raman spectroscopy by Horiba Jobin Yvon at room temperature. The current voltage response was characterized using Keithly to evaluate the resistance properties of the sample.

3. Results and Discussions

Figure 1 shows XRD spectra for all ZnO. It confirmed that all ZnO is a wurtzite structure with no impurities. From the XRD data, crystalline size of ZnO was calculated using Scherrer's equation [7]:

$$D = \frac{k\lambda}{\beta cos\theta} \quad (1)$$

Where k is plank constant and λ is wavelength of Kα of xray tube at 1.542 Å, β is full width at half maximum (FWHM) and θ is Bragg angle. The calculation considering an average of highest peak at (100), (002) and (101) for all ZnO. From the calculation, the crystallite size for all ZnO is as shown in table 1.

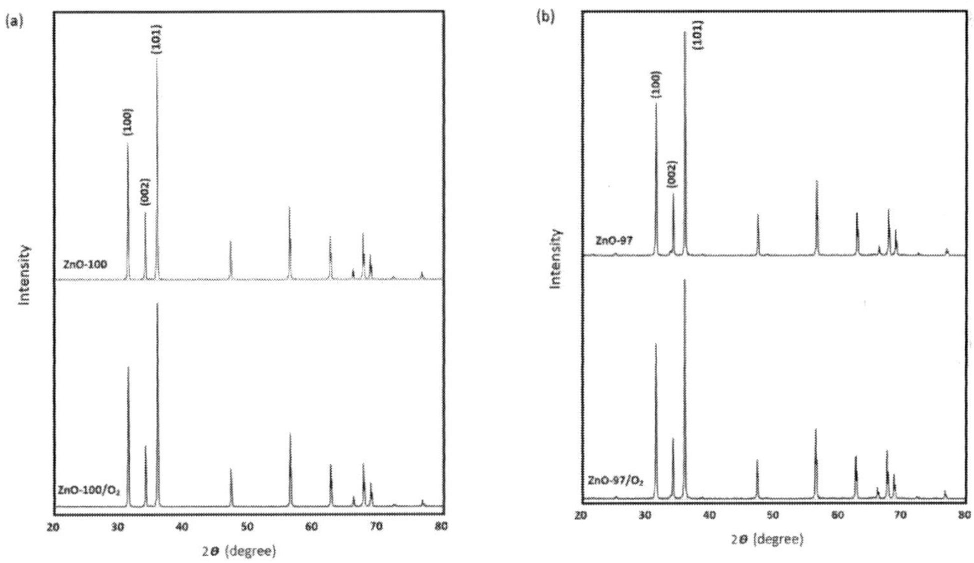

Figure 1. XRD patterns for (a) ZnO-100 and ZnO-100/O_2 ; (b) ZnO-97 and ZnO-97/O_2.

Table 1. The average crystallite size for all ZnO.

Sample	Crystallite size (nm)
ZnO-100	33.034
ZnO-100/O_2	41.972
ZnO-97	14.285
ZnO-97/O_2	13.054

The crystallite sizes of ZnO-100 and ZnO-100/O_2 discs were 33.0.34 nm and 41.972 nm respectively, showing increment in crystallite size. This result implies that oxygen annealing at 700°C did not alter the structural orientation of the sample but may modify the surface condition of the ZnO discs. Meanwhile, the crystallite size for ZnO-97 and ZnO/O_2 discs were 14.285 nm and 13.054 nm. It shows that by adding kaolin to ZnO can decrease the crystallite size of ZnO disc. Although XRD patterns in all ZnO are similar, but the calculated average crystallite size of all ZnO types showed some changes. This result implies that FWHM of peaks (100), (002) and (101) are significantly change.

Figure 2 shows the Raman spectra of all ZnO. It shows phonon modes of A1 (TO), E1 (TO), E2 (high), E1 (LO) is viable. Other than that, the raman mode is silent and could not be detected. The Raman shift for each ZnO is shown in Table 2 below. For A1(TO) the slightly blueshift of wavenumber occurred after doped with kaolin. On the other hand, E2 (high) for all ZnO are similar. Yoshikawa et al describe the decreased of E2 (high) phonon frequency is attribute to better crystallinity and particle size dependent [8]. This statement is contradicted with E2 (high) in Table 2, whereby the values are similar for all ZnO. This condition may attribute to insignificant effects of oxygen annealed on doped kaolin. Nevertheless, XRD calculation showed significant different in crystallite size, and Raman spectrum should be decreasing in E2(high) frequency according to Yoshikawa et al. But in this report, the E2(high) is not dependent on crystallite size. It may have influenced by the adsorption of oxygen molecules on ZnO surface by which it did not affect the phonon vibration.

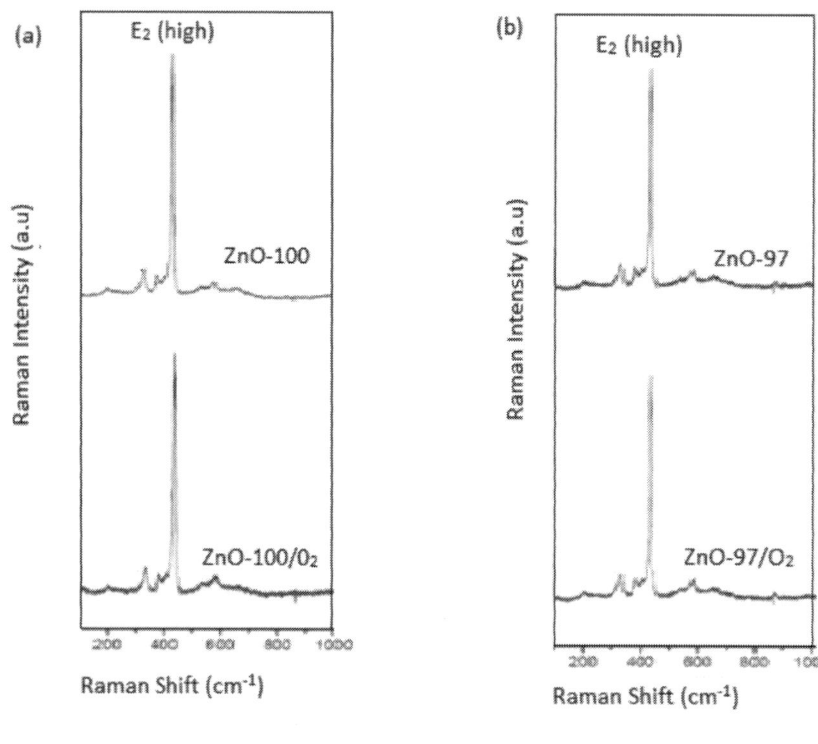

Figure 2. Raman spectra of (a) ZnO-100 and ZnO-100/O_2; (b) ZnO-97 and ZnO-97/O_2.

Table 2. Raman shift for all ZnO.

Sample	A_1 (TO)	E_1 (TO)	E_2 (high)	E_1 (LO)
ZnO-100	332.907	379.010	437.293	583.249
ZnO-100 O_2	332.355	378.462	437.293	584.481
ZnO-97	330.152	380.102	437.293	584.310
ZnO-97 O_2	330.704	379.557	436.751	584.841

Figure 3 shows the current-voltage curve of all ZnO. The resistance measured on the ZnO disc surface revealed that ZnO-97 had lower resistance (2 MΩ) compared to ZnO-100 (6 MΩ). On the other hand, annealed ZnO-97/O_2 had a lower resistance (0.8 MΩ) compared to annealed ZnO-100/O_2 (30 MΩ). The Kaolin doped ZnO showed lower resistance because kaolin contained Al and Si elements and caused atomic substitution [9]. Doping with Al and Si (in Kaolin) into ZnO crystal lattice increases free charge carriers (electrons) thereby reducing resistivity. According to Maldonado et.al, by computer simulation, the electron jump produced free electron in conduction band thus increase the conductivity and decreased the resistance. [9]. Moreover, the oxygen annealing at 700 °C effects has lower the oxygen vacancies concentration by oxygen adsorption [10] on the ZnO disc surfaces.

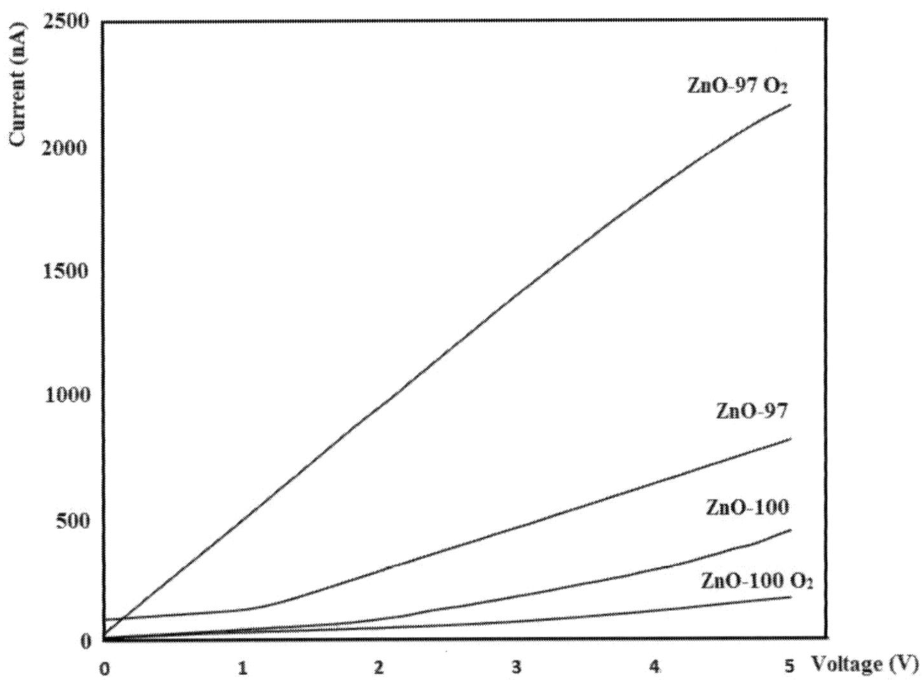

Figure 3. Current-voltage response for all ZnO dics.

4. Conclusion

In conclusion, the ZnO disc doped with kaolin have affected the physical and electrical properties. The changes are beneficial as it provides lower resistance to the ZnO disc. This advantage can be used in sensory application such as UV sensor or as dopant in metal oxide varistor as it can alter the breakdown voltage according to desired application.

Acknoledgement
The authors would like to acknowledge the FRGS grant [203/PFIZIK/6711605] and grant [304/PFIZIK/6313209] for financial support. Moreover, the author is grateful to Solid State Laboratory, School of Physics and Institute of Nano-optoelctronics Research and Technology (INOR).

References

[1] J Sun, F Juan, H Huang, J Zhao, Z Hu, X Zhang and Y Wang 2010 Fast response ultraviolet photoconductive detectors based on ga-doped zno films grown by radio-frequency magnetron sputtering *Appl. Surf. Sci.* 257 921-24
[2] O Lupan, L Chow, G Chai, L Chernyak and O Lopatiuk 2008 Focused ion beam fabrication of zno nanorod-based uv photodetector using in-situ lift-out technique *Phys. Stat. Sol. (a)* 205 2673-78.
[3] D C Reynolds, D C Look, B Jogai, C W Litton, T C Collins, W Harsch and G Cantwell 1998 Neutral donor bound excitation complexes in zno crystals *Physical Rev.* 57 12151-55
[4] S K Sahoo, S Parveen and J J Panda 2007 The present and future of nanotechnology in human health care *Nanomedicine* 3 20-31
[5] M Premanathan, K Karthikeyan, K Jeyasubramaniam and G Manivannan 2011 Selective toxicity of zno nanoparticles towards gram-positive bacteria and cancer cell by apoptosis through lipid peroxidation *Nanomedicine* 7 184-192
[6] S Mahmud 2011 One-dimensional growth of zno nanostructures from large micro-particles in a highly rapid synthesis *J. Alloys Compd.* 509 4035-40
[7] B M Han, S Chang and S Y Kim 1999 Chopping effect on the crystallinity of zno prepared by r.f. planar magnetron sputtering method *Thin Solid Films* 338 265-68

[8] M Yoshikawa, K Inoue, T Nakagawa, H Ishida and N Hasuike 2008 Characterization of zno nanoparticles by resonant raman scattering and cathodoluminescence spectroscopies *Appl. Phys. Lett.* 92 113115
[9] F Maldonado and A Stashans 2010 Al-doped zno : electronic, electrical and structural properties, *J. Phys.Chem. Solids* 71 784-87
[10] D Chakraborti1, G R Trichy1, J T Prater and J Narayan 2007 The effect of oxygen annealing on zno : cu and zno : (cu,al) diluted magnetic semiconductors *J. Phys. D Appl. Phys.* 40 24

IOP Conf. Series: Journal of Physics: Conf. Series **1083** (2018) 012007 doi:10.1088/1742-6596/1083/1/012007

Structural Properties of Cullet-Paper Ash-Kaolin Clay Ceramic

K.A.Samah[1,a*], M.R.Sahar[2,b] and M.Yusop

[1*,2]Advance Optical Material Research Group, Department of Physics, Faculty of Sciences, Universiti Teknologi Malaysia, 81310 UTM Skudai, Johor, Malaysia

[a*]khamisah86@gmail.com, [b]mrahim057@gmail.com

Abstract. This study focused on the structural properties of (x) Paper ash-(80-x) Cullet-20 Kaolin clay ceramic that are synthesized by solid state reaction due to the cullet content. The crystallinity and structure were determined using X-Ray diffraction (XRD), Fourier transform infrared spectroscopy (FTIR) and Raman spectroscopy. The major phase of the ceramic was Wollastonite (CaSiO3) while the minor phase consisted of Dicalcium Silicate (Ca2SiO4) and Quartz (SiO2). The FTIR spectra of the S2 sample exhibited broad peak near to 1020cm^{-1} corresponding to the stretching vibration of Si-O-Si linkage. The changes of Raman spectra linewidth of the S2 sample were observed near 575cm^{-1}, 480cm^{-1} and 968cm^{-1} which attributed to the Si-O-Si stretching respectively.

1. Introduction

Currently, there are two important composition based on natural recourses namely alumino-silicate-glass (ASG) and calcium-magnesium-alumino-silicate(CMAS) systems which have contributed to the excellent insulating materials. For instance, the main ingredients in cullet glass and paper ash contains ceramic network for ASG and CMAS:SiO_2, CaO, Na_2O and Al_2O_3. Thus, it is feasible to use cullet glass and paper ash as a raw materials to develop the ceramic matrices [1]. In addition, diopside, wollastonite, and melilite can be produced from these wastes which exhibits superior mechanical, erosion, thermal and electrical properties to the parent glass [2].

Ceramic from silicate waste consists of structural units of the SiO_4 tetrahedron. It has been reported that, the SiO_4 tetrahedron can be treated as rigid units which can be linked together by their corner to form pairs, rings, chains, sheets or frameworks [3]. On the basis of IR study, a particular structure of silicate is located around 400 to 1100 cm^{-1} which relates to bending or stretching vibration of both bridging Si-O-(Si) and terminal Si-O bond [4]. Infrared and Raman spectroscopy are well known tools to investigate the arrangement of the structural units.

Studies on cullet glasses have been introduced and grown due to their interesting behavior to improve the ceramic matrixes by incorporation of paper ash and Kaolin clay in these ceramics. Thus, in the present study, the cullet was used as host to vitrify paper ash and kaolin clay. In this study, X-ray diffraction spectra (XRD), Fourier transform infrared spectra (FTIR) and Raman spectroscopy have been used to investigate the structural aspect of cullet with paper ash and kaolin clay. The series of (x) Paper ash-(80-x) Cullet-20 Kaolin clay ceramic are prepared by using the sintering technique. The role of the cullet glass and paper ash on structural properties of ceramic has been studied.

Content from this work may be used under the terms of the Creative Commons Attribution 3.0 licence. Any further distribution of this work must maintain attribution to the author(s) and the title of the work, journal citation and DOI.
Published under licence by IOP Publishing Ltd

2. Experimental

Compositions of (x)paper ash-(80-x) cullet-20 Kaolin clay, where $(10 \leq x \leq 30)$ weight % were successfully prepared using modified solid-state reaction method. A proportional amount (in size 100 m) of cullet, paper ash and Kaolin clay were mixed together using ball milling for 2 hours for homogenization. Nominal composition of ceramic samples in weight percent has been enlisted in Table 3.1. These samples were then pressed into pallet using Palletizing Press Machine (HERZOG) at 40kN pressure. After that, it was sintered at 900 °C for 2 hours. Finally, ceramic samples were cut and polished for the structural measurements. The X-ray diffraction (XRD) pattern was obtained using PAN analytical X'Pert Pro with Cu Ka monochromatic radiation source at 40 kV in the 2θ range of 10-80°. The Fourier Transform Infrared (FTIR) spectra of fine powder of ceramic mixed with KBR was obtained by a Perkin Elmer Spectrum One FTIR spectrometer. The Raman spectra was obtained using a Nd:YAG laser (excitation wavelength 532nm) that operates in the range of 400 – 4000cm^{-1}.

Table 3.1 The nominal composition of ceramic samples in weight percents

Sample No.	Composition (wt%)		
	Cullet powder	**Paper Ash**	**Kaolin Clay**
S_1	50	30	20
S_2	55	25	20
S_3	60	20	20
S_4	65	15	20
S_5	70	10	20

3. Results and Discussion

3.1 X-Ray Diffraction (XRD)

The XRD pattern of all ceramic samples containing Cullet, Paper ash and Kaolin clay has been shown in Fig. 1 which contain multiple diffraction peaks owing to their ceramic nature. A strong peak of Wollastonite ($CaSiO_3$) representing the major phase of that ceramic phase. Minor phase of γ-Dicalcium silicate ($Ca_2(SiO_4)$) and Quartz (SiO_2) with low intensity peaks was also observed in XED pattern. The occurence of these phases was expected since the abundance of SiO_2, Al_2O_3, and CaO in the raw material of these ceramic [5]. The formation of Wollastonite phase is probably due to coalescence particle of CaO from paper ash and SiO_2 from cullet glass and Kaolin clay [6],[7]. XRD pattern also shows that as the cullet content exceeds 55wt% (after sample 2), the intensity of wollastonite peaks decreases while the intensity of γ- Dicalcium silicate increase. Thus, some structural changes from wollastonite to γ- Dicalcium silicate and Quartz(SiO_2) were observed as cullet content exceeds from 55wt% .

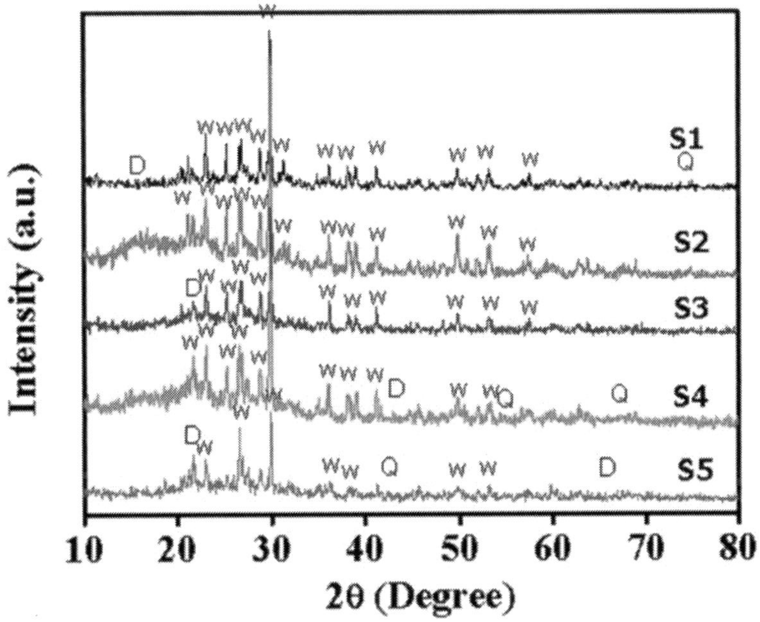

Fig. 1. XRD spectra of cullet-paper ash-kaolin clay ceramic (W=wollastonite, Q = quartz, and D = γ-dicalcium silicate).

3.2 FTIR spectra

The structures of the obtained polycrystalline ceramic samples were also characterized using FTIR and Raman spectroscopy. These two techniques are IR-active, Raman-active or both are complementarily when react with varies vibration modes frequency in molecule. Generally, there are two types of vibration frequency known as bending or tortion vibration which have in-plane and out-of plane bending and stretching vibration. Normally, the change in the inter-atomic distance along bond axis contribute to tortion vibration. The in-plane bending of tortion vibration can be separated into two groups like scissoring and rocking relying on the atoms swing opposite or in the same directions, respectively. Then, when the atoms swing up and down it can be classified into wagging and twisting vibrations for out-of plane bending of tortion vibration. Furthermore, the angle change between bonds form bending vibration [8]. FTIR and Raman analyses yield the signature effect of cullet glass and paper ash on the vibration modes of ceramic structure.

The FTIR spectra of prepared ceramics in the range of 400-1600cm^{-1} are shown in Fig 2. The strong absorption spectra in the range 400-800 cm^{-1} correspond to the bending vibration, while, weak absorption spectra in the range 900 -1100 cm^{-1} is due to stretching vibration of both bridging Si-O-(Si) and terminal Si-O bond [9], [4]. The absorption spectra appeared around 714-716 cm^{-1} is assigned to the stretching bridging Si–O(Si) with the presence of 3-membered ring and is referred as wollastonite band [10]. The observed absorption spectra around 472-475 cm^{-1} is allocated to the triple degenerated, stretching and bending vibration modes of the tetrahedron SiO_4 [11]. Additionally, the absorption peaks at 472, 902, 964, and 1015 cm^{-1} confirm the existance of wollastonite phase in the structure. Interestingly, the IR spectra for samples with addition of cullet glass in the ceramic samples are in close resemblance to each other. It was also observed that the intensity of absorption spectra (900-1100 cm^{-1}) is suppressed with further addition of cullet glass. Since the intensity of an absorption bands depends on the change in the dipole moment which contribute to the bond length and the charge difference between two atoms. Thus, more the cullet lesser is the electronegativity between Si-O-Si stretching vibration [12]. The wavenumber of the peaks shows an incremental shift from S1 to S2 at peak 472, 714, 902 and 1015 cm^{-1}. However, the wavenumber remain constant with more cullet concentration. Such phenomenon indicates that the degree of polymerization improve with addition of cullet at 50 wt%[13]. Further addition of cullet inhibits the ability of the sample to be polymerized [13], [14].

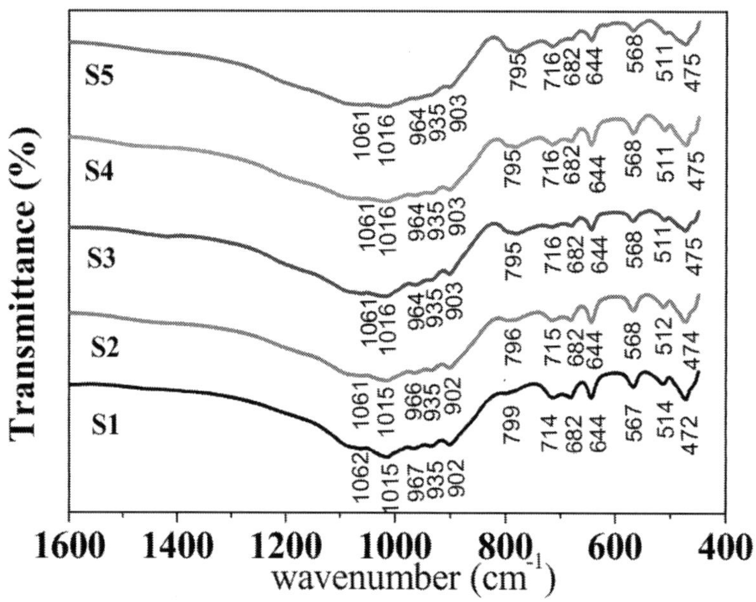

Fig. 2. FTIR spectra of cullet-paper ash-kaolin ceramic.

3.3 RAMAN spectra

The Raman spectra of all ceramic samples presented in Fig. 3. are selected to display in the range of 300 – 1200 cm^{-1}. Each Raman spectrum is exhibiting three Raman bands centered at 480 cm^{-1}, 577 cm^{-1}, and 968 cm^{-1}, respectively which are attributed to the fundamental vibrations of structural units/ lingkages in typical Si-O structure [1]. As reported by Kingma, the peaks around 480 cm^{-1} and 575 cm^{-1} are the symmetric Si-O-Si stretching-bending mode [13, 14]. More specifically, the intensity of peak at 480 cm^{-1} increases as cullet content is increased due to the strengthening of Si-O-Si bonds. The peak around 635cm^{-1} which clearly seen in sample 2 and 4 correspond to the rings of SiO_4 tetrahedra [3]. Whereas, the peaks between 960-981cm^{-1} are attributed to the Si-O stretching with 2 non bridging oxygen atoms per silicon(Q^2) [3]. Furthermore, the peak in all ceramic samples around 968 cm^{-1} are ascribed to the Ca^{2+} cation as non-briding Si-O stretching (Q^2) atoms in the structural unit with reduction of paper ash [3],[16].

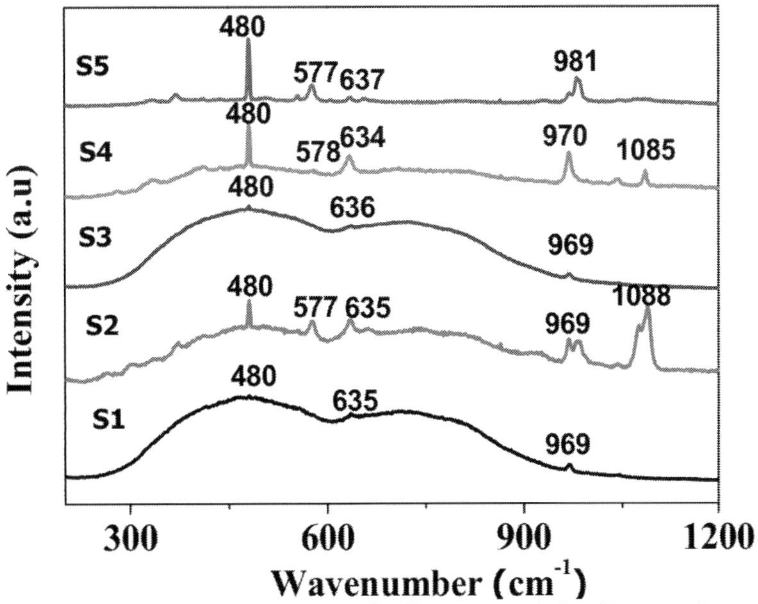

Fig. 3. RAMAN spectra of cullet-paper ash-kaolin ceramic.

4. Conclusion

The (*x*) Paper ash-(80-*x*) Cullet-20 Kaolin clay ceramic have been successfully prepared by solid state reaction method. The variation of cullet content that strongly affect the ceramic structure are clearly revealed from FTIR spectra consisting of vibrational mode of Si-O-Si stretching units. The participation of cullet glass in ceramic matrix modifies the ceramic forming network and also responsible for changing the structure and geometry of ligands. Raman peaks around 480 cm^{-1} and 575 cm^{-1} are assigned as the symmetric Si-O-Si stretching-bending mode, while the peak around 968 cm^{-1} referred to non-briding Si-O stretching (Q^2) atoms. The obtained results suggest that our method by varying the concentration of cullet glass may constitute a basis for significant enhancement in the structural properties for potential application in electronic devices.

Acknowledgments

The authors gratefully acknowledge the financial support from ministry of higher education, Universiti Teknologi Malaysia via Grants of Vot: 08H42 and 16H41, and 13J81

References

[1] Sahar M R, Hamzah K, Rohani M S, Samah K A and Razi M M 2011 *Phys. Procedia* 22 119–24
[2] Agathopoulos J M F F S, Tulyaganov D U, Ventura J M G, Kannan S, Saranti A, Karakassides M A 2006 *J. Non. Cryst. Solids* 352 322–28
[3] Salh R 2011 *In Crystalline Silicon - Properties and Uses*, Rijeka: InTech 135–72.
[4] Senthil Kumar R and Rajkumar P 2014 *Infrared Phys. Technol* 67 30–41
[5] Tarvornpanich T, Souza G P and Lee W E 2005 *J. Am. Ceram. Soc* 88 1302–08
[6] Okada K, Toya T, Kameshima Y, and Nakajima A 2004 *J. Waste Manag. Japan* 11–19
[7] Lu J, Lu Z, Peng C, Li X and Jiang H 2014 *J. Mater. Chem. Phys* 148 449–56
[8] Azmi S A M, Sahar M R, Ghoshal S K, and Arifin R 2015 *J. Non. Cryst. Solids* 411 53–58
[9] Sitarz M, Mozgawa W and Handke M 1997 *J. Mol. Struct* 404 193–97
[10] Paluszkiewicz C, Blazewicz M, Podporska J, and Gumuła T 2008 *J. Vib. Spectrosc* 48 263–68

[11] Gendron-Badou A, Coradin T, Maquet J, Fröhlich F, and Livage J 2003 *J. Non. Cryst. Solids* 316 331–37
[12] Anigrahawati P and Sahar M R 2017 *J.Solid State Phenom* 268 177–80
[13] Jiang S, Feng H, Chuqiao Y, Lan H, Junlin X, Hu Y, Xiaoqing L 2017 *J.Mater. Chem. Phys* 197 57–64
[14] Atalay S,Adiguzel H I, and Atalay F 2001 *J. Mater. Sci. Eng. A* 304 796–99
[15] Kingma K J and Hemley R J 1994 *J.Am. Mineral* 79 269–73
[16] Nguyen T P and Lefrant S 1985 *J.Solid State Commun* 57 235–36
[17] Manara D, Grandjean A and Neuville D R 2009 *J. Am. Mineral* 94 777–84
[18] Retsinas A, Kalampounias A G and Papatheodorou G N 2016 *J. Phys. Chem. Solids* 99 19–24

Influence of Microstructural Effect on Microvickers Hardness Properties of SiO_2-Na_2O-CaO (SNC) Waste Based Glass-ceramic

Nur Fadilah Baharuddin Pallan[1,2], N F Adnin[1,2], K A Matori[1,2], Mansor Hashim[1], F M Idris[1], I R Ibrahim[1], S N A Rusly[1], M Z A Khiri[1,2], Rodziah Nazlan[1], N Hapishah Abdullah[1], M S E Shafie[1], R S Aziz[2], Z N Alassan[2], N Zainuddin[3], N Faizah Baharuddin Pallan[4], Noraziah Ahmad[4], Maria Nuid[4]

[1]Institute of Advanced Technology, Universiti Putra Malaysia, 43400 Serdang, Selangor, Malaysia
[2]Physics Department, Faculty of Science, Universiti Putra Malaysia, 43400 Serdang, Selangor
[3]Chemistry Department, Faculty of Science, Universiti Putra Malaysia, 43400 Serdang, Selangor
[4]Department of Environmental Engineering, Faculty of Civil Engineering, Universiti Teknologi Malaysia, 81310 Skudai, Johor, Malaysia

nurfadilahbaharuddin@gmail.com, khamirul@upm.edu.my

Abstract. There are a lot of waste materials consist of silicate based such as coal combustion ash, slag from steel production, fly ash, mud, as well as glass cullet or mixtures to produce glass-ceramics. This research work using clam shell (CS) ash and soda-lime-silica (SLS) waste glass powder for fabricating novel SiO_2-Na_2O-CaO (SNC) glass-ceramic. The samples were composed of SLS (50%), Na_2CO_3 (30%), and CS (20%) in weight percentage via conventional melt-quenching technique and solid-state sintering technique. The samples were investigated via X-Ray Diffractometer (XRD), Field emission microscope (FESEM), and microvickers hardness tester. The samples were sintered at 550-950 °C to investigate the influence of microstructural effect on microvickers hardness properties at applied force 0.5 and 1.0 kgf. The optimal Vickers hardness properties at sintering temperature 850 °C due to high crystallization of SiO_2 phase from the residual glass and CaO content enhanced the viscosity flow, high compactness of particles arrangement and densification of sample.

1. Introduction

Glass-ceramics are composite materials consist of combination of residual amorphous phase and crystalline phase. Since the glass-ceramic is used in a large scale application, the materials technology is evolving rapidly. In this study, the mechanical properties of polycrystalline SNC glass-ceramic were determined by phase crystallization and microstructural properties at different sintering temperature. Generally, the mechanical properties of SNC system are less studied by previous researchers. The mechanical behaviour on evolving microstructure of the parent glass with fully crystallized phase have been explained and compared in previous published paper [1-3]. The crystal size plays an important role in preventing the propagation of the cracks in the structure of silicate glass-ceramics with fine

Content from this work may be used under the terms of the Creative Commons Attribution 3.0 licence. Any further distribution of this work must maintain attribution to the author(s) and the title of the work, journal citation and DOI.
Published under licence by IOP Publishing Ltd

grained microstructure increasing the hardness values [4]. The decrement in glass transition temperature influenced the decrement in microvickers hardness properties of bioactive glass [5]. The microhardness of glass-ceramics generally increased with the crystallization amount, smaller crystalline grains as well as formation of fine microstructure [6]. The mechanical properties of the glass-ceramic are influenced mainly by the evolution changes in microstructure properties [7]. The aim of present research is to investigate the phase composition, crystal system, morphological properties and mechanical properties with the low production cost by using SLS waste glass and CS in the SNC glass-ceramic. All the investigations were carried out to understand the microstructural-mechanical relationships at different applied loads dependence of sintering temperature.

2. Materials and methods

The SNC compositions used SLS (SiO_2), Na_2CO_3, and CS ($CaCO_3$) as the starting materials. The SLS waste glass and CS were plunged and milled by using mill jar for 24 hours to obtain fine powder. The glass-ceramic compositions were composed of 50 wt% SLS, 30 wt% Na_2CO_3, and 20 wt% CS was prepared using the melt-quenching and solid-state sintering. The powders were weighed accurately using an electronic balance and are mixed together by ball milling process at 700 rpm for 24 hours to improve homogeneity of the powders. The batch formulation of SNC composition was melted at 1400 °C. Then, the melted SNC composition was rapidly quenched in water to obtain the glass frits. The SNC glass frits were dried for one day at room temperature. After that, the SNC glass frits were crushed, dry milled for 24 hours and sieved at 45 μm. Finally, the prepared SNC glass-ceramic sample was pressed at 3 tonnes and sintered in air at 550-950 °C for 3 hours. The phases crystallization were investigated through XRD diffractometer (PANalytical X'Pert Pro PW3050/60) with Cu-Kα radiation (λ = 1.54060 Å) in the range of 20° - 80° by a 2θ scan mode and the XRD results were analyzed by using PANalytical X-Pert High ScorePlus Programme. The micrograph of samples that were coated with gold (Au) was observed by Nova Nano 230 Field Emission Scanning Electron Microscope (FESEM). The microvickers hardness tester was used to measure Vickers hardness (HV) values from the symmetrical indentation diagonal length of samples at applied force 0.5 kgf and 1.0 kgf.

3. Results and discussion

Figure 1 shows XRD pattern of the synthesized SNC glass–ceramic samples at different sintering temperatures. The broad peak in XRD pattern corresponds to low crystalline structure. Before sintering, sample was observed to have amorphous phase coexist with some crystalline phases at room temperature. The main phase formed before sintering belongs to $Ca_2Na_2Si_2O_7$ (Ref. code: 98-006-9558) which is in the monoclinic crystal system and SiO_2 phase formed as the secondary peak. Segnit (1953) obtained $Ca_2Na_2Si_2O_7$ compound in the parts of ternary system which is richer in CaO and Na_2O [8]. The compositions of $Ca_2Na_2Si_2O_7$ in SiO_2-Na_2O-CaO system were discovered by Kahlenberg and Hösch (2002) [9]. When sintering temperature at 550 °C, new phases were detected which were belong to Na_2CaSiO_4 and Ca_2SiO_4. The existence of SiO_2 phase from the residual glassy phases together with $Ca_2Na_2Si_2O_7$ phase remains as the temperature increased. When sintering temperature increases at 650–950 °C, the intensity increases and 2θ became broader with the appearance of new $Ca_3Na_6Si_6O_{18}$ (Ref. code: 98-002-1442) crystalline phase. These phase to be a mixture of a solid solution with Na_2CaSiO_4 and Ca_2SiO_4 phases transformed to individual compound, $Ca_3Na_6Si_6O_{18}$ phase which is hexagonal crystal system.

Figure 2 shows unit cell volume (Å^3) of SNC at different sintering temperature. The unit cell volume was calculated manually by multiplication of lattice parameters *a*, *b*, and *c* value. The lattice parameters *a*, *b*, and *c* value were obtained from the XRD programme analysis. The highest unit cell volume indicated un-sintered sample at room temperature. When SNC was sintered from 550-950 °C, unit cell volume increases from 415.161 - 1453.323 Å^3. The lowest unit cell volume was corresponding to sample at 550 °C due to decrement of crystallization phase and low intensity of crystallization peak as observable in Figure 1. The unit cell volume decreased at 850 °C and 950 °C correlated with increasing intensity peak of $Ca_2Na_2Si_2O_7$ phases which corresponding to high content of CaO and residual glassy phase of SiO_2. High content of CaO and SiO_2 induced high viscosity in the SNC system which leads towards high densification of sample as shown in Figure 3(d) and (e).

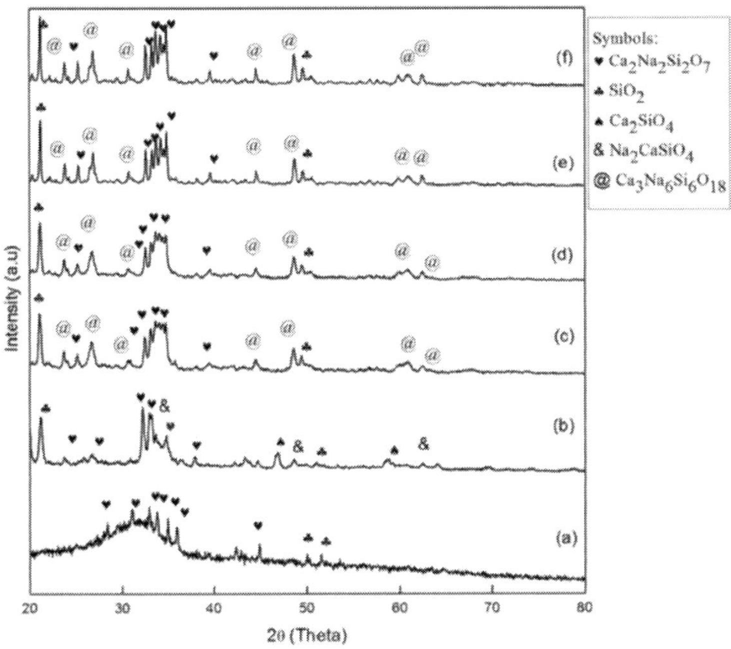

Figure 1. X-ray diffraction pattern of SNC samples (a) before sinter, (b) 550 °C, (c) 650 °C, (d) 750 °C, (e) 850 °C and (f) 950 °C.

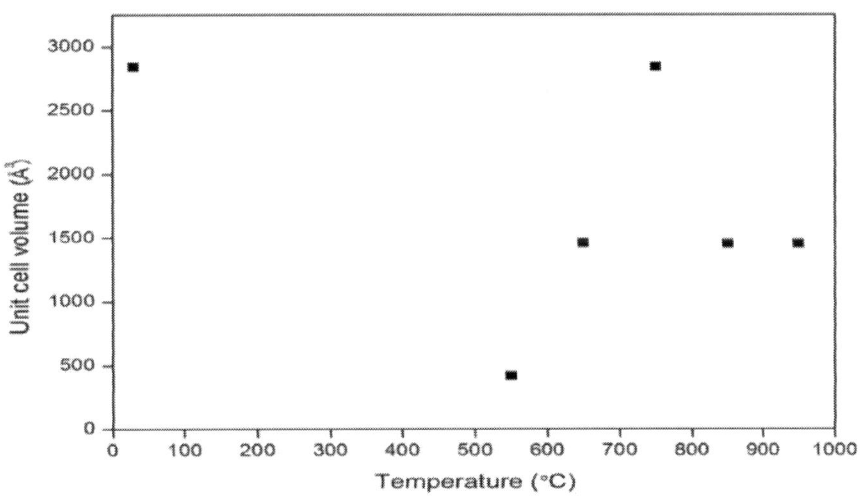

Figure 2. Unit cell volume of SNC versus sintering temperature.

Figure 3 shows the results of SNC FESEM microstructural at different sintering temperatures by using magnification 2000×. Generally, microstructural evolution was caused with increasing the sintering temperatures. The microstructures observed at 550–750 °C showed the increment of grain growth, while the grain boundary and porosity decrease. Samples at 850 °C and 950 °C showed the increment of grain, while grain boundary decreases induced the grain to combine each other to form agglomerate, less coalescence and porosity decreased. Sintering temperature increases affect the decrement in level of porosity of SNC microstructural. The grain in samples 3 (a) and (b), were appeared with small white spot interstitial remaining glassy phase of SiO_2 which is not fully homogeneous below 750 °C. Meanwhile, in samples 3 (c)-(e) the white spot decreased when it was sintered at 850-950 °C. All SNC samples show the presence of SiO_2 phase from the residual glass as detected by XRD analysis as shown in Figure 1 could enhance the viscous flow and densification of sample at high crystallization temperature. The grain increased as the sintering temperature increased due to the grain growth of major phase crystallization and agglomeration of particles at 850-950 °C. Therefore, SNC produced fine and uniform grain due to high crystallization of SiO_2 and high Na_2O content that act to homogenize the SNC composition as depicted in Figure 3(d) and (e).

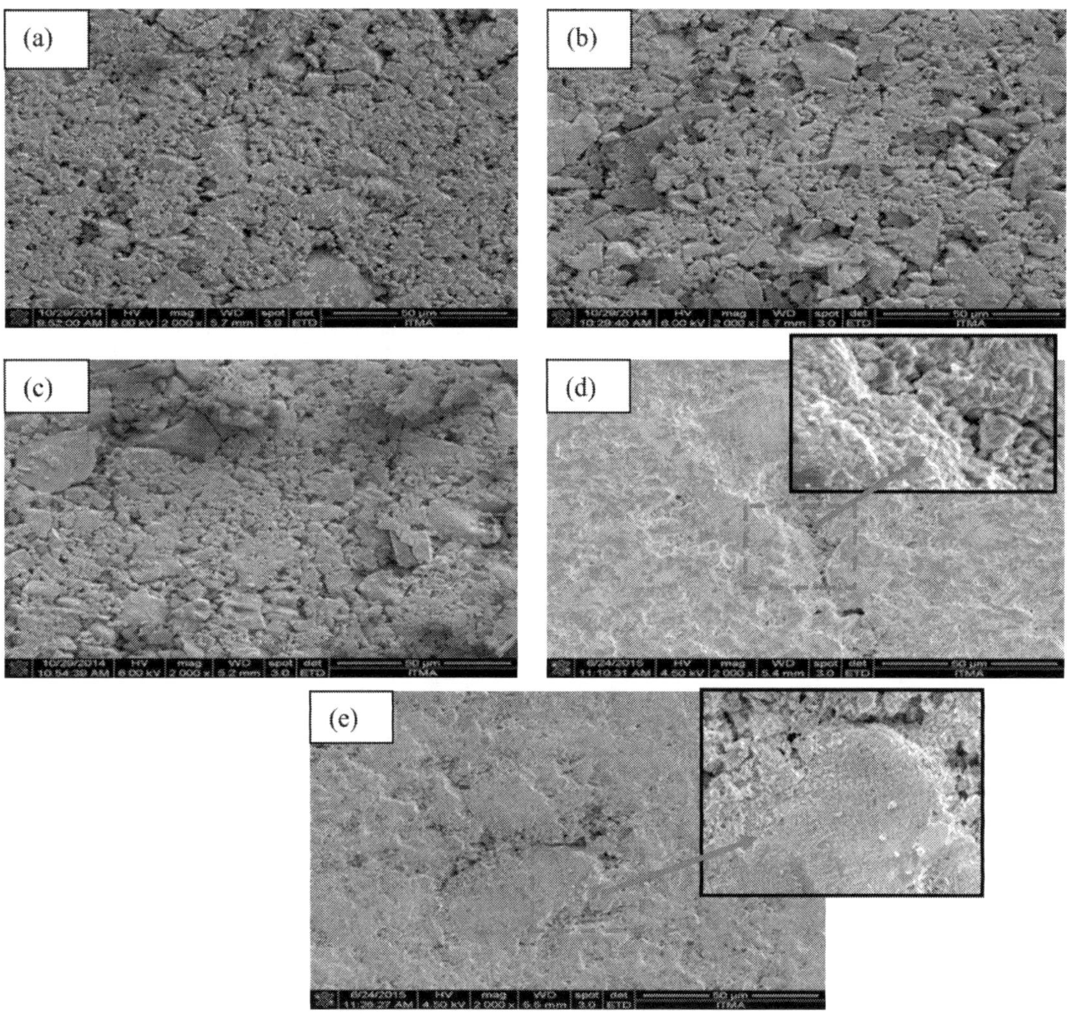

Figure 3. SEM micrographs of SNC at (a) 550 °C, (b) 650 °C, (c) 750 °C, (d) 850 °C, and (e) 950 °C

Microvickers hardness tester is used to measure hardness and the symmetrical indentation diagonal length of samples [10]. The applied force subjected to loads at 0.5 kgf and 1.0 kgf was chosen to evaluate the sample hardness that can withstand with the microvickers indentation at high and low applied force. Figure 4 demonstrates microvickers hardness (HV) value of SNC at different sintering temperature. It is apparent that both loads resulted in different shapes of Vickers hardness were corresponding to decrement of symmetrical indentation diagonal length on the sample at applied force 1.0 kgf while at 0.5 kgf the symmetrical indentation diagonal length on the sample increased. The HV graph trend at 0.5 kgf showed an increment from 189 HV to 322 HV. On the contrary, when sintering temperature was elevated at 950 °C, the hardness value dramatically decreased to 228 HV. The HV value decreased as the sintering temperature increased at 950 °C possible related with high intensity peak of SiO_2 phase as can be seen in Figure 1. The sample hardness properties at applied force 1.0 kgf obtained the highest value, 277 HV at 550 °C subsequently, drop to 238 HV at 950 °C. Both applied force have the highest HV value at the same sintering temperature, 850 °C. The HV value of SNC samples at 550 – 950 °C from waste-based glass-ceramic not achieved the hardness of SNCP system in the range of 458–600 HV [11-12].

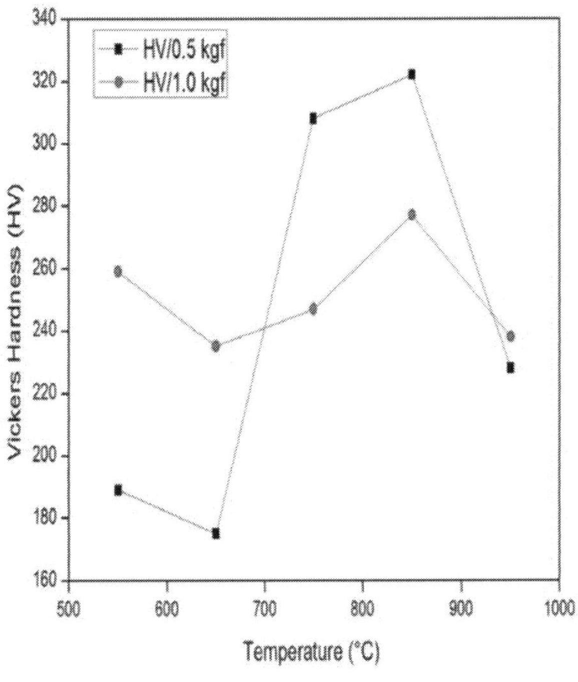

Figure 4. Microvickers hardness of SNC at 0.5 kgf and 1.0 kgf dependence of sintering temperature

4. Conclusion

In this study, the XRD characterization indicated the intensity phase crystallization of $Ca_2Na_2Si_2O_7$, SiO_2, and $Ca_3Na_6Si_6O_{18}$, was increased in correlated with agglomeration of particles and grain growth of major phase crystallization and at 850-950 °. It can be concluded that the SNC microstructural properties at sintering temperature 850 °C had high content of CaO and SiO_2 from the SNC phases crystallization induced high viscosity which leads towards high densification, reduction of porosity, and high compactness of particles arrangement influenced to the highest mechanical microvickers hardness value, 322 HV at applied force 0.5 kgf. The microvickers hardness properties above 950 °C affect to poor mechanical properties which was above high glass crystallization temperature.

Acknowledgements
The authors acknowledge the financial support from Universiti Putra Malaysia and Ministry of Education Malaysia (MOHE) for MyBrain15 sponsorship.

References

[1] Freiman S W and Hench L L 1968 *Journal of American Ceramic Society* 51(7) 382–387.
[2] Sambell R A J, Bowen D H, and Philips D C 1972 *Journal of Materials Science* 7 663–675.
[3] Thompson J Y 1995 *Ph.D. thesis* University of Florida.
[4] Sohn S B, Choi S Y, and Lee, Y K 2000 *Journal of Materials Science* 35 4815.
[5] O'Donnell M D 2011 *Acta Biomaterialia* 7(22) 64-69.
[6] Torres F J and Alarcon J 2004 45 347.
[7] Henry J and Hill R G 2003 *Journal Non-Crystalline Solids* 319 13–30.
[8] Segnit E R 1954 *Journal of the American Ceramic Society* 37(6) 273-277.
[9] Kahlenberg V and Hösch A 2002 217 155–163.
[10] Wachtman J B,Cannon W R, and Matthewson M J 2009 *John Wiley & Sons, Technology & Engineering.*
[11] Thompson I D and Hench L L 1998 *Proceedings of the Institution of Mechanical Engineers, Part H* 212(2) 127-136.
[12] Zanotto E D 2010 *American Ceramic Society Bulletin* 89 19–27.

The Effect of Sintering Time on The Microstructural and Nonlinear Electrical Properties of Zn-V-Mn-Nb-Nd-O Low-Voltage Varistor Ceramics

N H Isa[1*], R S Azis[1,2] and N K Saat[1]

[1]Department of Physics, Faculty of Science, Universiti Putra Malaysia, 43400 UPM Serdang,Selangor, Malaysia
[2]Institute of Advanced Materials, Universiti Putra Malaysia, 43400 UPM Serdang,Selangor, Malaysia

Abstract. There is lacking of study on the prolonged sintering time effect on the microstructural and nonlinear electrical properties of ZnO-V_2O_5-MnO_2-Nb_2O_5-Nd_2O_3-based low-voltage varistor ceramics sintered at 900 °C. The aim of this study is to investigate the effect of sintering time from 120 to 210 mins on the microstructural and nonlinear properties of the ceramics. It is expected that its properties is diminished as sintering time is increased. The sample was prepared by solid-state method and sintered at 900 °C. The XRD diffraction peaks detected V_2O_5 as secondary phase. The ZnO peak shifted to the low diffraction angle of 34.37° and increased the interplanar space slightly to 2.61 Å. The average grain size decreased to 2.67 μm and density decreased to 4.74 g/cm^3 as sintering time increased to 180 mins. However, the presence of liquid phase at 210 min increased both value to 3.63 μm and 5.00 g/cm^3. The nonlinear α value decreased to 5.74 since the barrier height decreased to 0.633 eV as sintering time increased. The value of leakage current density J_L increased to 0.54 mA/cm^2 as sintering time increased up to 180 mins but decreased to 0.42 mA/cm^2 at 210 mins. The breakdown field E_{1mA} value also decreased from 76.07 to 66.03 V/mm as sintering time increased.

1. Introduction

Zinc oxide (ZnO)-based low voltage varistors are semiconducting ceramic devices used to protect low voltage electrical devices from harmful overvoltage. The low voltage devices without varistor had risk being exposed to overvoltage transient repeatedly and have a high chance to damage. The ZnO-based varistor is produced by sintering ZnO powder containing additives of small amount such as vanadium (V), bismuth (Bi), manganese (Mn), and niobium (Nb) [1,2]. The nonlinear properties of ZnO-based varistors ceramics are ascribed to double Schottky barrier formed at active grain boundaries [3,4]. Double Schottky Barrier model proposed that the bulk of material behaves as an n-type semiconductor and the interface behaves as a p-type semiconductor [5].

Study shows that the prolonged sintering time from 1h to 3h deteriorates the varistor properties of the ZnO-based ceramics substituted with dysprosium oxide (Dy_2O_3) and yttrium oxide (Y_2O_3) where it reduces the nonlinear properties [8,9]. The low-voltage varistor based on Zn-V-Mn-Nb-O ceramics substituted with 0.03 mol% of Gd_2O_3 and Nd_2O_3 sintered at 900 °C shows good nonlinear α value which are 9.91 and 9.94 and low E_{1mA} value which are 88.48 and 75.05 V/mm [6,7]. The $Zn_2Mn_2Nb_2O_9$ phases was found to enhance the nonlinear properties in ZnO-based ceramics [6,7].

Content from this work may be used under the terms of the Creative Commons Attribution 3.0 licence. Any further distribution of this work must maintain attribution to the author(s) and the title of the work, journal citation and DOI.
Published under licence by IOP Publishing Ltd

However, there is lacking of study on the prolonged sintering time effect on the microstructural and nonlinear electrical properties of ZnO-V_2O_5-MnO_2-Nb_2O_5-Nd_2O_3-based low-voltage varistor at 900 °C. The previous study of ZnO- V_2O_5 ceramics investigating the sintering time effects in range from 60 to 480 min [6,7,10]. The aim of this study is to investigate the effect of sintering time on the microstructural and nonlinear electrical properties of the ceramics. The prolonged sintering time was expected to diminish the microstructural and nonlinear electrical properties of the ceramics. In this paper, the effect of sintering time from 120 to 210 min on the microstructural and electrical properties of ZnO-V_2O_5-MnO_2-Nb_2O_5-Nd_2O_3-based low voltage varistor ceramics have been discussed.

2. Materials and Methods

The sample was fabricated according to (97.40-*x*) ZnO (0.5) V_2O_5 (2) MnO_2 (0.1) Nb_2O_5 (0.03) Nd_2O_3 from high-purity reagent-grade raw materials (> 99.9%, Alfa Aesar). All the units were used in the bracket were in mole percentage. Raw materials were ground using ball milling with zirconia balls with distilled water for 24h. The slurry was dried at 70 °C for 12h before pre-sintered at temperature of 800 °C for 2h. Then, the binder polyvinyl alcohols 1.75 wt% was added and sieved through the 75 µm mesh screen. The powder from each ceramic combination was pressed into pellets with 10 mm diameter and 1 mm in thickness at a pressure of 4 tonne/m^2. The pellets were sintered at a temperature of 900 °C in air for 120, 150, 180 and 210 mins with heating and cooling rates of 5 °C/min. The crystalline phase was identified using Cu Kα radiation (λ = 1.540598 Å) with PANalytical X'Pert. XRD software X'Pert high score software Pro PW3040/60) was used to identify secondary phase and *d*-spacing. The *d*-spacing value was measured using Bragg law. The average density (ρ_{avg}) of sintered sample was determined using an electronic densimeter (Alfa Mirage, Model MD-300S), working based on the Archimede's principle.

$$\rho = \frac{m}{v} = \left(\frac{W_{air}}{W_{air} - W_{water}} \right) \times \rho_{water} \tag{1}$$

where w_{air} is a weight of sample in air, w_{water} is a weight of sample in water, and ρ_{water} is a density of water. The value of ρ_{water} is about 1.00 gcm^{-3}.The surface microstructures were examined by a scanning electron microscope (SEM, model: LEO 1455 VPSEM). One side of the samples was lapped and ground with SiC paper, and then polished with 0.3 µm Al_2O_3 powder to make a mirror-like surface. The average grain size (*D*) was determined by the linear intercept method [11].

The current density (*J*) – electric field (*E*) characteristics at room temperature were measured using the voltage source-measure (Keithley Model 2410) unit to obtain their non-linear coefficient (α). All samples were coated with silver conductive paint and cured at 550 °C for 10 min to make the electrodes. The α value was determined through the expression (2):

$$\alpha = \frac{\log J_2 - \log J_1}{\log E_2 - \log E_1} \tag{2}$$

where E_1 and E_2 are the electric fields corresponding to J_1= 1.0 mA/cm^2 and J_2=10 mA/cm^2, respectively. The breakdown electrical field (E_{1mA}) was measured at 1 mA/cm^2 in the current density and the leakage current density (J_L) was measured at 0.8 E_{1mA}. The barrier height ϕ_B is estimated according to (3)

$$J = AT^2 \exp\left(\beta E^{1/2} - \phi_B\right)\left(k_B T\right)^{-1} \tag{3}$$

where k_B is the Boltzmann constant (8.167×10^{-5} eV/K), A is the Richardson's constant (30 A/cm^{2}K^2) for ZnO, T is the absolute temperature, β is a constant related to the relation as $\beta \sim (r\omega)^{-1}$, where r is grains per unit length and ω is the barrier width [12].

3. Results and Discussion

Figure 1 shows the XRD patterns of ZnO-based varistor ceramics sintered at different sintering time. The major diffraction peaks belong to ZnO (ref. No. 98-006-5190) as the primary phase and the minor diffraction peaks the V_2O_5 (ref. No. 98-010-7061) and $Zn_2Mn_2Nb_2O_9$ (ref. No. 98-004-3631) as secondary phase as the XRD pattern is matched to ICSD database. The V_2O_5 liquid phase detected in all sintering time with very low-intensity peaks. The ZnO peak shifts to the low diffraction angle from 34.4962 to 34.3789° and increases the d-spacing space slightly gain from 2.6000 to 2.6080 Å as the sintering time is increased.

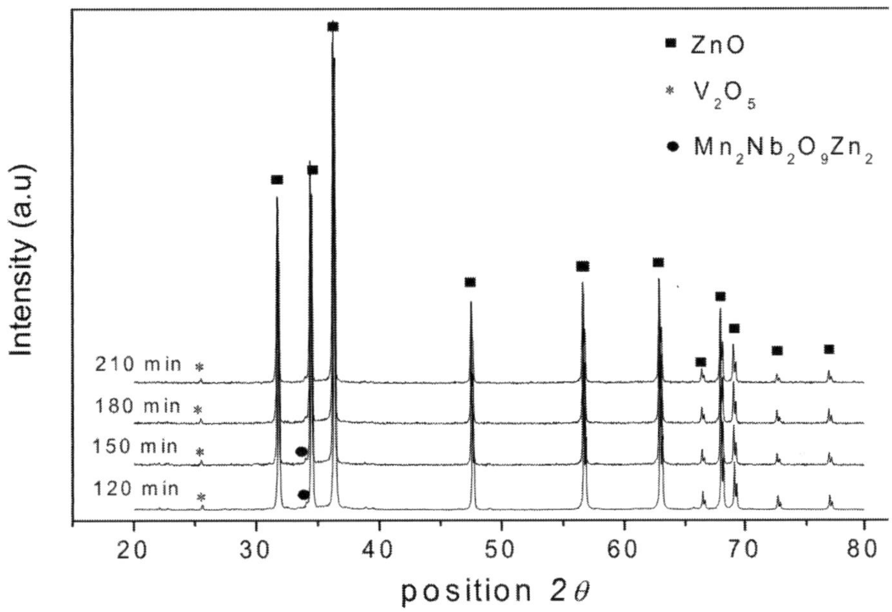

Figure 1. XRD patterns of ZnO-based varistor sintered at different time

Figure 2 shows the microstructure characterization of ZnO-based varistor ceramics at varied sintering time. The micrograph shows the large grains dispersed in a small grains at 120 mins. The disappearance of $Zn_2Mn_2Nb_2O_9$ phase at 150 mins inhibit the abnormal grain growth at the grain boundary by leaving the pores at the grain boundary. The pores are shrunk at 180 mins but the number of grain boundary is increased as Nd_2O_3 and MnO_2 act as grain inhibitor in the Zn-V-Mn-Nb-Nd-O ceramics [7, 13]. The liquid phase filled up the pores at the grain boundary is clearly observed at 210 mins as a result of segregation of Vanadium towards the grain boundary. The average grain size (D) decreases from 3.54 to 2.67 μm with increasing sintering time up to 180 mins due to pores but increases to 3.63 μm at 210 mins due to liquid phase. The pores also reduces the density (ρ) from 5.02 to 4.74 g/cm^3 with increasing sintering time up to 180 mins and but increases to 5.00 g/cm^3 at 210 mins due to liquid phase. The liquid phase act as a sintering aid that enhances the grain growth and densification by a solution and re-precipitation of ZnO [9, 14]. Assuming all additives disappear except V element, the grain boundary containing V element shown in Fig 3 confirmed the existence of V_2O_5 phase at 210 mins.

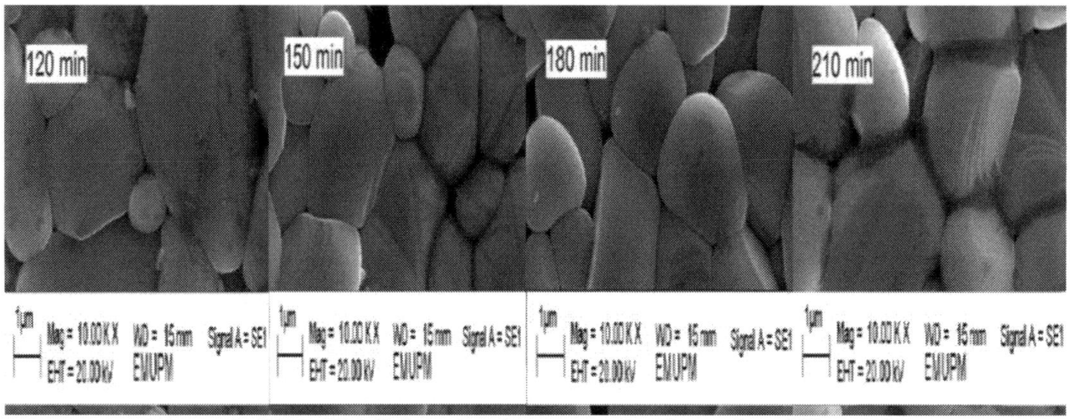

Figure 2. SEM of ZnO-based varistor ceramics sintered at different time

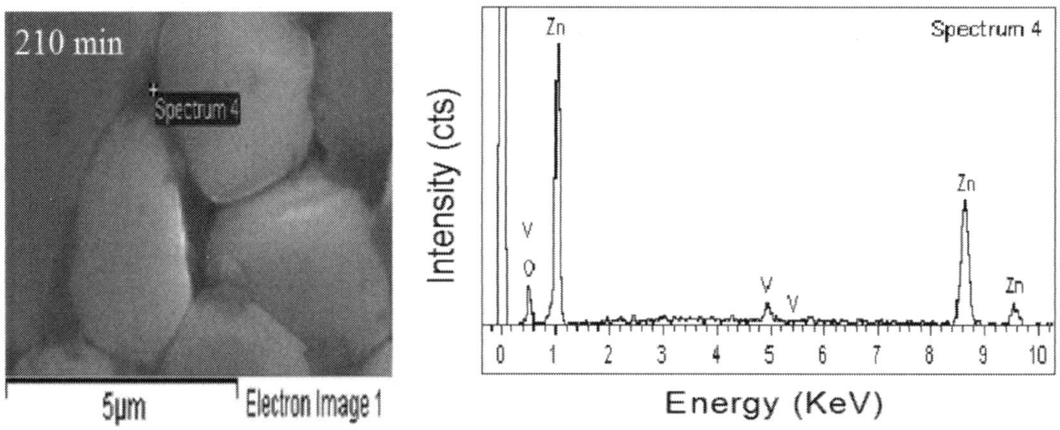

Figure 3. SEM attached with EDX of ZnO-based varistor ceramics sintered at 210 mins

The sintering time also affect the nonlinear electrical properties of ZnO-based varistor ceramics. Figure 4 shows the nonlinear region of *J-E* curves shifted to the lower electrical field region with increasing sintering time. The nonlinear α value decreased from 9.94 to 5.21 at 150 mins as the barrier height decreased from 0.849 to 0.625 eV might due to formation of pores at the grain boundary. This indicates the pores did not contribute to the build-up of barrier height. The leakage current density J_L increased from 0.45 to 0.46 mA/cm^2 as sintering time increased. The pores on the microstructures surfaces affecting the electrical conduction might contributes increment of J_L [15]. However, the pores shrunk at 180 mins causes the barrier height increased to 0.644 eV and the nonlinear α value increased to 5.80, but the J_L increased to 0.54 mA/cm^2. Assuming all additives disappear at the grain boundary might generates conduction carriers in the ZnO grain and thus weaken the insulating grain boundary [16]. Adversely, the barrier height decreased again to 0.633 eV at 210 mins and therefore, the nonlinear α value decreased to 5.74. It has been well accepted that the oxygen species promote density of interfaces states, thus increases the barrier height [17]. Assuming the liquid phase containing Vanadium might have promoted the oxygen vacancies V_O formation and depleted the O species. Thus, it reduces the density of interface state and hence decreases the barrier height. The behavior of nonlinear α and barrier height dependence on the sintering time is shown in Figure 5.

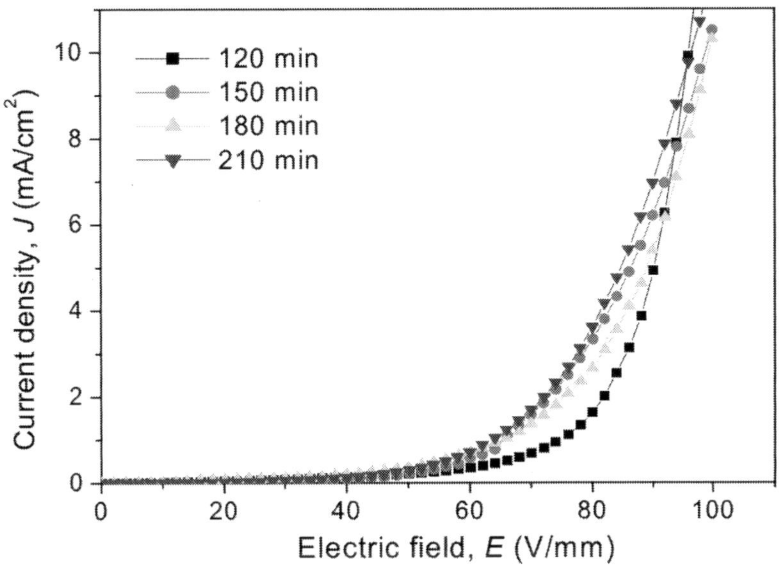

Figure 4. *J-E* curve of ZnO-based varistor sintered at different sintering time.

The J_L value also decreased to 0.42 mA/cm^2 at 210 mins. The liquid phase at the grain boundary might control the generation of conduction carrier in the ZnO grain and thus strengthen the insulating grain boundary [10]. The breakdown field E_{1mA} value also decreased from 76.07 to 66.03 V/mm as sintering time increased. This reduction is due to pores and liquid phase formation. Table 1 shows the value of position 2θ, *d*-spacing, density (ρ), average grain size (D), breakdown electrical field (E_{1mA}), nonlinear coefficient (α), leakage current density (J_L) and barrier height (Φ_b) of ZnO-based varistor ceramics sintered at varied sintering time.

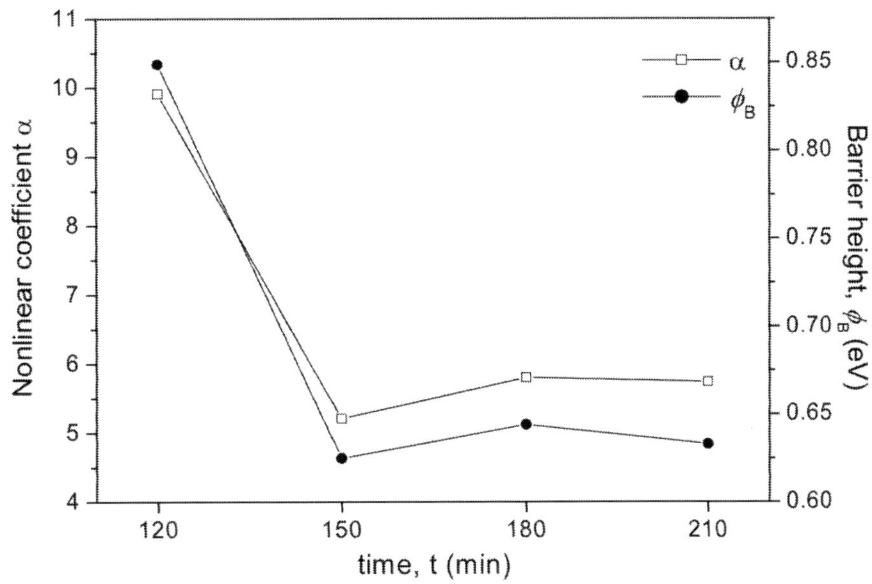

Figure 5. Nonlinear α and barrier height (Φ_b) dependence on sintering time

Table 1. Position 2θ, *d*-spacing, density (ρ), average grain size (D), breakdown electrical field (E_{1mA}), nonlinear coefficient (α), leakage current density (J_L) and barrier height (Φ_b) of ZnO-based varistor ceramics sintered at varied sintering time.

Sintering time (min)	Position 2θ (°)	*d*-spacing (Å)	ρ (g/cm^3)	D (μm)	α	Φ_b (eV)	E_{1mA} (V/mm)	J_L (mA/cm^2)
120	34.4962	2.6000	5.02	3.54	9.94	0.849	76.07	0.45
150	34.4163	2.6059	4.91	3.08	5.21	0.625	68.03	0.46
180	34.3939	2.6075	4.74	2.67	5.80	0.644	66.07	0.54
210	34.3789	2.6080	5.00	3.63	5.74	0.633	66.03	0.42

4. Conclusion

The effect of sintering time on the nonlinear coefficient of Zn-V-Mn-Nb-Nd-O varistor ceramics was studied systematically. The prolonged sintering time disrupts the microstructure and nonlinear electrical properties as expected. The pores were formed which retarded the densification which causes of decreasing in density and inhibited the grain growth which causes of decreasing in average grain size. The pores did not contribute to the build-up of barrier height which lead to decrement in the nonlinear α value. It also affecting the electrical conduction which contributes increasing in J_L value up as the sintering time increased to 180 mins. The liquid phase at 210 mins promoted the densification which causes of increase in density and grain growth which lead to increase in the average grain size with increasing sintering time. It also strengthens the insulating grain boundary which causes decrease in J_L value but promoted the V_O formation and depleted the O species, thus reduced the barrier height to 0.633 eV and consequently decreased the nonlinear α value to 5.74. The E_{1mA} value decreased from 76.07 to 66.03 V/mm as sintering time increased.

Acknowledgements

The authors are grateful to the Universiti Putra Malaysia for supporting this work under the Universiti Putra Malaysia GRANT No. GP-IPS/2016/9493000 and FRGS/2016/5524942.

References

[1] Matsuoka M 1971 *Jpn. J. Appl. Phys.* 10 736
[2] Kim ED, Kim CH and Oh MY 1985 *J. Appl. Phys.* 58 3231
[3] Cheng P, Li S and Alim MA 2007 *Phys. Stat. Sol.* 204 887
[4] Nahm CW and Shin BC 2003 *Mater. Lett.* 57 1322
[5] Mantas PQ and Baptista JL 1995 *J. Europ. Ceram. Soc.* 15 605
[6] Isa NH, Zakaria A, Azis RS and Rizwan Z 2017 *Solid State Phenom.* 268 181
[7] Isa NH, Zakaria A, Azis RS and Abdullah WRW 2017 *Dig. J. Nanomater. and Biostruct.*, 12 821
[8] Nahm CW 2004 *Mater. Lett.* 58 3297
[9] Nahm CW and Shin BC 2004 *Ceram. Int.* 30 9
[10] Tsai JK and Wu TB 1996 *Mater. Lett.* 26 199
[11] Wurst JC and Nelson JA 1972 *J. Am. Ceram. Soc.* 55 109
[12] Li C, Wang J, Su W, Chen H, Wang W and Zhuang D 2001 *Phys. B*, 307 1
[13] Chen CS 2003 *J. Mater. Sci.* 38 1033
[14] Zahid R, Azmi Z and Sabri MGM 2011 *Int. J. Mol. Sci.*, 12 1625
[15] Li BQ and Lu X 2011 *Transp Porous Med* 87 179
[16] Bai SN and Tseng TY 1992 *J. Electron. Mater.* 21 1073
[17] Abdullah WRW, Zakaria A and Ghazali MSM 2012 *Int. J. Mol. Sci.* 13 5278

Acoustical and Physical Characteristic of a New Blood Mimicking Fluid Phantom

Ammar A. Oglat [1, a], MZ Matjafri1, [b, *], Nursakinah Suardi[1, c], Mostafa A Abdelrahman[3, d], Mohammad A. Oqlat [2, d], Ahmad A. Oqlat [3, f], Raed Abdalrheem[1, g], Almutairi A.M [1, h]

[1] Department of Medical Physics and Radiation Science, School of Physics, Univirsti Sains Malaysia, 11800 Penang Malaysia
[2] Department of Biological Sciences, School of Science, Yarmouk University, Irbid Jordan
[3] Faculty of Medicine, Jordan University of Science and Technology, Irbid Jordan

mjafri@usm.my

Abstract. Offer a blood-mimicking fluids (BMF) for test object of medical ultrasound diagnostic instruments in this article is done. Both the acoustical (speed of sound and attenuation) and physical (density and viscosity) properties were defined and specified in the International Electrotechnical Commission standard and are given as a constant value. However, the viscosity of human blood depends greatly on the shear rate of the blood. In accurate studies that use a specified flow phantom, the BMFs with the appropriate viscosity are required. Furthermore, since the BMF considered as Newtonian, the viscosity it can has an arbitrary viscosity (more than 4.1 mPa.s). This article displays a general overview about the effect of density and molecular weight of dextran which is a component of mixture fluid that was used in BMF preparation. For this reason, we replaced the dextran 185000D (0.90 g/ml) with dextran 150000D powder material (0.25 g/ml). To produce liquids with proper viscosity in unit of mPa.s, we have prepared a new ternary fluid made of water, glycerol and dextran 150000D powder material.

1. Introduction

Using medical Doppler ultrasound technique for measuring the blood flow depends on the changing or variation of ultrasound frequency that reflected from movable particles of blood during the flow. The variation of frequency known as the Doppler shift or frequency shift and increases with increasing the speed of blood flow [1]. The achievement of Doppler medical ultrasound instrument can be estimated with several test objects [2-5]. The test object is also known as a flow phantom which has been used in clinical research field. Both the acoustical and physical properties blood mimicking fluid (BMF) must closely match to the human blood to make the examination meaningful.

The International Electrotechnical Commission (IEC) was used as a standard and constant values of the acoustical and physical properties of Blood Mimicking Fluid for flow phantoms Table 1. [2, 6]. Several various Blood Mimicking Fluids have been reported [2, 6-10]. Some scattered particles were added to the Blood Mimicking Fluid to produce scattering influence of blood cells, also the variating of the characteristics based on if the scattered particles deposit down or float.

Content from this work may be used under the terms of the Creative Commons Attribution 3.0 licence. Any further distribution of this work must maintain attribution to the author(s) and the title of the work, journal citation and DOI.
Published under licence by IOP Publishing Ltd

Traditional BMFs have contained scattered particles material such as nylon powder (Orgasol), polystyrene microspheres, starch, Sephadex, and silicon powder in the fluid to form BMF which was used for different purposes. However, the densities of these scattered particles should closely match the densities of mixture fluids to prevent float or deposit down of those particles and then to allows keep neutrally buoyant of scattered particle materials inside this liquid.

Table 1. Specifications of the BMF defined as the IEC standard. f is the acoustic frequency (Hz).

Attenuation coefficient [(dB/cm/MHz)]	<0.1 × 10^−4 × f
Speed of sound (m/s)	1570 ± 30
Density (×10^3 kg/m^3)	1.050 ± 0.040
Viscosity (×10^-3 Pas)	4.0 ± 0.4

2. Materials and Methods

In this research study, our target was to prepare a new BMF with proper speed of sound, attenuation, and density that defined and known in the IEC standard [2], in addition to the arbitrary viscosity since the BMF not depends on shear rate. Water, glycerol and dextran 150000D dispersed with nylon 12 particles were used to obtain the BMF.

However, in this experiment research, the concentration of glycerol and dextran is changed several times to get good acoustical and physical properties which known in the IEC standard. In this research study, three different experiments were made. In the first experiment, both the dextran 150000D concentration and the glycerol concentration were increased in the mixture with small ratios. The second experiment was done in the same way of the first experiment but with changing the items' concentration to obtain the suitable speed of sound and density which should be equivalent to IEC standard, the mixture made of the dextran 150000D concentration with the same amounts of the glycerol concentration. Finally, the glycerol was increased and fixed dextran 150000D concentration in some mixture samples and then fixed the glycerol and increase the dextran concentration in the other mixture. However, to make sure that the values of the sample number 3 in Table 2 are the optimum values for preparing the BMF. The changes of the items' concentrations around this sample were done in Table 3.

Applying of Scheffé equation for several reasons. Firstly, because the mixture made of three different components (water, glycerol and dextran) for the new BMF. Secondly, the density of the liquids have no linear relationship between each other in several experimental results [25, 26] for designing tests with a mixture fluid. The function of the three different components M represented by three different factors are S1, S2, and S3. Equations 1 and 2.

$$M = H_1S_1 + H_2S_2 + H_3S_3 + H_{12}S_1S_2 + H_{13}S_1S_3 + H_{23}S_2S_3 + H_{123}S_1S_2S_3 \quad (1)$$
$$S1+S2+S3=1 \quad (2)$$

Where S1, S2, and S3 are weight fractions. And, H1, H2, H3, H12, H13, H23, H123 are constants. There are seven coefficients (H1-H123) are unknown. In this case, the naming suggested by Scheffé equations (seven equations) will help us to find the values of the seven unknown coefficients since the Scheffé principle depends on seven equations with seven coefficients. The research study data M for seven group of values of S1, S2, and S3 can be utilized to design the BMF. The answer of M to ingredients of S1, S2, and S3 are given by the research study data displayed in Table 4.

Through the exchange M= Vo or M= Po, With the substitution M Vo or M= P, the general equation given by Equation 2 can be rearranged. Where the Vo and Po are the speed of sound and the density of the mixture fluid samples, respectively. The typical formula determined by Equation 1, can be rearranged. The amounts of Vo and Po for various group of values of S1, S2 and S3 are given by the research study data displayed in Table 3. Where S1, S2 and S3 are the components of water, glycerol, and dextran 150000D as wt% in the sample, respectively.

The BMF items of glycerol, dextran 150000D, water, and nylon were used in the research study experiments are 99% pure, supplied by Sigma Aldrich Company. The water was distilled utilizing a quartz distiller. The dynamic viscosities of the items were measured applying an electronic rotational viscometer in mPa.s, displayed in Tables 3 and 4. The temperature readout during measurement was 37 ◦C like human body temperature. The densities of the materials were measured applying density meter DMA35 with an accuracy of 0.001. The speed of sound in the materials was specified by the pulse-echo PE method by applying A-scan GAMPT ultrasonic device with uncertainty of ±3 m/s, and this uncertainty was estimated from repeated measurements of the reference distilled water scan in the Plexiglas and it was ±0.01 m/s as a random uncertainty and ±3.0 m/s as a systemic uncertainty. The temperature of the samples item when measuring the densities, speed of sound and viscosity was preserved within ±1 °C.

3. Results and Discussion

In the first experiment, when both the dextran 150000D concentration and the glycerol concentration were increased with slightly ratios, the speed of sound of mixture solution decreased and the density decreased too (proportional relationship) but there are no suitable result values between the density and speed of sound of the mixture fluid like values defined in the IEC standard. The reason of decreasing the speed of sound that it is directly proportional to the density, and the density of mixture decreases due to the glycerol concentration was more than the dextran concentration and the dextran density much lower (five times) than glycerol density. At the second experiment, the speed of sound of mixture solution decreased and the density decreased too, but also there was a mismatch between the speed of sound and the density according to IEC standard. The main cause of mismatch is the densities of items, such as glycerol has density more than blood density about 0.2 g/ml and the dextran 150000D has density less than the blood density about -0.8 g/ml. Thus, the minus value more than the positive value. At the final test, we noted that both the speed of sound and the density of the mixture solution increased (proportional relationship) when the glycerol increased and fixed dextran 150000D concentration, whereas the speed of sound and density decreased when we fixed the glycerol concentration and increased the dextran concentration Table 2, because the dextran density is much lower (five times) than glycerol density and vice versa. However, because the first two experiments were not a suitable to be mixture fluid for preparing BMF. Thus, we chosen the results of the final one experimental research study since it was a suitable result of the mixture solution for preparing BMF. Sample number 3 at Table 2 shows the BMF that has a proper speed of sound density and viscosity. However, the density of a suitable mixture fluid at sample 3 was 1.020 g/ml, speed of sound was 1581 m/s and the dynamic viscosity was 4.4 mPa.s. The viscosity of ternary mixture was increased gradually due to used dextran in all mixture with high molecular weight 150000D. And, when the molecular weight increases, the viscosity will increase directly.

Table 2. Constitutions and physical properties of water, glycerol and dextran 150000D solutions at change the mixture fluid ratios (temperature: 37.0 °C).

Number of samples	Water Wt%	Glycerol Wt%	Dextran 150000D Wt%	density (g/ml)	speed of sound (m/s)	viscosity (mPas)
1	100	0	0	0.998	1508	0.89
2	95	4.5	0.5	1.01	1522	3.9
3	90	9.5	0.5	1.02	1581	4.4
4	88	11.5	0.5	1.03	1610	5.4
5	86	12	2	1.012	1578	7.4
6	85	12	3	1.001	1530	10.9
7	83	12	5	0.97	1485	16

The research study results by using Scheffé equation displayed in Table 2 give the amounts of V_o, P_o, S1, S2, and S3 in Equation 1, and we get the following:

S1	S2	S3	S12	S13	S23	S123	Seven Unknown	Density g/ml
1.000	0.000	0.000	0.000	0.000	0.000	0.000	H1	0.988
0.950	0.045	0.005	0.043	0.005	0.0002	0.0002	H2	1.01
0.900	0.095	0.005	0.086	0.005	0.001	0.001	H3	1.02
0.880	0.115	0.005	0.101	0.004	0.001	0.001	H4	1.03
0.860	0.120	0.020	0.103	0.017	0.002	0.002	H5	1.012
0.850	0.120	0.030	0.102	0.026	0.004	0.003	H6	1.001
0.830	0.120	0.050	0.010	0.042	0.006	0.005	H7	0.970

Due to the number of formulas is like the number of unknown factors, thus the seven formulas are easy to solve together. Also, can specified H1, H2, H3, H12, H13, H23, and H123 relating to the density. So, the values of H1, H2, H3, H12, H13, H23, and H123 relating to the density are 0.998, 0.962, 0.933, 0.923, 0.900, 0.830, and 0.800, respectively. Thus, the standard deviation is ± 0.20. In other words, the different results between measurements by the equipment and Scheffé equation because the components of the mixture and density of the liquid don't have linear relationship between each other g/ml.

In addition. The data that resulted from the research study can be specified that H1, H2, H3, H12, H13, H23, and H123 are relating to the speed of sound. As V_o and P_o are known in the IEC standard, we can utilize Vo = 1581 m/s and P_o = 1.02 g/ml. The graph with the relation between the speed of sound and density in the dextran 150000D and glycerol aqueous solution are displayed in Figure. 1. The lines of the density P_o and speed of sound V_o intersect in the S1, S2, and S3 plane at values of S1, S2, and S3 for that the fluid meets both the speed of sound and density needs of the BMF. However, medical ultrasound technique produces Doppler spectral waveforms by the scattering of red blood cells inside a vessel. To mimic and simulate the scattering influence by red blood cells in the blood, scattered particles material of the BMF is dispersed with nylon 12 particles. Nylon 12 particles supplied by Sigma Aldrich were utilized in the experimental study. The density of the nylon 12 particles corresponds completely with the amount known in the IEC standard. Thus, the neutral buoyancy effect of the nylon 12 particles is guaranteed in the fluid with a density of 1.02 g/ml. Moreover, the particle size (diameter) of nylon 12 in this experiment is 5-10 μm.

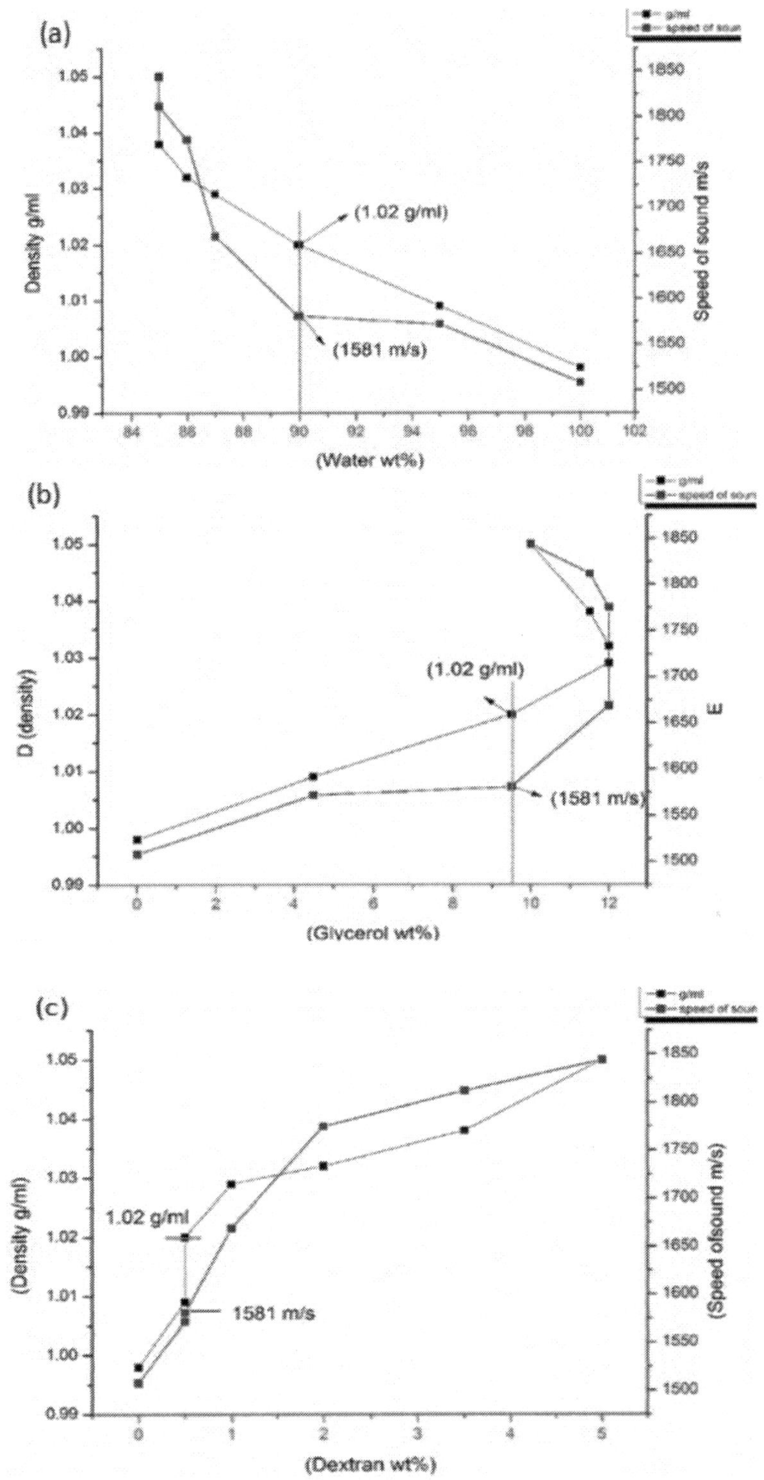

Figure 1. Suitable speed of sound and density of fluid mixture of a BMF at different concentrations by wt% of (a) water (b) glycerol and (c) dextran 150000D, respectively.

Moreover, the values that resulted from mixture fluid in Table 3 to make sure that sample number 3 in Table 2 is the optimum specimen, were indicated that the sample number 3 is the optimum sample of mixture fluid to mixing with nylon particles for preparing BMF, see Table 3.

Table 3. Constitutions and physical properties of water, glycerol and dextran 150000D solutions at change the mixture fluid ratios (temperature: 37.0 °C), around sample number 3 in Table 2.

Number of samples	Water Wt%	Glycerol Wt%	Dextran 150000D Wt%	density (g/ml)	speed of sound (m/s)	viscosity (mPas)
1	100	0	0	0.998	1508	0.89
2	95	4.5	0.5	1.01	1522	3.9
3	93	6.5	0.5	1.014	1543	4.1
4	91	8.5	0.5	1.018	1567	4.3
5	90	9.5	0.5	1.02	1581	4.4
6	89	10.5	0.5	1.025	1594	4.9
7	88	11.5	0.5	1.03	1610	5.4

Three different types of ternary aqueous solution were used for preparing the mixture fluid (water (90.0 wt %), glycerol (9.5 wt %), and dextran 150000D (0.5 wt %)) of BMF. The research study data, speed of sound and density of mixture fluids were 1581 m/s and 1.02 g/ml, respectively. The values were corresponded well with the theoretically calculated results. The BMF was prepared with both the density and speed of sound agreed in IEC standard. However, the items densities play an important role in the mixture fluid to get a suitable speed of sound and density, while the molecular weight plays an important role in mixture fluid viscosity.

4. Conclusion

New blood-mimicking fluid with the density (1.02 g/ml), viscosity (4.4 mPa.s), and speed of sound 1581 have been wanted for test object of medical ultrasound diagnostic instruments. the Results from the initial search study pointed out that we can prepare the fluid mixture of BMF from water, glycerol, and dextran. As a conclusion, the acoustical features of this BMF was agreed the IEC standard. Moreover, changing the dextran sort affects the physical and acoustical features of the fluid mixture.

References

[1] Smith, H.-J., *Quantitative Doppler Flowmetry I. Construction and Testing of a Duplex Scanning System.* Acta Radiologica Diagnosis (Sweden), 1984. 25(4): p. 305-312.

[2] Yoshida, T., K. Sato, and T. Kondo, *Blood-mimicking fluid using glycols aqueous solution and their physical properties.* Japanese Journal of Applied Physics, 2014. 53(7S): p. 07KF01.

[3] Browne, J., et al., *Assessment of the acoustic properties of common tissue-mimicking test phantoms.* Ultrasound in medicine & biology, 2003. 29(7): p. 1053-1060.

[4] Law, Y., et al., *On the design and evaluation of a steady flow model for Doppler ultrasound studies.* Ultrasound in medicine & biology, 1989. 15(5): p. 505-516.

[5] Hoskins, P.R., *Simulation and validation of arterial ultrasound imaging and blood flow.* Ultrasound in medicine & biology, 2008. 34(5): p. 693-717.

[6] Ramnarine, K.V., et al., *Validation of a new blood-mimicking fluid for use in Doppler flow test objects.* Ultrasound in medicine & biology, 1998. 24(3): p. 451-459.

[7] Tanaka, K., et al., *Blood-Mimicking Fluid for Testing Ultrasonic Diagnostic Instrument.* Japanese Journal of Applied Physics, 2012. 51(7S): p. 07GF18.

[8] Oates, C., *Towards an ideal blood analogue for Doppler ultrasound phantoms.* Physics in medicine and biology, 1991. 36(11): p. 1433.

[9] Samavat, H. and J. Evans, *An ideal blood mimicking fluid for doppler ultrasound phantoms.* Journal of Medical Physics, 2006. 31(4): p. 275.

[10] Hoskins, P., T. Loupas, and W. McDicken, *A comparison of the Doppler spectra from human blood and artificial blood used in a flow phantom.* Ultrasound in medicine & biology, 1990. 16(2): p. 141-147.

[11] Scheffé, H., *Experiments with mixtures.* Journal of the Royal Statistical Society. Series B (Methodological), 1958: p. 344-360.

[12] Cornell, J.A., *Experiments with mixtures: designs, models, and the analysis of mixture data.* Vol. 895. 2011: John Wiley & Sons.

Ex-situ Generation of the Gold Nanowire Networks Bovine Serum Albumin Bio-Conjugated System Using Pulsed Laser Ablation in a Harsh Environment

S.A. Abdulateef[1, 3], A.F. Omar[1], M.Z. Mat Jafri[1], Naser M. Ahmed[1], Azman Seeni[2] and E. A. Kabaa[1]

[1] School of Physics, Universiti Sains Malaysia, Penang, Malaysia.
[2] Advanced Medical and Dental Institute, Universiti Sains Malaysia, Penang, Malaysia.
[3]College of Medicine, Al Iraqia University, 1556 Baghdad, Iraq.

Sinan1974@yahoo.com, mjafri@usm.my

Abstract. The experimental conditions were investigated to fabricate gold nanowire networks (AuNNWs) by pulsed laser ablation in saline solutions. The study revealed that the production of alternative nanoparticles (or aggregates) or nanowire networks at the fundamental wavelength of a Nd:YAG laser (1064 nm) depend on the medium media surrounding Au nanoparticles during the process of synthesis. The aggregates and precipitates of the gold nanoparticles (AuNPs) in saline solutions were caused by the harsh environment of the anions with Na+ cations in the solution. The ex-situ was used for the preparation of bovine serum albumin (BSA) bio-conjugates. Enhanced colloidal stability was dramatically achieved when BSA was adsorbed onto Au nanowire networks. The optical extinction spectra of the colloids in the UV/vis region were obtained to evaluate the structure of the dispersed Au phase. Transmission electron microscopy (TEM) was applied to visualize the size and morphology of the colloidal particles. Au nanowire networks compatibility can be prepared and applied in real environments.

1. Introduction

Noble metal nanocrystals have attracted considerable attention from a fundamental point of view and for technological applications mainly due to the position of their surface plasmon resonance lying in the visible spectral range. Moreover, the strong dependence of the electronic, optical, catalytic, and magnetic properties of the nanoparticles on their shape [1] and size [2] provides opportunities of creating nanostructures for specific applications by way of controlling the preparation conditions to produce nanoparticles having the desired shape and size.

The profile of the optical transmission spectrum of the corresponding colloid obtained in the UV/Vis range allows their indirect assessment. For example, spherical nanoparticles (up to a certain size) are characterized by a single surface plasmon resonance (SPR) peak in their extinction spectrum. The appearance of two peaks arising from the transverse and longitudinal SPRs is observed in extended shapes, e.g., spheroids or nanorods [3]. The positions of these peaks in the extinction spectrum depend on the aspect ratio of the nanostructure considered [4]. The existence of aggregates or nanowires in the colloid influences the corresponding extinction spectrum in a specific way. One-dimensional nanostructures excite substantial interest regarding the fundamental properties of surface plasmon

Content from this work may be used under the terms of the Creative Commons Attribution 3.0 licence. Any further distribution of this work must maintain attribution to the author(s) and the title of the work, journal citation and DOI.
Published under licence by IOP Publishing Ltd

polaritons (SPPs) propagation in the nanoscale structured matter [5]. The gold nanowires exhibit excellent electric-conductivity, chemical and thermal inertness, and biocompatibility. Their properties depend on substances absorbed on the surface. Pulse Laser ablation (PLA) technique has attracted increasing interest and is extensively employed to synthesize new novel materials [6]. It has inherent advantages compared to the different chemical methods used, especially when biological applications are considered. Bovine serum albumin (BSA) often serves as a model protein for nano-particle–protein studies, as it shares 98% of its amino acid sequence with its human variant [7] Although this protein is negatively charged at physiological pH, it also contains 11 hydrophobic binding domains, 66 positively charged lysines, and an unpaired sulfhydryl group (cysteine) [8].

When a saline solution has a NaCl concentration less than that of blood plasma or bodily fluids, AuNP cations form a complex with Cl- anions, resulting in aggregation and precipitation of the AuNPs. Hence, it is difficult to maintain a stable dispersion of colloidal AuNPs within any saline solution for a relatively long time. This is because particle agglomeration and sedimentation may affect the biodistribution of the material, which may lead to harmful side effects [9].

A large variety of chemical methods for fabricating gold nanowires have been reported in the literature. A general disadvantage of the suspensions produced by these methods is the presence of residual ions as byproducts and their potential toxicity which precludes the biological applications [10-12]. In this study, we used the method of nanosecond laser ablation of an Au disc immersed in a harsh environment to produce contamination-free nanowire networks.

2. Materials and Methods

AuNNWs were prepared via the laser ablation technique in two different solutions. The first was a solution of deionized distilled water (DDW) and highly concentrated NaCl (150 mM), while the second contained similar components, but after PLA process directly the addition of BSA (375 μM). Both solutions were prepared at room temperature.

To synthesize AuNNWs, a gold target was placed at the bottom of a 10-mL quartz vessel filled with 3 mL of both solutions independently. The target was irradiated with a Q-switched Nd: YAG laser operated with a peak emission wavelength of 1064 nm, a repetition frequency of 5 Hz, and a pulse duration of 6 ns. A schematic diagram of the pulse laser ablation in liquid (PLAL) system used in this experiment is illustrated in Figure 1.

The focusing lens with a focal length of 100 mm was used to focus the laser beam on the gold target. The energy per pulse 650 mJ was used to ablate the Au target for 4 min in both solutions.

The optical properties of the solution of synthesized colloidal AuNPs were characterized using an ultraviolet/visible (UV-vis) spectrophotometer (Ocean Optics; SD1024DW, USA). Transmission electron microscopy (TEM, FEI CM 12, USA) operated at 120 kV were employed was to study the structural properties of the samples.

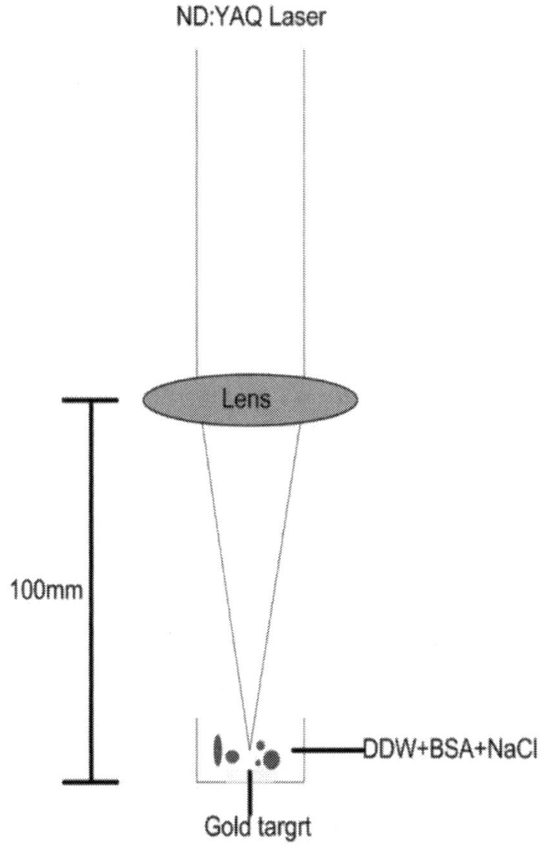

Figure 1: A schematic diagram of the PLAL system

3. Results and discussion

To ensure sample reliability, tests were performed to characterize the samples. After the laser ablation of the Au target in the DDW with NaCl, and ex-situ BSA-conjugated AuNNWs in saline solutions. Some changes in the coloration of the solutions were observed; the solution changed from colorless to transparent red. The change in color could be an indication of the production of colloidal AuNPs. Notably, after the ablation with laser fluences to obtain AuNPs in the 150 mM NaCl solutions, the color of the solution changed to different hues of color [13].

3.1 Absorption spectra

UV-vis absorption spectroscopy is the most commonly used technique to study the optical properties of AuNNWs, determine the relationship between the laser parameters and productivity, and evaluate the AuNNP-corona interaction. Furthermore, absorption spectroscopy is faster than other standard optical technologies. In colloidal GNNWs, the SPR band and its intensity depending on the size, shape, and distribution of the AuNNWs. It has been noticed after performing the ablation process to get Au NPs in 150 mol of NaCl solution, that decrease the values of the intensity; the solution color changed into transparent very light purple. Then, after an hour AuNPs started to aggregate and precipitate at the bottom of the sample container. Colloidal AuNNWs precipitate in NaCl solution at a physiological pH. The SPR absorbance is at a maximum at a maximum wavelength $\lambda max = 530$ nm and decreases with NaCl concentration. Colloidal AuNPs were aggregated and precipitated to the bottom of the sample container at room temperature. Figure 2 (solid line) shows the SPR of colloidal AuNNWs.

The ex-situ AuNP–BSA system, the absorption spectrum was acquired after laser ablation and the addition of BSA at the physiological concentration (375 μM) after AuNPs adsorb BSA.

The solution color was transparent very light blue and insignificant color changes even after long period of time.Therefore, we believe that AuNNWs in this state was stable. The SPR absorption spectrum decreases, broadening the plasmon band and a red-shift of the peak occurs as a result of plasmon coupling with AuNP aggregates or bigger size. See figure 2 (dash line).

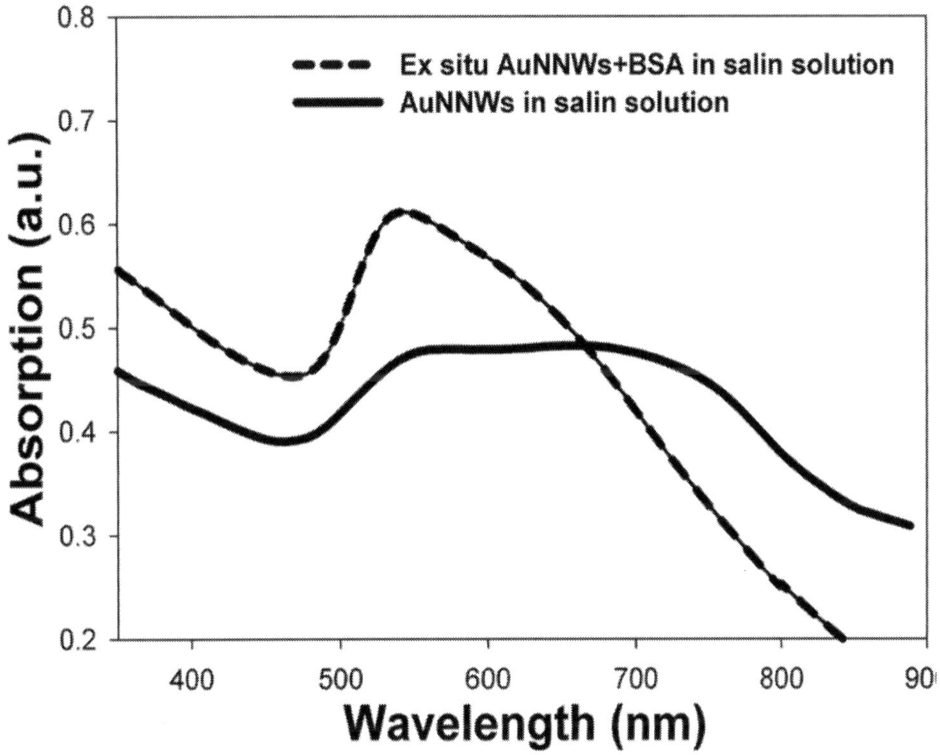

Figure 2: UV-vis spectra of a colloidal dispersion containing particles synthesized by means of laser ablation of an Au target in in DDW + NaCl solution (solid line) and DDW+BSA+NaCl solution (dash line).

3.2. TEM observations

TEM was employed to measure the sizes, size distributions, and shapes of the AuNNWs. The TEM image shows an increment of aggregation and agglomeration along with larger particles. This is indicated by the larger particles, agglomeration, and aggregation in the NaCl physiology concentration, which was confirmed by the TEM image shown in figure 3 (a).

When the ex-situ BSA conjugated with the AuNNWs, it control and stop aggregation. BSA can stable the sample which can be attributed to the insignificant aggregation and intensive stability and build chains of AuNNWs shows figure 3 (b).

IOP Conf. Series: Journal of Physics: Conf. Series **1083** (2018) 012011 doi:10.1088/1742-6596/1083/1/012011

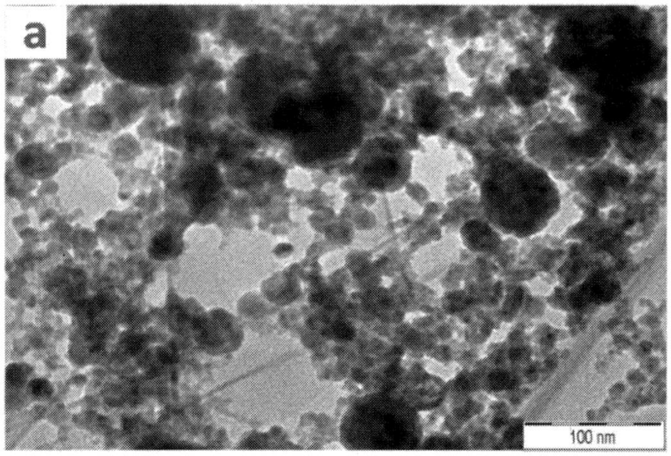

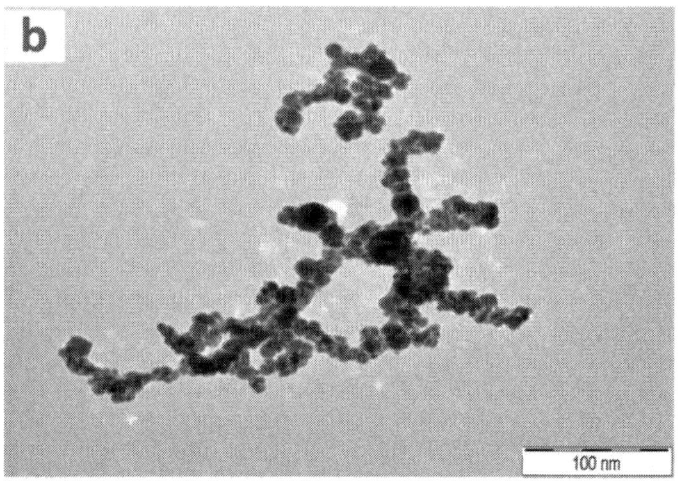

Figure 3: Transmission electron microscope images of AuNNWs prepared in (a) DDW + NaCl solution and (b) DDW+BSA+NaCl solution.

4. Conclusion

The study emphasized the important role of the self-absorption in creating AuNNWs by PLA in a Harsh Environment. Prepared AuNNWs by PLA technique in a harsh environment was investigated. The experimental results revealed that it is difficult to maintain colloidal AuNNWs for a long time in a solution containing only NaCl. After PLA process directly addition BSA to the NaCl solution was found to be an effective approach to producing AuNNWs within a short period in such harsh environments. Ex-situ technique in producing AuNNWs conjugated with BSA was successfully achieved by pulsed Nd:YAG laser ablation of the gold target in saline solution. The TEM image showed the development of investigating BSA effects on size, morphology, growth, and composition. The importance of the method of pulsed laser ablation for producing contamination-free AuNNWs suitable for medical and research aims was thus confirmed.

References

[1] Jin. R, Cao. Y, Mirkin. Kelly C.A, K.L., Schatz G.C and Zheng J.G., Science 294 (2001) 1901–1903.

[2] Alivisatos. A.P, J. Phys. Chem. 100 (1996) 13226–13239.
[3] Perez-Juste. J, Pastoriza-Santos. I, Liz-Marzan. L.M and Mulvaney. P, Coordin. Chem. Rev. 249 (2005) 1870–1901.
[4] van der Zande. B.M.I, Bohmer. M.R, Fokkink. L.G.J and Schonenberger. C, Langmuir 16 (2000) 451–458.
[5] Schider. G, Krenn. J.R, Hohenau. A, Ditlbacher H, Leitner A, Aussenegg F.R, Schaich W.L, Puscasu I, Monacelli B and Boreman G, Phys. Rev. B 68 (2003) 155427.
[6] San. N, Kurşungöz. C, Tümtaş. Y, Yaşa. Ö, Ortaç. B and Tekinay. T, Particuology, 17 (2014) 29-35.
[7] Kan. C, Cai. W, Li. C, Zhang. L and Hofmeister. H, J. Phys. D. Appl. Phys, 36 (2003)1609–1614.
[8] Tarhoni. M. H, Lister. T, D. Ray. E, and Carter W. G, Biomarkers 13(2008) 343–363.
[9] Albanese, Alexandre, and Warren CW Chan. ACS nano 5.7 (2011) 5478-5489.
[10] Pileni. M.P, J. Phys. Chem. 97 (1993) 6961–6973.
[11] Huang C.J, Chiu P.H, Wang Y.H, Yang C.F and Feng S.W, J. Colloid Interf. Sci. 306 (2007) 56–65.
[12] Hansen. C.R, Westerlund F, Poulsen K.M, Ravindranath R, Valiyaveettil. S and Bjornholm T, Langmuir 24 (2008) 3905–3910.
[13] Abdulateef, S. A., Omar. A. F, Jafri. MZ Mat, Ahmed. Naser M, and Seeni Azman Journal of Cluster Science 28 (2017) 3193-3207.

Ultraviolet Protection Properties of Commercial Sunscreens and Sunscreens Containing Zno Nanorods

Nur Athirah Ahmad Zaki[a] *, Shahrom Mahmud[b] and Ahmad Fairuz Omar[c]

School of Physics, Universiti Sains Malaysia, 11800 Pulau Pinang, Malaysia

[a]athirahzaki92@gmail.com, [b]shahromx@usm.my, [c]fairuz_omar@usm.my

Abstract. Zinc oxide nanorods were used as the main ultraviolet agent in the development of sunscreens. Four commercial sunscreens containing organic and inorganic UV agents were studied for their UV performances and SPF determinations. From FESEM and TEM micrographs of ZnO powder, the dominant structure was rod structure with width range 61-70 nm. Sunscreens containing different concentrations of ZnO nanorods were characterized using UV-Vis spectrophotometer and X-ray diffraction (XRD). The absorbance spectra for commercial sunscreens showed a good protection in range of UVB (280-320nm), while sunscreens containing 20-40% (weight percentage) of ZnO nanorods showed a greater UV absorbance in UVA and UVB (280-400 nm). The strong UVA absorbance of ZnO nanorod can offer extra UV protection for commercial sunscreens. The average crystallite sizes of the sunscreen samples calculated from the XRD spectra were in the range 45.27 to 47.36 nm. In vitro sun protection factor (SPF) method was carried out to determine the SPF value of sunscreen specimens. SPF values of sunscreens containing ZnO nanorods were 29.50 to 43.48, whereby these SPF values indicated more than 93% of UVB blocking capability. The SPF values found of commercial sunscreens were slightly different from the SPF values in the product labels. In conclusion, incorporation of ZnO nanorods into sunscreen can greatly enhanced the UV blocking performance of sunscreens.

1. Introduction

The destruction of ozone layer of the atmosphere led more ultraviolet to penetrate in the earth and causes health issues, such as sunburn, aging, skin reddening, acne allergies and even skin cancer [1]. The use of sunscreen is one of the most popular approach to protect skin against UV radiation. The introduction of zinc oxide nanoparticles (ZnO NPs) contributed to higher UV protection efficacy due to higher surface area to volume ratio and appear transparent to the skin surface compared to ZnO bulk particles [2]. Besides, they offer a broad spectrum protection besides very low of skin irritation and sensitization, inertness of the ingredients and low penetrate skin [3], [4]. Sun protection factor (SPF) is the universal parameter for describing the efficiency of sunscreen products, which can draws attention of consumers in terms of reduce UV-induced erythema [5]. It is measured by both in vivo [6] and in vitro SPF determination methods [7]. However, due to high cost and time consumption of in vivo methods, the SPF of sunscreens was determined by in vitro of ultraviolet spectrophotometry, which substitutes in the in-vivo method [7]–[9]. UV-vis spectroscopy was employed to measure absorbance at 290 to 320 nm and a SPF was calculated using the absorbance measured and Mansur mathematical equation [10].

Content from this work may be used under the terms of the Creative Commons Attribution 3.0 licence. Any further distribution of this work must maintain attribution to the author(s) and the title of the work, journal citation and DOI.
Published under licence by IOP Publishing Ltd

In this study, sunscreen contained different concentrations of ZnO were prepared. Four commercial sunscreens from different brands and different values of SPF were purchased. We report the UV performances of sunscreen contained ZnO and commercial sunscreen. The determination of SPF value for both samples were determined using an in-vitro UV spectrophotometry method.

2. Experimental details

A commercial pharma grade ZnO powder was used in this work. The ZnO was synthesized using French process [11]. In this study, sunscreen creams were made from mineral oil and beeswax, while ZnO was used as the main UV agent. The sunscreens were produced with different weight percentage (wt %) of 5, 10, 15, 20, 25, 30, 35 and 40 of ZnO, respectively. Other than these higher concentrations were not produced because it will cause the texture of sunscreen to be more harder and appear white pigment on the skin [12]. Besides that, four different brands of commercially available sunscreens were purchased in the drugstore in Penang. Morphology and composition of the ZnO powder were characterized using field-emission scanning electron microscope (FESEM-EDS, model: JSM – 6460 LV) and Phillips CM12, transmission electron microscopy (TEM). The structural and crystallinity properties of sunscreen samples were characterised by X-ray diffraction (XRD, model: Panalytical X'pert Pro MRD Pw3040). The absorption spectra sunscreen containing ZnO and commercial sunscreens were obtained from Varian Cary 5000 UV/Vis/NIR spectrophotometer.

Sun protection factor of sunscreen contained ZnO and commercial sunscreen, were measured using in vitro UV spectrophotometry method [9]. All sunscreen contained different concentration of ZnO and commercial sunscreen (1.0 ± 0.05 g) were weighed individually, transferred to 100 mL volumetric flask and diluted with 40% ethanol and 60% water solution. Followed by 5 minutes of ultrasonication, the solution was filtered through cotton filter and producing 10 mL filtered solution. A 5.0 mL of aliquot was transferred to 50 mL volumetric flask and diluted with volume 40% ethanol and 60% water solution. The absorbance spectra of aliquot prepared were obtained from 290-320 nm at 5 nm interval. The measurements were taken three times and the determination of SPF values by application of Mansur mathematical equation in Equation (1);

$$SPF = CF \times \int_{290}^{320} EE\ (\lambda) \times I\ (\lambda) \times Abs\ (\lambda) \qquad (1)$$

where *EE* is erythema effect spectrum; *I* is solar intensity spectrum; *Abs* is absorbance of sunscreen and *CF* is correction factor (=10). The absorbance spectra were expressed as mean ± SEM, then multiplied with the respective EE (λ) × I (λ) constants in Table 1. Their summation $\sum_{290}^{320} EE\ (\lambda) \times I\ (\lambda) \times Abs\ (\lambda)$ were taken and multiply with the correction factor (10).

Table 1: Product function used in the calculation of SPF [10].

Wavelength (nm)	EE × I
290	0.0150
295	0.0817
300	0.2874
305	0.3278
310	0.1864
315	0.0839
320	0.0180

3. Results and discussion

From FESEM micrograph in Figure 1(a), several morphologies were identified in ZnO powder: rods, slabs, tripods and irregular-shaped particles. Among these structural morphologies, nanorods were dominant as observed in TEM image (shown in Figure 1(c)). The rod structures had width 61-70 nm (Figure 1(d)) was determined from TEM micrograph. The FESEM-EDS Figure 1(b) shows the composition percentage of ZnO powder. The relative ratio of O: Zn in the atomic and weight percentages measurement were 0.74 and 0.18, respectively. This result shows content of oxygen atom was lower than zinc. The excess zinc could be attributed to intrinsic defects in crystalline ZnO including oxygen vacancies and zinc interstitials [13]. As a result, sunscreen contained ZnO nanorods as a main active UV filter. ZnO nanorods possess unique optical properties, where their absorption very high in UV regions that able to prevent UV rays from penetrating the skin [14].

XRD pattern of the sunscreen contained ZnO nanorods and commercial sunscreens are shown in Figure 2(a-b). The highest peaks including (100), (002) and (101) were selected for average size analysis. Figure 2(a) shows the higher the concentration of ZnO nanorods in the sunscreen, exhibited sharp and greater peak intensities. It clearly revealed that the ZnO nanorods diffraction peaks matched very well with the hexagonal wurtzite phase. Since the XRD do not have any additional peaks apart from the peaks characteristic for ZnO, suggesting that the formed product contained high-quality ZnO nanorods. The average particles size of the ZnO nanorods in the sunscreen were measured using Scherrer equation in the range of 45.27 to 47.36 nm, respectively. Figure 2(b) shows the presence of inorganic compounds such as ZnO. However, the percentages of these compound contained in the product were not labelled in the product labels; hence, it is difficult to determine their amount in the products. No peak indicated TiO_2 particles was found in the spectra of sunscreen A even though TiO_2 was label-printed to be one of the contents. Meanwhile, sunscreen B shows a very weak intensity of peak at 25.34° indicating the presence of TiO_2 particles found in the sunscreen. By contrast, the other two sunscreens showed high peak intensities allowing average of particles sizes to be calculated. Average zinc oxide particles sizes of 13 nm and 14 nm were found in sunscreen C and sunscreen D, respectively. The sharp intensity corresponding to TiO_2 particles in sunscreen D shows crystallite sizes below 20 nm. TiO_2 particles in the sunscreen might possessed anatase, rutile or brookite crystalline structures [3].

Figure 3(a-b) show the absorption spectra for the sunscreen contained different concentration of ZnO and commercial sunscreen. The observation of absorbance spectra shows higher concentration of ZnO, as main active UV filter offered higher protection over UVA (320-400 nm) and UVB (280-320 nm) regions. The highest absorption at ultraviolet region (376.3 nm) was achieved for sunscreen containing 40 wt% of ZnO nanorods. Commercially available sunscreen A, B, C and D contained both organic and inorganic UV filters and the percentages of these filters were not labelled in the packaging of the product. The absorbance spectra showed the same trend, which is exhibited good absorbance against UVB region and low absorbance approached UVA region. It was emphasized that the homemade ZnO sunscreens in this work offered much bigger protection in the UVA regions than other commercial sunscreens. With the nanorod shape, ZnO particles possessed very large surface-to-volume ratio that can offer much larger optical interaction in absorbing both UVA and UVB radiation. The absorbance spectrum of commercial sunscreen might be influenced by several factors affecting the action against UV rays such as combination of UV filters, concentration of sunscreen and type of emulsion [13-14].

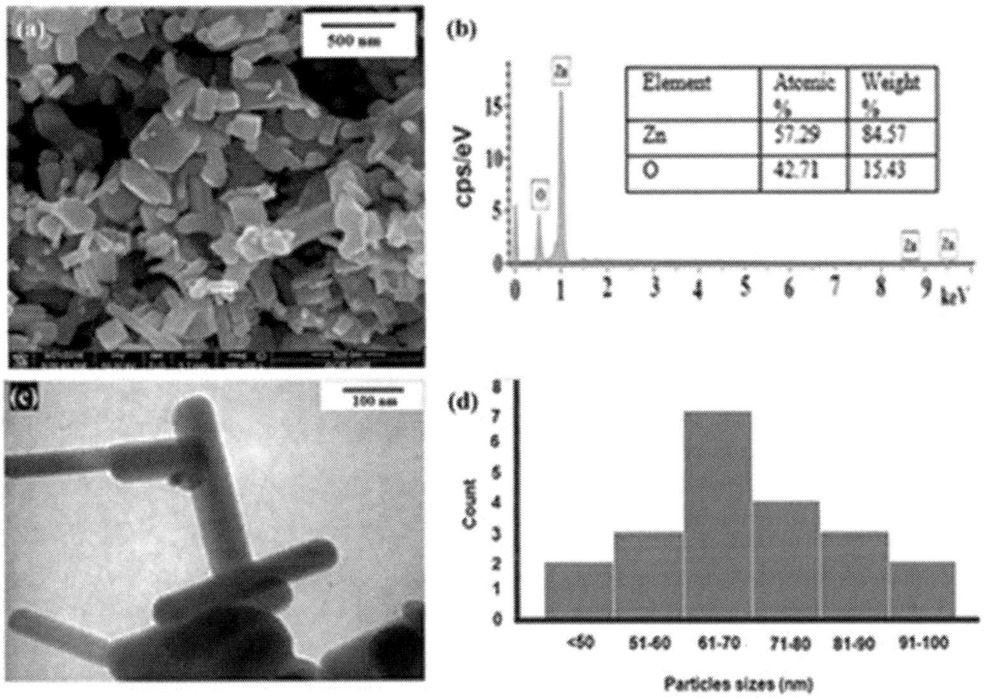

Figure 1: (a) FESEM micrograph; (b) FESEM-EDS; (c) TEM image; (d) particle distributions of ZnO powder.

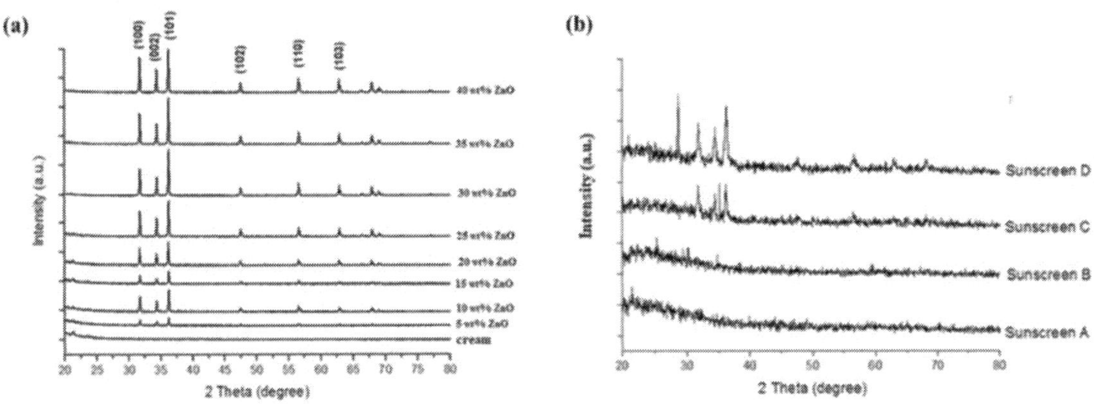

Figure 2: X-ray diffraction of (a) sunscreen contained ZnO nanorods and (b) commercial sunscreens.

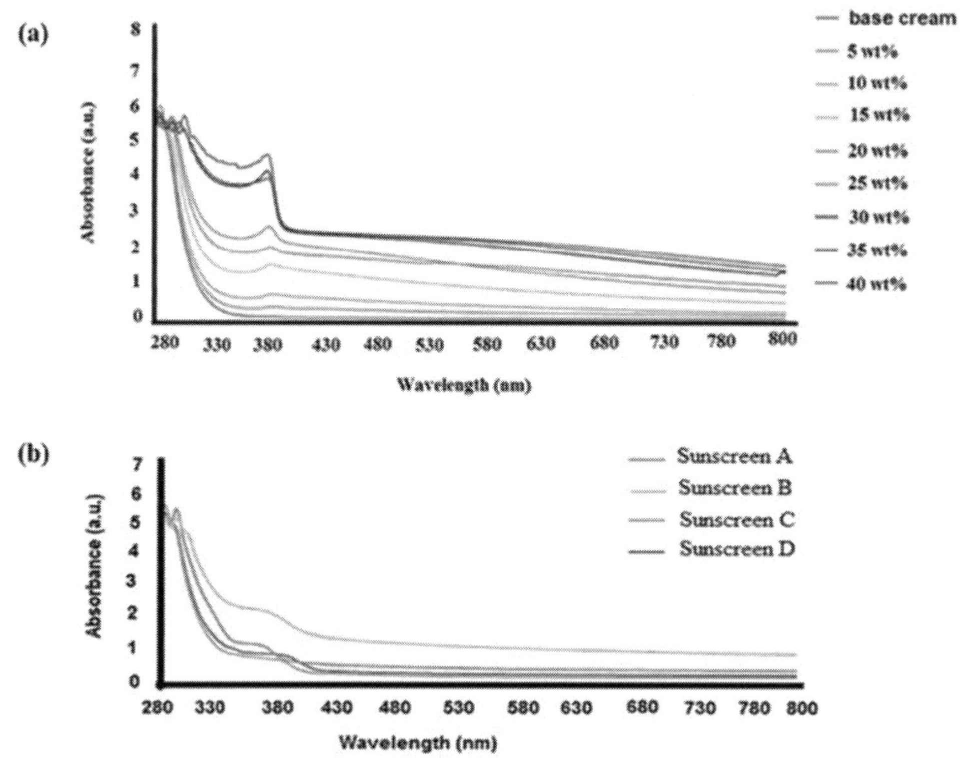

Figure 3: Absorbance spectra of (a) sunscreen contained different concentration of ZnO; (b) commercial sunscreen.

SPF values of sunscreen contained different concentration of ZnO nanorods were determined and tabulated in Table 2, whereby the percentage amount of ZnO nanorods in the sunscreen played an important role in determining the sunscreen SPF. The higher the ZnO amount, the higher the SPF value of the sunscreen. SPF values of sunscreen containing 5 to 40 wt% of ZnO showed very good protection, where more than 93% of UVB blocking. SPF value of commercial sunscreens measured in this study showed sunscreen A and sunscreen B exhibited lower SPF than the SPF value in the product label. This observation was probably because of higher SPF values are more difficult to measure [17]. However, sunscreen C and D exhibited higher SPF value than their labelled SPF. These two sunscreens possessed ideal SPF value for adequate UV protection [18].

Table 2: In-vitro SPF value of sunscreens.

Sun protection factor (SPF)			
Sunscreen contained ZnO	**SPF value**	**Commercial sunscreen**	**SPF value**
Base cream (mineral oil and beeswax)	11.51	**Sunscreen A (SPF60)**	51.04
5 wt%	29.50		
10 wt%	36.06	**Sunscreen B (SPF50)**	48.52
15 wt%	37.03		
20 wt%	40.37	**Sunscreen C (SPF30)**	42.48
25 wt%	40.16		
30 wt%	42.40	**Sunscreen D (SPF24)**	26.26
35 wt%	42.77		
40 wt%	43.48		

4. Conclusion

ZnO nanorods as active ingredient showed excellent UVA and UVB absorbing performances in sunscreen. Sunscreen incorporated ZnO nanorods increased the SPF value and achieved greater UV protections. Commercial sunscreens exhibited various UV performances due to many chemical and physical compound in the formulations and mostly protects in UVB regions with little protection in the UVA region. SPF value of sunscreen must reflect their protection against UV rays and helps consumer to get adequate protections under the sun.

Acknowledgement

This work was supported by Universiti Sains Malaysia through FRGS grant (203/PFIZIK/6711491) and USM Physics School Postgraduate Fund (308/AIFIZIK/415403). We acknowledge assistance from USM electron microscopy unit and NOR laboratory.

References

[1] B. Xu and Z. Cai, "Trial-Manufacture and UV-Blocking Property of ZnO Nanorods on Cotton Fabrics," vol. 108, pp. 3781–3786, 2008.
[2] L. Chuo *et al.*, "Journal of Environmental Chemical Engineering Structural morphology and in vitro toxicity studies of nano- and micro-sized zinc oxide structures," *Biochem. Pharmacol.*, vol. 3, no. 1, pp. 436–444, 2015.
[3] T. G. Smijs and S. Pavel, "Titanium dioxide and zinc oxide nanoparticles in sunscreens: Focus on their safety and effectiveness," *Nanotechnol. Sci. Appl.*, vol. 4, no. 1, pp. 95–112, 2011.
[4] E. B. Manaia, R. C. K. Kaminski, M. A. Corrêa, and L. A. Chiavacci, "Inorganic UV filters," *Brazilian J. Pharm. Sci.*, vol. 49, no. 2, pp. 201–209, 2013.
[5] S. Schalka and V. M. S. Dos Reis, "Sun protection factor : meaning and controversies.," *An. Bras. Dermatol.*, vol. 86, no. 3, pp. 507–515, 2011.
[6] U. V. V. Spectrometry and F. Padera, "Sunscreen Testing According to COLIPA 2011 / FDA Final Rule 2011 Using UV / Vis LAMBDA Spectrophotometers," pp. 1–9, 2011.
[7] V. Sudhahar and V. Balasubramanian, "Sun production factor (SPF) determination of marketed sunscreen formulation by In-Vitro method using UV-VIS spectrophotometer," vol. 5, no. 6, pp. 119–122, 2013.
[8] E. A. Dutra, D. Almança, E. R. M. Kedor-, M. Inês, and R. Miritello, "Determination of sun protection factor (SPF) of sunscreens by ultraviolet spectrophotometry," *Brazilian J. Pharm. Sci.*, vol. 40, no. 3, pp. 381–385, 2004.
[9] M. Zarkogianni and N. Nikolaidis, "Determination of Sun Protection Factor (SPF) and Stability of Oil-in-Water Emulsions Containing Greek Red Saffron (Crocus Sativus L .) as a Main Antisolar Agent," vol. 3, no. 7, pp. 1–7, 2016.
[10] H. J. Hussein and M. K. Ismail, "Measuring the Sun Protection Factor of Sunscreens," *World J. Pharm. Sci.*, no. 8, pp. 1–3, 2016.
[11] S. Mahmud, "One-dimensional growth of zinc oxide nanostructures from large micro-particles in a highly rapid synthesis," *J. Alloys Compd.*, vol. 509, no. 9, pp. 4035–4040, 2011.
[12] T. A. Egerton and I. R. Tooley, "UV absorption and scattering properties of inorganic-based sunscreens," pp. 117–122, 2012.
[13] L. C. Ann *et al.*, "Effect of surface modification and UVA photoactivation on antibacterial bioactivity of zinc oxide powder," *Appl. Surf. Sci.*, vol. 292, pp. 405–412, 2014.
[14] L. Guo, Y. L. Ji, H. Xu, and P. Simon, "Regularly Shaped , Single-Crystalline ZnO Nanorods with Wurtzite Structure," pp. 14864–14865, 2002.
[15] S. Seité *et al.*, "A full-UV spectrum absorbing daily use cream protects human skin against biological changes occurring in photoaging," *Photodermatol Photoimmunol Photomed*, vol. 16, no. 4, pp. 147–155, 2000.
[16] G. Engla and M. A. Khan, "Studies on sun protection factor of sunscreen cosmetic formulations for their selection and use," vol. 3, no. December, pp. 45–47, 2012.

[17] S. Miksa, D. Lutz, C. Guy, and E. Delamour, "Sunscreen sun protection factor claim based on in vivo interlaboratory variability," pp. 541–549, 2016.
[18] B. L. Diffey, "Sunscreens and melanoma: the future looks bright.," *Br. J. Dermatol.*, vol. 153, no. 2, pp. 378–81, 2005.

Optimisation of DECT Spectral Separation for Quantitative Evaluation of Clinical Contrast Agent

Mohammed Y.M Abujazar[1,*], Rafidah Binti Zainon [2], and Kurama M.B [3], Elaine Salmi

[1] School of Physics USM, 11800 Pulau Pinang, Malaysia
[2] institut perubatan dan pergigian termaju, Bandar Putra Bertam, Kepala , Pulau Pinang
[3]A-5-6 Menara Indah jalan 9 Taman TAR, Ampang 50400 Selangor

*muh.jazar@gmail.com

Abstract: This study is focusing on the evaluation and optimization of the spectral separation of clinical contrast agents in DECT imaging. Based on previous literatures, the two most important methods to evaluate the DECT systems are the evaluation of the DE ratio and the contrast to noise ratio. In the first study, abdomen CT phantom was used with an insertion of three contrast media: iodine, gadolinium and milk. The phantom was scanned using abdomen routine protocols for both single and dual energy CT. The scanner was then switched from DECT to SECT to scan the phantom with single energy ranges such as 80 kV, 100 kV, 120 kV and 140 kV. The MATLAB was used to fuse the images by weighted averaging 30 % low energy and 70 % high energy. The collected data were analysed using the Weasis software. Statistical analyses were then performed to compare the contrast to noise ratio with different tube voltage combinations. The MATLAB was again used to fuse the images. The results showed that 80/140 kV is the best spectral separation for evaluation of the clinical contrast medium. In addition, the results displayed significant CNR effect ($p < 0.05$) for all DECT images. The dual energy technique improved the contrast to noise ratio and dual energy ratio for the DECT images.

1. Introduction

The first CT imaging device was created in 1972 by G. N. Hounsfield [1, 2]. CT is considered as a non-invasive diagnostic tool that allows for 3D visual reconstruction. The first DECT device was introduced in 1987 [3]. Then DECT has been introduced and depends on using two spectra; a dual-energy CT is typically performed using 80 and 140 kV spectra [3-5]. The dual-energy CT was developed to provide the ability of a CT to differentiate between different contrast media. Thus, DECT improves the ability to distinguishes different material [6, 7].The contrast improvement depends on using different X- ray spectra for high and low energy spectra of the evaluated contrast materials [5, 8, 9]. Contrast Medias are radioopaque materials that are not transparent to X-rays. Contrast Medias that have high atomic number (Z) and high density will incline to absorb X-rays better. The X-ray absorption coefficient (μ) relationship is perhaps best expressed in the formula below [2]:

$$\mu \approx \frac{\rho Z^4}{AE^3} \quad (1)$$

Content from this work may be used under the terms of the Creative Commons Attribution 3.0 licence. Any further distribution of this work must maintain attribution to the author(s) and the title of the work, journal citation and DOI.
Published under licence by IOP Publishing Ltd

IOP Conf. Series: Journal of Physics: Conf. Series **1083** (2018) 012013 doi:10.1088/1742-6596/1083/1/012013

where "ρ" is the density, "Z" is atomic number, "A" is atomic mass and E the X-ray energy.
X-Ray computed tomography contrast agents can be administrated orally for G.I.T examination or intravenously for angiography examination. The contrast media that are containing atoms of high atomic number such as gadolinium and iodine are commonly used in diagnostic imaging department to highlight the abnormal lesions better. This study evaluate and optimization the spectral separation of clinical contrast agent in computed tomography imaging

2. Methods

2.1 Phantom design

Abdomen phantom fabricated from perspex. The phantom have five inserts, which two inserts blocked and filled with water. Remain two inserts filled with known concentrations and volume of iodine and gadolinium and the last insert was filled with milk. The Insert number 1 contained diluted iodine (Atomic number = 53) with concentration 10.34 mg/ml. Insert number 2 contained diluted Gadolinium (Atomic number = 64) with concentration 0.313 M and insert number 3 with Milk. All the contrast agents had the same volume 100 ml.

2.2 Image acquisition

The phantom was positioned on the CT table with the aid of laser light. It was immobilized with form pad to maintain the stability of the phantom in fixed place without movement. The phantom was scanned with SECT technique at different single tube voltage 80, 100, 120 and 140 kV. Also scanned by DECT technique by common tube voltages of 80/140 kV. The CT value difference between the water that surrounds the clinical contrast materials and the water inside the inset was less than 1.5 HU. Measurements for contrast to noise ratio were performed with different X-ray tube voltage combinations for low and high energy.

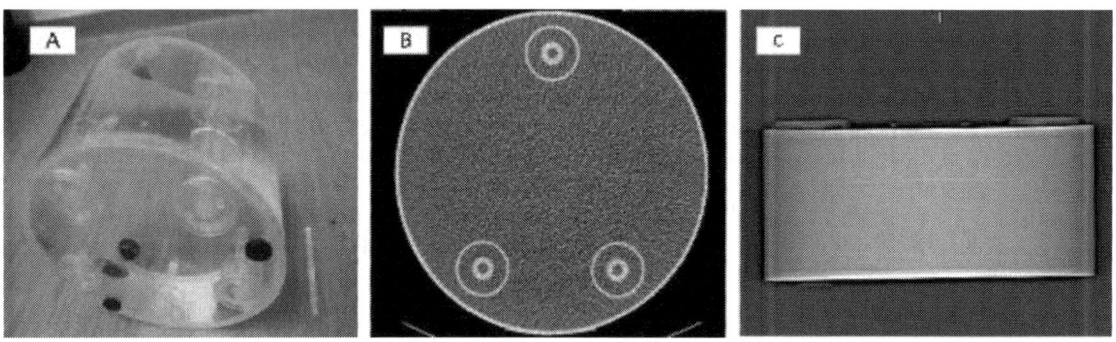

Figure 1: (A) Image of the circular abdomen phantom of diameter 32 cm used to simulate abdomen patients. (B) CT image for the phantom after filled with water to take the background. (C) Tomography image to accurate scan for the phantom.

Table 1. DECT abdomen routine protocol parameters.

Parameters	Abdomen protocol DECT
mAs for 80 [kV]	343 [mA]
mAs for 140 [kV]	189 [mA]
Pitch factor	0.35 and 1.2 [mm]
Slice thickness	0.6 and 10 [mm]
Scan mode	Spiral
Exposure Time per Rotation	0.5 [s]

2.3 Technique for Reconstruction of DECT data

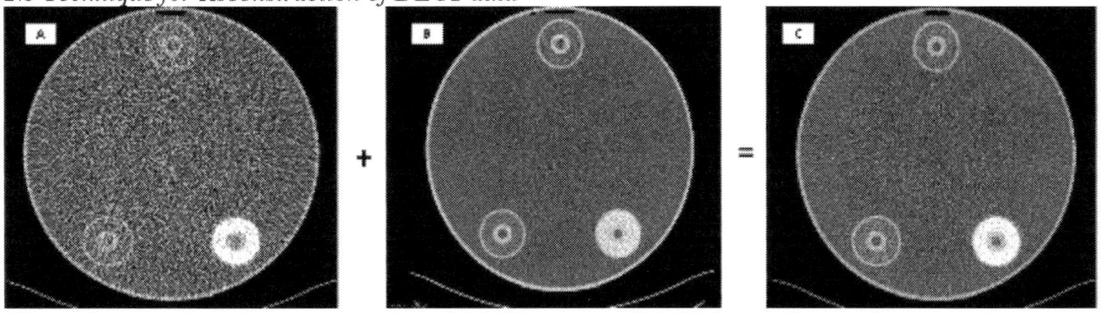

Figure 2: Technique for reconstruction of DECT images. (A) CT image for the phantom at low energy tube voltage about 80 kV and (B) phantom scanned at 140 kV and (C) the fused image at 80/140 kV.

In our research, The IMMA phantom scanned at several single energy CT were 80, 100, 120 and 140 kV. The images were reconstructed using FBP, and the data were analyzed. The typical fusing technique was performed with weighting factor of 0.3 [10-13]. The 0.3 weighted- average data set are 30% from 80 kV data and 70% from 140 data. As found in this experiment, this weighting factor minimizes the image noise and improved the contrast to noise ratio in VNC images.

2.4 Measurements of contrast noise ratio (CNR) and DE ratio

Using MATLAB program, the attenuation coefficient for X-ray photons was measured for each contrast media in Hounsfield unit (HU) with different pitch value and tube voltage. This measurement was performed by drawing a circular ROI at the areas of interest. The area of the ROI was 60 mm^2 and was maintained constant for all analysis to decrease bias in measurement [14].

$$\text{Contrast noise ratio} = \frac{ROI\ c.m - ROI\ b}{SD\ b} \qquad (2)$$

where; ROI_m is the mean attenuation of the contrast media and ROI_b is the mean attenuation in background from water. SD_b is the image noise, defined by the standard deviation of the pixel values from HU in the ROI. DE ratios of the three contrast agents were measured by dividing the Hounsfield unit (HU) of the low-energy such as 80 kV by mean Hounsfield unit (HU) for high energy such as 140 kV. Where low energy tube voltage is 80, 100 kV and high 120 or 140 kV [5, 15].

3. Results and Discussion

The measured CNR in the combined images at fixed slice thickness is 0.6 mm and minimum pitch value 0.35 and the maximum pitch value 1.2 mm. In addition different X-ray tube voltage combinations as drawn in Figure 4. For all the VNC images the CNR level was higher at 80/140 kV compared with the other tube voltage combination. Gadolinium enhancement in the fused images ranges from 16 to 23 HU for the different pitch value and tube voltage combination. However, the HU for iodine enhancement ranges from 179 to 214 HU. Similarly, the CNR is significantly higher with low pitch compared to high pitch value ($P<0.05$) (Figure 4). The CNR for all the contrast media, are significantly higher at 80/140 kV ($P < 0.05$). The highest standard deviation of all contrast agents in the fused images was higher at 80/120 kV.

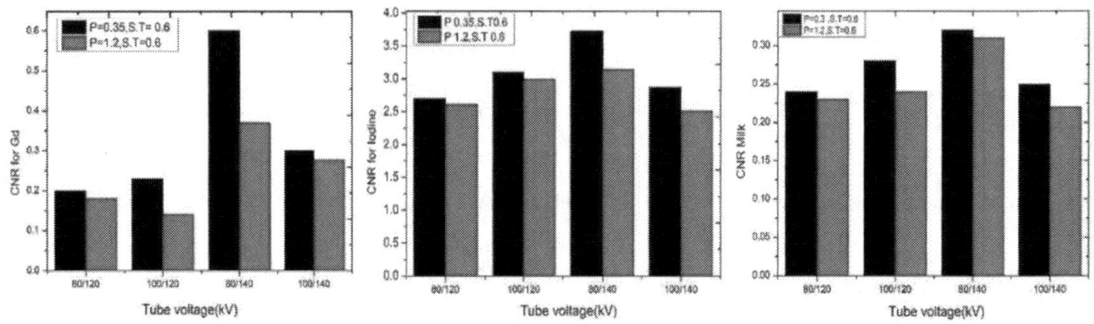

Figure 3: Comparing CNR values between different tube voltage combinations for gadolinium, iodine and milk.

The measured DE ratios for the different contrast media are presented in matrix color as shown in Figure 5. The values ranged from 1.18 to 3.4, 1.18 to 2.1 and 1.1 to 2.3 for gadolinium, iodine and milk, respectively. The combination voltage of 120/140 kV has the lowest DE ratios, while 80/ 140 kV has the highest DE ratios (Figure 5).

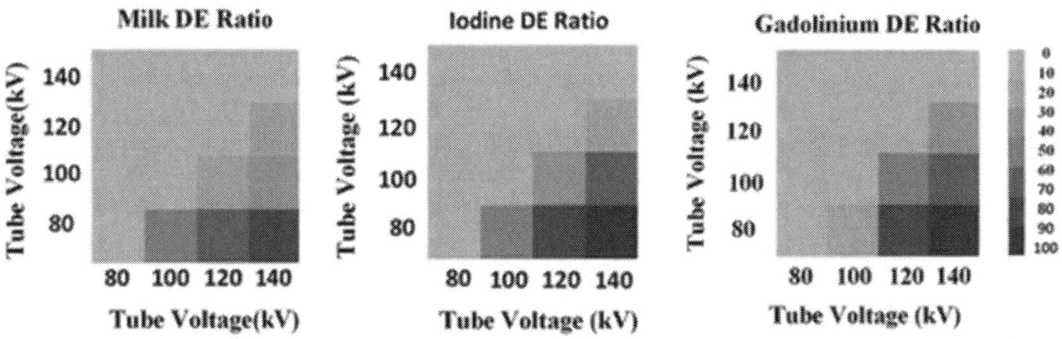

Figure 4: Matrix color at different single energy CT energy to measure DE ratio.

In this study, the performance of DECT at different X-ray tube voltage combination with constant slice thickness and different pitch values was assessed. Similarly, the CNR for the different contrast agent within the abdomen CT phantom was evaluated (Figure 3). There is apparent and essential difference in CNR between the different tube voltage combinations. The combination voltage of 80/140 kV shows the highest CNR for all the contrast agents, this is as a result of combining images with the highest enhancement that is 80 kV and highest penetration obtained at 140 kV. At low tube voltage (80 kV) the effective energy is similar to the k-edge of iodine and Gd and hence, better enhancement. However, using 80 kV alone will have a limitation of not penetrating the body adequately. The 140 kV will have high energy to penetrate the body but with low contrast. The combination of these two energies will strike a balance between contract and penetration and hence, give a better image with high CNR. This is inline, which study conducted by Graser, A., et al 2009 and they mentioned that 80 and 140 kV tube voltage , lead to optimum maximum material differentiation for materials [16]

Results from Figure 4 shows that 80/140 kV demonstrate the best DE ratio for milk, iodine and Gd, respectively. Gd and iodine are used for IV contrast. However, iodine show superior DE ratio compared to Gd. This is because the attenuation is directly related to the concentration of the contrast media and Gd has a low concentration. Therefore, it is less distinguished with the surrounding background. On the other hand, iodine has high concentration and is more distinguished with the surrounding environment. Milk has shown attenuation difference with the surrounding background. However, it has low attenuation difference compared to iodine. This might be because milk constitute of fat, which makes it have little attenuation.

On the contrary, 120/140 kV tube voltage gives the lowest DE ratio. It accounts for low DE ratio because of the high energy and will lead to more penetration with low attenuation compared to the low energy such as 80 kV. Similarly, the small pitch value shows a high CNR. This might be as a result of the fact that there is more radiation deposited in each slice with low pitch because of the small movement in table per gantry rotation and hence, more CNR.

4. Conclusion

Out of the tube voltage combinations, the highest CNR fused images were obtained using 80/140 kV for Gd, iodine and milk, respectively. This indicate a better image quality. Similarly, low pitch value has improved CNR compared to a high pitch in all the contrast media. Results from the study showed that the DE ratios is higher at 80/140 kV for the different contrast media. This optimizes spectral separation of the various contrast media used.

Reference

[1] Lindsten, J.E., *Physiology Or Medicine: 1981-1990*. 1993: World scientific.
[2] Lusic, H. and M.W. Grinstaff, *X-ray-computed tomography contrast agents*. Chemical reviews, 2012. 113(3): p. 1641-1666.
[3] Krauss, B., et al., *The importance of spectral separation: an assessment of dual-energy spectral separation for quantitative ability and dose efficiency*. Investigative radiology, 2015. 50(2): p. 114-118.
[4] Karcaaltincaba, M. and A. Aktas, *Dual-energy CT revisited with multidetector CT: review of principles and clinical applications*. Diagnostic and Interventional Radiology, 2011. 17(3): p. 181.
[5] Krissak, R., et al., *Gold as a Potential Contrast Agent for Dual-Energy CT*. Advances in Molecular Imaging, 2013. 3(04): p. 37.
[6] Avrin, D.E., A. Macovski, and L. Zatz, *Clinical application of Compton and photo-electric reconstruction in computed tomography: preliminary results*. Investigative radiology, 1977. 13(3): p. 217-222.
[7] Yeh, B.M., et al., *Dual-energy and low-kVp CT in the abdomen*. American Journal of Roentgenology, 2009. 193(1): p. 47-54.
[8] Johnson, T.R., et al., *Material differentiation by dual energy CT: initial experience*. European radiology, 2007. 17(6): p. 1510-1517.
[9] Kruger, R., S.J. Riederer, and C. Mistretta, *Relative properties of tomography, K-edge imaging, and K-edge tomography*. Medical physics, 1977. 4(3): p. 244-249.
[10] Tawfik, A.M., et al., *Dual-energy CT of head and neck cancer: average weighting of low-and high-voltage acquisitions to improve lesion delineation and image quality—initial clinical experience*. Investigative radiology, 2012. 47(5): p. 306-311.
[11] Tawfik, A., et al., *Image quality and radiation dose of dual-energy CT of the head and neck compared with a standard 120-kVp acquisition*. American Journal of Neuroradiology, 2011. 32(11): p. 1994-1999.
[12] Wichmann, J.L., et al., *Virtual monoenergetic dual-energy computed tomography: optimization of kiloelectron volt settings in head and neck cancer*. Investigative radiology, 2014. 49(11): p. 735-741.
[13] Albrecht, M.H., et al., *Assessment of an advanced monoenergetic reconstruction technique in dual-energy computed tomography of head and neck cancer*. European radiology, 2015. 25(8): p. 2493-2501.
[14] Razak, H.R.A., S.M.S.S. Rahmat, and W.M.M. Saad, *Effects of different tube potentials and iodine concentrations on image enhancement, contrast-to-noise ratio and noise in micro-CT images: a phantom study*. Quantitative imaging in medicine and surgery, 2013. 3(5): p. 256.

[15] Boll, D.T., et al., *Renal stone assessment with dual-energy multidetector CT and advanced postprocessing techniques: improved characterization of renal stone composition—pilot study.* Radiology, 2009. 250(3): p. 813-820.

[16] Graser, A., et al., *Dual energy CT: preliminary observations and potential clinical applications in the abdomen.* European radiology, 2009. 19(1): p. 13.

Diode Dosimetric Characteristics Assessment for In-Vivo Dosimetry in Radiotherapy Treatment

R Marzuki[1], A A Rahman[1], I S Mustafa[1] and A N Shabandi[2]

[1]School of Physics, Universiti Sains Malaysia Main Campus, 11800 Minden, Penang, Malaysia.
[2]National Cancer Institute, No. 4, Jalan P7, Presint 7, 62250, Putrajaya, Malaysia

rffadli@yahoo.com

Abstract. Diodes are radiation dosimetry system that has been applied widely in radiotherapy treatment and very useful for in-vivo dosimetry (IVD). It is duly important to understand the characteristic of detector before it can be used in clinical treatment. In this study, characteristics of diode model IBA dosimetry EDP-15^{3G} diode were analysed and compared to calibrated ionization chamber. 6 MV linear accelerator (Linac) was used with different field sizes (FS), Linac monitor unit (MU), tray field (TF), wedge field (WF) and source to surface distance (SSD). Actual diode thickness, diode reproducibility, perturbation effect and diode effect with radiation beam angle and temperature were performed. From the determination of diode dosimetric characteristic, a clinical treatment planning system (TPS) model Elekta Monaco and calibrated ionization chamber (IC) were used to verify the accuracy of diode reading in different setup conditions. Results of study showed wedge, SSD, angle of incidents and temperature were affected the accuracy of diode reading. Phantom measurement and simulation of patient treatment were less than 3%.

1. Introduction

IVD is an important method to monitor and compare dose measured from the detector such as diode with the theoretical values, as calculated by TPS or manual calculation. IVD is also a tool for the detection of systematic errors and prevention of unintended exposures to patients.

A silicon diode dosimeter is a p-n junction. The diodes are produced by taking n-type or p-type silicon and counter-doping the surface to produce the opposite type material. These diodes are referred to as n–Si or p–Si dosimeters, depending upon the base material [1]. An n-type diode is formed by doping acceptor impurities into a region of n-type silicon. A p-type diode is formed by doping donor impurities into a p-type substrate [2]. This study was aimed to study diode characteristics and to understand its capabilities and limitation when it is used as IVD in radiotherapy treatment. This characterization will give guidance on correction method and evaluate its efficacy.

2. Material and Method.

2.1. Diode Characteristics.

Measurements were performed using 6 MV photon beams of Linac model Elekta Synergy, Sweden. The beams was calibrated using an IC (FC65-G Farmer Type IBA dosimetry, Germany) placed at 10

Content from this work may be used under the terms of the Creative Commons Attribution 3.0 licence. Any further distribution of this work must maintain attribution to the author(s) and the title of the work, journal citation and DOI.
Published under licence by IOP Publishing Ltd

cm depth in a water phantom according to IAEA TRS-398 protocol [3]. The IC dosimetry system was calibrated at the Secondary Standard Dosimetry Laboratory (SSDL), Malaysian Nuclear Agency.

Preliminary testing for diodes was perturbation effect. Diode perturbation effect was study using a 3D water phantom system, Model Blue Phantom2, IBA Dosimetry, Germany with stereotactic field diode (SFD). Inline and crossline profile was study at depth of maximum dose (d_{max}), 5 cm, and 10 cm depth as shown in Figure 1. SFD diode was chosen due to sensitivity with small field.

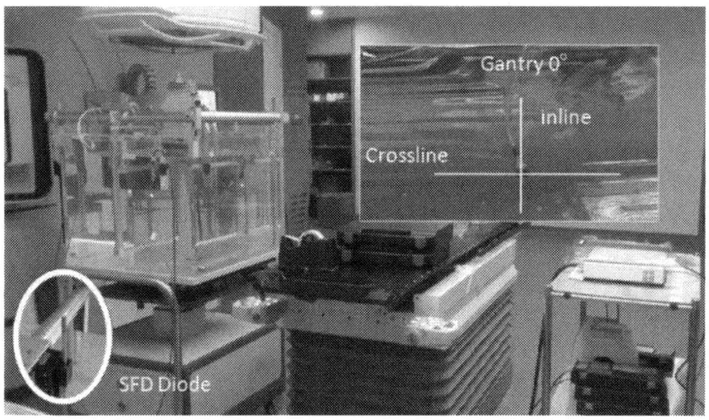

Figure 1. 3D water phantom for inline and crossline profile measurements.

Protocol for diode dose entrance calibration was based on IAEA [4], Estro [5] and AAPM TG-62 [2]. Solid Water (Gammex, Wisconsin, USA) was used for the calibration of the IVD system with solid phantom to water correction factor. Three Diodes were taped onto the surface of the phantom. Diode perturbation effect results were taken into account during calibration. Diodes are relative dosimeters and therefore, these need calibration to present the dose at the d_{max}. This system calculated linear factor between dose and ADU unit (diode reading).

The IVD system used in this study consisted of Model DPD-12pc (electrometer, IBA Dosimetry, Germany) and p-type diodes (EDP-15^{3G}, IBA Dosimetry, Germany). The build-up cap material of the diodes is made of stainless steel with epoxy encapsulation that is equivalent to 15 mm water-equivalent thickness and primarily designed for 6 to 12 MV photon [6]. Actual thickness of diode was study using 10 MV photon. Solid Phantom slabs of 1 mm, 2 mm and 5 mm are used to cover the diodes dose till maximum is reached. Air gaps are eliminated by using a perforated slab for diode. Diode reading with solid phantom to water correction factor was plotted and compared to d_{max} (22 mm) water depth for 10 MV, which was based on baseline value during commissioning. Fluctuate reading from linac also taken into account by using IC, where it was placed at the head of gantry.

Reproducibility was determined according to the procedure stated in the ESTRO Booklet [7]. Signal stability and leakage were measured as outlined in the IAEA Human Health reports No. 8 guidelines [8]. Correction factors (dose non-linearity of the response (different MU), field size and angle of incidence, SSD correction, wedge correction) for entrance setup for all diodes were determined according to literature [8]. Fluctuate reading from linac taken into account by using IC where it was placed at the head of gantry for angle of incidence correction factor. Relation diode with temperature was studied according to the procedure defined in the literature [9,10].

2.2. Phantom measurements.

Measurements were made in a Thorax Phantom Model CIRS 002LFC (30 cm long x 30 cm wide x 20 cm thick) [11]. The phantom is constructed of proprietary tissue equivalent epoxy materials. Tissue equivalent interchangeable rod inserts accommodate IC allowing for point dose measurements.

The experiments were performed for single field (200 Monitor Unit (MU), 100 cm source to surface distance (SSD) with different field size (FS)), wedge field (300 MU, 100 cm source to axis distance (SAD), with different FS), two field (Anterior Posterior/Posterior Anterior (APPA), 150 MU for AP and PA, 100 cm SAD with different field size) and lateral opposed (400 MU, 100 cm SAD, with different FS). The phantom measurements were compared with TPS model Monaco, Elekta, Sweden, Ver. 5.10 and calibrated IC.

Patient simulation treatment was done using Alderson Rando Phantom. TPS was planned as standard patient treatment as shown in Table 1. Measurement was performed as planned and diodes reading were compared with TPS.

Table 1. Simulation of patient treatment using.

No.	Case name and prescription	Setup	Treatment field	Field Size (cm)
1	Spine, Single field 20 Gy/5 fraction (#) to 6 cm depth	SSD 100cm	Posterior Anterior	8x12
2	Ca Rectum, 3 fields 45 Gy/ 25# to isocentre	SAD 100cm	Anterior Posterior	14x17
			Right Lateral	14x17
			Left Lateral	14x17
3	Ca Endometrium, 4 fields 46 Gy / 23# to isocentre	SAD 100cm	Anterior Posterior	13x14
			Posterior Anterior	13x14
			Right Lateral	13x14
			Left Lateral	13x14
4	Whole brain, 2 fileds 20 Gy / 5# to 100% isodose line	SAD 100cm	Right Lateral	22x18
			Left Lateral	22x18
5	Nasopharyngeal Cancer Facial :40 Gy / 20# to 100% isodose line (IL) Neck :40 Gy / 20# to d_{max}	SAD 100cm	Right Lateral facial	14x10
			Left Lateral facial	14x10
			Anteriror Neck Right	10x7
			Anteriror Neck Left	10x7
6	Rt Chest Wall 40 Gy / 15# to 100% isodose line	SAD 100cm	Right Mediastinum	8x16
			Right Lateral	8x16
7	Lt Chest Wall and supraclavicular (SCF) Chest : 40 Gy / 15# to 100% IL SCF : 40 Gy / 15# to d_{max}	SAD 100cm	Left Mediastinum	8x15
			Left Lateral	8x15
			Anteriror SCF	9x8

3. Results and Discussions

3.1. Diode Perturbation

Inline profile shown that diodes perturbations is at d_{max}, 5, 10 cm were 5.4%, 6.1% and 5.8%, respectively and crossline were 6.1%, 6.5% and 5.6%, respectively. It was also observed that the cable is affecting the amount dose at 1.9%. Measurement of perturbation effect for 5 cm was higher 0.5% compared to manufacturer technical specification 5-6%. This study will preserve as guidelines towards the effect of diode to the treatment area and frequency of diodes measurements. Furthermore, 1.3 cm square field area was affected throughout the measurement.

3.2. Actual thickness

A study was performed and confirmed that build-up cap material of the EDP-15^{3G} diodes for all diodes were 1.5 cm. These results were follow manufacturer technical specification. Thicknesses of diodes are equivalent with d_{max} for 6 MV, which make it suitable with the amount of energy.

3.3. Reproducibility, Signal stability and leakage.

Two types of reading measurements were performed, consisted of i) 20 readings at the same days, and ii) 15 readings at different days with 1 day gap, observed that the standard deviation (SD) was less 0.1 % and 1.0%, respectively. These results are within the acceptable limits and follows IAEA guidelines

[8]. Signal stability after radiation was observed for 5, 10, 15 and 20 minutes. Results presented in figure 2 shown that Diode 1 (D1) and Diode 2 (D2) increase linearly with time, while Diode 3 (D3) do not have any significant changes. Signal stability is crucial for longer treatment time, especially in total body irradiation. Signal does not exceed 1% for 20 minutes is acceptable. Furthermore, leakage measurement was negligible.

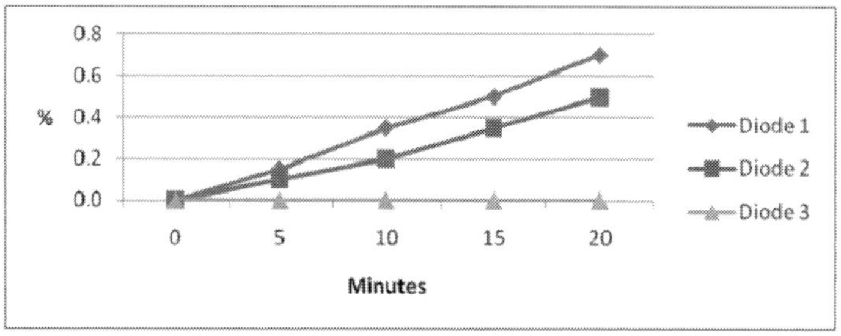

Figure 2. Signal stability after radiation.

3.4. Correction factors

Correction factor for different dose rate, MU, field size (symmetry and asymmetry jaw) and tray is less than 1.0%. Correction factor are in range 0.99 – 1.01, which is considered can be ignore and no need to use it. Based on the analysis performed, the diode sensitivity was affected by angle, wedges and SSD. Result for axial and tilt angle of incidence were illustrated in figure 3. Referring to the vendor technical specification, angle of incidence (axial and tilt) are supposed to be less than 2% for angle ± 45°. Therefore, it was found that only diode 3 has confirmed within the specification requirement. The measured angle for diode 1 and 2 were less than 2% at angle ± 41° and ± 37°, respectively.

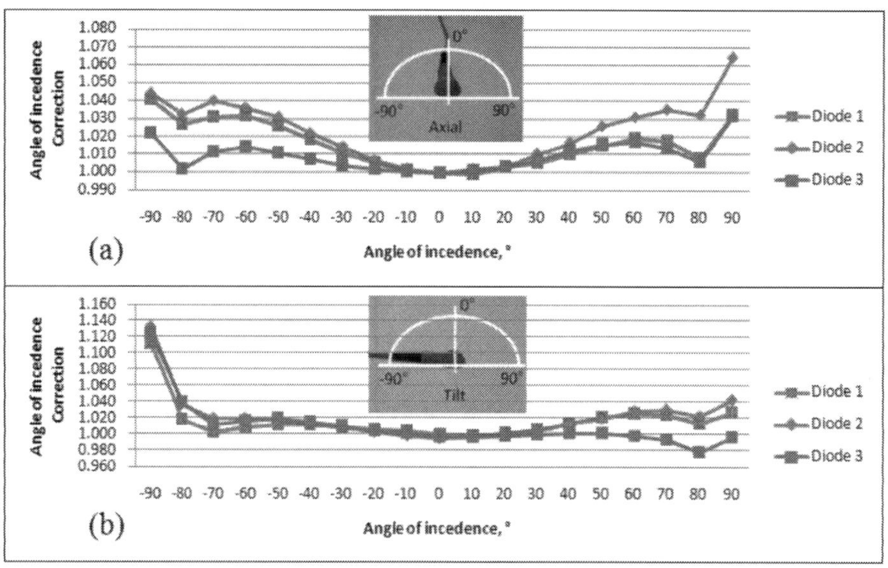

Figure 3. Correction factor for different angle (a) axial and (b) tilt.

Elekta Linac (used in this study) is equipped with motorised wedge. It has single physical wedge angle of 60° and attach permanently in the gantry head. Different wedge angle can be produced by TPS by combination of physical wedge and open field. The results were shown in Figure 4(a). Wedge correction factor has increased proportionally by wedge angle. Wedge correction factor was calculated from the average reading of 3 different field sizes (5 x 5 cm, 10 x 10 cm and 15 x 15 cm). Diodes also affected by different SSD, which was shown in figure 4(b). SSD correction has increased proportionally by increased SSD.

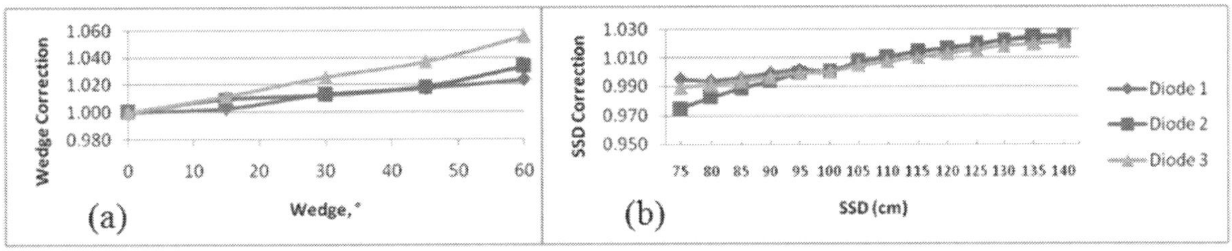

Figure 4. Correction factor for (a) wedge and (b) SSD.

Analysis of diodes with temperature deviation showed that diode is sensitive with temperature. Diode produced signal or readout even though there are no radiations occurring in the circumference. Analysis was performed with temperature 25-34°C. Results showed that higher temperature will increase the readout. Temperature correction factor can be avoided when insulator, such as paper was placed between diode and patient.

Based on the diode characteristics analysis, it was concluded that only three factors (angles wedge and SSD) were applied for correction. The placement of diode orientation should be recorded to ensure the angle correction factor is correct.

3.5. Thorax phantom measurements

For diode verification, thorax phantom was used in this study to measure dose with IC. The phantom consisted of lung and spine tissue, which is very useful to verify the accuracy of TPS algorithm [12]. Manual calculation was done in order to get the value for d_{max}.. The results of mean percentage difference and standard deviation for individual diode measurement besides having comparison with IC and TPS are tabulated in Table 2. Mean percentage between ionization chamber reading and TPS for single, wedge, APPA and Lateral Opposed were 0.17 (Std Dev 1.45), 1.58 (Std Dev 0.6), 1.83 (Std Dev 0.8) and 0.66 (Std Dev 0.62), respectively.

Table 2. Mean percentage different and standard deviation between diode vs ionisation chamber and diode versus TPS.

Plan	Diode	Diode vs Ionization chamber		Diode vs TPS	
		Mean %	Std Dev	Mean %	Std Dev
Single field	D1	0.20	1.65	0.36	0.42
	D2	0.78	1.43	0.94	0.32
	D3	1.21	1.22	1.38	0.33
	Average	0.73	1.43	0.89	0.36
Wedge field	D1	1.39	0.55	2.99	0.05
	D2	1.03	0.71	2.62	0.12
	D3	1.05	1.00	2.64	0.41
	Average	1.15	0.75	2.75	0.19
APPA	D1	-0.02	0.97	-1.85	0.33
	D2	-0.44	0.99	-2.26	0.40
	D3	0.43	1.04	-1.41	0.45
	Average	-0.01	1.00	-1.84	0.39
Lateral Opposed	D1	-1.01	0.51	-1.67	0.40
	D2	-0.89	0.48	-1.55	0.48
	D3	0.15	0.48	-0.51	0.44
	Average	-0.58	0.49	-1.24	0.44

3.6. Aldersons Rando Phantom results
Analysis with rando phantom was performed in order to simulate the real clinical cases. Different prescription, body area, treatment method and treatment parameter were chosen in order to evaluate accuracy of diode. Generally, all measured datas were having less than 3% of acceptable limit (mean percentage 0.3, Std Dev 1.9). Results are tolerable based on LINAC acceptable daily output tolerance, which is ± 3% [13].

4. Conclusion
In this study, EDP-15^{3G} diode was sensitive with angle of incidence, wedge, SSD and temperature. Each diode has their own characteristics and should analyse individually to avoid any inaccurate data. From the patients simulation results, it can be concluded that by applying proper calibration and correction factors, ± 3% reading able to achieve.

Acknowledgment
We would like express our gratitude to the Director General of Health Malaysia for his permission to publish this article. Our sincere thanks also goes to Dr. Asmayani binti Khalib, Director National Cancer Institute (NCI) Malaysia, Dr. Gerard Lim Chin Chye, Head of Department, Radiotherapy and Oncology NCI and Medical Physicist team who provide access to the linac and associated facilities. The first author would like to thank Ministry of Health, Malaysia for the scholarship given under *Hadiah Latihan Persekutuan.*

References
[1] Izewska J and Rajan G 2005 Radiation dosimeters *Radiation Oncology Physics : A Handbook For Teachers And Students / [Technical ed. Podgorsak E.B]* (Vienna: International Atomic Energy Agency) pp 71–99
[2] AAPM 2005 *Diode In Vivo Dosimetry For Patients Receiving External Beam Radiation Therapy Report of Task Group 62 of the Radiation Therapy Committee* (Madison: Medical Physics Publishing)
[3] IAEA 2000 *IAEA Technical Report Series No.398* (Vienna: International Atomic Energy Agency)
[4] IAEA 2013 Development of Procedures for In Vivo Dosimetry in Radiotherapy *IAEA Reports Series*
[5] Van Dam J and Marinello G 2006 *Methods for in vivo dosimetry in external radiotherapy* (Brussels: Estro)
[6] Scanditronix Wellhofer 2000 *The accurate and reliable choice for patient dose verification: 3G-pSi in vivo detectors* ed A Salih (Schwarzenbruck, Germany: Scanditronix Wellhofer)
[7] Huyskens D ., Bogaerts R, Verstraete J, Loof M, Nystrom H, Fiorino C, Broggi S, Jornet N, Ribas M and Thwaites D I 2001 Practical guidelines for the implementation of in vivo dosimetry with diodes in external radiotherapy with photon beams (entrance dose), Booklet No. 5 *Estro*
[8] IAEA 2013 *IAEA Health Report No. 8, Development Of Procedure For In Vivo Dosimetry In Radiotherapy* (Vienna: International Atomic Energy Agency)
[9] Van Dam J, Leunens G and Dutreix A 1990 Correlation between temperature and dose rate dependence of semiconductor response; influence of accumulated dose *Radiother. Oncol.* 19 345–51
[10] Ozleyis Altunkok S, Tuncel N and Ucar N 2015 Investigation of temperature dependence of semiconductor detectors used in medicine for radiation measurements *EDP Sci.* 100, 4001 1–3
[11] http://www.cirsinc.com/products/all/12/imrt-thorax-phantom/
[12] IAEA (International Atomic Energy Agency) 2004 Technical reports series no. 430. Commissioning and quality assurance of computarized planning systems for radiation treatment of cancer
[13] Klein E E, Hanley J, Bayouth J, Yin F F, Simon W, Dresser S, Serago C, Aguirre F, Ma L, Arjomandy B, Liu C, Sandin C and Holmes T 2009 Task group 142 report: Quality assurance of medical acceleratorsa *Med. Phys.* 36 4197–212

Chlorophyll *a* Concentration of Fresh Water Phytoplankton Analysed by Algorithmic based Spectroscopy

Fairooz Binti Johan[1, a *], Mohd Zubir Bin Mat Jafri[1,b], Lim Hwee San[1,c], Wan Maznah Wan Omar[2,d] and Tan Chun Ho[1,e]

[1]School of Physics, Universiti Sains Malaysia, 11800 Minden, Penang, Malaysia
[2]School of Biological Science, Universiti Sains Malaysia, 11800 Minden, Penang, Malaysia

[a*]fairoozjohan@yahoo.com.my, [b]mjafri@usm.my, [c]hlim@usm.my, [d]wmaznah@usm.my, [e]chunho86@hotmail.com

Abstract. Phytoplankton are microscopic single-celled plants that play an important role in the ecosystem as a major primary producers through photosynthesis. The main objective of this study is to investigate properties of reflection and chlorophyll *a* concentration of phytoplankton in fresh water. The reflection of phytoplankton is taken using spectroradiometer to observe the relationship between the reflectance and the wavelengths of phytoplankton. Tasik Harapan is selected as a study area. This lake is located in Universiti Sains Malaysia (USM), Penang, Malaysia. In this study, the water samples collected are filtered using phytoplankton net and analyzed in the laboratory to determine the reflectance and chlorophyll *a* concentration of phytoplankton. The water samples taken were prior to culture in the medium known as Bold's Basal Medium (BBM) before the reflections are taken so that accurate reflection measurements of phytoplankton can be acquired. Two spectrometers are used in this study, firstly, spectroradiometer, which was used to measure the reflectance of phytoplankton and secondly, spectrophotometer, which was used to measure the concentration of chlorophyll *a*. The algorithms that related between chlorophyll *a* concentration and reflectance of phytoplankton are used. In this study, three regions also are focused, which are the red band, green band and blue band. These bands are related in analyzing phytoplankton. Reflectance of each band specified referred to the concentration of chlorophyll *a* for calibration algorithm. Finally, three wavelengths of 438 nm, 550 nm and 675 nm were selected. The selection of these three wavelengths were found to be strongly correlated to phytoplankton and chlorophyll *a*. The water sampling for validation was also taken from the same lake and was analyzed using the same algorithm. The best results are obtained in this study, which are evidenced by the good correction coefficient, R^2 in the analysis results using the developed algorithm.

1. Introduction

Fresh water is a fundamental resource for human life, and the services provided by surface fresh water ecosystems underpin global water security, food security and economic productivity. Fresh water is a natural resource that is vital to the ecosystem life, whereas good water quality is required for many ecosystems [1] and rich with a variety of flora and fauna, including phytoplankton as a life resource. Everyone knows, phytoplankton give more advantages in today's technology, such as the measurement of phytoplankton biomass is important in aquatic ecology studies. Among the commonly discussed

Content from this work may be used under the terms of the Creative Commons Attribution 3.0 licence. Any further distribution of this work must maintain attribution to the author(s) and the title of the work, journal citation and DOI.
Published under licence by IOP Publishing Ltd

factors that influenced biomass loss are light, temperature, nutrient and control algal growth. The impacts of one factor are dependent on other the factors, for example, the nutrient loading effect on phytoplankton abundance in ecosystems is dependent on other factors including light availability and sedimentation [2]. Consequently, the study on phytoplankton communities and water quality become popular since the last decades [3]. Phytoplankton are microscopic single-celled plants and plays a major role as dominant primary producer of plankton community through photosynthetic activities in aquatic ecosystems. Phytoplankton have a few groups, such as green algae, cyanobacteria, red algae, diatom, dinoflagellates and yellow-green algae.

Spectroscopy is used in studying of absorption, emission and other radiation (ie. scattering) by matter, which is related to the processes on the wavelength of the radiation. In addition, spectroscopic techniques are widely used in almost all technical fields of science and technology including in phytoplankton study. However, spectroscopic techniques are very sensitive and must be set up carefully to study to get an accurate result [4]. The spectral reflectance in certain wavelength range corresponding to the chlorophyll *a* concentration is identified according to the optical properties of phytoplankton [5]. Spectral reflectance is measured when imaging is done under different conditions including sediment, pigment, chlorophyll of water, depth, distribution and optical properties of the medium [6].

Spectroscopy has a various type and each type shows pictures and characteristics spectrum of matter differently. The variation in intensity of the radiation as a function of the wavelength or frequency showing in a graph is called a spectrum. The different regions of the electromagnetic spectrum, such as ultraviolet (UV), visible (VIS) and infrared (NIR) are dependent on the matter characteristics. In the UV region, the wavelength ranges from 200 to 400 nm and in the VIS region, the ranges of wavelength between 380 until 750 nm, whereas in the NIR region, the wavelength ranges from 750 to 2500 nm. In phytoplankton, many previous study are focused on the visible (VIS) and near infrared (NIR) regions, which could be used in vegetation studies [7].

The existing of the chlorophyll *a* concentration in phytoplankton and the phytoplankton spectra reflectance is related to the optical activity of the pigment, the composition and appearance of the algal cells. Chlorophyll, in various forms, is bound within the living cells of phytoplankton found in surface water. Chlorophyll is an important pigment in marine and freshwater algal species and has always been used as an indicator for phytoplankton biomass as well as bio-production calculation in water bodies [8]. The measurement of algal biomass is important in aquatic studies and is commonly estimate the concentration of chlorophyll *a* in fresh water with 90% acetone method using spectrophotometer [9]. This is because spectrophometry is the classical method of determining the quantity of chlorophyll in surface water. The main objective of this study is to investigate properties of reflection and chlorophyll *a* concentration of phytoplankton in fresh water.

2. Methodology

2.1 Sample data collected

In this study, the 40 samples were taken in the morning about 10 a.m. These samples were taken using the phytoplankton net in the filtrate 20 litres of lake water for each sample. The samples were taken randomly but still around the entire study area. The water sample was filled into plastic bottles of 500 ml and stored in an ice box containing ice packs to ensure the conditions of phytoplankton remained good and would not die. After the samples were taken, the samples were brought to the plankton lab for analysis.

2.2 Study Area

Tasik Harapan is chosen as the study area because it was suitable for studying phytoplankton chlorophyll concentration in term of size, condition and colour of the lake. . This lake is located in Universiti Sains Malaysia (USM). Furthermore, the location is also near to make observations. The proposed algorithm has been tested on the lake. This lake has an apparent distribution of phytoplankton chlorophyll concentration and is suitable to be studied. Tasik Harapan is one of the lakes that exists in this university. The area of Tasik Harapan is about 8000 m^2.

2.3 Cultured Method

In this study, culture methods were also needed so that when the reflection reading of phytoplankton were taken, reflectance was shot exactly on phytoplankton rather than the other solid. The medium used in culturing is called Bold's Basal Medium (BBM). This medium was rotated in an autoclave to sterilize equipment and supplies. After preparing the samples were done, these samples were stored in cultured room.

2.4 Optical Instrument and Software

Reflectance of phytoplankton was taken using ASD spectroradiometer (handheld2) with height from sample to spectroradiometer of 30 cm and 1 degree FOV. The reflectance radiation wavelength regions to be recorded range from 325 to 1075 nm and was supported by tripod stand. Each sample was taken ten readings. An average of ten measurements was recorded for each sample. All the results were analyzed by using ViewSpec Pro software. After reflectance was taken, the sample was taken rapidly for analysis to measure the concentration of chlorophyll *a*.

2.5 Extract and Measurement Chlorophyll a Concentration

In this study, the samples were extracted by using a solution of 90% acetone. After extracting chlorophyll *a*, the concentration of chlorophyll *a* is measured using spectrophotometry method using model HITACHI U-1900 UV-Visible spectrophotometer with selected wavelengths at 750 nm, 664 nm, 647nm and 630 nm and 90% acetone was used as a blank in the spectrophotometer [10]. The concentration of chlorophyll *a* was calculated using formula by Jeffrey and Humphrey, 1975 [11]. The absorbance at 750 nm was subtracted from those three wavelengths to give the turbidity-corrected value.

2.6 Regression Algorithm

Equations for different wavelength such as (1) and (2) are needed to assess the content of chlorophyll concentration.

$$C_a = a_0 + a_1R_1 + a_2R_2 + a_3R_1R_2 + a_4R_1^2 + a_5R_2^2 + a_6 R_1^2R_2 + a_7R_1R_2^2 + a_8R_1^2R_2^2 \quad (1)$$

$$C_a = e_0 + e_1R_1 + e_2R_2 + a_3R_3 + e_4R_1R_2 + a_5R_1R_3 + a_6R_2R_3 + a_7 R_1^2 + a_8R_2^2 + a_9R_3^2 \quad (2)$$

Equation (1) is an algorithm model for two wavelengths. The algorithm for three wavelengths such as equation (2) can produce with enlarged equation (1), where, $j = 0,1,2,...$is the constant of the equation. It can be empirically determined using regression analysis.

3. Result and Discussion

3.1 Reflectance spectra of phytoplankton

Figure 1 shows the six spectral reflectance graphs of randomly picked samples using spectroradiometer for the study area at Tasik Harapan, Universiti Sains Malaysia (USM), Penang. In general, the graph shown by the six samples during the study has similar patterns. This could probably because many factors such as distribution, chlorophyll, composition, species, pigment, condition, and measurement instrument while taking the reading. For graph of reflectance against wavelength, there are two obvious peaks in the green (500-600 nm) and NIR (700 nm) regions due to several factors, and high absorption in the blue (400-500 nm) and red regions. Besides, a broad reflectance in this region (500-600 nm) is because of minimum absorption by phytoplankton pigment. Phytoplankton pigments absorb blue and red light, but have minimum absorbance in green light. In the red region, wavelength near 675~677 nm has the lowest reflectance and highest absorbance. In this region the spectra reflectance of phytoplankton was the lowest which also corresponded to high absorption by chlorophyll.

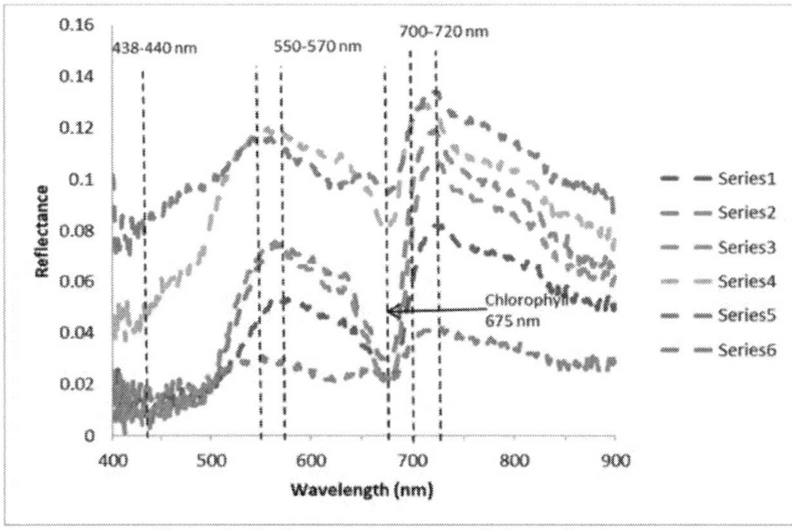

Figure1: Reflectance spectra of phytoplankton

3.2 Correlation between Absorbance and Chlorophyll a Concentration

Figure 2(a), 2(b) and 2(c) show the scatter plot of the absorbance and concentration of chlorophyll *a* value for phytoplankton. The absorbance values increased with concentration of chlorophyll *a*. The increasing value was experienced by all the wavelengths. Figure 2 (a), 2(b) and 2(c) also show that these three wavelengths which were 630, 647 and 664 nm produced the higher R^2. The values of R^2 for absorbance for these three wavelengths were 0.9196, 0.8895 and 0.9984. The correlation of these three wavelengths are also quite high and all these three wavelengths are located in the red region. The concentration of chlorophyll *a* for these 40 samples were produced from by using spectrophotometry method. This instrument was chosen to measure the actual values for chlorophyll *a* because spectrophotometry is the classical method of determining the quantity of chlorophyll in surface water. From Figure 2(a) and 2(b), the data were not as precise as desired on the points obtained for the trend line analysis compared to Figure 2(c) the all almost closed to the line and the value of R^2 was near to 1. It was assumed that the inconsistencies in the data were due to errors in the dilution of the stock solution when making the phytoplankton samples. By the way, these results indicate still can be accepted because the values of R^2 more than 80%.

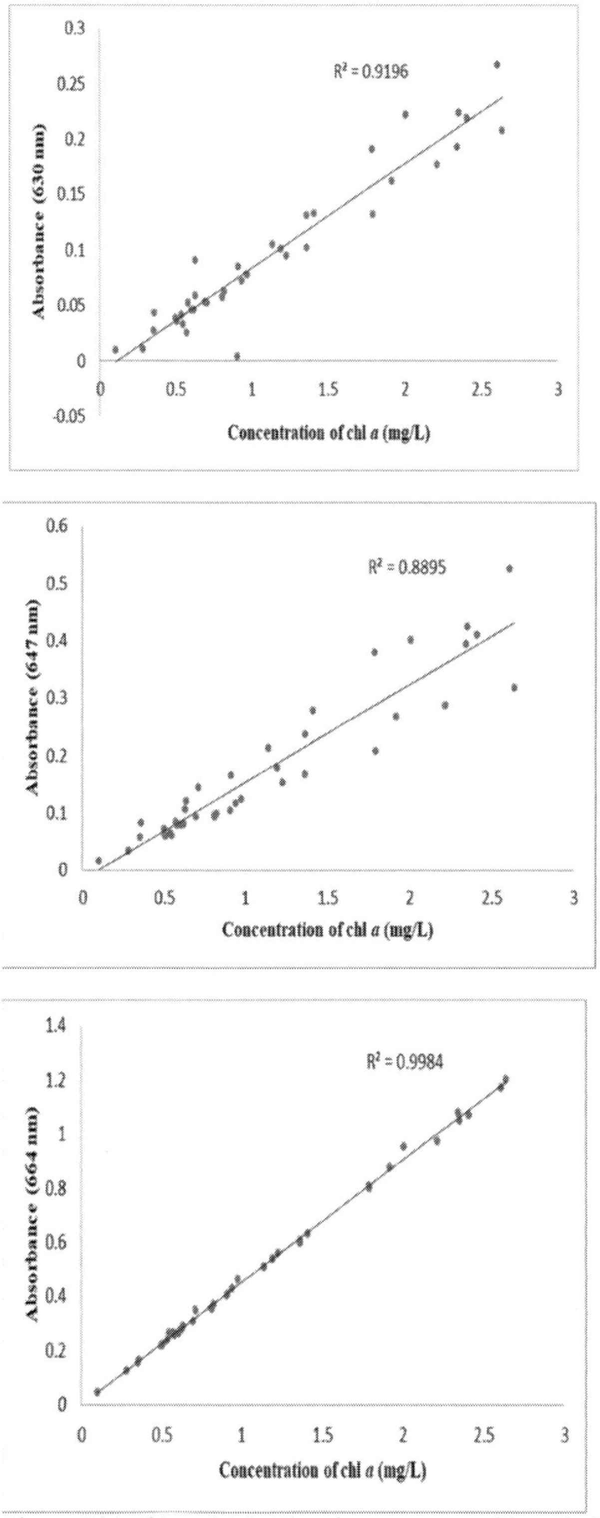

Figure 2(a), (b) and (c): Absorbance versus chlorophyll *a* concentration for 630 nm, 647 nm and 664 nm.

3.3 Correlation between Reflectance and Chlorophyll a Concentration

Figure 3(a), 3(b) and 3(c) shows the scatter plot between chlorophyll *a* concentration and reflectance of phytoplankton at wavelength 438, 550 and 675 nm. The graph patterns for these two variable relationships are polynomial pattern. The scattering response of chlorophyll *a* concentration is decreased with increased value of reflectance spectra. At the beginning, the graph is steeper and gradually approaches horizontal line. This result indicates that the values of reflectance change with different concentration of chlorophyll *a* for phytoplankton. From the result, the graph pattern between these two variables in this study is quite similar with the previous research, Babin, 2003 [12]. The chlorophyll *a* concentration content of phytoplankton ranges from 0.1 to 2.7 mg/L. The points were not closed to the line because of this the correlation coefficient, R for two variables is low and cannot show a relationship between these two variables is perfect. This is due to dependent parameter like instrument setting, chlorophyll *a* content and water samples. Chlorophyll *a* concentration could be influenced by a few factors such as light, nutrient, total phosphorus (TP), total nitrogen (TN), pH, water flow, water temperature and others.

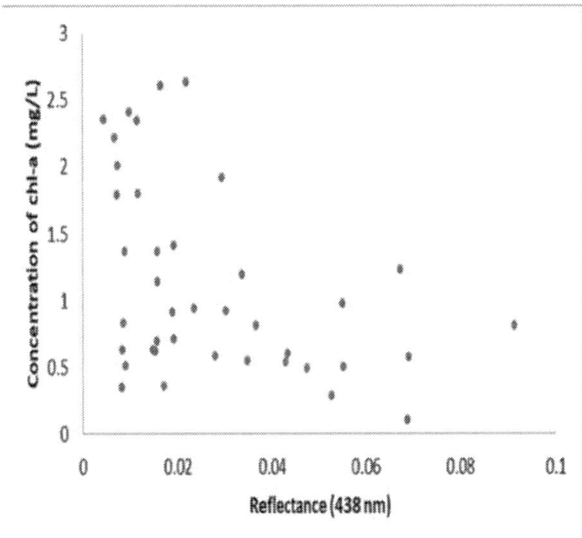

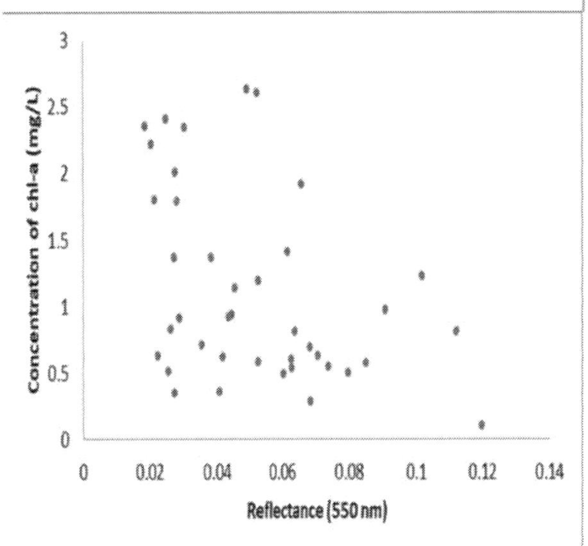

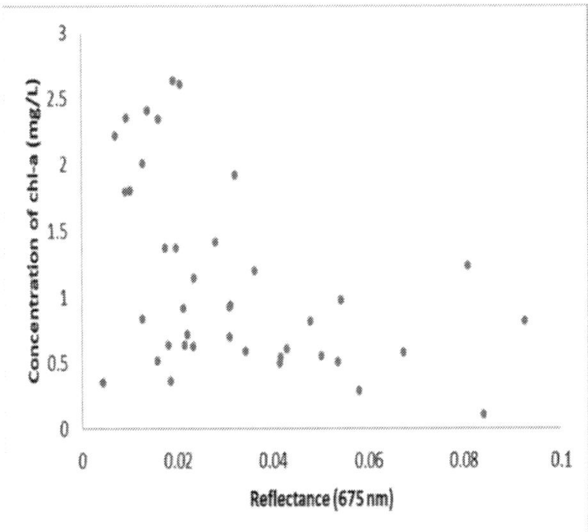

Figure 3(a), (b) and (c): Concentration of chlorophyll *a* versus reflectance for 438 nm, 550 nm and 675 nm

3.4 Validation The Concentration Of Chlorophyll a from The Algorithms Developed.

Figure 4(a) and 4(b) show the measurements of predicted and actual concentration of chlorophyll *a* for two and three wavelengths involving three selected wavelengths which are 438, 550 and 675 nm. The algorithm that used to measure the chlorophyll *a* concentration of phytoplankton for two wavelengths and three wavelengths as shown in Equation (4.1) and (4.2) with the correction coefficient, R^2 of 0.753 and 0.879. Meanwhile, the value of R^2 for validation are 0.768 and 0.856. The R^2 values are quite high. The algorithms are efficient in measuring concentration of chlorophyll *a* for phytoplankton. Compared to two wavelengths, three wavelengths are more efficient for measuring concentration of chlorophyll *a* based on R^2 value.

$$\textit{Concentration of chl a}\ (\text{mg/L}) = 1.67 + 299\, R_{438} - 33.1\, R_{675} - 7217\, R_{438}R_{675} - 14022\, R_{438}^2 - 973\, R_{675}^2 + 373702\, R_{438}^2R_{675} + 112440\, R_{438}R_{675}^2 - 3317051\, R_{438}^2R_{675}^2 \quad (4.1)$$

$R^2 = 0.753$

Validation

$R^2 = 0.768$

$$\textit{Concentration of chl a}\ (\text{mg/L}) = 3.45 + 66.2\, R_{438} - 100\, R_{550} - 3.9\, R_{675} + 5349\, R_{438}R_{550} - 16643\, R_{550}R_{675} - 12682\, R_{438}R_{675} + 3077\, R_{438}^2 + 5209\, R_{550}^2 + 15992\, R_{675}^2 \quad (4.2)$$

$R^2 = 0.879$

Validation

$R^2 = 0.856$

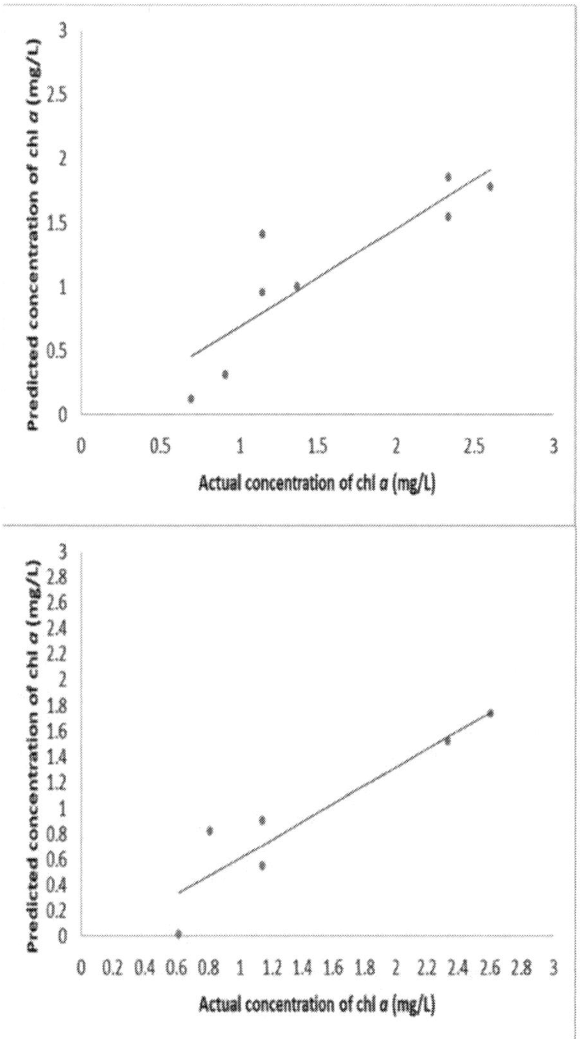

Figure 4(a) and (b): Relationship between predicted and actual concentration of chlorophyll *a* for two and three wavelengths

4. Conclusion

In this research, the concentration of chlorophyll *a* was measured using a spectrophotometry technique. This method was adopted because it is the classic method and high accuracy, but the costs incurred are quite expensive and taken a long time to measure the chlorophyll *a* concentration. The result shown the graph between the concentration of chlorophyll *a* and observance of phytoplankton have straight line with high value of R^2. Besides, the results also shown the graph between the concentration of chlorophyll *a* and reflectance of phytoplankton for three wavelengths 438 nm, 550 nm and 675 nm. The concentration value of chlorophyll *a* is decreased with reflectance, but the correlation coefficient, R for two variables are lower. This is because the result depends on instrument setting, chlorophyll *a* content and water samples. Chlorophyll content is influenced by a few factors, such as nutrient, phosphorus, pH, water flow, water temperature and others

Generally, the spectral reflectance of phytoplankton in this study showed similar pattern. The result also was shown two obvious peaks which were in green and NIR regions. The spectra composed of high absorbance in the blue region near to 440 nm and low absorbance with high reflectance in the green region part of spectrum (550~570 nm). There was a high absorbance (low reflectance) in the red region (~675 nm). Meanwhile, in the NIR region the reflectance increased and decreased slowly.

A simple method to detect and calculate the chlorophyll *a* concentration of phytoplankton in the fresh water was introduced. It based on the spectroradiometer in the calculation of the concentration of chlorophyll *a*. The developed algorithms in this study were created based on the concept of water reflectance properties and can be measured the concentration of chlorophyll *a* in the study area with acceptable accuracy by the value of R^2. This is evidenced in the concentration of chlorophyll *a* measurement and also used as validation in the Tasik Harapan. The methods of chlorophyll *a* concentration measurement are introduced in this study was used to determine the distribution of phytoplankton in fresh water with more ease, simpler saving a cost and time. The results of studies performed in this research achieve the objectives.

Reference

[1] Meybeck, M. & R. Helmer, (1996) An introduction to water quality. In Chapman, D. (ed), Water quality assessments, 2nd ed. Taylor & Francis, New york: 1-22.

[2] Richardson, K., & Jorgensen, B. B. (2013). Eutrophication: Definition, History and Effect *Eutrophication in Coastal Marine Ecosystems (pp. 1-9): American Geophysical Union.*

[3] Stomp , M., Huisman, J., Mittelbach, G. G., Litchman, E., & Klausmeier, C. A. (2011). Large-scale biodiversity patterns in freshwater phytoplankton. *Ecology*, 92(11), 2096-2107. doi: 10.1890/10-1023.1

[4] Stoner,J.O.(2015)"Spectroscopy"http://global.britannica.com/EBchecked/topic/558901/spectroscopy

[5] Johan, F. Jafri, M.Z., Lim., & Wan Maznah W.O. (2015, 24-25Nov 2015). *Estimating the Chlorophyll a Concentration of Phytoplankton from an Empirical Analysis* in Smart Instrumentation Measurement and Application (ICSIMA), IEEE 3rd International Conference on 3015, doi: 10.1109/ICSIMA.2015.7559027.

[6] Peddle, D. R., Peter White, H., Soffer, R. J., Miller, J. R., & LeDrew, E.F. (2001). Reflectance processing of remote sensing spectroradiometer data. *Computers & Geosciences, 27(2), 203-213. doi: 10.1016/S0098-3004(00)00096-0.*

[7] Johan, F., Jafri, M.Z., Lim, H.S., & Sim, C. K. (2013, 1-3 July 2013). *Preliminary study: Spectral reflectance properties of microalgae in freshwster* in Space Science and Communication (IconSpace), IEEE Internation Conference on 2013, pp. 337-340.

[8] Duan, H., Ma, R., Zhang, Y., Loiselle, S. A. Xu, J., Zhao, C., Shang, L. (2010). A new three-band algorithm for estimating chlorophyll concentration in turbid inland lakes. *Environmental Research Letters, 5,* 044009.

[9] Johan, F., Jafri, M.Z., Lim, H.S., & Wan Maznah, W.O. (2014, 9-12 Dec. 2014). *Laboratory measurement: Chlorophyll-a concentration measurement with acetone method using spectrophotometer in Industrial Engineering and Engineering Management (IEEM), IEEE International Conference on 2014, pp. 744-748.*

[10] Jeffrey, S. W. (1975). Preparation and some properties of crystalline chlorophyll c_1 and c_2 from marine algae. Biochim. biophys. Acta 279, 15-33.

[11] Jeffrey, S.W., and G.F. Humphrey. (1975). New spectrophotometric equations for determining chlorophylls a,b, c_1 and c_2 in higher plants, algae and natural phytoplankton. Biochem. Physiol. Pflanzen. 167: 191-194.

[12] Babin, M., Stramski, D., Ferrari, G.M., Claustre, H., Bricaud, A., Obolensky, G., & Hoepffner, N. (2003). Variations in the light absorption coefficients of phytoplankton, nonalgal particles. And dissolved organic matter in coastal waters around Europe. *Journal of Geophysical Research:Oceans, 108 (C7), n/a-n/a. doi 10.1029/2001jc000882.*

A Computerized Time Domain and Spectral Analysis System for Acoustic Characterization of Tissue Mimicking Materials

Anis Nazihah Mat Daud[1], Md Supar Rohani[1,*] and Rosly Jaafar[2]

[1]Department of Physics, Faculty of Science, Universiti Teknologi Malaysia, 81310 Skudai, Johor, Malaysia.
[2]Department of Physics, Faculty of Science and Mathematics, Universiti Pendidikan Sultan Idris, 35900 Tanjong Malim, Perak, Malaysia.

*mdsupar@utm.my

Abstract. In this study, an integrated computerized system for acoustic characterization of tissue mimicking materials (TMMs) was developed. It is used to determine the longitudinal velocity and acoustic impedance of TMMs for time domain analysis and frequency dependent phase velocity and attenuation coefficient of TMMs for spectral analysis. It utilizes the pulse echo immersion technique and consists of a pulser/receiver generator, a transducer as both transmitter and receiver, a digital oscilloscope and a personal computer installed with a custom-developed program. The program is developed on LabVIEW 2012 platform and its graphical user interface consists of three parts; signal acquisition and selection panel, signal display panel and acoustic properties panel. Its performance is validated using KB-Aerotech Alpha R Series transducer with 5.0 MHz center of frequency and RIGOL DS2202 2 Channel 200 MHz 2 GSa/s digital oscilloscope for poly(methyl methacrylate) (PMMA) sample with (1.000±0.050)cm thickness at $(24.0 \pm 0.1)\,^{\circ}C$ water temperature. Findings indicated that the developed system produced accurate and consistent results less than 2.00% error compared to reference values. The developed computerized system offers a user-friendly interface, fast results display and convenient connection.

1. Introduction

Tissue mimicking materials (TMMs) refer to materials that mimic the acoustic properties of human tissue [1, 2]. The transmissions of acoustic signals through TMMs are usually used to determine the longitudinal velocity, acoustic impedance and attenuation coefficient at the center value of frequency. The signals also can be analyzed to determine their phase velocity and attenuation coefficient over a certain range of frequency. The frequency dependent phase velocity is important to determine the acoustic dispersion characteristics of TMMs, while the frequency dependent attenuation coefficient is essential to determine the effect of scattering and absorption with the change of frequency [3, 4].

The common techniques to measure the acoustic properties of TMMs are ultrasonic insertion technique (UIT) and pulse echo immersion technique (PEIT). In 2013, Nazihah, *et al.* developed a computerized acoustic characterization system based on UIT [5]. However, the developed system was limited to determine the longitudinal velocity, acoustic impedance and attenuation coefficient of TMMs at the center value of frequency. Furthermore, UIT is adapted from through transmission technique and it requires two identical transducers aligned in a line facing each other [6]. Therefore,

Content from this work may be used under the terms of the Creative Commons Attribution 3.0 licence. Any further distribution of this work must maintain attribution to the author(s) and the title of the work, journal citation and DOI.
Published under licence by IOP Publishing Ltd

the sample should be carefully inserted between the transducers to ensure that the sample's surface is perpendicular with respect to both transducers [6–8].

In this study, PEIT was employed to develop a computerized acoustic characterization system of TMMs. It is adapted from pulse echo technique (PET), which consists of the ultrasound pulse transmission with and without sample immersed in the distilled water [9]. It is used to determine the longitudinal velocity and acoustic impedance of TMMs for time domain analysis and frequency dependent phase velocity and attenuation coefficient of TMMs for spectral analysis.

2. Acoustic Characterization System

The acoustic characterization system employs PEIT as shown in Figure 1. An electrical pulse was produced by a pulser/receiver generator (Olympus Panametrics NDT model 500 PR) and the transducer was excited to convert the electrical pulses into mechanical vibration and create an ultrasound pulse. The pulse traveled through water and reflected back to its original path at water-reflector interface for the first measurement and at water-sample and sample-reflector interfaces for the second measurement. The reflected pulse was detected by the same transducer and was reconverted into electrical pulse. Then, the pulse was amplified and conditioned using pulser/receiver generator and was monitored using RIGOL DS2000 Series digital oscilloscope. The oscilloscope was connected to personal computer via Transmission Control Protocol/Internet Protocol (Versatile Instrument Control Protocol) control or TCPIP (VICP) control. Both water and sample signals were analyzed using custom-developed program, which was installed on personal computer. The transducer was positioned at 5.00 cm from glass reflector and the water temperature was measured using thermometer.

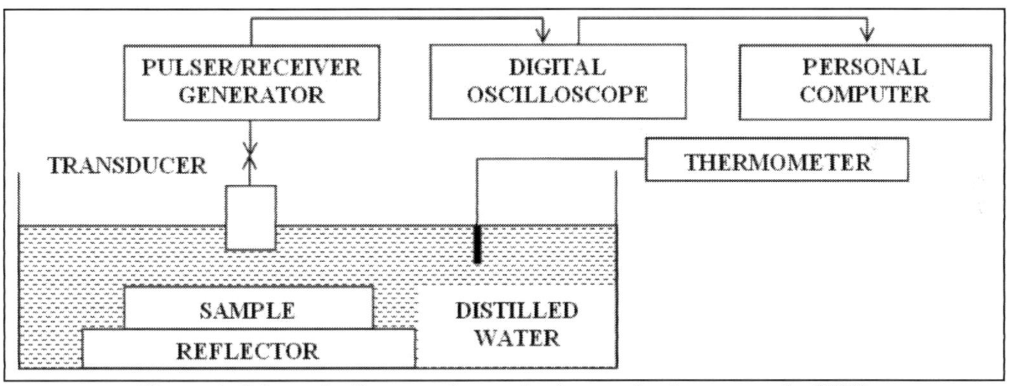

Figure 1. Schematic diagram of acoustic characterization system for PEIT.

2.1 Acoustic Properties

There are four acoustic properties of sample are involved in this study; longitudinal velocity, acoustic impedance, frequency dependent phase velocity and frequency dependent attenuation coefficient. The longitudinal velocity in sample, v_s, is calculated from the longitudinal velocity in distilled water, v_w, time of reflected pulse at water-reflector interface for the water signal, t_w, time of reflected pulse at sample-reflector interface for the sample signal, t_s, and thickness of sample, d_s, using Equation (1) [9].

$$\frac{1}{v_s} = \frac{1}{v_w} - \frac{t_w - t_s}{2d_s} \tag{1}$$

The longitudinal velocity in distilled water, v_w, is calculated from the measured temperature of distilled water, T_w, using Equation (2) [5–7, 10].

$$v_w = \begin{array}{c} 1.40238744\times10^3 + 5.038836171T_w - 5.81172916\times10^{-2}T_w^2 \\ +3.34638117\times10^{-4}T_w^3 - 1.48259672\times10^{-6}T_w^4 + 3.16585020\times10^{-9}T_w^5 \end{array} \tag{2}$$

The acoustic impedance of sample, Z_s, is calculated from the density of sample, ρ_s, and longitudinal velocity in sample, v_s, using Equation (3).

$$Z_s = \rho_s\, v_s \tag{3}$$

Meanwhile, the frequency dependent phase velocity of sample, $v_s(f)$, is calculated from the longitudinal velocity in distilled water, v_w, frequency, f, time of reflected pulse at water-reflector interface for the water signal, t_w, time of reflected pulse at sample-reflector interface for the sample signal, t_s, phase of reflected pulse at water-reflector interface for the water signal, ϕ_w, and phase of reflected pulse at sample-reflector interface for the sample signal, ϕ_s, and thickness of sample, d_s, using Equation (4).

$$\frac{1}{v_s(f)} = \frac{1}{v_w} - \frac{1}{2d_s}\left[(t_w - t_s) - \left(\frac{\phi_w - \phi_s}{2\pi f}\right)\right] \tag{4}$$

The last acoustic property, the frequency dependent attenuation coefficient of sample, $\alpha_s(f)$, is calculated from the amplitude of reflected pulse at water-reflector interface for the water signal, $A_w(f)$, amplitude of reflected pulse at sample-reflector interface for the sample signal, $A_s(f)$, and thickness of sample, d_s, using Equation (5) [9].

$$\alpha_s(f) = \frac{20\log_{10}\left[\dfrac{A_w(f)}{A_s(f)}\right]}{2d_s} \tag{5}$$

2.2 Graphical User Interface (GUI)

The graphical user interface (GUI) of the custom-developed program was designed on the LabVIEW 2012 platform as shown in Figure 2. It consists of signal acquisition and selection panel, signal display panel and acoustic properties panel.

IOP Conf. Series: Journal of Physics: Conf. Series **1083** (2018) 012016 doi:10.1088/1742-6596/1083/1/012016

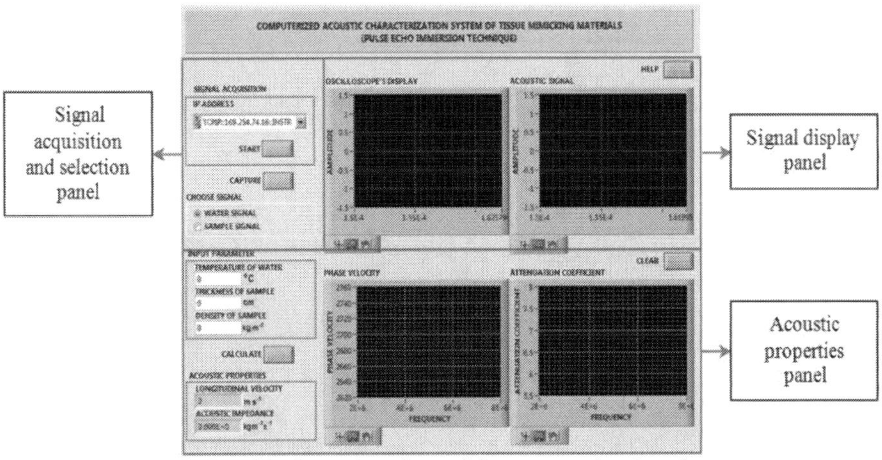

Figure 2. Graphical user interface of the custom-developed program.

The signal acquisition and selection panel consists of two sections (SIGNAL ACQUISITION and CHOOSE SIGNAL), and two buttons (START and CAPTURE). It is used to acquire the acoustic signal from digital oscilloscope and select the appropriate signals to represent water and sample signals. The second part, signal display panel consists of HELP button, OSCILLOSCOPE'S DISPLAY and ACOUSTIC SIGNAL graphs. It is used to display the acoustic signal from oscilloscope and the selected water and sample signals. The last part, acoustic properties panel consists of two sections (INPUT PARAMETER and ACOUSTIC PROPERTIES), two graphs (PHASE VELOCITY and ATTENUATION COEFFICIENT) and two buttons (CALCULATE and CLEAR). It is used to calculate and display the values of LONGITUDINAL VELOCITY and ACOUSTIC IMPEDANCE and the graphs of PHASE VELOCITY and ATTENUATION COEFFICIENT over a certain frequency range.

2.3 Operation Procedure

To acquire the acoustic signal, the user has to enter IP address and press START button and the signal is displayed on OSCILLOSCOPE'S DISPLAY graph. If the program is not connected to oscilloscope, the user has to enter IP address and press START button again. Then, the displayed signal is determined to represent the signal travel through water or immersed sample in water by choosing WATER SIGNAL or SAMPLE SIGNAL in CHOOSE SIGNAL section and press CAPTURE button. Hence, the signal is immediately displayed on ACOUSTIC SIGNAL graph and this procedure is repeated to acquire another signal. Once both signals are available to be analyzed, the user has to enter the values of TEMPERATURE OF WATER, THICKNESS OF SAMPLE and DENSITY OF SAMPLE in INPUT PARAMETER section and press CALCULATE button. Then, the values of LONGITUDINAL VELOCITY and ACOUSTIC IMPEDANCE are calculated and displayed in ACOUSTIC PROPERTIES section. At the same time, the values of PHASE VELOCITY and ATTENUATION COEFFICIENT of sample over a certain frequency range are calculated and displayed in PHASE VELOCITY and ATTENUATION COEFFICIENT graphs. The flow chart of operation procedure to use the custom-developed program is shown in Figure 3.

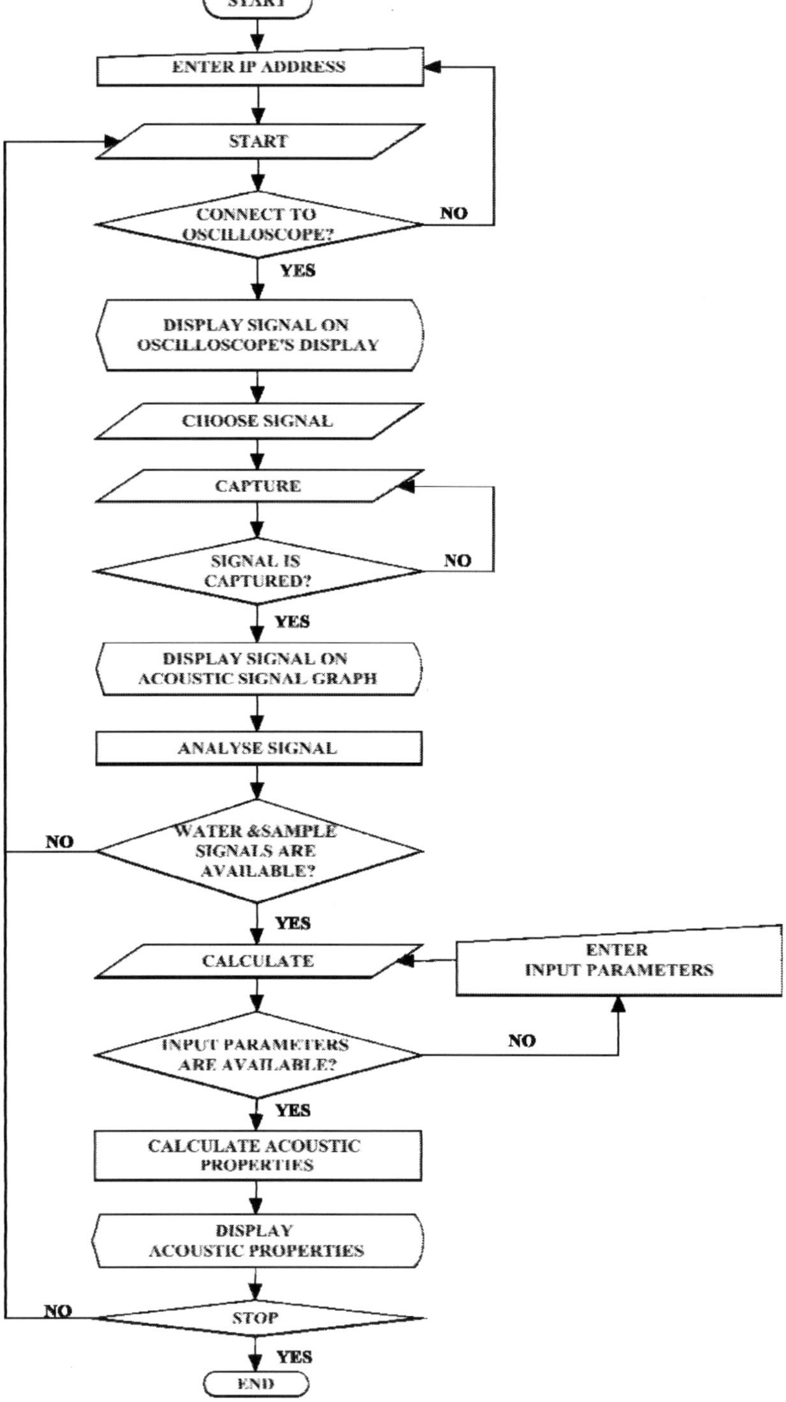

Figure 3. Flow chart of operation procedure to use the custom-developed program.

3. System Accuracy and Consistency

The accuracy of developed system was validated using KB-Aerotech Alpha Series transducer with 5 MHz center frequency and RIGOL DS2202 2 Channel 200 MHz 2 Gigasamples per second (GSa/s) digital oscilloscope by comparing the data obtained from the developed system and standard UIT with reference values. Meanwhile, its consistency was determined by repeating five measurements for the same sample. The experiments were carried out in distilled water at $(24.0 \pm 0.1)^{\mathrm{O}}\mathrm{C}$ and poly(methyl methacrylate) (PMMA) was used as sample since its acoustic properties was consistent within small range of temperature of distilled water [11]. The dimension of sample was $8.000 \times 4.000 \times (1.000 \pm 0.050)\ \mathrm{cm}^3$ and its density was $(1180 \pm 10)\ \mathrm{kg\ m}^{-3}$. The comparison between the results of acoustic properties obtained from the developed system and standard UIT with reference values are summarized in table 1.

Table 1. Comparison between the results of acoustic properties obtained from the developed system and standard UIT with reference values.

Acoustic Properties	Reference Values	Developed System		Standard UIT	
		Values	%	Values	%
$v_s\ (\mathrm{ms}^{-1})$	2730[a]	2742 ± 27	0.44	2747 ± 27	0.62
$Z_s\ (\times 10^6\ \mathrm{kg\ m}^{-2}\ \mathrm{s}^{-1})$	3.22[a]	3.235 ± 0.041	0.46	3.241 ± 0.042	0.65
$v_s(f)\ (\mathrm{ms}^{-1})$	2753[b]	2747 ± 40	0.22	2742 ± 59	0.40
$\alpha_s(f)\ (\mathrm{dB\ cm}^{-1}\ \mathrm{MHz}^{-1})$	$0.94f$[b]	$(0.923 \pm 0.044)\ f$	1.81	$(0.987 \pm 0.098)\ f$	5.00

[a] Reference values [12]
[b] Reference values [4]

Table 1 indicated that the obtained results from the developed system and standard UIT were within 1.81% and 5.00% errors compared to reference values, respectively. The errors were probably due to the inaccurate measurement of water temperature since both developed system and standard UIT involved the water temperature measurement in determining the acoustic properties of sample. In addition, the errors were also probably caused by the inhomogeneity of materials, which could occur during manufacturing processes [13]. However, the small percentage error confirmed that the acoustic signals were more accurately analyzed by the developed system compared to standard UIT in order to determine the acoustic properties of sample. The results from the developed system and standard UIT were also consistent within 4.77% and 9.93% from the average values of longitudinal velocity, acoustic impedance, frequency dependent phase velocity and frequency dependent attenuation coefficient of sample, respectively. It indicated that the results obtained from the developed system were more consistent compared to standard UIT since the sample's surface should be perpendicular with respect to single transducer for developed system compared to both transducers for UIT [6–8].

In addition, the frequency dependent phase velocity and attenuation coefficient of PMMA obtained from the developed system and standard UIT were in good agreement with previous studies [4, 14]. The phase velocity of PMMA was approximately constant with the increasing frequency. Thus, its phase velocity was independent with the change of frequency. It proved that PMMA has nondispersive medium characteristics of acoustic wave [15]. Meanwhile, the attenuation coefficient of PMMA was linearly increased as the frequency increased. It indicated that the effect of scattering and absorption phenomena in PMMA were greater at high frequency compared to low frequency as the ultrasonic wave propagates in the sample [16].

4. Conclusion

An integrated computerized system to determine the acoustic properties of tissue mimicking materials (TMMs) for pulse echo immersion technique was successfully developed in this study. It is used to determine the longitudinal velocity and acoustic impedance of TMMs for time domain analysis and frequency dependent phase velocity and attenuation coefficient of TMMs for spectral analysis. It consists of a pulser/receiver generator, a transducer as transmitter and receiver, a digital oscilloscope and a personal computer installed with custom-made developed program. It produces reliable and consistent results less than 2.00% error. The developed system offers a user-friendly interface, fast results display and convenient connection as it is connected with digital oscilloscope via TCPIP (VICP) control.

Acknowledgement
This research was financially support by MyBrain15 scholarship and grant GUP/MOHE Vot. 4F897, 08H16 and 4F412 of Universiti Teknologi Malaysia from Ministry of Higher Education. We also would like to acknowledge the facilities support from Faculty of Science, Universiti Teknologi Malaysia.

References

[1] Zell K, Sperl J I, Vogel M W, Niessner R and Haisch C 2007 Acoustical properties of selected tissue phantom materials for ultrasound imaging, ***Phys. Med. Biol.*** 52 N475–84.
[2] Maggi L E, Von Kruger M A, Pereira W C A and Monteiro E E C 2009 Development of silicon-based materials for ultrasound biological phantoms, ***IEEE Int. Ultrason. Symp.*** 0483 1962–5.
[3] Lee K I, Humphrey V F, Kim B-N and Yoon S W 2007 Frequency dependencies of phase velocity and attenuation coefficient in a water-saturated sandy sediment from 0.3 to 1.0 MHz *J. Acoust. Soc. Am.*121 2553–8.
[4] He P 1999 Experimental verification of models for determining dispersion, *IEEE Transactions on Ultrasonics, Ferroelectrics, and Frequency Control* 46 706–4.
[5] Nazihah M D, Kadri S, Yaacob M I H and Rosly J 2013 Computerized acoustical characterization system of medical phantoms *AIP Conf. Proc.* 1528 406–11.
[6] Mat Daud A N, Rohani M S and Jaafar R 2017 Acoustic characterisation of Konjac Glucomannan gel as a medical phantom *Solid State Phenomena* 268 379–83.
[7] Mat Daud A N, Jaafar R, Ayop S K, Yaacob M I H and Rohani M S 2017 Elastic constant determination of hardwoods using ultrasonic insertion technique *Ultrasonics* 75 194–8.
[8] Fahr A 2013 *Aeronautical Applications of Non-Destructive Testing* (Pennsylvania: DEStech Publications Incorporated).
[9] Cournane S, Cannon L M, Browne J E and Fagan A J 2010 Assessment of the accuracy of an ultrasound elastography liver scanning system using a PVA-cryogel phantom with optimal acoustic and mechanical properties *Phys. Med. Biol.*55 5965–83.
[10] Maggi L E, Souza A B B, Ichinose R M, Pereira W C A, von Kruger M A and Costa-Felix R P B 2011 The importance of expression of uncertainty of acoustical parameters of ultrasonic phantoms *J. Phys. Conf. Ser.* 279 1–6.
[11] Carlson J E, van Deventer J, Scolan A and Carlander C 2003 Frequency and temperature dependence of acoustic properties of polymers used in pulse-echo systems *IEEE Symposium on Ultrasonics* 1 885–8.
[12] MacIntire P (Ed.) 1991 *Nondestructive Testing Handbook Second Edition: Volume 7 Ultrasonic Testing* (United States of America: American Society for Nondestructive Testing).
[13] Lane C 2014 *The Development of a 2D Ultrasonic Array Inspection for Single Crystal Turbine Blades* (Switzerland: Springer International Publishing).
[14] Bauer A Q, Marutyan K R, Holland M R and Miller J G 2008 Negative dispersion in bone: the role of interference in measurements of the apparent phase velocity of two temporally overlapping signals *J. Acoust. Soc. Am.* 123 2407–14.
[15] Ida N 2015 *Engineering Electromagnetics* (Switzerland: Springer International Publishing).

IOP Conf. Series: Journal of Physics: Conf. Series **1083** (2018) 012016 doi:10.1088/1742-6596/1083/1/012016

[16] Cobbold R S C 2007 *Foundations of Biomedical Ultrasound* (New York: Oxford University Press).

T2* Relaxation of Agar Gel With and Without The Presence of Tumor-Like Structure as Obtained from Resting State Fmri Sequence Protocol

A N Yusoff [1, a *], N S Abdul Rashid[1, b] and S Usman Ali[1, c]

[1]Diagnostic Imaging & Radiotherapy Program, School of Diagnostic & Applied Health Sciences, Faculty of Health Science, Universiti Kebangsaan Malaysia, 50300 Jalan Raja Muda Abdul Aziz, Kuala Lumpur, Malaysia

[a]nazlimtrw@ukm.edu.my, [b]syera_eyra16@yahoo.com, [c]syeera94@yahoo.com

Abstract. The objective of this study was to determine the change in signal-to-noise ratio (SNR) as a function of TE of agar gel phantom with and without the presence of tumor-like structure, using a resting state fMRI sequence protocol. The phantoms were prepared using a conventional method. *Citrus hystrix* DC (kaffir lime) was used to mimick tumor and was submerged in the agar gel. The phantoms were scanned using a 3T-MRI system. An rsfMRI sequence protocol (TR = 2000 ms; TE = 51 ms – 115 ms) was used in obtaining the SNR at several TEs. In addition, a standard MRI phantom as control was also scanned using the same sequence protocol. For the prepared phantoms, the SNR was obtained from several 7-mm diameter ROIs at a distance 2, 4 and 6 cm from the centre of the image and was averaged. For the standard MRI phantom, similar ROI diameter was used for SNR determination but for three different slice thicknesses; 3, 5 and 10 mm. The MRI, agar gel and agar gel with kaffir lime phantoms show T2* relaxation that fits the exponential equation $S \propto e^{-TE/T2^*}$. The average T2* for the standard MRI phantom is 86.7 ± 0.1 ms. The average T2* for agar gel phantom is 70.1 ± 0.1 ms while T2* is shortened to an average of 54.8 ± 0.1 ms for agar gel phantom with kaffir lime. This indicate an increase in local magnetic field inhomogeneity in the vicinity of the tumor. The relaxation is thus faster causing a drop in the magnetic resonance signal intensity. The agar gel and agar gel with tumor mimicking phantom are suitable for a long TE rsfMRI scans.

1. Introduction

Resting state fMRI (rsfMRI) is an fMRI experiment conducted on patients while they are at rest to investigate, among others, the functionality of the default mode network (DMN) [1]. The output from rsfMRI is very much depends on the sequence protocol in particular the time to echo (TE) and the transversal relaxation of a tissue (T2*). The amount of T2* weighting that is seen in an image depends on TE [2], thus making it an important controlling parameter in obtaining sufficient amount of signal from an fMRI experiment.

T2* value of a substance or tissue can be determined from the SNR vs. TE plot and the equation $SNR \propto e^{-TE/T2^*}$ [3]. The exponential curve can be fitted to the equation to determine T2*. By knowing the value of T2*, a suitable TE for an fMRI experiment or QC can be chosen. Conventionally, suitable

Content from this work may be used under the terms of the Creative Commons Attribution 3.0 licence. Any further distribution of this work must maintain attribution to the author(s) and the title of the work, journal citation and DOI.
Published under licence by IOP Publishing Ltd

TE is about the T2* of tissue or substance. Newly developed phantom would also be possibly tested and investigated.

This study implemented T2* determination on MRI phantom, agar gel phantom and agar gel phantom with the presence of tumor-like structure represented by fresh *Citrus hystrix* DC (kaffir lime). Consistency of the SNR value by using the TE determined in the present study will be the main focus of our subsequent study.

2. Methodology

2.1 Preparation of agar gel phantom

An amount of 50 g dry agar gel powder (Merck) and 10 g preservative thimerosal powder (Fisher Scientific) were mixed with 1000 m*l* distilled water in a 1-*l* container at 5% concentration. Magnetic stirring technique was used to homogenize the solution for 30 minutes, while the solution was increasingly heated on the hot plate. The solution complete transformation from the sol state to molecular state was indicated by the change in its color from murky to crystal clear, which occurred at its boiling temperature of 100°C. The solution was continued to be heated for 10 minutes to ensure a complete transformation and subsequently poured into a cylindrical plastic container (12 cm diameter and 12 cm height) and was let to gradually cool down to room temperature to form the jelly phantom structure. For the phantom that contained tumor-like structure, a fresh *Citrus hystrix* DC (kaffir lime) was submerged into the solution. Its position is controlled so that it will be at the center of the plastic container longitudinally and axially as the solution hardened.

2.2 Magnetic resonance imaging scans

Magnetic resonance imaging (MRI) scans on the phantoms were conducted in the Department of Radiology, Hospital Canselor Tuanku Muhriz using a 3-T Siemens Magnetom Verio MRI system. The phantom was placed vertically at the iso-centre of the magnet bore inside the head coil, which was used to deliver the radiofrequency (RF) pulses. The resting state functional MRI (rsfMRI) protocol implementing gradient echo – echo planar imaging (GRE-EPI) sequence was used to obtain axial T2* weighted images of the phantoms at TR = 2000 ms and several TE values (TE = 51, 55, 61, 65, 71, 75, 81, 85, 91, 95, 101, 105, 111 and 115 ms). Other imaging parameters are number of slice = 29, field of view (FOV) = 200 × 200 mm, flip angle (α) = 90°, matrix size = 64 × 64, slice gap = 1 mm. For comparisons and validation, resting state fMRI protocols were also implemented on a standard MRI phantom using the above parameters but for different slice thickness; 3 mm, 5 mm and 10 mm.

2.3 SNR determination

The middle slice obtained from each measurement (or each TE) of the phantoms was chosen for SNR evaluation [4]. Three circular shaped regions of interest (ROIs) were drawn on the phantom image and the average signal intensity (I_p) was measured from the three ROIs. The intensity of the background noise (I_b) and its standard deviation (σ_b) were measured from a large ROI drawn outside the phantom image. For a good averaging, each signal intensity measurement was done three times. The SNR were than calculated from SNR = $(I_p - I_b)/\sigma_b$. This procedure was repeated for all TEs. For the standard MRI phantom, this SNR determination was done for three different slice thickness; 3 mm, 5 mm and 10 mm for validation purpose. For the agar gel phantom, SNR was determined at three different radii (2, 4 and 6 cm) from the center of the phantom image.

2.4 T2 determination*

The T2* value for the standard MRI phantom and for each of the agar gel phantom was then determined from the SNR vs. TE plot from which the equation SNR $\propto e^{-TE/T2^*}$ was fitted with the experimental data using Microsoft Excel trend line function. The equation of the trend line will then be compared to SNR $\propto e^{-TE/T2^*}$ to determine the T2* value of the standard MRI and agar gel phantoms.

3. Results

Figure 1 shows the exponential decrease of SNR as a function of TE for the standard MRI phantom. The exponential model has been fitted as closely as possible to the experimental data as indicated by the trend line and the respective equation. Generally, SNR increases with slice thickness at all TEs but decreases with the increase in TE for a fixed thickness. By comparing the exponential equations with SNR $\propto e^{-TE/T2^*}$, it can be shown that T2* values are 76.9 ± 0.1 ms, 83.3 ± 0.1 ms and 100.0 ± 0.1 ms for slice thicknesses 3 mm, 5 mm and 10 mm, respectively. Thus, the average T2* for the standard MRI phantom is 86.7 ± 0.1 ms, slightly larger than its standard manufactured value which is 80 ms. These results validate the technique to be used for agar gel phantom.

Likewise, Figure 2 shows a similar trend of exponential SNR decrease as a function of TE for agar gel phantom. The trend line for each plot shows a good fit of the exponential model to the experimental data. SNR is about the same for 2 cm and 4 cm distances at all TEs, but decreases markedly at 6 cm. The trend line equations obtained for each plot indicated T2* values of 66.7 ± 0.1 ms, 66.7 ± 0.1 ms and 76.9 ± 0.1 ms for radial distance of 2 cm, 4 cm and 6 cm, respectively. An average T2* value for agar gel phantom is 70.1 ± 0.1 ms.

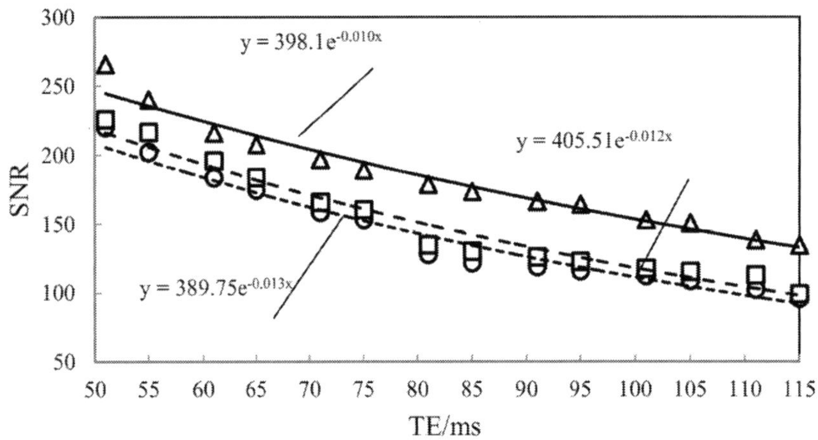

Figure1. The plots of SNR vs. TE for the standard MRI phantom at different slice thicknesses; ○: 3 mm, □: 5 mm and Δ: 10 mm

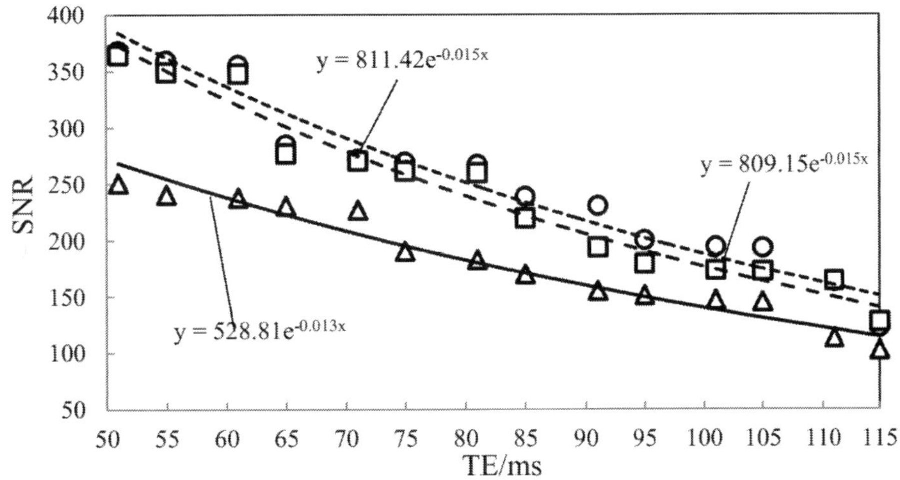

Figure 2. The plots of SNR vs. TE for the agar gel phantom at different radial locations; ○: 2 cm, □: 4 cm and Δ: 6 cm from the center of the phantom.

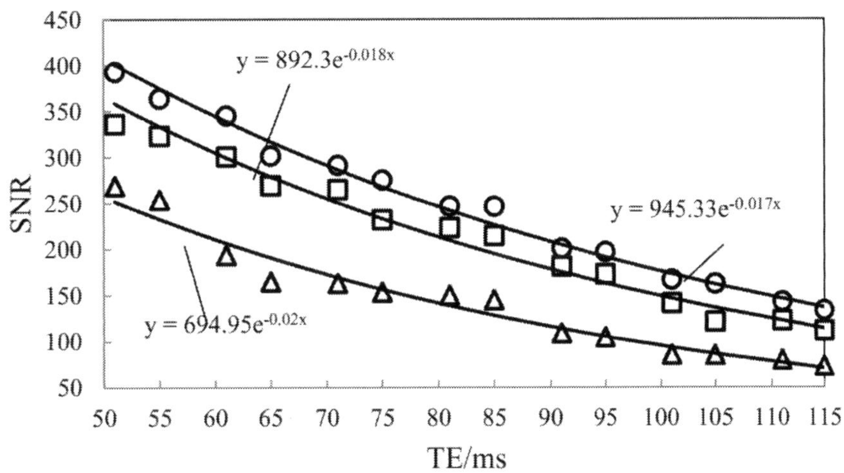

Figure 3. The plots of SNR vs. TE for the agar gel phantom in the presence of tumor-like structure at different radial locations; ○: 2 cm, □: 4 cm and Δ: 6 cm from the center of the phantom.

Figure 3 is the behavior of SNR similar to that shown in Figure 2 but in the presence of kaffir lime as tumor like structure, being longitudinally and axially submerged in the agar gel phantom. Similar change in SNR with distance can be observed. The T2* values determined from fitting the exponential model to the experimental data are 58.8 ± 0.1 ms, 55.6 ± 0.1 ms and 50.0 ± 0.1 ms for radial distances of 2 cm, 4 cm and 6 cm, respectively. The average T2* of the agar gel phantom in the presence of kaffir lime is 54.8 ± 0.1 ms which is lower than that obtained in the absence of the tumor like structure.

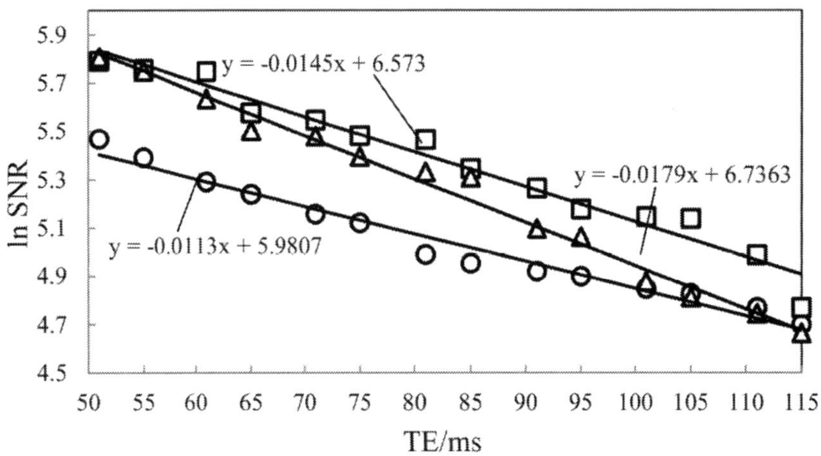

Figure 4 Natural logarithm of SNR vs. TE for the average of SNR shown in figure 1 – 3; ○: standard MRI phantom, □: agar gel and Δ: agar gel with tumor-like structure.

Figure 4 shows another way of evaluating the change of SNR with TE, hence determination of T2* value. The plots are the *ln* of the average of SNR of Figure 1, Figure 2 and Figure 3 as a function of TE. For comparisons, it can be seen from the slope of the lines that a faster exponential decrease is indicated by a higher slope which is shown by the agar gel phantom with kaffir lime, while the MRI standard phantom has a slower relaxation. Obviously, the presence of tumor like structure has the effect of decreasing the T2* value of the agar gel. From the slopes, it can be shown that T2* values for the phantoms under study are 88.5 ± 0.1 ms, 69.0 ± 0.1 ms and 55.9 ± 0.1 ms for the standard MRI, agar gel and agar gel with kaffir lime, respectively, which are about the same with that obtained from the exponential equations.

4. Discussion

Studies on other gelling materials as an alternative to agar gel for MRI phantom fabrication has been done [5–7] to investigate their homogeneity, gel stability, reproducibility and time consumption in preparation. Two important quantities of interest are the T1 and T2 relaxation times, which were obtained from their respective curves. In Hellerbach et al. [5], it is shown that Carbomer-980 and Carbopol-974P were two promising novel phantom materials. These gelling agents are readily available, inexpensive and easy to handle given that thermal treatment is not required. In Christoffersson et al. [6] it is shown that The T1 and T2 values of the gel can be varied independently by changing the relative amounts of magnetic additives (nickel) and agarose. In addition, a novel tissue-equivalent MRI phantom based on carrageenan gel was fabricated [7]. Carrageenan gel is an ideal solidifying agent for making large, strong and flexible phantoms in a wide variety of shapes. In this study, agar gel was used as opposed to gelling materials in [5–7] due to its flexibility, availability and cheaper cost.

The SNR obtained from the standard MRI phantom, agar gel and agar gel in the presence of kaffir lime mimicking phantom follows an exponential decrease as a function of TE. As TE increases, more signals are lost due to the duration of spin-spin interactions that are allowed to occur. All hydrogen spins in the MRI and agar gel phantoms are subjected to the same constant external magnetic field (of 3T in this study). However, despite the homogeneity of the 3-T magnetic field upon the application of shimming, spin-spin interactions caused by local magnetic field inhomogeneity of a material or tissue still persist. Protons will still be experiencing differences in their local magnetic fields caused by differences in their surroundings. Some protons precess faster than the other. A larger difference between the fastest and the lowest precession frequency of the protons indicates a larger inhomogeneity and will enhance a larger amount of spin-spin interaction, hence, a shorter T2*.

T2* relaxation is slower (or longer T2*) for the standard MRI phantom. The standard MRI phantom is fluid in nature and has smaller molecules and higher mobility causing a longer T2*, which is 86.7 ± 0.1 ms on average as compared to 70.1 ± 0.1 ms for agar gel and 54.8 ± 0.1 ms for agar gel in the presence of kaffir lime. Materials with higher mobility, such as fluid will increase local magnetic field homogeneity causing a small difference between the precessional frequencies among the protons. On the contrary, gelling materials such as agar gel contains larger molecules thus will create a larger local magnetic field inhomogeneity that will cause spin-spin interaction to complete in a shorter duration. This is more prominent in the presence of kaffir lime. Kaffir lime has caused disturbance in the local field homogeneity of the field distribution in the gel phantom resulting in a faster drop in SNR as a function of TE.

The three highest chemical compounds in kaffir lime peel oil are Sabinene (4-methylene-1-(1-methylethyl)bicyclo[3.1.0]hexane), Limonene (1-Methyl-4-(1-methylethenyl)-cyclohexene) and β-Pinene (6,6-Dimethyl-2-methylidenebicyclo[3.1.1]heptane) [8] all with chemical formula of $C_{10}H_{16}$ which contained about 71.8% of its total content as determined by gas chromatography (GC) and gas chromatography–mass spectrometry (GC-MS) [8]. It is believed that an abundant amount of hydrogen atom in kaffir lime has caused disturbance in the distribution of magnetic field lines in the agar gel surrounding the kaffir lime which in turn introducing magnetic field inhomogeneity that has caused a decrease in T2* value of the phantom.

5. Conclusion

Due to the local inhomogeneity of the magnetic field introduced by the gelling materials, the T2* values are shortened and the effect is larger in the presence of kaffir lime. This could be due to the increasing number of hydrogen atom resulting in a higher degree of field inhomogeneity. The fabricated phantoms are suitable to be used with long TEs resting state fMRI protocols.

Acknowledgement

Authors would like to thank Sollahuddin Hj. Omar for his assistance in the preparation of the agar gel phantom and Mohd Nor Affendi for his assistance in the MRI scans. Authors would also like to express their gratitudes to Biomedical Science Program for the permission to use the laboratory for the preparation of agar gel phantom and Department of Radiology, Hospital Canselor Tuanku Muhriz for the permission to use the MRI system.

References

[1] Di X and Biswal B B 2014 *Neuroimage* 86 53
[2] Tofts P S 2009 Presented at the International Seminar on Magnetic Resonance in Medicine 18 April 2009 Hawaii
[3] Bernstein M A, King K F and Zhou X J 2004 *Handbook of MRI Pulse Sequences* (Amsterdam: Elsevier Academic Press)
[4] Ahmad Nazlim Yusoff 2007 *Pengantar Pengimejan Resonans Magnet* (Bangi: UKM Press)
[5] Hellerbach A, Schuster V, Jansen A, Sommer J 2013 *Plos One* 8(8) e70343
[6] Christoffersson J O, Olsson L E and Sjöberg S 1991 *Acta. Radiol.* 32(5) 426
[7] Yoshimura K, Kato H, Kuroda M, Yoshida A, Hanamoto K, Tanaka A, Tsunoda M, Kanazawa S, Shibuya K, Kawasaki S and Hiraki Y 2003 *Magn. Reson. Med.* 50 1011
[8] Kasuan N, Muhammad Z, Yusoff Z, Rahiman M H F, Taib M N and Haiyee Z A 2013 *Mal. J. Anal. Sci.* 17(3) 359

The Effect of Chenodeoxycholic Acid (CDCA) in Mangosteen (Garcinia mangostana) Pericarps Sensitizer for Dye-Sensitized Solar Cell (DSSC)

Mashasriyah Ismail[1], Norasikin Ahmad Ludin[1 *], Norul Hisham Hamid[2], Mohd Adib Ibrahim[1], Kamaruzzaman Sopian[1]

[1]Solar Energy Research Institute (SERI), Universiti Kebangsaan Malaysia, 43600 UKM Bangi, Selangor, MALAYSIA
[2]Faculty of Forestry, Universiti Putra Malaysia, 43400 UPM Serdang, Selangor, MALAYSIA

*sheekeen@ukm.edu.my

Abstract. Dye-sensitized solar cell (DSSC) is a third generation of solar cell, which has been pursued by numerous researcher to increase its efficiency, stability, and compatibility with our surrounding. One of advancement is a usage of a natural product or organic material to replace an artificial sensitizer such as ruthenium sensitizer, which is expensive, difficult to synthesize and has its own noxiousness. A natural product such as mangosteen (Garcinia mangostana) pericarps is easily obtained from eaten mangosteen fruits. Mangosteen fruits are mainly grown in Southeast Asia such as Malaysia and certain tropical areas around the globe. Mangosteen pericarps gave a promising performance in natural dye research however unable to compete with artificial due to low anchor groups such as carboxyl and hydroxyl. In this study, chenodeoxycholic acid (CDCA) was used to help increase the device performance. The concentration of CDCA in the sensitizer used in this study are 0.10 mM, 0.50 mM, 0.75 mM, 1.00 mM, 1.50 mM and 2.00 mM. There are several characteristics are studied such as UV-VIS spectrometer, cyclic voltammetry (CV), Ffourier-transform infrared spectroscopy (FTIR) and current-voltage measurement (I-V). Mangosteen pericarps sensitizer itself applied on DSSC have efficiency of 0.36% with short-circuit current density (J_{sc}) = 1.00 mA/cm^2, open circuit voltage (V_{oc}) = 0.64 V and fill factor (FF) = 56.75%. The highest performance of DSSC with mangosteen sensitizer and CDCA are at 0.56% with short-circuit current density (J_{sc}) = 1.40 mA/cm^2, open circuit voltage (V_{oc}) = 0.65 V and fill factor (FF) = 62.03%. The increment of device performance shows that CDCA helps to increase mangosteen sensitizer performances by adding additional -O-H into the sensitizer which increases the ability to anchor onto the TiO_2 surface.

1. Introduction

From 1991, DSSC device has become an interest by numerous researchers when Gratzel and O'Regan were able to create an alternative device of silicon solar cells with an efficiency of 12% in a diffuse light [1]. A standard DSSC consist of photoanode, counter electrode, dye sensitizer, and electrolyte which each of them played a very critical role related to each other. As for this study, the sensitizer will be main focus, which it helps to harvest photons and producing photo-excited electron at the surface of photoanode [2]. A fully efficient sensitizer are needed to have several properties such as intense

Content from this work may be used under the terms of the Creative Commons Attribution 3.0 licence. Any further distribution of this work must maintain attribution to the author(s) and the title of the work, journal citation and DOI.
Published under licence by IOP Publishing Ltd

absorption within visible light range (400–700 nm), absorbed or anchored robustly onto surface of photoanode thin film, high extinction coefficient, a stable oxidation form for regeneration phase, capable to have ~108 turnovers correspond to 20 years of cell operation and possess more negative lowest unoccupied molecular orbital (LUMO) level than conduction band (CB) of the semiconductor and also more positive highest occupied molecular orbital (HOMO) level than the redox potential of the electrolyte [3]. Throughout years there are several different types of photosensitizer was created and studied such as ruthenium (II) dye, porphyrin dye, coumarin and natural products sensitizer [4]. Natural photosensitizer was obtained from natural products such as fruits, leaves, seeds, and trees [5]. Natural sensitizer was selected based on its anthocyanin, chlorophyll, and carotenoids, which helps to give the colour to the products and consists carboxyl and hydroxyl groups [6]. Those groups help the sensitizer molecule to anchor on to the surface of a semiconductor. There are different types of natural products were studied to be able replacing the artificial sensitizer because of its rarity elements, poisonous, and expensive processes such as perilla and crocetin [7]. The natural photosensitizer is easy to be developed using cold extractions and the products easily found in our surrounding without damaging our planets. Mangosteen (Garcinia mangostana) fruit is easily found in a tropical place or Southeast Asia such as Malaysia and Indonesia. Mangosteen fruit itself has a structure which consists of aril and pericarps. Mangosteen pericarps were used in this study due to its infamous high performances due to the colour of purple provided by anthocyanin group. However, it has a staggering development such low efficiency of 1.17% at the active area of 0.2 cm^2 compare to artificial sensitizer due to lacking the amount the anchoring groups, dye aggregation and limited absorption light range [8]. There are several advancements were studied such as purifications, combining different types of dyes and adding a catalyst to boost performances [6]. CDCA was used as a catalyst to increase the amount of carboxyl and hydroxyl groups within mangosteen photosensitizer in order to improve the co-absorbance properties and chemical structure. In this study, it is also expected that the catalyst will reduce dark current, enhancing V_{oc} and weakened dye aggregation.

2. Experimental

2.1. Materials

In this study, we obtained mangosteen dye extract from a dried pericarp of consumed mangosteen immersed in methanol 99.8% (JT. Baker) solvent. The extract was separated into six different dark glass bottles and added chenodeoxycholic acid (CDCA) powder (Solaronix) with concentrations of 0.10 mM, 0.50 mM, 0.75 mM, 1.00 mM, 1.50 mM and 2.00 mM respectively. DSSC fabrication consists of fluorine doped tin oxide (FTO) glass, titanium oxide (TiO_2) paste (Dyesol), platinum paste (Dyesol), iodine electrolyte (Solaronix).

2.2. Preparation of mangosteen dye sensitizer

Mangosteen pericarps are obtained from a consumed mangosteen fruits and cleaned thoroughly using tap water which to make sure no residue of mangosteen fruits. The pericarps were dried under sunlight for two to four days after cleaning process. The drying process was performed to make sure the pericarps does not contain any water or moist. The dried pericarps were turned into powder using pester and mortar, grinder, and blender. The processed pericarps powder was soaked in methanol solvent with a ratio of 1:10. This process was called as cold extraction. The first extraction of dye from the powder was taken durations of 24 hours [9]. After the extraction was filtered and separated from pericarps residue, the extractions were repeated two times in durations of three days. The repeating process of the extraction was conducted to make all the dye was fully extracted from the powder. The filtered extracted solutions were needed to have suitable dye colour, which is concentrated using rotary evaporator and able to be applied as a sensitizer for DSSC.

2.3. Fabrication of DSSC

A standard DSSC structure was used as a device to study the sensitizer with CDCA performance as catalyst. It consists of substrate glasses, photoanode, counter electrode, sensitizer, and electrolyte. In this study, one-sided FTO glasses was used as a substrate for both photoanode and counter electrode. The photoanode was using FTO glass with a resistivity of 15 Ω/square and counter electrode using FTO glass with a resistivity of 8 Ω/square. All FTO glasses in this study was cut into dimension of 3 cm x 2 cm using a diamond cutter and cleaned using ultrasonic bath using solvents of dilute acetone, dilute ethanol and DI water in sequence. The cleaning process was done in a duration of 10 minutes for each solvent then the substrates were dried using air dryer or nitrogen gas. For photoanode, initially, an area of 1 cm x 1 cm was created using a scotch tape and the first layer of TiO_2 (absorption layer) was deposited using a doctor-blade technique. The first layer of photoanode was annealed for 30 minutes at 450°C. Then by using the same technique and annealing process, on top of the annealed absorption layer, the second layer of TiO_2 (scattering layer) was deposited. A finished product of annealed photoanode was treated inside $TiCl_4$ solution for 10 minutes at 70°C then rinsed thoroughly with water. The photoanode was heated inside the furnace at 100°C then immediately immersed inside the mangosteen sensitizer dyes with CDCA for 24 hours. A different amount of CDCA powder was dissolve inside the dyes with each concentration of 0.10 mM, 0.50 mM, 0.75 mM, 1.00 mM, 1.50 mM and 2.00 mM. As for counter electrode, the substrate was deposited with platinum paste using doctor-blade technique and annealed at 450°C for 30 minutes. The annealed counter electrode and the dye-immersed photoanode was sandwiched together with adhesion (Surlyn) using heat press machine. Finally, iodine electrolyte was inserted inside the sandwiched device then sealed using a polymer glue.

2.4. Characteristics and device performances

The characteristics of mangosteen dye sensitizer with CDCA was studied using fourier transform infrared spectroscopy (FTIR), UV-Vis spectrometer, photoluminescence (PL) and cyclic voltammetry. FTIR spectra measured with Perkin Elmer Spectrum 400 FT-IR/FT-NIR & Spotlight 400 imaging systems on wavenumber 4000 to 650 cm^{-1} [10]. Perkin Elmer UV-Vis spectrometer Lambda 35 was used to study the absorption strength at a certain wavelength range of 350 to 600 nm. Highest occupied molecular orbital (HOMO) and lowest unoccupied molecular orbital (LUMO) levels can be studied and calculated from data collected in photoluminescence and cyclic voltammetry (CV). Equation (1) was used to calculate HOMO, LUMO and an energy gap (E_g), where h is Planck's constant (6.62607004 × 10^{-34} m^2 kg s^{-1}), c is the speed of light (3.00 x 10^8 ms^{-1}) and the wavelength, λ is the highest peak in PL. FLS920 Edinburgh Instrument was used to measured PL with excitation value of 400 nm [11]. From CV, the redox characteristic of the sensitizer with CDCA can be observed and investigated by using METROTHM AUTOLAB and NOVA software. An electrochemical cell was constructed for CV measurement consists glossy carbon as the working electrode (WE), Ag/AgCl immersed in 3 M NaCl as reference electrode (RE), platinum as the counter electrode (CE) and dye sensitizer consist of 0.1 M lithium perchlorate, 99+% from ACROS ORGANICS acted as an electrolyte. An empirical equation Bredas et al [2] was used to calculate the HOMO energy level, which is -4.4 eV is the value of ferrocene. The value of LUMO level can be calculated by inserting the value of HOMO level and E_g in Equation (3).

$$Eg = \frac{hc}{\lambda} \tag{1}$$

$$E\,(HOMO) = -e\,[E_{ox}{}^{onset} + 4.4] \tag{2}$$

$$E\,(LUMO) = E\,(HOMO) + Eg \tag{3}$$

A DSSC device performances can be measured and study using current-voltage (*I-V*) characteristics. A Keithley 2400 source meter and solar simulator from SAN-EL ELECTRIC was used in the *I-V* measurement. The luminosity of solar light is 1000 Watts/m^2 measured with the daystar meter. The performance that can be observed from *I-V* measurement is the efficiency of the cell, which can be calculated from the equation.

$$\eta = \frac{Voc \times Jsc \times x\, FF}{Pin} \quad (4)$$

3. Results and discussions

3.1. Electrochemical properties

The properties of mangosteen sensitizer were changes when added CDCA as a catalyst. Numerous chemical compound count was increased as in FTIR studied. A strong and broad signal was identified at 3320 cm^{-1} and 1026 cm^{-1} in stretch vibrations manners, which represent O-H and C-O compound, respectively. There are increment amount of O-H and C-O from mangosteen sensitizer itself to added CDCA sensitizer. It shows the sensitizer helps to increase anchoring groups, hence improved the bonding process between TiO_2 and sensitizer. Photon spectra also studied using UV-Vis and PL measurements to determine the absorption spectra and Eg. The absorption spectra was observed within 400 to 600 nm with blue-shift behaviour. The highest peak, which is around 488 nm was observed from PL spectra. This value was substituted in the equation (1) to determine E_g, which is 2.54 eV, respectively. Both UV-Vis and PL measurements have increment intensity of peaks regards on the amount added catalyst but there are no additional peaks was found. CDCA helps broadening the wavelength spectra, where it changes the compound inside the mangosteen dye in methanol solvent. HOMO and LUMO levels were calculated from $E_{oxidation}{}^{onset}$ obtained from extrapolated CV graph and E_g from PL spectra using equation (1), (2), and (3). LUMO level played a role between CB of TiO_2, which it needed to be higher than -4 eV to -4.3 eV [12]. The redox potential is in the range of -4.6 to -5 eV which the HOMO level needed to be lower from the potential. Mangosteen itself has E(HOMO) and E(LUMO) at -4.81 eV and -2.27 eV respectively. E(HOMO) and E(LUMO) for photosensitizer with the catalyst amount of 2 mM CDCA is at -4.97 eV and -2.43 eV, respectively. The different gap between LUMO and CB of TiO_2 are needed to be more than 0.2 eV, which show the mangosteen capable to have high efficiency. However, large gap also can contribute with energy losses, then affect the performances. HOMO level for both mangosteen sensitizer and the sensitizer with CDCA however at the same level at the redox potential range, which explains low performances caused by poor dye regenerations [11].

3.2. Photovoltaic properties

The mangosteen dye with CDCA as a catalyst was applied as a photosensitizer for DSSC and its photovoltaic properties was observed and studied. One of the measurement was used is current-voltage measurement, which is able to obtain the whole device efficiency. Based on Figure 1 and Table 1, it shows that CDCA fully acted as a catalyst at the concentration of 1.00 mM, 1.50 mM, and 2.00 mM. The efficiency for concentration 0.10 mM, 0.50 mM, and 0.75 mM is at 0.34%, 0.37% and 0.36%, which are lower than the efficiency for mangosteen sensitizer by itself as DSSC sensitizer at 0.38%. This is due to the amount of CDCA was small and does not have enough amount of anchoring group to help increase the device performances. It also reduces the short circuit current density (J_{sc}), which is probably due to lack of anchoring group, which reduces the electrons diffusion from dye to TiO_2 surface. However, the concentration at 1.50 mM and 2.00 mM of CDCA have higher efficiency at 0.44% and 0.56%, respectively. Compared to mangosteen sensitizer, its own as DSSC photosensitizer. As the efficiency increases, the J_{sc} also increase, which shows the higher amount of electron can transfer from the sensitizer to TiO_2 surface. All the concentrations do not have a significant change for open circuit voltage (V_{oc}), which shows the CDCA does not improve the recombination activity performance [13].

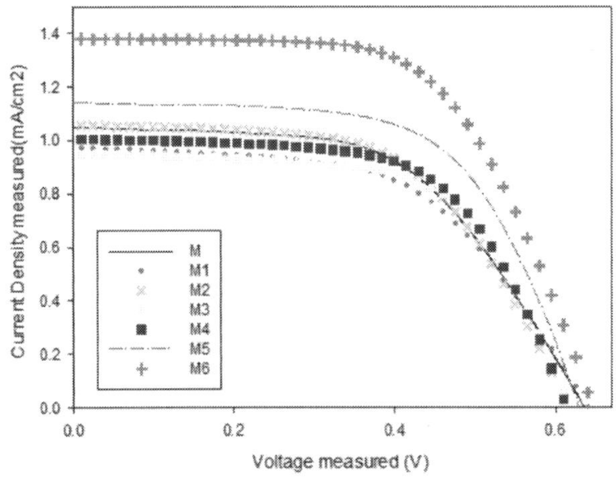

Figure 1. Current-voltage characteristics graph.

Table 1. Photoelectrochemical parameters for mangosteen sensitizer with CDCA.

Sample	Sensitizer	Voc(V)	Jsc (mA/cm^2)	FF (%)	η (%)
M	Mangosteen	0.64	1.03	57.32	0.38
M1	Mangosteen+0.10 mM CDCA	0.64	0.93	56.75	0.34
M2	Mangosteen+0.50 mM CDCA	0.62	0.99	59.52	0.37
M3	Mangosteen+0.75 mM CDCA	0.62	0.95	64.28	0.36
M4	Mangosteen+1.00 mM CDCA	0.62	1.05	62.93	0.41
M5	Mangosteen+1.50 mM CDCA	0.63	1.14	63.32	0.44
M6	Mangosteen+2.00 mM CDCA	0.65	1.40	62.03	0.56

4. Summary

In this study, CDCA was used as a catalyst to increase DSSC performance by adding dye sensitizer. The CDCA helps to introduced more anchoring group inside dye molecule which increases electron transfer between the photosensitizer and TiO_2 surface. It also to reduce the dye aggregation occurs on the surface on TiO_2, which also affect the charge transfer performance. The highest efficiency is at 0.56% for CDCA concentration of 2.00 mM with J_{sc} = 1.40 mA/cm^2 compare to mangosteen sensitizer itself only have the efficiency of 0.38% with J_{sc} = 1.03 mA/cm^2. The catalyst for dye sensitizer shows improvement of the DSSC performance. However, the efficiency for the natural dye not capable to surpass the artificial sensitizer. Another advancement for dye sensitizer can be studied in future by increasing the amount of CDCA inside sensitizer, purification of natural dye sensitizer or add nobel metal to co-sensitized with the photosensitizer.

5. Appendices

Equation Symbol	Definition
h	Planck's constant ($6.62607004 \times 10^{-34}$ m^2 kg s^{-1})
c	Speed of light (3.00×10^8 ms^{-1})
λ	The wavelength of the highest peak in PL
E(HOMO)	Energy value at highest occupied molecular orbital
E(LUMO)	Energy value at lowest unoccupied molecular orbital
η	The efficiency of device performance
V_{oc}	Open circuit voltage occurred at current = 0 A
J_{sc}	Short circuit current density measured at voltage = 0 V
FF	A measure of series resistance and junction quality of the cell
P_{in}	The input power of incident light 1000 Watts/m^2

Acknowledgments
The authors are grateful for a support from Solar Energy Research Institute (SERI), Centre for Research and Instrumentation Management (CRIM), The National University of Malaysia (UKM) and research grant DIP-2016-025.

References
[1] O'Regan B and Grätzel M A low-cost, high efficiency solar cell based on dye-sensitized colloidal TiO2 films
[2] Nuran A, Zulkifili B, Kento T, Daiki M and Fujiki A 2015 The Basic Research on the Dye-Sensitized Solar Cells (DSSC) 3
[3] Mehmood U, Rahman S, Harrabi K, Hussein I A and Reddy B V S 2014 Recent Advances in Dye Sensitized Solar Cells *Adv. Mater. Sci. Eng.* 2014 1–12
[4] Ye M, Wen X, Wang M, Iocozzia J, Zhang N, Lin C and Lin Z 2014 Recent advances in dye-sensitized solar cells: from photoanodes, sensitizers and electrolytes to counter electrodes *Mater. Today* 18 155–62
[5] Saxena S and Raja A S M 2014 *Natural Dyes : Sources , Chemistry , Application and Sustainability Issues*
[6] Shahid M, Shahid-ul-Islam and Mohammad F 2013 Recent advancements in natural dye applications: a review *J. Clean. Prod.* 53 310–31
[7] Adedokun O, Titilope K and Awodugba A O 2016 Review on Natural Dye-Sensitized Solar Cells (DSSCs) 2 34–41
[8] Desalegn J. Godibo, a Sisay T. Anshebo b and T Y A 2015 Dye Sensitized Solar Cells Using Natural Pigments from Five Plants and 26 92–101
[9] Isah Kimpa M, Momoh M, Uthman Isah K, Nawawi Yahya H and Muhammed Ndamitso M 2012 Photoelectric Characterization of Dye Sensitized Solar Cells Using Natural Dye from Pawpaw Leaf and Flame Tree Flower as Sensitizers *Mater. Sci. Appl.* 3 281–6
[10] Priyadharsini P and David S T 2014 Natural Dyes and Their Ft-Ir Spectroscopy Studies 8 15–8
[11] Safie N E, ludin N A, Hamid N H, Sepeai S, Teridi M A M, Ibrahim M A, Sopian K and Arakawa H 2017 Energy levels of natural sensitizers extracted from rengas (Gluta spp.) and mengkulang (Heritiera elata) wood for dye-sensitized solar cells *Mater. Renew. Sustain. Energy* 6 9227–43

[12] Li Y, Ku S H, Chen S M, Ali M A and AlHemaid F M A 2013 Photoelectrochemistry for red cabbage extract as natural dye to develop a dye-sensitized solar cells *Int. J. Electrochem. Sci.* 8 1237–45

[13] Piers R. F. Barnes,a Assaf Y. Anderson,a Mindaugas Juozapavicius,a Lingxuan Liu,a Xiaoe Li,a Emilio Palomares b A F and B C O 2011 Factors controlling charge recombination under dark and light conditions in dye sensitised solar cells 3547–58

IOP Conf. Series: Journal of Physics: Conf. Series **1083** (2018) 012019 doi:10.1088/1742-6596/1083/1/012019

The Gafchromic EBT3® film as alpha detector

A S Mustaqim[1*], S Kandaiya[2], N N A Razak[2], and N Z Yahaya[1*]

[1]School of Distance Education, Universiti Sains Malaysia
[2]School of Physics, Universiti Sains Malaysia

[*]shafawiahmad@yahoo.com; norzakiah@usm.my

Abstract. Film dosimetry is developed as a powerful tool for radiotherapy treatment verification and quality assurance. This project was done to study the response of Gafchromic EBT3® film for alpha particle detection. Gafchromic EBT3® film was exposed to alpha particle of energy 5.4 MeV from Am-241 source for different exposure time after one of the 125 µm polyester layer was removed. Then, the film was scanned with a flatbed scanner (Epson V700). From the results obtained, the maximum alpha particle absorbance was measured within the red light wavelength which was between 620 nm - 750 nm (alpha energy of 5.4 MeV). The absorbance also increased linearly with the exposure time. The exposure of film at different heights of pipe washer indicated that the film was energy dependent.

1. Introduction

The concept of radiochromic film was first proposed by McLaughlin and Chalkley (1965). With the recent improvements in the technology associated with the production of these films, it has been widely used in radiotherapy and recently in brachytherapy dosimetry. The vital advantages of radiochromic film as dosimeter are high spatial resolution, tissue equivalence, large dynamic dose range, relatively low spectral sensitivity variation, insensitive to visible light and no chemical processing is needed. Gafchromic EBT3® film (figure 1) consists of a single active layer, nominally 28 µm thick, containing the active component, marker dye, stabilizers, and other additives giving the film its low energy dependence. The yellow marker dye is less sensitive to visible light which decreases the light sensitivity. The active layer is sandwiched between two of 125 µm transparent polyester substrates. The polyester substrate has a special surface treatment containing microscopic silica particles that maintain a gap between the film surface and the glass window in a flatbed scanner [1].

Content from this work may be used under the terms of the Creative Commons Attribution 3.0 licence. Any further distribution of this work must maintain attribution to the author(s) and the title of the work, journal citation and DOI.
Published under licence by IOP Publishing Ltd

IOP Conf. Series: Journal of Physics: Conf. Series **1083** (2018) 012019 doi:10.1088/1742-6596/1083/1/012019

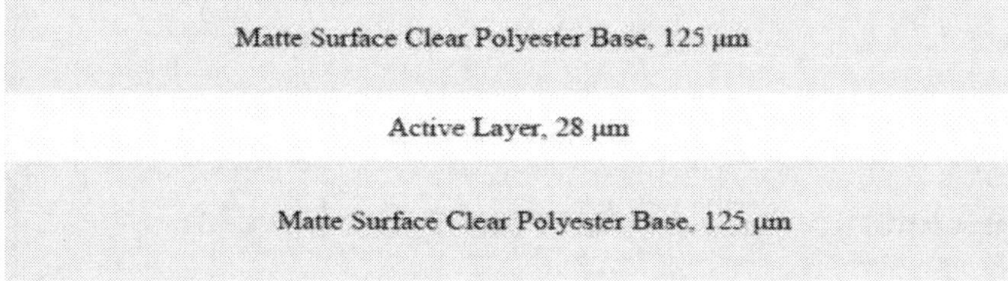

Figure 1: Gafchromic EBT3® film properties [2]

The Gafchromic EBT3® is primarily used for photon dosimetry and is insensitive to alpha particles. Therefore, in this work, one of the protective polyester sheaths was removed to make the sensitive dye layers directly exposed to the alpha particles. This constitutes the basic building block of a novel alpha detector.

2. Materials and Methods

In this study, the Gafchromic EBT3® film was prepared for alpha detection by removing one of the transparent polyester substrates using a razor knife. The procedures were as follow:

2.1. Film Preparation Procedure

The film was cut into 3 x 2 cm^2 using a very sharp blade to peel one of the polyester layer [3]. The tip edge of the knife was inserted into the film until it peeled a bit of the polyester as shown in figure 2. Then, the polyester layer was peeled carefully to prevent film damage.

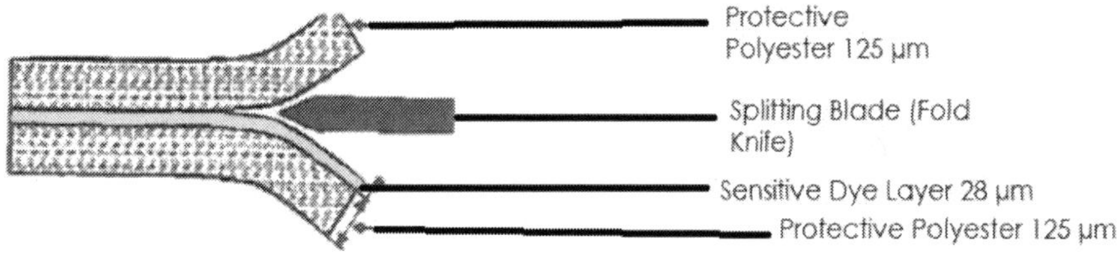

Figure 2: Schematic diagram of the cross section of Gafchromic EBT3® [3]

2.2. Film Irradiation

2.2.1. Using Different Time Period

The alpha emitting radiation source (Am-241) was placed on a pipe washer 2 mm height to prevent any direct contact of source with the film as shown in figure 3. It also prevented the crosswire from touching the film when the source Am-241 was placed on it which could affect the reading of pixel value.

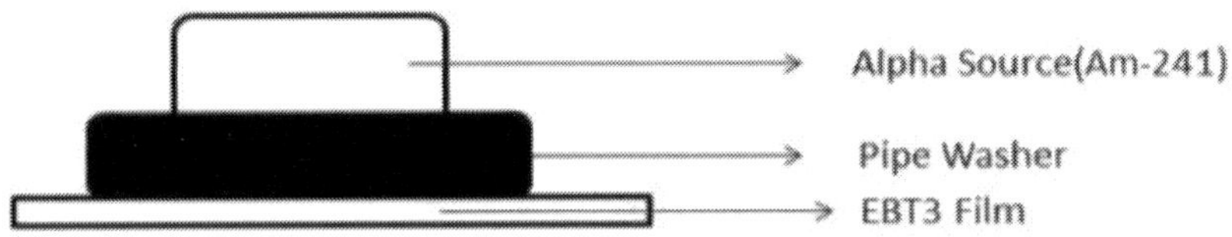

Figure 3: Schematic diagram of film exposure with alpha source (Am-241)

The initial time was set at twenty minutes. After twenty minutes, the film was observed for any colour changes of the film. This procedure was repeated at different exposure time of 1, 2, 4, 6, 8, 16 and 24 hours using different films.

2.2.2. The Irradiation of Source at Different Film Distance

In this work, the different height of pipe washer from the radiation source (Am241) was used to observe the colour changes on the film. The number of pipe washers was increased from 4.20 mm to 16.20 mm as shown in figure 4 to increase the distance of source to the film. The exposure time was kept constant for one hour throughout the experiment.

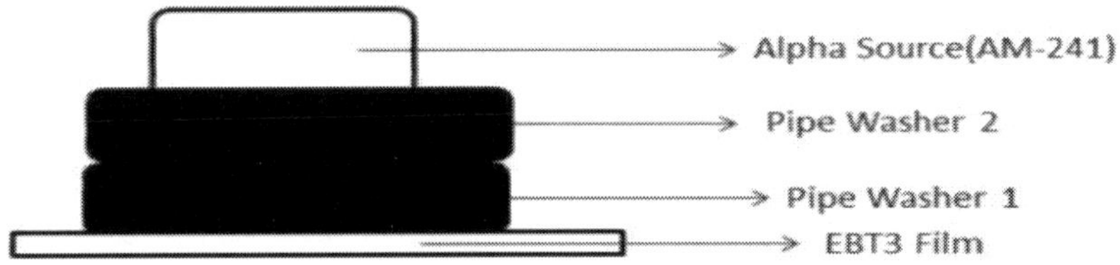

Figure 4: Schematic diagram of film irradiation with Am-241 at increase distance of the pipe washer

2.3. Scanner and Scanning Procedure

2.3.1. Scanning Procedure
The film was scanned using The Epson Perfection™ V700 Photo scanner and the transmission mode was selected which was suitable for film scanning. The image was saved as Tagged Image File Format (tiff) image files. All scanned images were analyzed using ImageJ software in terms of pixel value.

2.4. Spectrophotometry
The unexposed and exposed film were placed in the different holder as shown in figure 5, where the unexposed film acteds as "base" and exposed film acted as "sample". The wavelength was set from 200 to 800 nm and all obtained data were saved in ".BSW" and ".csv" format.

Figure 5: The film attached to the film holder

3. Results and Discussion
3.1. Measurement of Optical Density
The pixel values of the film (extracted from the ImageJ) was converted into net optical density (netOD) using the equation 1 [4]. The figure 6 shows the response of the EBT3® film at three colour channels (red, green and blue). This can be considered as multichannel analysis. According to the figure 6, the red channel shows the highest sensitivity than the green and blue channels. Therefore, the red channel was selected for whole image analysis.

$$\mathrm{netOD} = \log_{10}\left(\frac{I_{unexp}}{I_{exp}}\right) \quad (1)$$

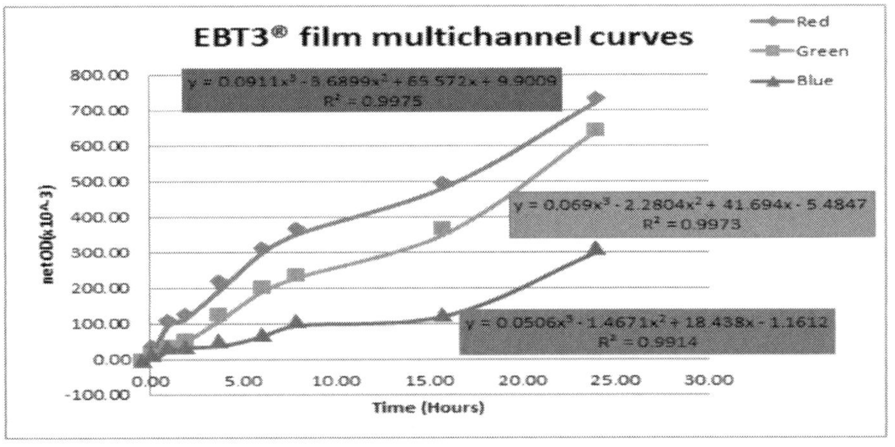

Figure 6: EBT3® film multichannel curves

3.2. *Estimation of Energy Dependence on the Film using Distance*

According to the figure 7, as the distance of the source to the film was increased, the pixel value also increased. According to the inverse square law, the energy of the alpha particle is reduced when the height of pipe washer is increased (increased the distance of source to the film). When the distance reached at 16.20 mm, the pixel value of film was equal to the background reading. Thus, it can be concluded that the alpha particles cannot reached the film exceeding 16.20 mm.

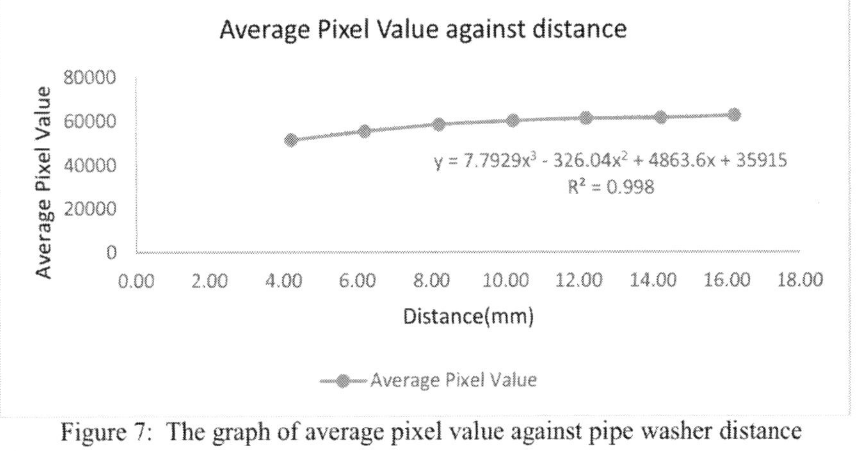

Figure 7: The graph of average pixel value against pipe washer distance

3.3. Absorbance and wavelength of the film

A spectrophotometer device (The Agilent technologies, Cary series UV-Vis-NIR) was used in this study to read the absorbance and the wavelength of the film. From figure 8, it shows that the value of the wavelength was between 636 to 638 nm.

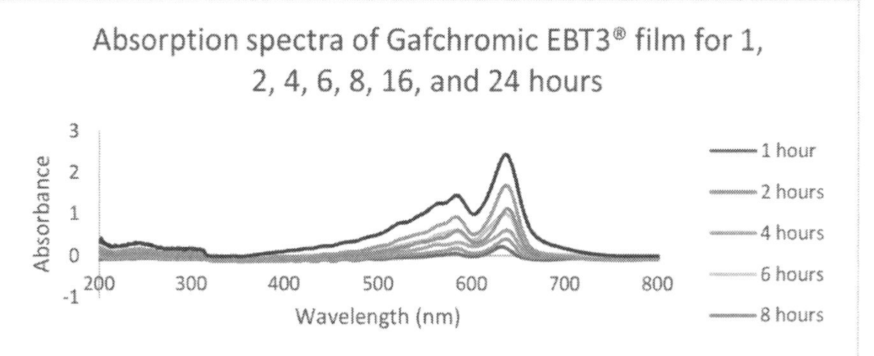

Figure 8: Absorption spectra of Gafchromic EBT3® film for 1, 2, 4, 6, 8, 16, and 24 hours

From figure 8 it can be concluded that the wavelength of 633.00 to 638.00 nm for 1,2,4,6,8,16, and 24 hours were in red colour wavelength [8].

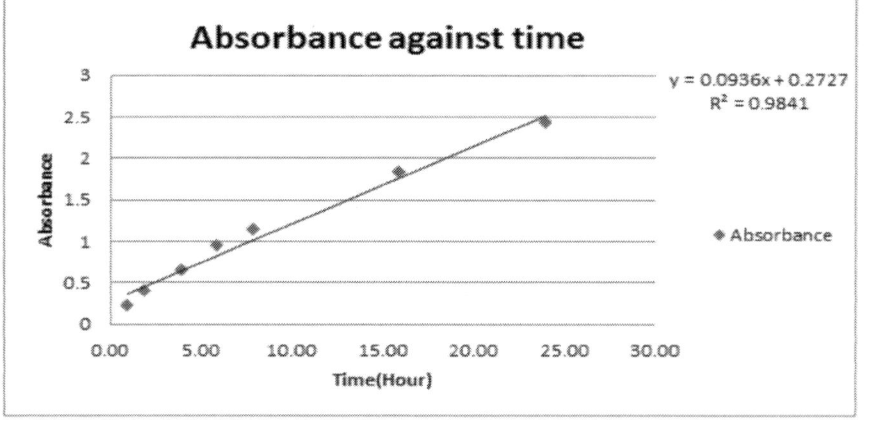

Figure 9: The absorbance at different exposure time

Figure 9 shows that as the time was increased, the absorbance of the Am-241 source on the films was increased linearly with regression coefficient of 0.9841.

4. Conclusion

In this study, one of the 125 μm polyester layer of the EBT3® film was removed resulting in the EBT3 film sensitive to alpha particles (Am-241 source). The film's wavelengths obtained were within the range of 633 nm - 638 nm.

5. Acknowledgement

This work was funded under Fundamental Research Grant Scheme, FRGS, Ministry of Higher Education (203/PJAUH/6711489) and Universiti Sains Malaysia for Short Term Grant (304/PFIZIK/6313323).

6. References

[1] V. Casanova Borca *et al.*, "Dosimetric characterization and use of GAFCHROMIC EBT3 film for IMRT dose verification.," *J. Appl. Clin. Med. Phys.*, vol. 14, no. 2, p. 4111, 2013.

[2] A. Inc, "Gafchromic ™ Dosimetry Media, Type Ebt-3."

[3] B. Mukherjee, Y. H. Gholami, U. Bhonsle, R. Hentschel, and J. Khachan, "A unique alpha dosimetry technique using Gafchromic EBT3® film and feasibility study for an activity calibrator for alpha-emitting radiopharmaceuticals," *Br. J. Radiol.*, vol. 88, no. 1056, 2015.

[4] L. Menegotti, A. Delana, and A. Martignano, "Radiochromic film dosimetry with flatbed scanners: A fast and accurate method for dose calibration and uniformity correction with single film exposure," *Med. Phys.*, vol. 35, no. 7, p. 3078, 2008.

[5] A. Aydarous and M. El Ghazaly, "Characterization of HD-V2 Gafchromic Film for Measurement of Spatial Dose Distribution from Alpha Particle of 5 . 5 MeV," vol. 7, no. 7, pp. 1174–1176, 2013.

[6] S. Devic *et al.*, "Precise radiochromic film dosimetry using a flat-bed document scanner," *Med. Phys.*, vol. 32, no. 7, pp. 2245–2253, 2005.

[7] H. Alnawaf, P. K. N. Yu, and M. Butson, "Comparison of Epson scanner quality for radiochromic film evaluation.," *J. Appl. Clin. Med. Phys.*, vol. 13, no. 5, p. 3957, 2012.

[8] T. Bruno, *CRC handbook of spectroscopic correlation charts*. Boca Raton Fla.: Taylor & Francis, 2005.

Microwave Dielectric Properties for Detection of 'Harumanis' Mangoes Ripeness

N Z Yahaya[1], N S Ayob[2], R M Ramli[2], N A N A Razak[2], N Z N Azman[2], H A Rahim[3]

[1] Physics Section, School of Distance Education, Universiti Sains Malaysia, 11800 USM, Penang, Malaysia
[2] School of Physics, Universiti Sains Malaysia, 11800 USM, Penang, Malaysia
[3] School of Computer and Communication Engineering, Universiti Malaysia Perlis (UniMAP), Pauh Putra, 02600 Arau, Perlis, Malaysia

norzakiah@usm.my

Abstract. The qualities of mango are affected by moisture content and storage conditions on fruits stability. Thus, the dielectric properties of maturity stages of mango were studied in this work. The dielectric constant and loss factor were measured using the commercial probe (Agilent 85070E) at microwave frequencies range (0.2 – 5 GHz) whilst the moisture contents were obtained using the standard oven drying method. Comparison results were obtained between the different dielectric and moisture content to predict ripeness of mango. Moisture content values for of mangoes decreased with ripening time. The decreasing tendency was mainly attributed to increased ε' and ε" values. The prediction mango ripeness can be obtained from measured values of the dielectric constant which obtained ε' values for unripe (19 – 24), while both ripe and overripe mangoes has similar ε' range values (16 – 21) at microwave range (0.3 GHz to 5 GHz).

1. Introduction

'Harumanis' is a mango clone and known as one of the best mango breeds in Malaysia. 'Harumanis' mango which is only available once a year in the season will give the best mango qualities in terms of shape, texture of content, taste and demand in the market. However, it must be had challenges to produce high qualities of Harumanis mango which is greatly dependent on many parameters such as ripeness, shape, size, and is influenced by other factors such as harvesting time. Hence, it's important to harvest fruit at the right maturity stage to suit the purpose.

Traditionally, the method of determining the ripeness of a fruit including mango fruit is done manually by human grader which is dependent on hands-on experience or visual observations. Obviously, this is the easiest and fastest way to determine the ripeness of fruits but the level of accuracy in deciding whether that fruit is unripe, ripe or overripe is low. Thus, it affects the quantity and quality of the mango fruit that can be marketed. Furthermore, the problem with the traditional method is the inconsistency of choosing the ripeness of the mangoes in the large scale. It is difficult for humans to maintain their focus when dealing with a large quantity. This method also inconsistent and subjective in nature because each grader has different techniques of grading the mango [1–4]. Besides that, either the mango is unripe, ripe or overripe, the skin colour of Harumanis mango is

Content from this work may be used under the terms of the Creative Commons Attribution 3.0 licence. Any further distribution of this work must maintain attribution to the author(s) and the title of the work, journal citation and DOI.
Published under licence by IOP Publishing Ltd

normally green and hard for people to identify the ripeness of mango by traditional way method due to the same colour of mango. Sometimes the result obtained from traditional method is not so accurate. Therefore, this problem is can solved by using non-destructively method for determination ripeness of mango is mostly required.

Several types of mango ripeness sensor have been establish such as acoustic Resonance Spectroscopy [4], Near Infrared (NIR) [5–7], electronic nose [8,9], RGB (3-channel of photodiode sensitive to the blue, green and red regions of the spectrum), color sensor and fuzzy technique [2]. These techniques were concern on detection of aroma, color and firmness but none studied on moisture content and dielectric properties to detect the mangoes ripeness. Thus, this paper deals with dielectric properties measurement to detect the ripeness of mangoes. Dielectric properties are described by the complex permittivity, ε* of a given material, expressed in equation (1)

$$\varepsilon^* = \varepsilon' - j\varepsilon'' \quad (1)$$

where the real part ε' is the dielectric constant is related to the material to store electrical energy, while dielectric loss factor, ε" is defined as the imaginary of the ε', usually, it is significantly smaller than ε'. ε" is used to measure of how dissipative or lossy a material is to an external electric field. ε" is also included the effects of both dielectric loss and conductivity [10].

2. Materials and methods

2.1. Sample preparations

Harumanis mangoes were acquired from a local farm in Perlis, Malaysia. Immature fruits (unripe) were picked from 1 tree only and considered as at 0th day for storage at room conditions. The ripeness of samples were collected and the softness of fruits were manually checked by farmer. The samples measurements were taken on 0th day , 4th day and 8th day of storage for experiments and can be categorized into 3 stages of ripeness that are unripe, ripe and over ripe, respectively. The stages of ripeness were determined by measurement of dielectric properties and moisture content of samples that will be discussed in results and discussions section. Conventionally, it can be proved by checking the softness of fruits. The measurement is takes on different position on Part A, B and C (Fig. 1). However, only measurement on part A was concerned because the values of moisture content and dielectric properties were different at different parts of fruits due to inhomogeneous compositions.

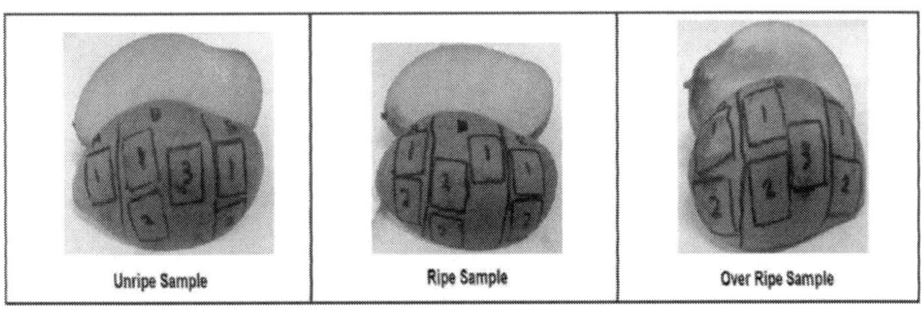

Figure 1. Ripeness stages of Harumanis mangoes.

2.2. Moisture content measurements

The actual moisture content in mango was determined using conventional, standard an oven drying method. Each sample was dried in an oven at 70 °C for 24 hours. The dried samples were allowed to cool at room temperature 25 °C before weighing for dry sample. The data weight of mangoes with different position (stage: unripe, ripe and over ripe) is taking before and after drying process. The percentage moisture content is calculated refer to the data taken. The percentage moisture content is calculated by using equation (2)

$$\text{Moisture content } (\%) = [(m_{wet} - m_{dry})/\, m_{wet}] \times 100\% \quad (2)$$

where m_{wet} and m_{dry} are the initial and final mass before and after drying.

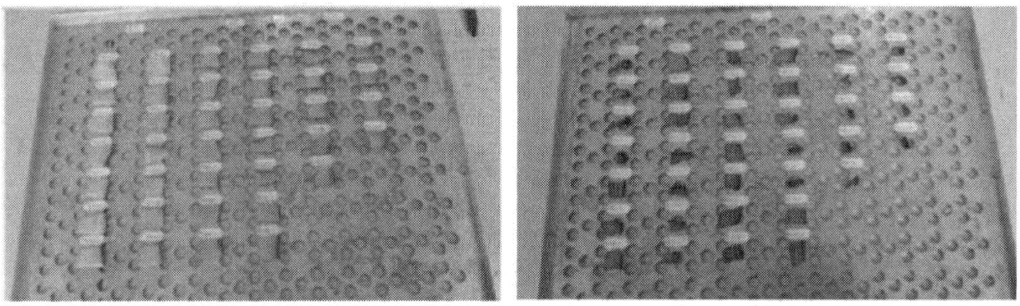

(a) (b)

Figure 2. Cutting sample (a) before and (b) after drying using a conventional drying oven

2.3. Dielectric properties measurements

The Agilent E8362B Professional Network Analyzer (PNA) was calibrated by implementing a full one port calibration procedure at the end of the coaxial cable. Open, short and load standards used in the calibration were automatically measured with the Agilent Electronic Calibration Kit (85052D) (Fig. 2). These entire requirements are need to avoid and reduce the losses in the coaxial cable. Besides that, this technique is used to determine the error terms, requires a test system consisting of a PNA, cables, and generally a test fixture, and is performed by sequentially making measurements using calibration standards.

The experimental set-up consists of a PNA and the commercial probe (Agilent 85070E). The measurement frequency range was set from 0.2 GHz to 5.0 GHz. The dielectric properties of mangoes were measured by connecting the dielectric probe to the cable.

3. Results and discussions

3.1. Relationship between moisture content and ripening time

Moisture content values for of mangoes decreased with ripening time (Table 1). Thus, the decreasing tendency was mainly attributed to increased ε' and ε" values (Fig 4 and Fig. 5), respectively. Two measurements were taken for each day to check the consistency values. Dielectric properties of mangoes were affected by ripening time until 8th day only. The result of moisture content of mangoes are remain same at 8th day and above, so the result at 10th day were not included in this paper. The best harvest time was obtained at 4th day. However, there are slightly different between moisture content for various ripeness stages. Thus, the ability of dielectric properties was studied to predict the ripening time of 'Harumanis' mangoes.

Table 1. Relationship between moisture content and ripening time

Ripeness stages	Ripening day	Moisture content (%)
Unripe	0	82.7
		82.6
Ripe	4	84.1
		85.1
Over ripe	8	85.5
		87.0

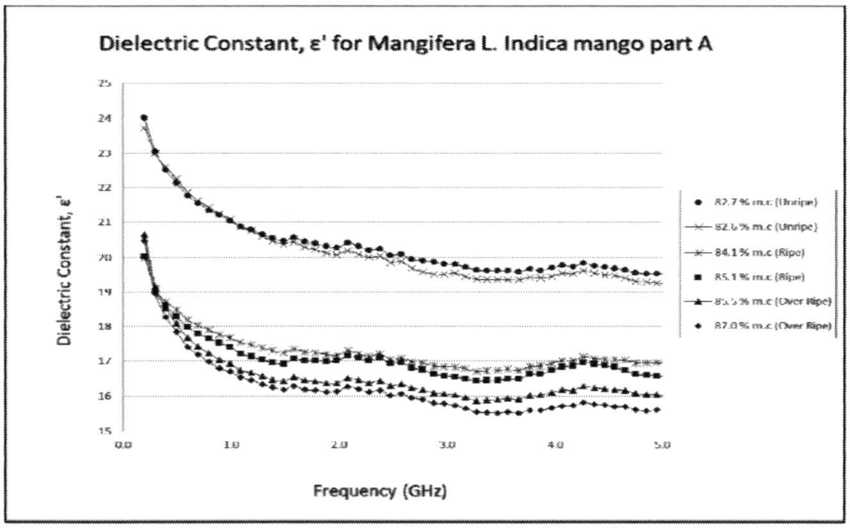

Figure 3. Dielectric constant, ε' for Mangifera Indica L. mango part A.

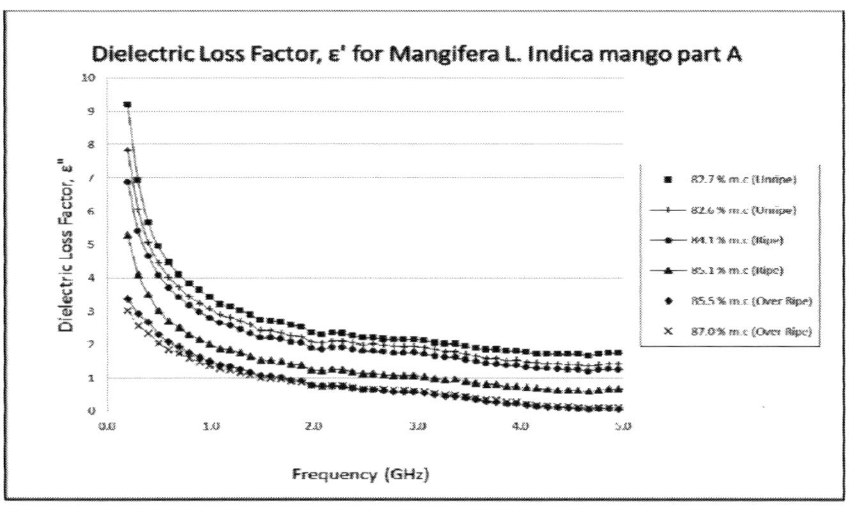

Figure 4. Dielectric Loss, ε" for Mangifera L. Indica mango for part A

Relationship between dielectric constant, ε' and loss factor, ε" with microwave frequencies (0.2- 5 GHz) for different ripening stages of Mangifera Indica L. mango are shown in Fig. 3 and Fig. 4, respectively. Both the dielectric constant and the loss factor decreased with increasing frequency at all moisture content. At frequency below 0.3 GHz, ionic conduction is the dominant loss mechanism. But, above than 0.3 GHz, dipole relaxation accounting for most of the energy loss that occurs. The critical edge frequency is at 0.3 GHz. The dielectric properties values for unripe (19 – 24) and for both ripe and overripe (16 – 21) can be obtained from Fig. 3. The value of overripe mango is closest to dielectric properties ripe mangoes values might due to water content in the sample is a strong polar solvent and it reorients in responses to change in the field polarity as a major component of the samples that contributes to the ε' [11].

4. Conclusion

Harumanis mangoes ripeness can be determined using the dielectric constant, ε' values which for unripe was 19 – 24 while for both ripe and overripe were 16 – 21. The 16-21 range is the best ε' value range for harvesting the Harumanis mangoes which is at 4th day of storage.

Acknowledgement

This work was funded under Fundamental Research Grant Scheme, FRGS, and Ministry of Higher Education (203/PJJAUH/6711489).

References

[1] Omid M, Khojastehnazhand M and Tabatabaeefar A 2010 Estimating volume and mass of citrus fruits by image processing technique *J. Food Eng.* 100 315–21

[2] Mansor A R, Othman M, Ahmad K A, Nazari M, Bakar A and Razak T R 2014 Fuzzy Ripening Mango Index Using RGB Color Sensor Model *J. Arts, Sci. Commer.* 118–23

[3] Patel A, Kadam P and Naik S 2015 Color , Size and Shape Feature Extraction Techniques for Fruits : A Technical Review *Int. J. Comput. Appl.* 130 6–10

[4] Raju P S, Jagannath J H, Ashok N, Gupta D K D and Bawa A S 2006 Nondestructive monitoring of ripeness in mango cultivars by acoustic resonance spectroscopy *Int. J. Food Prop.* 9 487–501

[5] Jha S N, Narsaiah K, Jaiswal P, Bhardwaj R, Gupta M, Kumar R and Sharma R 2014 Nondestructive prediction of maturity of mango using near infrared spectroscopy *J. Food Eng.* 124 152–7

[6] Jha S N, Jaiswal P, Narsaiah K, Gupta M, Bhardwaj R and Singh A K 2012 Non-destructive prediction of sweetness of intact mango using near infrared spectroscopy *Sci. Hortic. (Amsterdam).* 138 171–5

[7] Rungpichayapichet P, Mahayothee B, Nagle M, Khuwijitjaru P and Müller J 2016 Robust NIRS models for non-destructive prediction of postharvest fruit ripeness and quality in mango *Postharvest Biol. Technol.* 111 31–40

[8] Ali S B, Ghatak B, Gupta S D, Debabhuti N, Chakraborty P, Sharma P, Ghosh A, Tudu B, Mitra S, Sarkar M P, Bhattacharyya N and Bandyopadhyay R 2016 Detection of 3-Carene in mango using a quartz crystal microbalance sensor *Sensors Actuators, B Chem.* 230 791–800

[9] Lihuan S, Liu W, Xiaohong Z, Guohua H and Zhidong Z 2017 Fabrication of electronic nose system and exploration on its applications in mango fruit (M. indica cv. Datainong) quality rapid determination *J. Food Meas. Charact.* 11 1969–77

[10] Yahaya N Z, Abbas Z, Ibrahim N M, Hafizi M H M and Yahaya M Z 2014 Permittivity models for determination of moisture content in Hevea Rubber Latex *Int. J. Agric. Biol. Eng.* 7 48–54

[11] Sipahioglu O and Barringer S a. 2003 Dielectric Properties of Vegetables and Fruits as a Function of Temperature, Ash, and Moisture Content *J. Food Sci.* 68 234–9

IOP Conf. Series: Journal of Physics: Conf. Series **1083** (2018) 012021 doi:10.1088/1742-6596/1083/1/012021

Effect of Annealing Temperature on Growth Particles of YAG: Ce^{+3} Phosphor and White Light Chromaticity Values

H. R. Abd[1], Z. Hassan[2], N. M Ahmed[1], A. F. Omar[1], F. H. Alsultany[1]

[1]School of Physics, Universiti Sains Malaysia, 11800 USM, Penang, Malaysia
[2]Institute of Nano Optoelectronics Research and Technology (INOR), Universiti Sains Malaysia, 11800 USM, Penang, Malaysia

husnen1978@gmail.com, zai@usm.my

Abstract. In the present work white-emitting $Y_3Al_5O_{12}$:xCe^3 (x = 0.04) nanophosphor in the form of powder were synthesized by a microwave-induced combustion synthesis method (MW) using metal nitrates as precursors and urea as fuel. By covering blue light-emitting diodes (blue-LED, 445 nm) white light emission (WLED) was generated. The sintering temperature with fixed time (5 hours) for phosphor powder was optimized and found to be 1050 °C. The crystallinity structure, luminescent properties and chromaticity were characterized by X-ray diffraction (XRD), field emission-scanning electron microscope (FE-SEM), electroluminescence (EL) and standard CIE 1931 chromaticity diagram. The results show that the obtained YAG:Ce^{+3} phosphor sintered at 1050 °C has good crystallinity with pure phase, low agglomeration particles and strong yellow emission that offering daylight white LED with tuneable correlated color temperature (CCT) and a good colour rendering index (CRI) compared to those sintered at 950 °C, 850 °C and non-sintered phosphor powders.

1. Introduction

Phosphors are inorganic luminescent materials that absorb energy and re-emit this absorbed energy in the form of light [1]. The incorporation of the small amount of rare earth (RE) ions into host crystals came into existence as a promising method for regulating the optical and photoluminescence (PL) properties [2] White LEDs (WLEDs) are generated using near-ultraviolet or ultraviolet InGaN LEDs coated with different luminescent phosphors [3]. In general, phosphor such as $Y_3Al_5O_{12}$ (YAG) is considered inert host lattice that circumvent these issues. YAG is doped with a small amount of lanthanide element (activator) such as cerium (Ce). In the free Ce^{3+} ion, the electronic configuration is the simplest with one electron in the 4*f*-shell. This ion absorbs light and emits it through 4*f* to 5*d* transition levels and that emission was found in a wide range under excitation between UV and blue [4]. The general approach for producing WLEDs is to combine blue InGaN chips (450–470 nm) with yellow light-emitting $Y_3Al_5O_{12}$: Ce^{+3} (YAG: Ce^{+3}) phosphors [5].The crystal structure of the host material (i.e., YAG) is a cubic phase that belongs to the Ia 3*d* (230) space group with Y in 24(c) sites, AlO_6 octahedra in 16(a) sites, AlO_4 tetrahedra in 24(d) sites, and the remaining 96 oxygen atoms in (h) sites [6].
Many different methods were used for preparation of inorganic phosphors (YAG: Ce^{+3}). The most common method is solid-state reaction [7]. However, the solid-state reaction method suffers from several shortcomings such as the processing times are long and it requires a high sintering temperature (above 1800 ° C) to obtain the pure phase. On the other hand, researchers have investigated novel

Content from this work may be used under the terms of the Creative Commons Attribution 3.0 licence. Any further distribution of this work must maintain attribution to the author(s) and the title of the work, journal citation and DOI.
Published under licence by IOP Publishing Ltd

methods to obtain pure phase of YAG:Ce^{3+} at lower temperature, such as sol-gel/Pechini method (SG) [8], co-precipitation (CP) [9], hydrothermal (HT) [10], and combustion method (CS) [5].

High purity, homogeneous and small particle size powder and low-temperature synthesis was achieved in this study by using microwave-induced combustion synthesis method (MW) to prepare YAG: Ce^{+3} phosphor powder. The main feature achieved in this method is that the high temperature was generated instantly by exothermic reaction and not from an external source. Microwaves interact with the matter from bulk to the surface and the heat generation is governed by the rapid kinetics of the dipole moments of the molecules [11].

In this present work, Ce^{+3} doped YAG ceramics were obtained by using microwave oven method with fixed Ce^{+3} concentration (x=0.04) according to the formula, $Y_{(3-x)}Al_5O_{12}$: x Ce^{+3}. The luminescent intensity of the phosphor is a function of Ce^{+3}concentration. By increasing Ce^{+3}concentration the luminescent intensity increases until 4 mol% Ce^{+3} and then decreases due to non-radiative energy transfer. The heat treatment was achieved at different temperatures (850, 950 and 1050 °C) and fixed time (5 hours).

2. Experimental procedures and measurements

Stoichiometric amounts with high purity (Sigma-Aldrich, 99.99%) of $Y(NO_3)_3{\bullet}_6H_2O$, $Al(NO_3)_3{\bullet}_9H_2O$ and $Ce(NO_3)_3{\bullet}_6H_2O$ were needed to form $Y_{2.96}Ce_{0.04}Al_5O_{12}$ and mixed with an amount of urea $CO(NH_2)_2$ (Merck, purity ≥98%). The materials were dissolved in distilled water in 300ml capacity Pyrex beaker and stirred in order to obtain a clear solution. Then the urea was added to this solution. The solutions process offers many advantages over conventional methods in the synthesis of fine- grain powders and particularly phosphor materials. After stirring for 30 min at 60°C the solution was placed into a modified domestic microwave oven (900W, KOC-9Q5TR, Daewoo) for about 30 min. The solution boiled and liberated a large amount of gases such as N_2, CO_2 and H_2O. Finally, the powder was crushed with a mortar and pestle and placed in a ceramic boat to be heat-treated at different temperatures (850, 950 and 1050 °C) and fixed time (5 hours). It was previously determined that a heat treatment step is necessary to enhance luminescence and to remove the organic residues.

To generate WLED, the YAG: Ce nanopowder was mixed with two-part clear liquid epoxy (Devcon ® home, made in the USA with global materials) according to a volume ratio of 1: 2.5. Then, a thin phosphor layer (approximately 1 mm) was coated on the top surface of the InGaN LED chip (445 nm, 10 W, chip size 945±25μm×945±25μm, square chips, chips covered with transparent resin, EPILEDS Company) to obtain WLED as shown in figure 1. This method is low cost and simple.

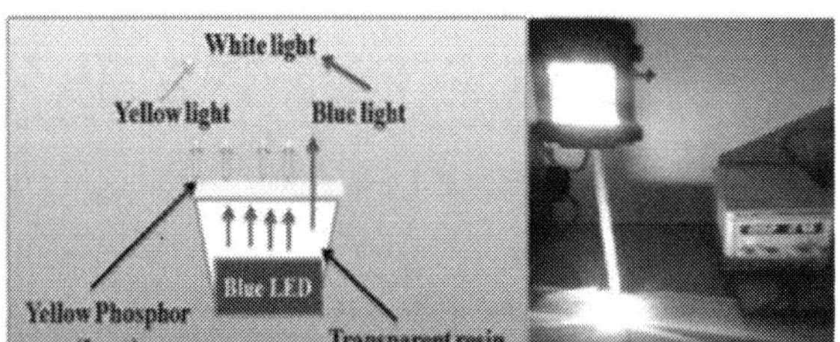

Figure 1. Schematic cross-section illustration and image of the WLED device made from a composite of nano-YAG: Ce and liquid epoxy.

The crystal phase of the YAG:Ce phosphors prepared was characterized by X-ray diffraction using a PANalytical X'Pert PRO MRD PW3040 with Cu Kα wavelength of 1.5443 Å. Powders morphology characterization was studied by using field emission-scanning electron microscope (FE-SEM; NOVA NANO SEM 450). Heat treatment was achieved by using a tube furnace (model: F21100). All measurements were performed at room temperature. The electroluminescence (EL) spectra,

International Commission on Illumination (CIE) chromaticity coordinates, color rendering index (CRI), and correlated color temperature (CCT) were measured using United Power Research Technology Corporation (UPRtek) model: MK350N PLUS spectrometer with spectral responses between 380 and 780 nm. The spectrometer was calibrated prior to sample testing. All luminance's measurements were performed in a dark room.

3. Results and discussions

The crystallographic purity phase of non-sintered YAG: Ce phosphor powder and the samples sintered at different temperatures (850 °C, 950 °C and 1050 °C) with the fixed time (5 h) was examined by X-ray diffraction (XRD) as shown in figure 2. It can be observed that the YAG: Ce powder matched closely with that of cubic YAG (JCPDS Card No. 01-072-1853 and JCPDS Card No. 00-033-0040) and no other peaks corresponding to the impurities such as YAP ($YAlO_3$, JCPDS Card No. 016-0219) and YAM ($Y_4Al_2O_9$, JCPDS Card No. 01-080-1695). Moreover, the diffraction peaks became narrower and more intense were observed for the powder sintered at 1050 °C.

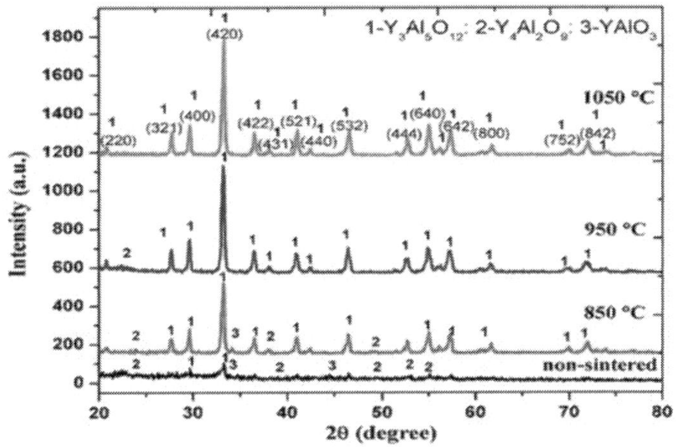

Figure 2. XRD patterns of non-sintered YAG: Ce^{+3} phosphor powder and sintered at various temperatures with fixed time (5 hours).

For the non-sintered powder, the YAG phase, other intermediate phases such YAM and YAP was observed at peaks 23.75 °, 37.94 °,44.65 °, 49.37 °, 53.12 °, 55.09 ° and 34.34 °, 44.60 ° respectively, and the intensity of these peaks decreases for the powders sintered at 850 °C and 950 °C but disappear at 1050 °C. This behavior is expected because higher temperature enhances ion diffusivity and influences the solubility of cerium ions in the YAG matrix [12].

The crystallite size of the YAG: Ce phosphor powder was determined from the full-width at half maximum (FWHM) in the (420) peak reflection using Scherrer formula $D= k\lambda/ \beta\cos\theta$ [13], where D is the crystallite size in nanometres, K is a geometric factor equal to 0.94, λ is XRD wavelength radiation (1.54056 Å for CuKα radiation), β is the peak width in radians at FWHM and θ is the diffraction peak position. According to Scherrer equation, the crystallite sizes of the non-sintered YAG: Ce^{+3} phosphor powder and sintered samples at different temperatures are presented in Table 1. Annealing of the YAG:Ce^{+3} powder increased the size of crystals and the roughness of particles. At high temperature, the particles are interconnected by necking while the particle size increased by crystal growth.

The lattice parameters for non-sintered powder and annealed at different temperatures are larger than the standard values (a = 12.008 Å for YAG, JCPDS No. 01-072-1853). This indicate the presence of disorder due to defects resulting from different radiuses between Ce^{+3} and Y^{+3} ions [14]. In addition, the lattice parameter value shows a small but clear contraction for samples annealed at 850 °C, 950 °C and 1050 °C. This can be owing to the lattice parameter to be a function of calcination temperature. However, increasing calcination temperature led to a decrease in a lattice parameter value. The shift of the position of diffraction peak to larger angle in the XRD patterns may be due to the larger ionic radius of Ce^{3+} ion than that of Y^{3+} ion, implying that the cerium incorporation into the YAG matrix was improved by sintering. All structural parameters values are illustrated in Table 1.

Table 1. The peak position, lattice parameter and grain size of YAG: Ce^{3+} phosphors for non-sintered and sintered at different temperatures for 5 hours.

Samples	Peak position 2θ (degree)	Lattice parameter a (Å)	Crystallite Size D(nm)
as- prepared	33.06	12.091	18.02
Sintered at 850 °C	33.09	12.061	25.90
Sintered at 950 °C	33.15	12.061	25.122
Sintered at 1050 °C	33.27	12.052	28.86

The morphology of the obtained YAG: Ce^{+3} phosphors powders were investigated by using FE-SEM (figures 3a-d). For the non-sintered powder, no specific shape is observed but only a smooth surface due to non-completed ignition for the phosphor powder. Significant changes of the particle shape were observed after powder was sintered at 850 °C and 950 °C, flaky and thin platelets they can be clearly seen, this reflects that the particles had been start growth . The powder sintered at 1050 °C shows a crispy foam structure. On the other hand, the cracks and pores formed due to escaping gases such as carbon and nitrogen during the combustion reaction were observed; this indicates that the sintering process had been locally completed. These type of structure are highly friable, facilitating easy grinding to obtain finer particles [15].

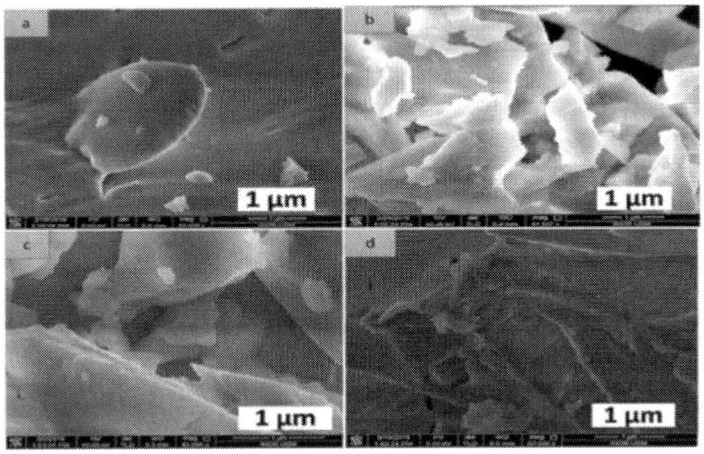

Figure 3. FE-SEM micrograph of YAG: Ce^{+3} powder: (a) non-sintered: (b) sintering at 850°C for 5h; (c) sintering at 950°C for 5h; (d) sintering at 1050°C for 5h.

The significant changes of the morphology powder after sintering is not surprising because the surface area of powders produced from sol is very high, surface tension increased rapidly with an increase of surface area, which caused shrinkage of the unit cell [16]. For sintered powders, the surface area decreases due to releasing of large number of gaseous products before 850 °C. It is known that the small surface area of materials induces roughness of the unit cell, which caused the unit cell to become larger. This is agree with the calculated lattice parameters using XRD data.

Figure 4 illustrates the electroluminescence (EL) spectra of the WLED devices without any heat treatment (non-sintered) and sintered at 850 °C, 950 °C and 1050 °C for 5 h with fixed current injection (10 mA). The short wavelength emission peaks (blue) which is attributed to the InGaN LED chips and the broad long-wavelength peaks originates from the Ce^{3+} inter-shell transition ($5d \rightarrow 4f$) in the YAG lattice. The highest emission intensity appears in the powder prepared at 1050 °C because of the improved crystallinity and the homogeneous distribution of Ce^{3+} ions in the YAG lattice. Thus, high absorption of blue light.

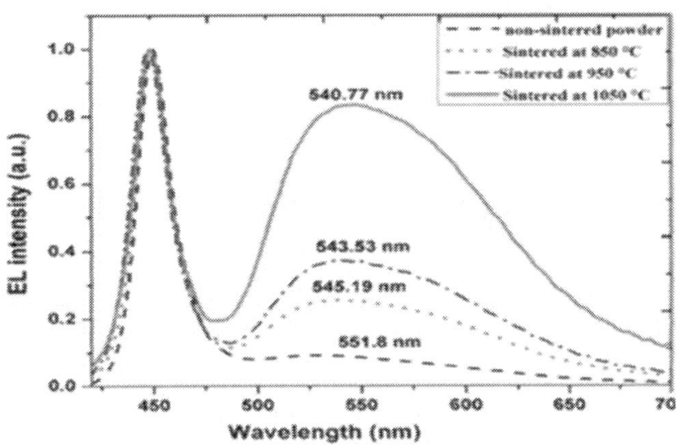

Figure 4. EL spectra of the WLED devices with non-sintered, and sintered at various temperatures samples for 5h.

The blue light emitted from the LED chips was not completely absorbed by the phosphors. This could be attributed to the low efficiency of the down conversion of phosphor or because of scattering of the powder phosphor particles [17]. It is well- known that down-conversion efficiency depends strongly on phosphor grain size.

The increasing sintering temperature tends to shift the emission peak to the blue region (from 545.19 nm to 540.77 nm) which indicates that the lowest 5*d* has a shift to a lower energy level. This blue shift originates from the change of the cell parameter (see Table I) that affects the crystal field around Ce^{+3}, which implies a smaller atomic spacing and a stronger ligands field of YAG [18], and the blue shift will affect the luminescence parameters values.

All of the luminescence parameters values such as emission coordinates (x, y), CCT and CRI with non-sintered, sintered at 850 °C, 950 °C and 1050 °C for 5 h samples are plotted in the standard CIE 1931 chromaticity diagram as shown in figure 5 a-d and listed in Table 2. Both x and y have increased and the colour coordinates shifted from blue to yellow-green, near the colour coordinate values of standard white light (0.333, 0.333) and offering daylight white LED when the sample are sintered at 1050 °C in comparison with the non-sintered and sintered at 850 °C and 950 °C samples. This finding confirms the Ce^{+3} incorporation and homogeneous distribution into the YAG matrix. For the sample sintered at 950 °C shows lower emission intensity, colorimetric coordinate (x,y) value decreases accompanied with an increase of CCT and presenting bluish white LEDs. The decreases of luminance parameters can be due to the thermal quenching from thermally activated crossover from the lowest excited $5d_{(1)}$ state to the $4f_1$ levels [19].

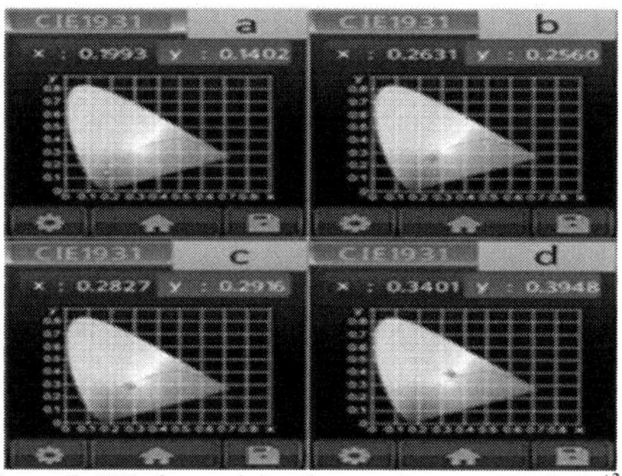

Figure 5. CIE chromaticity diagram for nano-YAG:Ce^{+3} phosphors excited at 445 nm: (a) non-sintered: (b) sintering at 850°C for 5h; (c) sintering at 950°C for 5h; (d) sintering at 1050°C for 5h.

The colorimetric coordinates (x,y) of the non-sintered sample were observed from blue region and no CCT and CRI can be detected. These findings confirm that the annealing step is necessary to improve luminance parameters. The low CRI value (< 75) is attributed to the lack of the red component [20]. Moreover, the decrease of CRI and CCT values were ascribed to the blue shifting in the long-wavelength EL peak (see figure 4). Table 2 summarizes the comparison of CIE colorimetric coordinates x, y, CCT and CRI for WLED devices.

Table 2. The CIE 1931 coordinate, color temperature (CCT) and color rendering index (CRI) of WLED devices with non-sintered and sintered samples at different temperatures during 5 hours.

Samples	luminance parameters		
	CIE (x,y)	CCT (Kelvin)	CRI
as- prepared	0.1993,0.1402	0	0.0
Sintered at 850 °C	0.2631,0.2560	18065	80.4
Sintered at 950 °C	0.2827,0.2916	9601	75.4
Sintered at1050 °C	0.3401,0.3948	5270	68.4

4. Conclusion

The nano-$Y_{2.96}$ Al_5 O_{12}: $Ce_{0.04}$ phosphor powders were successfully synthesized by microwave-induced solution combustion (MW) at different sintered temperatures for 5 hours. The results of these studies show that sintering powder at 1050 °C for 5 h markedly affects the structure, morphology, and luminescence properties of YAG:Ce phosphor. The XRD results indicate that cubic YAG:Ce pure phase is formed for powder sintering at 1050 °C. For non-sintered YAG:Ce phosphor powder and the powders sintered at 850 °C and 950 °C, the diffraction peaks indicate the formation of YAG, YAP, and YAM phases. Crystallite size was calculated using Scherrer formula. FE-SEM examination was carried out to study the grain growth process of the YAG:Ce^{+3} particles. The grain growth process is controlled via the sintering powder. The highest EL intensity was prepared from the powder sintered at 1050 °C that suggests the incorporation with best distribution of the Ce ion in the YAG host. In addition, a daylight WLED with a value approaching an ideal white light, low color temperature, and a suitable rendering index was recorded by controlling the powder sintering temperature.

Acknowledgements

The authors acknowledge the support from LRGS grant (203/CINOR/6720013) and Universiti Sains Malaysia.

References

[1] Lopez O A *et al* 1997 *J. Lumin.* 71 1.
[2] Kumar V *et al* 2016 *Mater.Des.* 93 203.
[3] Chang C J *et al* 2015 *J. Lumin.* 160 145.
[4] Silveirac L G D *et al* 2012 *Mater. Lett.* 89 86.
[5] Gupta K V K *et al* 2011 *Appl. Phys. B* 105 479.
[6] Muñoz G A B and Seijo L 2011 *J. Phys.Chem. A* 115 815.
[7] Shi Y *et al* 2012 *Opt. express* 20 21656.
[8] Chung D N *et al* 2014 *J. Nanomaterials* 2014 11.
[9] Zhou Y *et al* 2002 *Mater. Lett.* 56 628.
[10] Chen, X. P *et al* 2009 *J. Appl. Phys.* 106 063518.
[11] Mangalaraja R V *et al* 2009 *Powder Technol.*191 309.
[12] Abd H R *et al* 2017 *J. Alloys Compd.* 711 42.
[13] Suryanarayana C and Norton M G 2013 *X-ray diffraction: a practical approach* Springer Science and Business Media New York.
[14] Saladino M L *et al* 2008 *Opt. Mater.* 31 261.
[15] Zhang Y and Stangle G C 1994 J. *Mater. Res.* 9 1997.
[16] Park G S *et al* 2008 *Electrochem. Solid-State Lett.* 11 J23.
[17] Sheu J K *et al* 2003 *IEEE Photon. Technol. Lett.* 15 18.
[18] Abd H R *et al* 2017 J. Electron. Mater. https://doi.org/10.1007/s11664-017-5968-9 1.

[19] Ning L *et al* 2016 *J. Mater. Chem. C* 4 5214.
[20] Yuhua W *et al* 2015 *J. Rare Earths* 33 1.

Determination of Selectivity Coefficients of Sodium and Potassium Ion-Selective Electrode Using Porous Silicon N-Type (100) Based Extended Gate Field Effect Transistor

E. A. Kabaa [1, a *], Naser M. Ahmed [1,b], S. A. Abdulateef [1,c], Hasma A. Wahab [1, d]

[1] School of physics – Universiti sains Malaysia -11800- Penang-Malaysia

[a] eas14_phy050@student.usm.my, [b] Naser@usm.my, [c] sinan1974@yahoo.com, [d] farahatul3110@gmail.com

Abstract. Ion-selective electrode (ISE) was able to accurately detect the ions, precisely for laboratory analysis purposes. The selection of the material varies to select a particular ion, according to the electrode response, and the decisive factor for the optimal selection of ISE for specific ion is the selectivity coefficient. This study used porous silicon (PSi) (100) as an ISE based on extended gate field effect transistor (EGFET) to characterize the selection's preference for sodium (Na^+) as a target ion and potassium (K^+) is an interference ion. The potential response in the linear region of the EGFET sensor between the detection electrode (PSi) and a reference electrode Ag/AgCl was measured in NaCl, KCl, and a mixed solution of the two salts along ions' concentration (10^{-4}-1 M) under drain-source voltage 0.3 V at temperature room. The potentiometric selectivity coefficient was calculated by two methods: (1) separate solution method (SSM) at the same activity (0.1 M) and (2) two solution method (TSM). The SSM method, $K^{pot}_{Na,K}$ was 0.581, while the result of TSM method was 0.743. The results showed a slight superiority of Na^+ selectivity and acceptable interference resistant for K^+.

1. Introduction

Ion-selective electrode (ISE) was used in several fields such as: bio-chemistry, environmental analysis, medicine, and process control [1]. There are several factors that assess ISE performance, such as conditioning time, the ions concentrations in the electrolyte, the stability of the potential of an electrode, the ability of the ISE to identify between the primary and interfering ions (the selectivity) [2].

The selectivity coefficient can be defined as the responding degree of ion-selective electrode to a certain ion with respect to a reference ion, as predicted by Nikolsky-Eisenman [3]:

$$E = E_0 + 2.303\frac{RT}{z_A}log\left(a_A + K^{pot}_{A,B}\,a_B^{\frac{z_A}{z_B}}\right) \quad (1)$$

where: E_o is the potential at standard conditions, R is the universal gas constant, T is the absolute temperature, F is Faraday constant, z_A and z_B are the valences of ion A and B in aqueous solution, respectively, a_A and a_B are the activities of the dissolved ions.

Content from this work may be used under the terms of the Creative Commons Attribution 3.0 licence. Any further distribution of this work must maintain attribution to the author(s) and the title of the work, journal citation and DOI.
Published under licence by IOP Publishing Ltd

Nikolsky-Eisenman coefficient is considered a good selectivity characteristic. It can take notice as a conditional equilibrium constant, where its value is influenced by changing the membrane components and electrolytes concentrations. Nikolsky-Eisenman equation is accurate relationship for the development of electrode selectivities [4].

This study aims to evaluate porous silicon n-type (100) efficacy in sodium ion selecting toward the potassium ion.

2. Methodology

A n-type Si wafer (100) with Resistivity 0.697 (Ωcm) was cut with dimensions of (2.4 × 2.4) cm. Ethanol/HF mixture solution with a volume/volume ratio (4:1) was used to form a porous silicon (PSi) layer using photo-electrochemical etching under conditions: current density of 8.5 mA/cm2, anodization time 15 minutes, and under illumination of 100 W at room temperature, followed by cleaning with ethanol.

The two parts of an EGFET are a commercial MOSFET (CD4007UB) and the sensing component (PSi). The PSi-based EGFET chemical sensor was used as in Naser M. Ahmed et al. [5]. PSi was tested to analyze its responding to Na^+ and K^+, and investigate the selectivity of the porous structure for the two cations above. The sensing part was connected to the MOSFET's gate. The sensing membrane and an Ag/AgCl reference electrode were immersed in electrolytes along ion concentrations of NaCl, KCl, and a mixture solution of the two salts above (10^{-4} to 1 M) and subsequently connected to the measurement system. The response of the sensor was analyzed using a Keithley 2400 with Lab Tracer 2 software.

Field emission scanning electron microscope (FESEM, FEI Nova NanoSEM 450) was used to characterize the surface morphology of PSi layer at 10 keV.

2.1 The Potentiometric Selectivity Coefficient Calculations

The selectivity coefficient of PSI for the Na+ (primary ion) toward K+ (interfering ions) was calculated using two methods according to International Union of Pure And Applied Chemistry (IUPAC)[6].
The first method is using the separate solution method (SSM); This method is to measure the potential of a cell (*E*) for two separate solutions, one containing Na^+ and the other one containing K^+ at the same concentration. The potentiometric selectivity coefficient is calculated from the equation:

$$K_{Na,K}^{pot} = a_{Na}^{\left(1-\frac{z_{Na}}{z_K}\right)} e^{(E_K-E_{Na})\frac{F}{RT}} \tag{2}$$

and the second method is the two solution method (TSM) involves measuring potential of a solution of Na^+, and a mixed solution containing Na^+ and K^+. The potentiometric selectivity coefficient is calculated into the following equation:

$$K_{Na,K}^{pot} = a_{Na} \frac{\left(e^{\left(\frac{\Delta E z_A F}{RT}\right)}-1\right)}{(a_K)^{\frac{z_{Na}}{z_K}}} \tag{3}$$

where $\Delta E = E_{Na+K} - E_{Na}$.

3. Results and Discussion

Figure 1 shows the morphological properties of PSi layer by FESEM images. The pore diameter is between 1-3 μm, while the thickness of PSi layer is 55.37 μm. Furthermore, the pores of porous n-type (100) silicon are tubal in shape and characterized by a sponge-like structure.

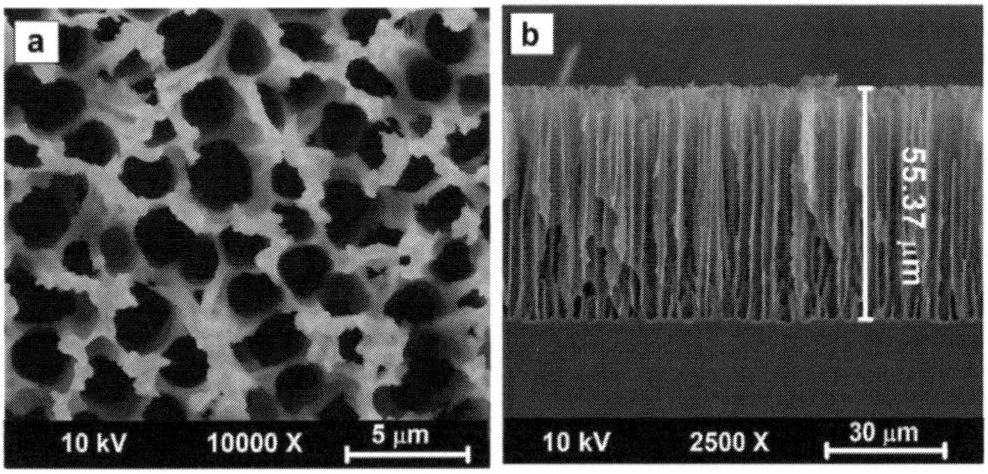

Figure 1. FESEM images for (a) the surface and (b) a cross-section of PSi (100) layer.

Based on the site-binding model, the negative sensing surface (n-type PSi) attracts positive ions (Na^+ and K^+). The positive charge of cations modifies the electrical conductivity of the sample by increasing the concentration of carriers (in case of n-type electrons) in FET channel, while addition of negative charge drains the channel due to a voltage difference between the working electrode (negative) and the cations. This behavior causes the drain-source current (I_{DS}) increase due to more cations concentration in the solutions. I_{DS} adjusts to the corresponding gate voltage with increasing cations concentration [7].

Figure 2 reveal a positive correlation between I_{DS} and reference voltage (V_{Ref}) in the linear region at V_{DS} = 0.3 V for Na^+ and K^+. The increase of V_{Ref} concomitantly increases the current due to increased dissociation of chloride ions; i.e., the rise in cations concentrations produces V_{Ref}. The set down of I_{DS} resulting from the decrease of Na^+ and K^+ concentrations can be attributed to the negative bias on the gate [8].

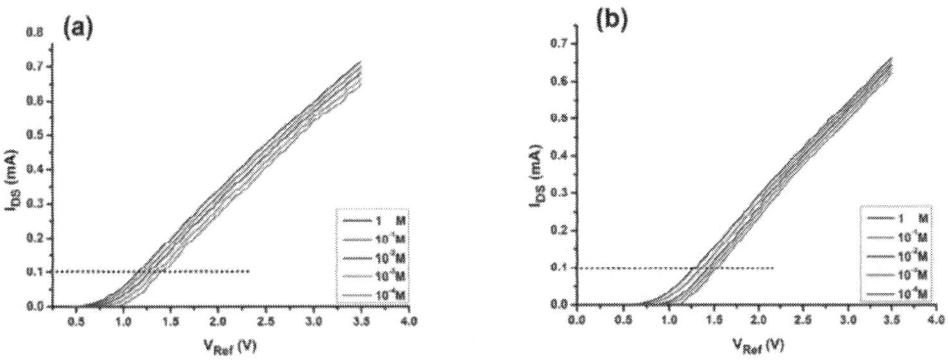

Figure 2: $I_{DS} - V_{Ref}$ characteristics of PSi based-EGFET for (a) Na^+ (b) K^+.

It has been extracted V_{Ref} values at I_{DS} = 0.1 mA, because this region has the best linear correlation between them. Fig.3 shows the relation between V_{Ref} and ion concentrations. Obviously, the sodium selectivity is very close to the potassium response, where the difference between the two values is only 0.3 mV/dec. They are values very close to the theoretical Nernst limit 59.2 mV/dec [5].

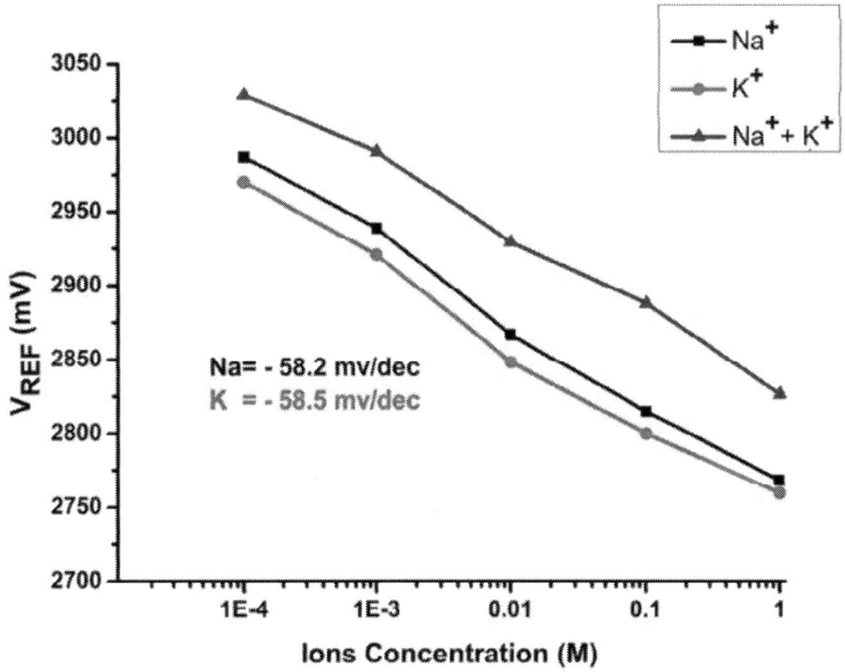

Figure 3: The relation between V_{Ref} and ion concentrations.

The Potentiometric Selectivity Coefficient for Na^+ toward K^+ was calculated by using Equation (2) according to (SSM) method at 0.1 M. the result showed that PSi has no high selectivity for Na^+ toward K^+. The value of the potentiometric selectivity coefficient is near 1, where it was 0.581. While by (TSM) method (Equation 3), the selectivity of PSi for Na^+ to K^+ became less in the mixture solutions to be closer to 1, the value was 0.743 [5]. A weak selectivity of PSi because the diameter of the pores is large allowing for free alignment of ions. Also, the standard potential value of Na and K is very close each other, which are (-2.71 V) and (-2.931 V), respectively. While for SiO_2 is (-0.91 V), this the reason caused PSi prefers Na^+ on K^+, because the deferent between SiO_2 and Na is less than between SiO_2 and K [9].

4. Conclusion

Experiments carried out to characterize PSi as ion-selective electrode based EGFET. The results showed good responding for Na^+ and K^+ 58.2 and 58.5 mV/dec respectively, values close to theoretical limit. The potentiometric selectivity coefficient was also calculated in two ways SSM and TSM. The calculations revealed a low PSi selectivity for Na^+ toward K^+, where the values of the potentiometric selectivity coefficient were close to one in both methods.

Acknowledgements

The authors are extremely grateful to Universiti Science Malaysia and the School of Physics for facilitating us in this study.

References

[1] N. He, S. Papp, T. Lindfors, L. Höfler, R.M. Latonen, R.E. Gyurcsányi, Pre-Polarized Hydrophobic Conducting Polymer Solid-Contact Ion-Selective Electrodes with Improved Potential Reproducibility. Analytical Chemistry, 89 (2017) 2598-2605.

[2] RP Buck, E. Lindner, Recommendations for nomenclature of ionselective electrodes (IUPAC Recommendations 1994). Pure Appl Chem., 66(1994) 2527–2536.

[3] C. Bieg, K. Fuchsberger, M. Stelzle, Introduction to polymer-based solid-contact ion-selective electrodes—basic concepts, practical considerations, and current research topics. Anal. Bioanal. Chem., 409 (2017) 45-61.

[4] E. Bakker, Selectivity of liquid membrane ion-selective electrodes. Electroanalysis, 9(1997) 7-12.

[5] N.M. Ahmed, E.A. Kabaa, M.S. Jaafar, A.F. Omar, Characteristics of extended-gate field-effect transistor (EGFET) based on porous n-type (111) silicon for use in pH sensors. Journal of Electronic Materials, (2017) 1-10.

[6] Y. Umezawa, P. Bühlmann, K. Umezawa, K. Tohda, S. Amemiya, Potentiometric selectivity coefficients of ion-selective electrodes. Part I. Inorganic cations (technical report). Pure Appl. Chem., 72 (2000 1851-2082.

[7] C.D. Fung, P.W. Cheung, W.H., Ko, A generalized theory of an electrolyte-insulator-semiconductor field-effect transistor. IEEE Transactions on Electron Devices, 33 (1986) 8-18.

[8] E.J. Guidelli, E.M. Guerra, M. Mulato, V2O5/WO3 mixed oxide films as pH-EGFET sensor: sequential re-usage and fabrication volume analysis. ECS J. Solid State Sci. Technol., 1 (2012) N39-N44.

[9] Petr Vanýsek, "Electrochemical Series", in Handbook of Chemistry and Physics: 92nd Edition (Chemical Rubber Company) (2011).

IOP Conf. Series: Journal of Physics: Conf. Series **1083** (2018) 012023 doi:10.1088/1742-6596/1083/1/012023

Spectroscopic Colour Evaluation under Different White Light Emitting Diode Illumination Angle

Wan Long Chiong, Ahmad Fairuz Omar, Mohd Zubir Mat Jafri and Chun Ho Tan

School of Physics, Universiti Sains Malaysia, 11700 Penang, Malaysia

wanlong3@hotmail.com, mjafri@usm.my

Abstract. Colour appearance can vary according to object material, geometry, surface characteristics and lighting conditions. In this study, a visible spectroscopy system using white LED as light source is used to measure the reflectance of colour paper at several different illumination angles of 15°, 30°, 45°, 60°, and 75°. In this experiment light was reflected from the surface of colour papers and was captured by spectrometer yielding in spectra ranging from 380 nm to 780 nm. Obtained spectra were plotted as a scatter plot to get the spectral gradient curves for each colour series. Peak wavelengths were also determined to check the shift in wavelength at different conditions. In this work influence of illumination angle on measurement of colour of a colour paper in spectral responsivity is showed. The best responsivity was obtained when light source of 5700 K white LED was illuminated at angle of 60° for blue and green colour series with spectral gradient values of 0.573 (blue glossy), 0.577 (blue non-glossy), 0.245 (green glossy) and 0.214 (green non-glossy) respectively while 3500 K white LED was found to be a better light source at 60° illumination angle in red colour series measurement with spectral gradients of 0.171 (red glossy) and 0.159 (red non-glossy). Choice of white LED in term of CCT was also found to affect the peak wavelengths of the colour as measured by spectrometer.

1. Introduction

Colour is defined to be the subjective appearance of light as detected by human eye and plays an important role in human visual experience [1]. While human vision system is more flexible as compared to instrument as eye is sensitive to geometric factors such as direction, pattern and shape of the object; however, the drawbacks are that evaluation is a subjective matter as different observers observe differently and might even change according to viewing conditions. Therefore, it is advantageous to use instrument as repeatable measurements can be made regardless time and place although only a specific attribute can be taken at a time [2].

In fact, colour cannot be measured as colour is just a perceptual quality and hence, colour measurement actually refers to measuring of optical properties of a sample such as reflectance, absorbance and transmittance [3]. Colour measurement using visible spectroscopy across 380 to 780 nm spectral range in the electromagnetic spectrum has been widely applied in the past years. A spectroscopic system consists of a light source, a spectrometer, and a fiber optic cable. Light interacts with sample and sends back portion of the light via fiber optic cable to be analyzed by spectrometer. Spectrometer is considered as the heart of whole instrumentation system. Fiber optic sensor collects portion of the electromagnetic radiation after interaction with internal structure of sample and then transfers it to the spectrometer. The device will split up the sample light beams into their spectral components and will

Content from this work may be used under the terms of the Creative Commons Attribution 3.0 licence. Any further distribution of this work must maintain attribution to the author(s) and the title of the work, journal citation and DOI.
Published under licence by IOP Publishing Ltd

digitize the signal as a function of wavelength and final data is observed in the form of a spectrum on a computer screen. The standard light source used in previous works was tungsten-halogen lamp with continuous spectrum from ultraviolet to infrared region in the electromagnetic spectrum which is comparable to the electromagnetic spectrum of Sun. Light Emitting Diode (LED) has been the rising star in lighting industry with higher efficiency, better light quality and lower energy consumption [4]. LED lighting market is growing continuously and will impact general lighting the most in 2020 [5].

Colour can be determined by the reflected light from a surface of an object, however, amount of light reflected from an object is regarded to vary according to the wavelength of the light, direction of the incident light on object (illumination angle) and direction from which the angle is viewed [2]. Therefore, this work is a preliminary study to determine the effect of illumination angle on colour measurement under two different white LED CCTs illuminations using visible spectroscopy.

2. Methodology

Experimental setup for this study consisted of a Jaz spectrometer (Ocean Optics Inc., Florida, USA), an optical fiber connecting spectrometer to computer and white LED lighting panels. The samples used for colour evaluation were monochromatic red, green and blue colour series papers in increasing hue from 20 to 255 printed on two types of paper; the glossy (G) and non-glossy (NG) ones. These colour papers were illuminated under white LEDs with different CCTs of 3500 K (D35) and 5700 K (D57) at illumination angles of 15°, 30°, 45°, 60°, and 75° respectively as shown in figure 1. The fiber optic probe was placed perpendicularly to sample at a fixed distance of 1.0 cm. White LED specifications are presented in table 1.

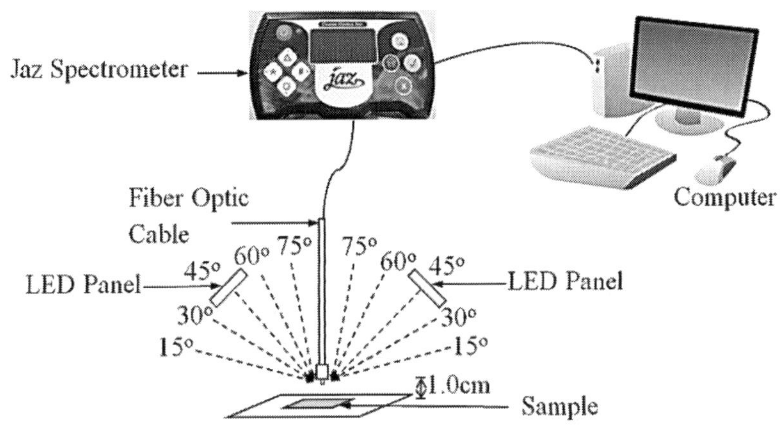

Figure 1. Experimental setup.

Table 1. White LED specifications.

	D35	D57
Correlated Colour Temperature (K)	3500	5700
Product Type	LCW W5AM-KXKY-409Q	LUW W5AM-KYLX-4E8G
LED Colours	Warm White	Cool White
LED Material	GaN	InGaN
Luminous Flux (lm)	71-97	82-130
Luminous Intensity(mcd)	19000-22000	22400-30200
CRI	80	70

Prior to measurement, Jaz spectrometer was first calibrated using tungsten- halogen lamp (ASD Inc., Colorado, USA) by illuminating it at an angle of 45° on a white diffuse reflectance standard (WS-1-SL) to obtain reference spectrum. The SpectraSuite software used equation (1) to produce the reflectance spectra:

$$R_\lambda = \frac{S_\lambda - D_\lambda}{W_\lambda - D_\lambda} \times 100 \tag{1}$$

where R_λ= the reflectance at the wavelength of λ
S_λ=the sample intensity at the wavelength of λ
D_λ=the dark intensity at the wavelength of λ
W_λ= the reference intensity at the wavelength of λ

The spectral reflectance was further analysed using spectral gradient where higher spectral gradient indicates better responsivity with smaller change in increasing hue of colour series. This technique is useful in reducing the influence of blue light from white LED itself rather than direct analysis by using raw spectra as raw spectra contain peaks from white LED which are unwanted in colour measurement. Peak wavelengths were also recorded to determine the shift in values due to difference in CCT of light source.

3. Results and Discussion

Figure 2 shows white LED spectra obtained when white LEDs were shone onto white diffuse reflectance standard at different illumination angles. Spectra showed that white LEDs used in this study were phosphor-converted white LEDs with combination of blue light and phosphor mixture. D57 had higher blue light composition which made it appear to be more 'bluish' compared to D35 which appears to look 'yellowish-orange' instead. Illumination angles showed direct influence on the spectral reflectance as 60° gave the highest reflectance followed by 75°, 45°, 30° and 15° for D35 while for D57 a decreasing reflectance order was observed for 60°, 45°,75°,30° and 15° respectively. Illumination angle of 75° was found to produce inconsistent results for different white LEDs despite spectral reflectance still being higher than 15° and 30°.

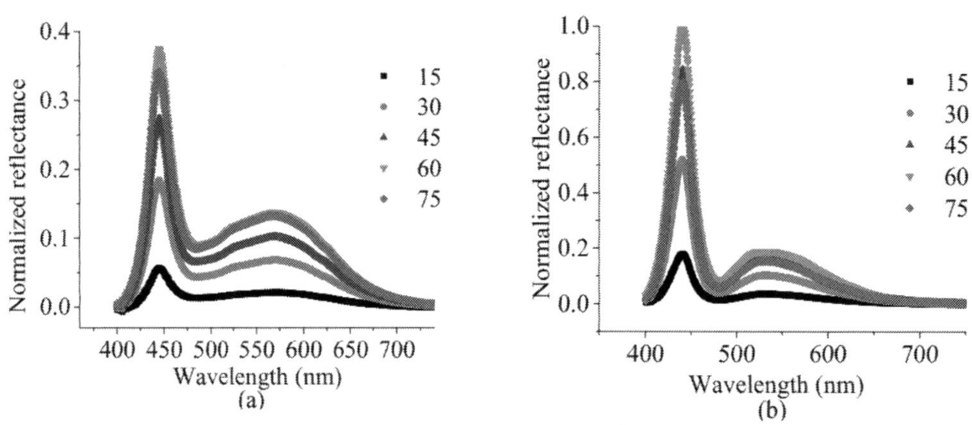

Figure 2. Spectra of white LED for (a) D35 and (b) D57 under different illumination angles.

Spectral gradient graphs of blue, green and red colour series for glossy (G) and non-glossy (NG) papers under D35 and D57 illuminations are shown in figure 3, 4 and 5. In general, 60° illumination angle gave the highest spectral gradient indicating best responsivity towards change in the constant colour series for both D35 and D57.

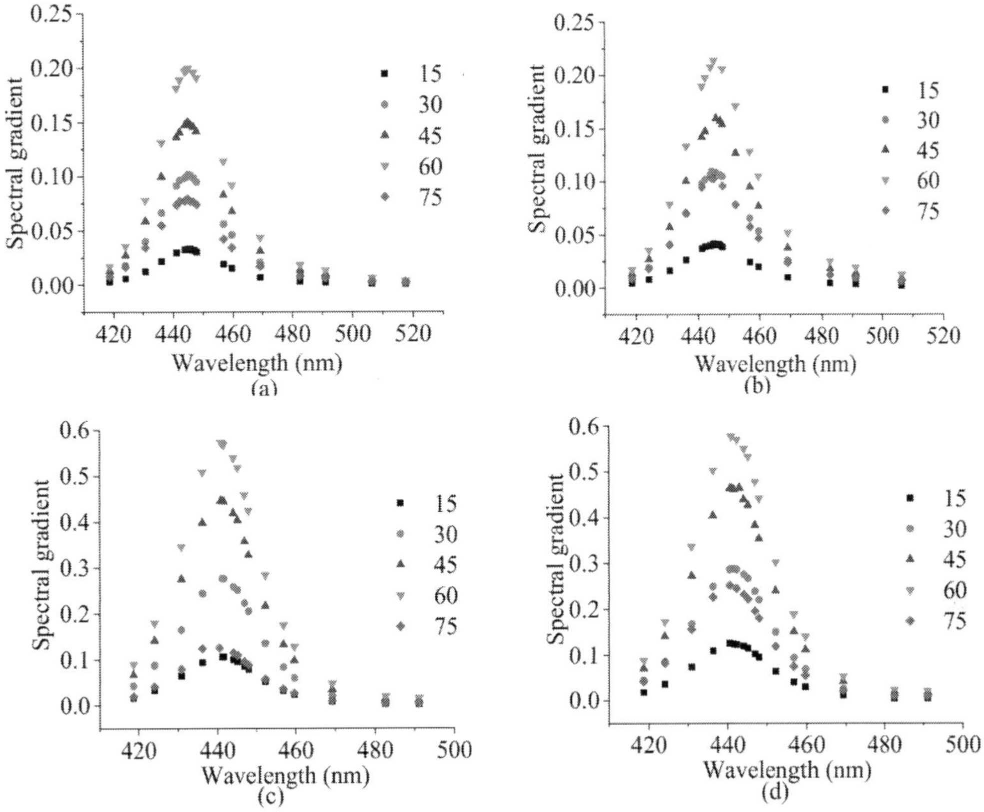

Figure 3. Blue colour series observation for (a) D35-G papers, (b) D35-NG papers, (c) D57- G papers and (d) D57-NG papers.

For blue colour series, the differences in spectral gradient of glossy and non-glossy results are not as significant as in case of green and red. This is due to the influence of blue light of white LED as reflectance measured by spectrometer is the result of interaction between colour paper and lighting which illuminates the samples. It might be a drawback for this type of white LED (phosphor-converted type) to act as light source in observing blue-coloured samples. For green colour series observation, there are peaks observed in spectra between 420 nm to 475 nm which actually are originated from the light sources (Refer figure 2). The overall patterns are discerned and close to the white LEDs spectra demonstrating influence of light sources on reflectance spectra and indirectly on the responsivity of the measured colour.

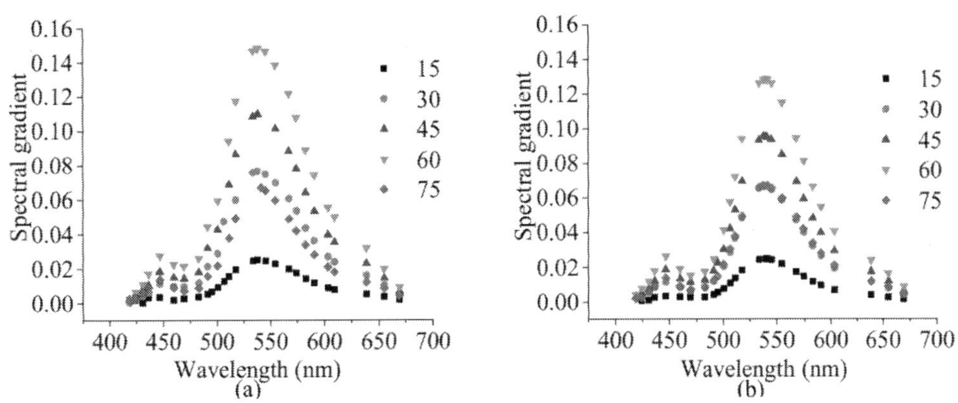

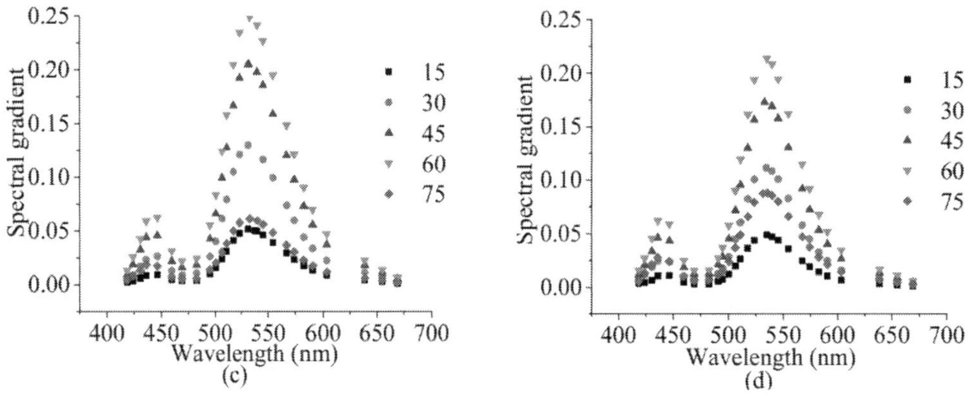

Figure 4. Green colour series observation for (a) D35-G papers, (b) D35-NG papers, (c) D57- G papers and (d) D57-NG papers.

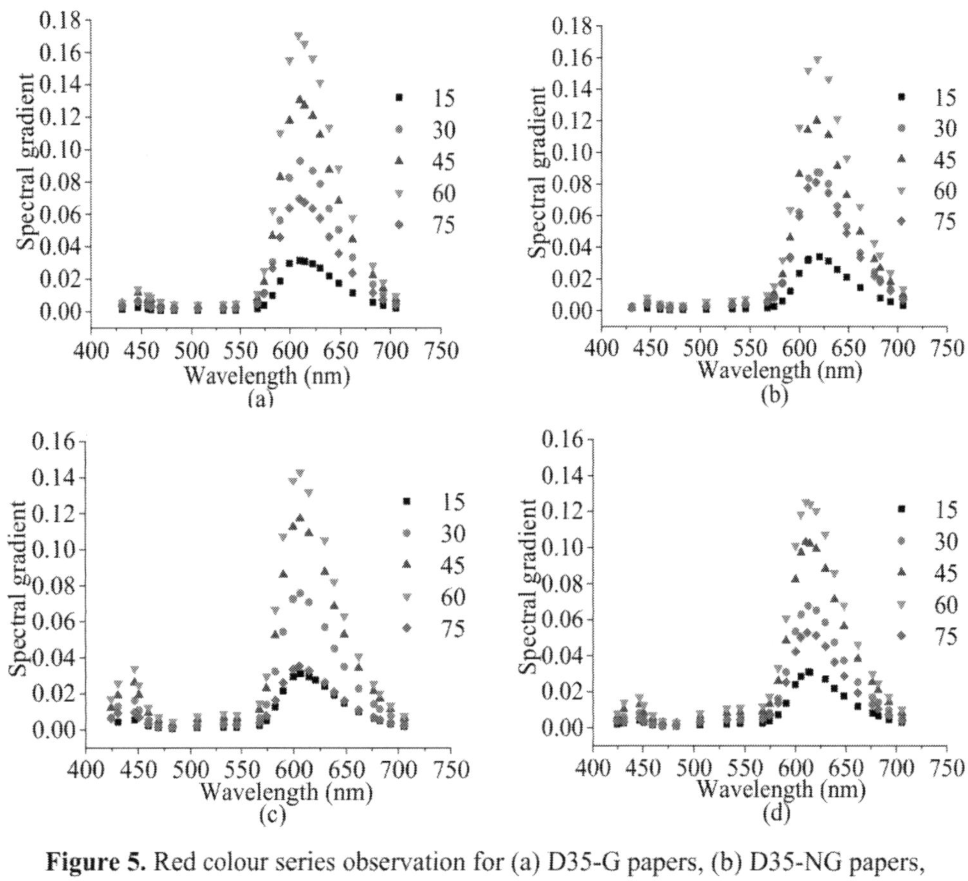

Figure 5. Red colour series observation for (a) D35-G papers, (b) D35-NG papers, (c) D57- G papers and (d) D57-NG papers.

Conversely, for red colour series, it is noteworthy that the spectral gradients for D35 are higher than for D57 unlike for blue and green colour observations. It can be deduced that D35 which is lower in CCT has improved observation for red colour series and strongly correlated such that warm white LED is a better light source in observing red-coloured samples.

Table 2 shows the peak wavelengths for spectral gradient curve for different illumination angles. The peak wavelengths observed to have similar values despite having different illumination angles for same type of paper and light source. It shows that illumination angle does not affect colour measurement of colour papers despite spectral gradient having significant differences such as in green colour series. The differences in peak wavelengths for same illumination angle and type of paper between lightings also show that LED CCT does affect colour appearance as measured by spectrometer especially for green colour series. Peak wavelengths for different types of colour paper under same lighting are compared next. For green colour series, the difference in values of D35 is smaller (±2 nm) than for D57 (±3 nm) but results are still acceptable as compared to red colour series results. It is found that the differences for both D35 and D57 are even bigger (±10 nm for D35 and ±6 nm for D57) indicating surface type of samples must be the same for measuring red-coloured samples in order to avoid this error.

Table 2. Peak wavelengths for colour series observation.

Colour Series	Illumination Angles	D35		D57	
		Glossy	Non-Glossy	Glossy	Non-Glossy
White Diffuse	15°	444.03		440.47	
	30°	444.74			
	45°	445.10			
	60°	445.10			
	75°	444.39			
Blue	15°	445.81	445.10	441.54	440.47
	30°	445.10	444.39	441.36	440.83
	45°	445.10	445.81	440.83	442.96
	60°	445.10	445.10	440.83	440.83
	75°	445.10	444.39	440.47	440.47
Green	15°	537.69	539.77	531.47	534.93
	30°	537.69	537.69	531.47	534.58
	45°	538.39	537.69	531.81	532.85
	60°	538.39	539.77	532.85	534.93
	75°	541.15	541.15	532.85	535.62
Red	15°	609.40	620.17	606.03	612.77
	30°	609.74	617.82	606.03	612.10
	45°	609.40	617.14	606.03	609.74
	60°	608.39	617.82	606.03	609.74
	75°	609.06	616.47	605.01	611.42

4. Conclusion

This work showed that illumination angle has direct influence on responsivity of colour measurement as measured by spectrometer even though peak wavelengths are found to be close to one another. It was also found that different CCT of white LED as light source does alter the colour measurement of samples as the peak wavelengths values have significant differences. Further study is required to discover methods to overcome the variance caused by white LED as light source in spectroscopy measurement.

Acknowledgments
This study was supported by Universiti Sains Malaysia Short Term Grant (Grant no. 304 / PFIZIK / 6313209). The first author thanks Ministry of Higher Education and Universiti Sains Malaysia for financial assistances under MyMaster and USM Fellowship respectively.

References

[1] R.J. Tilley 2011 *Colour and the optical properties of materials: an exploration of the relationship between light, the optical properties of materials and colour* (United Kingdom : John Wiley & Sons) p 1
[2] R. S. Hunter 1987 *The Measurement of Appearance* (New York: Wiley–Interscience)
[3] S. Johnsen 2016 *J. Exp. Biol.* 219 772-778

[4] A. De Almeida, B. Santos, B. Paolo, and M. Quicheron 2014 *Renew. Sust. Energ. Rev.* 34 30-48
[5] T. Baumgartner, F. Wunderlich, A. Jaunich, T. Sato, G. Bundy, N. Grießmann, J. Kowalski, S. Burghardt, and J. Hanebrink 2012 *Lighting the way: Perspectives on the global lighting market* (Europe: McKinsey & Company, Inc.) p 7

Effect of Annealing Temperature in ZnO For Photonic Applications

Nur 'Izazi Maad Tazri[1], Wan Maryam Wan Ahmad Kamil[1,*]

[1]School Of Physics, Universiti Sains Malaysia, 11800 USM, Penang, Malaysia

*wanmaryam@usm.my

Abstract: ZnO nanorods were synthesized on glass substrate by two step chemical bath deposition (CBD) method and annealed at different temperatures in argon environment for 1 hour. The structural and optical properties of the as-grown and annealed samples were investigated using X-ray diffractometer (XRD), UV-Visible-NIR spectroscopy, Raman spectroscopy and photoluminescence (PL) measurement. As grown sample has the largest crystallite size with a photonic gap of 90 nm whilst. Raman measurement shows sharp E_2 (H) peak typical for the wurtzite ZnO structure and PL spectra exhibit a strong UV emission peak at 380 nm for all samples. The ratio of UV to visible peak in the PL spectrum for as grown sample was the highest indicating lowest level of defects for these set of samples. This is consistent with the crystallinity of the samples. Results suggest potential of using CBD as a method of growth for synthesizing ZnO nanorods for nanophotonic devices.

1. Introduction

Zinc Oxide (ZnO) has generated great interest due to its direct wide band gap of 3.37 eV and large exciton binding energy of 60 meV with many applications including light emitting diodes [1], UV photodetectors [2], solar cells [3] and many more. In addition to that, due to its transparency and electro-optical and elasto-optical properties, ZnO is attractive for integrated photonic devices [4]. As nanotechnology blooms, investigations of optical properties of nanostructures have led to better design and control for nano scale devices.

Several techniques have been used to prepare ZnO nanostructures, such as molecular beam epitaxy (MBE) [5], chemical vapour deposition (CVD) [6], pulse laser deposition [7] , sol-gel method [8] and chemical bath deposition (CBD) method [9]. However, chemical bath deposition (CBD) method have many advantages, such as low cost and low working temperature [10]. Compared to other low cost method, CBD provides a more promising option due to its large scale controllability and have shown to be superior in producing good morphology of ZnO nanorods [11]. Based on previous studies, post annealing treatment in samples prepared by CBD has been reported using air flow [12], oxygen [13], nitrogen and even in vacuum [14] environment. However, the effect of annealing under argon environment no detailed reports of the growth characteristics is available, up to our knowledge.

In this paper, the properties of ZnO nanorods that prepared by a simple chemical bath deposition method for potential nanophotonic applications was addressed. The nanorods were synthesized using the CBD method at a fixed growth temperature of 96°C for 4 hours. A thin ZnO layer was deposited by using RF sputtering prior to growth and several parameters like molarity, growth temperature, time of annealing and gas for annealing has been fixed. The growth characteristics at different annealing

Content from this work may be used under the terms of the Creative Commons Attribution 3.0 licence. Any further distribution of this work must maintain attribution to the author(s) and the title of the work, journal citation and DOI.
Published under licence by IOP Publishing Ltd

temperature are analysed by scanning electron microscope (SEM) and X-ray diffraction measurements. Optical properties were measured by UV-Vis spectroscopy and photoluminescence measurements.

2. Method

Growth: Prior to growth, the glass substrates were pre-coated with 100 nm of ZnO using RF sputtering at room temperature in high purity of argon using a ZnO target with purity 99.999%. Then, all the samples except an as grown sample undergo annealing process at different temperatures of 350, 400, 500 and 600°C for 1 hour in argon environment. The rate flow for argon in a furnace is 3.0 slm. Then, all the seeded substrates were immersed into chemical solution consisting equi-molar of zinc nitrate hexahydrate and hexamethylenetetramine (HMT), dissolved in deionized water. The growth temperature was fixed at 96°C for the entire growth period of 4 hours. All samples undergo annealing for 1 hour in argon environment at different temperatures post growth.

Characterisation: The crystallinity of the ZnO nanorods was analysed from XRD measurement using a Cu Kα radiation source with wavelength of 1.5406 Å. Scherrer's formula was then used to determine the crystallite size. UV-Visible-NIR spectrophotometer was utilized to obtain the transmittance spectra and to determine the photonic gap of the samples. Raman measurement with spectrometer Horiba Jobin Yvon HR800 with Ar^+ ion laser as the excitation source operating at a wavelength of 514.55 nm and power of 20 mW was used to study the vibrational modes from the ZnO nanostructures. Meanwhile, photoluminescence (PL) measurement was carried out using a continuous wave He-Cd laser (325 nm, 20 mW) at room temperature to observe optical characteristics of the samples.

3. Result and Discussion

3.1 X-ray diffraction analysis

XRD spectra of the ZnO nanorods annealed at different annealing temperatures are shown in Figure 1. The as grown sample was included as reference. All peaks can be indexed to the hexagonal wurtzite structure with 2θ value of 34.4°, which is corresponding to the (002) plane. No diffraction peaks related to impurities were observed. The average particle size of all samples was calculated using the Debye-Scherer formula[15] :

$$D = \frac{\kappa\lambda}{\beta\, cos\theta} \qquad [1]$$

where κ is a constant with a value 0.9, λ is radiation source wavelength, β is full half-maximum (FWHM) intensity and θ is diffraction angle. The average particle size of all samples, calculated using equation (1) is approximately 48.99 nm, 46.28 nm, 43.75 nm, 41.95 nm and 44.05 nm for as grown, 350, 400, 500 and 600°C, respectively as shown in Table 1.

The lattice strain (ε) has been determined by using the tangent formula [15], whereas the *d*-spacing can be measured by the following relations [15]:

$$\varepsilon = \frac{\beta}{4\tan\theta} \qquad [2]$$

$$d_{hkl} = \frac{1}{\sqrt{\frac{4(h^2+k^2+hk)}{3a^2}+\frac{l^2}{c^2}}} \qquad [3]$$

where β is FWHM, d is interplanar distance, λ = 1.5406 Å is the wavelength of the X-ray radiation used, θ is the angle of the diffraction peak with respect to the orientation considered for calculation. The calculated values for each sample are tabulated in Table 1.

The as grown sample showed strongest intensity with smallest full-width at half-maximum (FWHM) value and largest crystallite size; indicating good crystallinity of the ZnO nanorods grown in the *z*-direction (growth direction). The spacing and lattice strain follows the same trend, whereby as annealing temperature increases, both lattice strain and *d*-spacing increases until the annealing temperature reach 600°C. Crystallinity starts to improve at this temperature indicating that a temperature above 600°C is needed to improve crystallinity in samples prepared by this method of CBD. However due to the limit of annealing soda lime glass substrates, it is not possible to maintain the glass structure at higher annealing temperatures. However, as grown samples already exhibit good crystallinity without the need of annealing treatment. Preheating the substrate, as done in this work, would also affect zinc acetate decomposition and crystal growth and may explain the reduced intensity from the XRD spectrum at the *c*-axis orientation obtained from annealed samples [16]

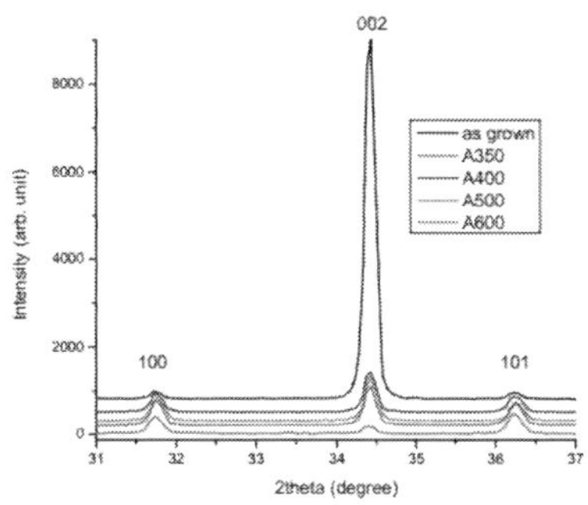

Figure 1: XRD spectra of all annealed samples in argon environment at different temperatures for 1 hour. Sample without annealing treatment also included in this measurement. Samples undergo annealing shows a polycrystalline structure whereas as-grown sample showed good crystallinity.

Table 1: Growth parameters of as-grown and annealed samples at different temperatures for 1 hour in argon atmosphere.

Temperature	2-theta (θ)	FWHM (°)	Average crystallite size (nm)	*d*-spacing (Å)	Lattice strain (ε)
As grown	34.417	0.158	48.99	2.6053	0.1318
A350	34.403	0.168	46.28	2.6066	0.1396
A400	34.418	0.177	43.75	2.6052	0.1476
A500	34.427	0.185	41.95	2.6048	0.1538
A600	34.418	0.176	44.05	2.6035	0.1465

3.2 UV-Visible Spectroscopy analysis

The UV-Vis transmission spectra of the ZnO nanorods are shown in Figure 2. The photonic gap, the region of zero light transmission, for all samples lie between 287 and 381 nm, which is typical for samples grow by CBD [17]. The widest photonic gap belongs to as-grown sample, whereby the photonic band gap is 94 nm. A summary has been included in Table 2.

A plot of $(\alpha hv)^2$ vs. hv as shown in Figure 2(b) was used to determine the optical band gap with relation

$$\alpha hv \approx (hv - E_g)^{m/2} \qquad [4]$$

where hv is the photon energy and m=1 for a direct transition. The optical band gap was found to be 3.21, 3.24, 3.26, 3.27 and 3.27 eV for as-grown, 350, 400, 500 and 600°C, respectively as summarized in Table 2. With the increase of annealing temperature, the red-shift of the band energy may be attributed to the increase of particle size[18].

The photonic gap provides information on the disallowed bands of wavelengths for photonic crystals and stops at the point where the threshold for photons to be absorbed is reached. This threshold gives the optical band gap of the sample and in our case, is shown to increase with increasing annealing temperature, indicating the threshold for photon absorption is higher at higher annealing temperatures. For nanophotonic devices, this would mean tuning the wavelength for photon absorption through annealing as different annealing temperatures showed different threshold (energy bang gaps) for photon absorption.

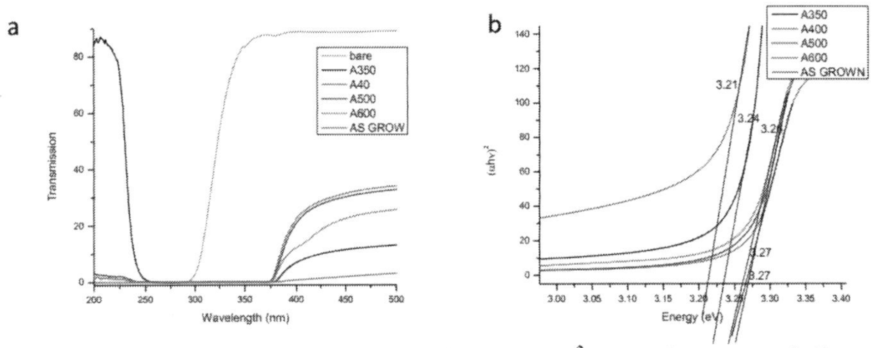

Figure 2: (a) The transmission spectra for all samples (b) $(\alpha hv)^2$ versus hv curves of all samples: used to determine the band gap energy of the material.

Table 2: Summarize Photonic band gap and band gap energy of all sample.

Sample	Photonic Band Gap (nm)	Band Gap Energy (eV)
As grown	94	3.21
A350	90	3.24
A400	85	3.26
A500	86	3.27
A600	87	3.27

3.3 Raman Spectroscopy

Figure 3 shows the Raman scattering spectra of the samples performed at room temperature. Spectra exhibit E_2 (H) mode peak around 438 cm^{-1} , which is typically the wurtzite hexagonal phase of ZnO and assigned to the optical phonon E_2 mode of the ZnO crystal [19]. A very weak and small band appeared at 331 cm^{-1} attributed to the second-order Raman scattering due to the zero boundary phonons, typically called E_{2H}-E_{2L} (multi-phonon process). Peaks at 380 cm^{-1} also can be seen in the spectra designated to the A_{1T} modes. From the spectra, E_2 has the strongest peak when anneal in 600°C compared to other samples which indicate sample had strong optical absorption, which is preferred for photonic applications. The broad phonon peaks located at 580 cm^{-1} corresponding to A_1 (LO) and E_1 (LO) modes is nominally attributed to defect, which is likely due to oxygen and zinc interstitials in our case.

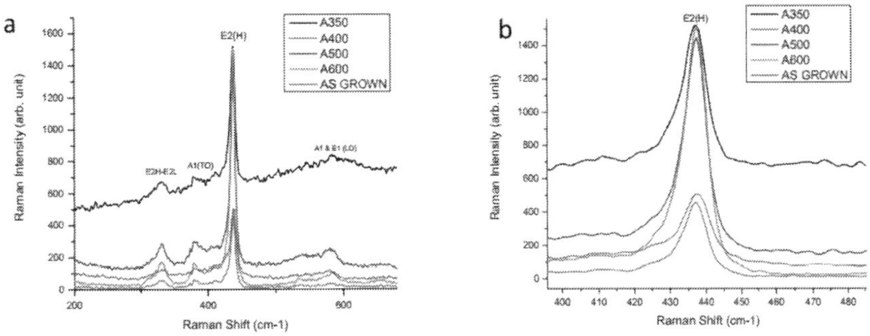

Figure 3 (a) Raman Spectra of all sample (b) zoomed image of the Raman spectra for the E_2(H) peak.

3.4 Photoluminescence Measurement

Figure 4 shows the PL spectra of the ZnO nanorods annealed at different temperatures. The spectra consist of a strong UV peak at 385 nm related to near-band edge emission and a broad deep-level emission (visible emission) peak in the visible range. The deep level emission mostly in the green and partly in the yellow and red spectral regions is generally attributed to deep level defects, such as O vacancies (V_o), Zn vacancies (V_{Zn}), interstitial O (O_i), interstitial Zn (Zn_i) and extrinsic impurities [20]. Shifts in the UV peak is consistent with shifts in the calculated optical band gap energy as in Table 2.

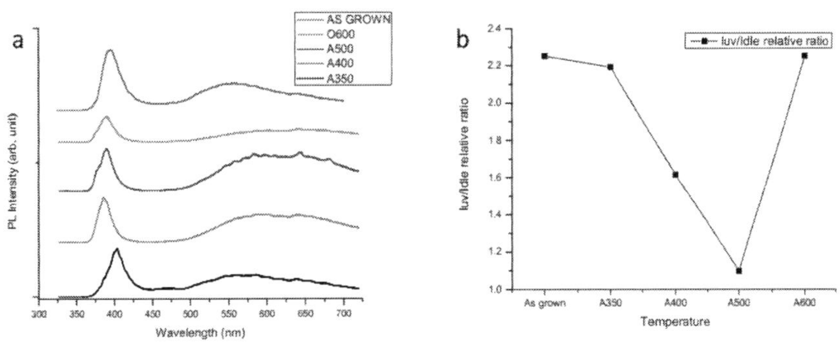

Figure 4 (a) PL spectra of all samples at room temperature and (b) relative ratio of the UV/DLE peak intensity obtained from the PL spectra.

The relative PL intensity ratio of UV emission to deep level emission for all samples, as shown in Figure 4(b) is roughly 2.25, 2.19, 1.62, 1.10 and 1.98 corresponding to as grown, 350, 400, 500 and 600°C, respectively. The highest ratio value for the as-grown and sample annealed at 600°C indicate lowest level of point defects in these samples. As temperature is introduced it is likely these defects move deeper into the structure, hence showing increased emission in the visible spectrum. However, 600 °C seems to be the recrystallization temperature for these samples as these defects are significantly reduced at this temperature. This result is consistent with the XRD analysis indicating increased crystallinity as these point defects are reduced.

4. Conclusion

ZnO nanorods were successfully grown on glass substrates by a simple chemical deposition method and undergo annealing process at different temperatures in argon environment for 1 hour. Sample annealed at 600°C shows the best sample for photonic applications with strong optical absorption as observed in the Raman spectra and lowest deep level associated defects despite reduced crystallinity compared to as-grown samples. Results also suggest 600°C as the minimum temperature for recrystallization in such a structure and annealing above this value is needed for ZnO nanorods prepared by CBD for potential nanophotonic applications.

Acknowledgement
We acknowledge partial funding from USM short term grant (304/PFIZIK/6312155), AKAUN SISWAZAH PPSF (308/AIFIZIK/415403) and bridging grant: (304.PFIZIK.6316077). We are also thankful to Prof Boon S Ooi for providing usage of their facility to conduct Photoluminescent measurements in Photonics lab, KAUST, Saudi Arabia

References

[1] J Dong, C Zhen, H Hao, J Xing, Z Fan, and Z Zhang 2014 *Jpn. J. Appl. Phys.* 53 055201.
[2] A Khayatian, M A Kashi, R Azimirad, S Safa, and S F A Akhtarian 2016 *Optik (Stuttg).* 127 4675–4681.
[3] J Huang, Z Yin, and Q Zheng 2011 *Energy Environ. Sci.* 4 3861–3877.
[4] L Znaidi, T Touam, D Vrel, N Souded, S Ben Yahia, and O Brinza 2012 *Acta Physica Polonica A.* 121 165–168.
[5] Y W Heo, M Kaufman, K Pruessner, D P Norton, M F Chisholm, and P H Fleming 2008 *Solid-State Electronics.* 47 2269–2273.
[6] Z Ye, T Wang, S Wu, X Ji, and Q Zhang 2017 *J. Alloys Compd.* 690 189–194.
[7] P S Shewale and Y S Yu 2016 *Journal of Alloys and Compounds.* 654 79–86.
[8] Y C Yoon, K S Park, and S D Kim 2015 *Thin Solid Films.* 597 125–130.
[9] N T Son, J-S Noh, and S Park 2016 *Appl. Surf. Sci.* 379 440–445.
[10] S Kahraman, H M Çakmak, S Çetinkaya, H A Çetinkara, and H S Güder 2013 *Int. J.Mater. Res.* 104 799–804.
11] T Lee, H Ryu, and W-J Lee 2014 *J. Alloys Compd.* 597 85–90.
[12] L L Yang, Q X Zhao, M Willander, J H Yang, and I Ivanov 2009 *J. Appl. Phys.* 105 053503.
[13] C M Mbulanga, Z N Urgessa, S R Tankio Djiokap, J R Botha, M M Duvenhage, and H C Swart 2015 *Phys. B Condens. Matter.* 480 42–47.
[14] S Erat, H Metin, and M Ari 2008 *Mater. Chem. Phys.* 111 114–120.
[15] S Shanmugan, Mohamed Mustaqim A B, and D Mutharasu 2015 *International Journal of Engineering Trends and Technology (IJETT)* 28 27–36.
[16] M Ohyama, H Kouzuka, and T Yoko 1997 *Thin Solid Films.* 306 78–85.
[17] W Q Lim, L K Sim, N Fazrina, and W Maryam 2017 *1st UKM-ISESCO-COMSATS Int. Work. Nanotechnol. Young Sci. AIP,* 1, 20002.
[18] R Bahramian, H Eshghi, and A Moshaii 2016 *Mater. Des.* 107, 269–276.
[19] Y J Xing, Z H Xi, Z Q Xue, X D Zhang, J H Song, R M Wang, J Xu, Y Song, S L Zhang, and D P Yu, 2003 *Appl. Phys. Lett.* 83 1689–1691.
[20] M Chen, Z Wang, D Han, F Gu, and G Guo 2011 *Sensors Actuators, B Chem.* 157 565–574.

IOP Conf. Series: Journal of Physics: Conf. Series **1083** (2018) 012025 doi:10.1088/1742-6596/1083/1/012025

Optical Properties and UV Sensing Response of Nitrogen-doped TiO_2 Thin Film by CVD

Siowwoon Ng[1,]*, Sau Siong Tneh[1], Fong Kwong Yam[1] and Zainuriah Hassan[2]

[1]School of Physics, Universiti Sains Malaysia, 11800 Penang, Malaysia
[2]Institute of Nano Optoelectronics Research and Technology (INOR), Universiti Sains Malaysia, 11800 Penang, Malaysia.

*sw.ng@live.com

Abstract. This work was demonstrated a cold-wall chemical vapour deposition (CVD) setup to prepare nitrogen (N)-doped titanium dioxide (TiO_2) thin film on glass substrates, which presents semi-transparent yellowish surface. The prepared N-doped TiO_2 film is homogeneous and possesses polycrystalline anatase structure. The transmittance measurement was carried out in the ultraviolet-visible spectral region to evaluate the film thickness, optical constants, and the optical band gap was determined by Tauc plot. A metal-semiconductor-metal (MSM) ultraviolet (UV) photodetector by the Pt/N-doped TiO_2 film has shown a positive response towards UV illumination.

1. Introduction

Ultraviolet (UV) radiation sources possess wide applications in various industries, for instance, astronomy, communication, memory storage, optoelectronic circuits, organic disinfection, food industry, and biological analysis [1–3]. As such, UV photodetection or monitoring becomes crucial to prevent excessive UV exposure, which may lead to negative effects, including premature aging, skin cancer, and cataracts [1,4]. Wide bandgap semiconductor based UV photodetector such as n-type titanium dioxide (TiO_2) is advantageous for its room-temperature operation and intrinsic visible blindness [1,5]. UV photodetector based on metal-semiconductor-metal (MSM) structure with two back-to-back Schottky barriers, is commonly adopted due to its high response, low cost, simple structure, and the ease of optoelectronics integration [6,7].

Titanium dioxide (TiO_2) has been extensively studied, primarily as a photocatalyst for photoelectrochemcial (PEC) water splitting [8], photo-decomposition of organic compounds [9,10], and dye-sensitized solar cells (DSSCs) [11]. Owing to its excellent physical, optical, chemical and thermal stability, TiO_2 is also applied in Li-ion batteries, supercapacitor, cell imaging, drug carrier, bio, chemical and physical sensors [12,13].

Commonly reported TiO_2 thin films preparation techniques include sol-gel spin coating, spray pyrolysis, magnetron sputtering, e-beam evaporation and chemical vapor deposition (CVD) [6,13–18]. CVD is certainly a versatile approach to achieve large surface area deposition with uniform coating and perform simultaneous doping in a single step. Among the non-metal (N, C, S, F) doping, N doping prevails in the modification of TiO_2 [19]. However, the major drawback of a CVD system is its complicated setup, high temperature and low vacuum operation. Thus, many efforts have been carried out to improve this aspect through the use of different Ti and dopant precursors [20].

Content from this work may be used under the terms of the Creative Commons Attribution 3.0 licence. Any further distribution of this work must maintain attribution to the author(s) and the title of the work, journal citation and DOI.
Published under licence by IOP Publishing Ltd

In the current work, N-doped TiO_2 thin film was prepared with high deposition rate. A customized and facile cold-wall CVD setup operating in atmospheric pressure with low deposition temperature is presented. The deposited film is investigated for its structural and optical properties, and applied as a MSM UV photodetector.

2. Experimental

Glass substrates were pre-cleaned and cut into desired sizes. Cold-wall CVD reactor was pre-heated and the substrates were placed in contact with the base of the reactor. The Ti and N dopant precursors were titanium tetrachloride ($TiCl_4$) and ammonium acetate (CH_3COONH_4), respectively, in 1:1 volume ratio. Air with controlled humidity of $\approx$ 80% was served as O_2 precursor and carrier gas. These precursors were simultaneously channeled into the reactor and horizontal laminar flow of air was introduced to initiate the chemical reactions. The deposition was carried out under atmospheric pressure with constant temperature lower than 500°C for 60 s.

The N-doped TiO_2 thin film morphology was imaged by field-emission scanning electron microscope (FE-SEM, Zeiss Leo Supra 50 VP) and the elemental composition was confirmed by energy dispersive X-ray spectroscopy (EDX). For structural analysis, the crystalline structure was verified by high-resolution X-ray diffractometer (HR-XRD, PANalytical X'Pert PRO MRD PW3040) with $CuK\alpha_1$ source of 0.154 nm wavelength. The optical transmittance spectrum was recorded by an ultraviolet-visible (UV-Vis) spectrophotometer (Agilent Cary 5000).

To evaluate the ultraviolet (UV) sensing response of the prepared film, 30 nm Pt was deposited by radio-frequency (RF) magnetron sputtering system (Auto HHV 500) at the two edges on the surface of the film as metal pad contact. A 6 W power UV lamp with peak intensity at λ = 372 nm was served as UV excitation source, maintained at a distance of 5 cm from the sensing film. A source measuring unit (SMU, Keithley 2400) was employed to supply a constant 8 V bias and to record the photoresponse profile.

3. Results and discussions

3.1. Morphology, composition and structural properties

The as-synthesized N-doped TiO_2 thin film is yellowish and semi-transparent as observed by naked eye, shown in Figure 1(a), and a transparent glass substrate is included for comparison. The yellowish film is in line with the nitrogen doped TiO_2 thin film or nanostructures reported in the literature. The higher color intensity suggests that a higher doping intensity because the color of a solid is determined by the position of its absorption edge [17].

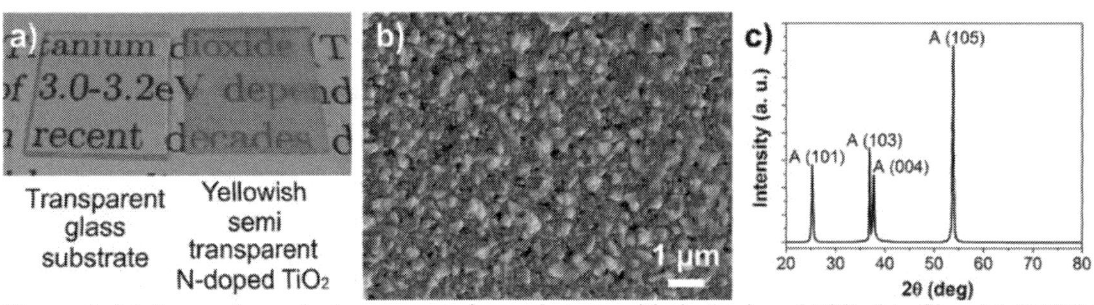

Figure 1. (a) Comparison of a transparent glass substrate with the N-doped TiO_2 thin film, (b) FESEM image and (c) XRD pattern of the N-doped TiO_2 thin film.

The FESEM image in Figure 1(b) shows that the entire coated surface is homogeneous. The atomic percentage from EDX analysis for Ti, O, Cl and Si elements are 35.08, 62.84, 0.60 and 1.48%, respectively. The tiny amount of Cl is originated from the $TiCl_4$ precursor and a small amount of Si is expected from the glass substrate. The rich amount of Ti and O affirms the formation of TiO_2 film, and no trace of N suggests that the TiO_2 film was appropriately doped. XRD pattern in Figure 1(c) shows four distinguishable peaks, and the peaks can be assigned to anatase (101), (103), (004) and (105) plane. This result shows that the N-doped TiO_2 thin film is composed of polycrystalline anatase and the film is readily crystalized even within 60 s of deposition.

3.2. Optical properties and band gap evaluation

The transmittance spectrum of the N-doped TiO_2 thin film is presented in Figure 2(a). The interference fringes of the incident beam on a homogeneously deposited film may result in adjacent maxima and minima in the transmittance spectrum, marked as the highly transparent region. These fringes may be utilized to calculate the optical constants such as refractive index, absorption and extinction coefficient based on Swanepoel's envelope method [21].

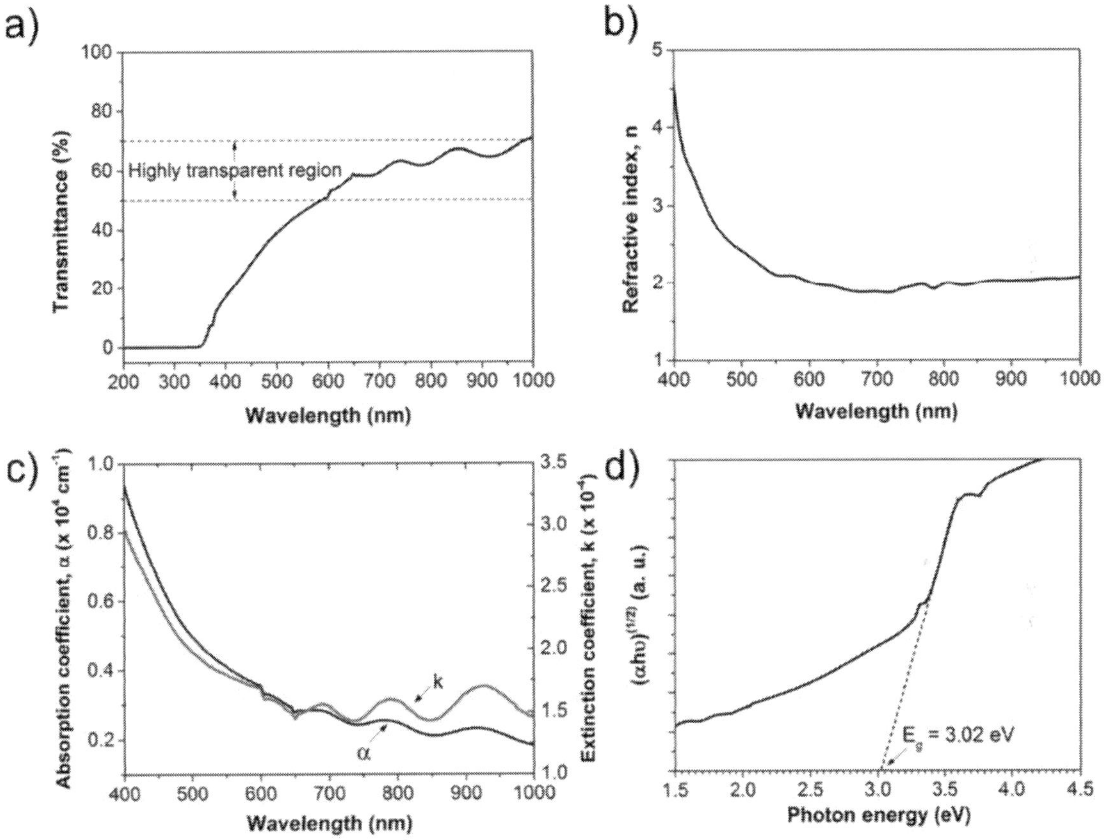

Figure 2. (a) Transmittance spectrum, (b) refractive index, (c) absorption and extinction coefficient as a function of wavelengths, (d) Tauc plot for the indirect allowed transition of the N-doped TiO_2 thin film.

Based on the envelope pattern in Figure 2(a), the transmittance at maxima, T_M and minima, T_m, the refractive index, n can be determined by Equation (1), where s is the refractive index of the glass substrate.

$$n=[N+(N^2-s^2)^{\frac{1}{2}}]^{\frac{1}{2}} \quad (1)$$

Where $N=2s\dfrac{(T_M-T_m)}{T_M T_m}+\dfrac{(s^2+1)}{2}$ and $s=\dfrac{1}{T_s}+\left(\dfrac{1}{T_s^2}+1\right)^{\frac{1}{2}}$, T_s is the transmittance of the substrate.

The thickness of the N-doped TiO_2 thin film, d evaluated through Equation (2) is found to be ≈ 1.9 μm thick. As a constant laminar flow was applied, the growth rate is deduced to be ≈ 30 nm/s.

$$d = \frac{\lambda_1 \lambda_2}{2(\lambda_1 n_2 - \lambda_2 n_1)} \tag{2}$$

Where n_1 and n_2 are refractive indices at two adjacent maxima (or minima) at λ_1 and λ_2, respectively. With the transmittance of the thin film, T_{film}, the absorption coefficient, α and extinction coefficient, k can be then obtained by Equation (3) and (4). The variations of n, α and k with wavelengths are plotted in Figure 2 (b)-(c).

$$\alpha = -\frac{1}{d}\ln T_{film} \tag{3}$$

$$k = \frac{\alpha\lambda}{4\pi} \tag{4}$$

Despite the thickness of the film was found to be nearly ≈ 1.9 μm, the semi-transparent film presents 50 – 70% of transmittance in the visible and near-infrared spectral region, shown in Figure 2(a). The refractive index changes from ≈ 4 to ≈ 2 with the increase of wavelength from 400 to 1000 nm in Figure 2(b). In particular, at 550 and 632 nm, the n is 2.1 and 1.96, respectively. These values are lower than several works reporting 2.3 to 2.5 [14,16]. The lower refractive index can be related to the deposition process, where only ≈ 20% of O_2 was presence in the air which was served as the O_2 precursor. This O_2 amount is significantly lower than the other works which employed high O_2 content precursor. This is in line with the observation by Stamate et al. [14], that the refractive index of TiO_2 films deposited by direct current magnetron sputtering is influenced by the gas flow rate consisted of Ar and O_2. At lower flow rate (lower O_2 content), the refractive index was recorded at 2.0 compared to 2.3 at a higher flow rate (higher O_2 content) [14]. Furthermore, the deposition was carried out in atmospheric pressure at temperature lower than 500°C with a total deposition time of 60 s. Such rapid growth process does not allow sufficient time and energy to break the oxygen O–O bond, compare to low-pressure deposition. While Ghamsari et. al [16] has also reported a refractive index <1.96 at 550 nm for their TiO_2 thin film fabricated through sol-gel method, it is obvious that the optical constants can be varied according to the fabrication process as well as the Ti and O_2 precursors. The calculated α and k values are in the order of 10^4 cm^{-1} and 10^{-4}, respectively, in the visible spectral region.

The α values in combination with Tauc plot, applying indirect allowed transition for anatase TiO_2, is used to evaluate the optical band gap of the thin film. The extrapolation of the linear portion of the curve to the energy axis in Figure 2(d) shows that the optical band gap energy for N-doped TiO_2 film is 3.02 eV. In the literature, the indirect band gap of undoped TiO_2 is reported as ≥3.2 eV [13]. It is apparent that the N doping has resulted in a reduced bandgap of ≈5% and has shifted the absorption edge towards the visible spectral region, which potentially absorbs more photons from the solar spectrum for enhanced photocatalytic activities.

3.3. *UV sensing response*

To evaluate the photo-electrical properties and demonstrate the practicality of the prepared N-doped TiO_2 thin film, The film was tested for its UV sensing response. The film has shown positive response towards UV illumination and the response under 8 V bias is depicted in Figure 3(a). Upon UV illumination, the film reacted almost instantaneously with a rise of current. However, when the illumination is removed, the film requires a longer duration to return to the base current. This slower recovery response is caused by the presence of carrier depletion layer at the film surface due to surface trap states [22,23].

Here, the Pt/N-doped TiO_2 interface forms a Schottky contact, due to the larger work function of Pt (≈5.65 eV) [7,23] compare to TiO_2 (≈5.1 eV) [24]. For larger work function, more electrons will be collected from N-doped TiO_2 to Pt upon UV illumination [23]. The UV sensing mechanism for MSM

N-doped TiO_2 follows that, the O_2 molecules capture free electrons from n-type TiO_2 (with excess electrons) and being adsorbed into the film, given in equation (5) [6,15].

$$O_2(g) + e^- \rightarrow O_2^-(ad) \tag{5}$$

Upon UV illumination, the photons excite the electrons in the valence band to conduction band and generate electron-hole pairs. The photogenerated electron-hole pairs are separated by the applied electric field in the depletion region, and drift to their respective regions [25]. The hole trapping states at the Pt/N-doped TiO_2 interface are partially occupied by the photogenerated holes, and a positively charged region is created. At the Pt interface, negative charges are induced to counter the positive charges at the TiO_2 interface. Thereby, a local electric field is produced to lower the Schottky barrier, shown in Figure 3(b). The photogenerated holes also drift to the film surface and discharge the adsorbed O_2^- ions, and release the O_2 from the surface, according to Equation (6) [6,15].

$$h^+ + O_2^-(ad) \rightarrow O_2(g) \tag{6}$$

The series of O_2 related hole-trap states prevent charge carrier recombination, and extend the charge carrier lifetime for higher photoresponse. In turn, the longer life time increases the recovery time when UV illumination is removed [26], observed in Figure 3(a).

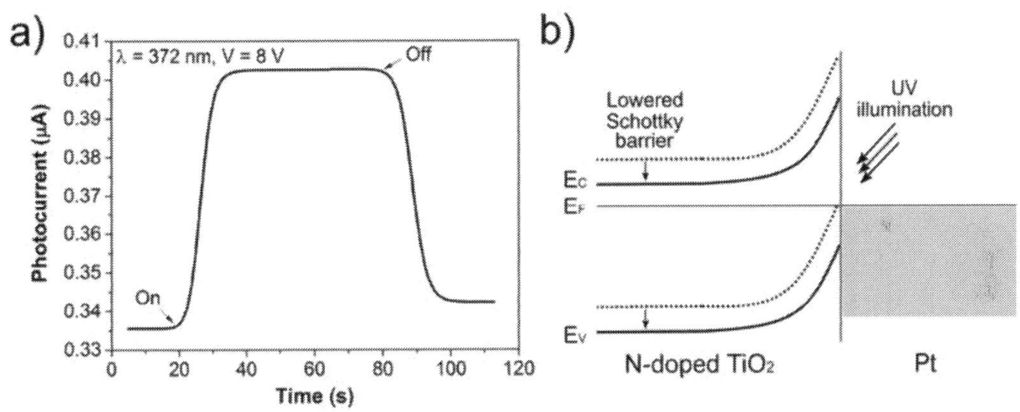

Figure 3. (a) UV detection response of N-doped TiO_2 thin film biased at 8 V (b) Schematic energy band diagram of Pt/N-doped TiO_2 upon UV illumination.

4. Conclusion

Semi-transparent yellowish surface of the as-synthesized film suggests that N-doped TiO_2 thin film was successfully grown on glass substrate by a cold-wall CVD setup. The as-deposited film presents polycrystalline anatase phases which were preferred for UV sensing purpose. The optical transmittance in the visible spectral region was 50 – 70% and the transmittance spectrum is utilized to evaluate the film thickness and optical constants based on Swanepoel's envelop method. The film thickness was 1.9 μm, as a result of 30 nm/s growth rate. At 550 nm, the refractive index, absorption and extinction coefficient were 2.1, 0.4 x 10^{-4} cm^{-1} and 2.0 x 10^{-4}, respectively. Through N doping, the optical band gap of anatase TiO_2 was reduced to 3.02 eV. A Pt/N-doped TiO_2 film was applied as MSM UV photodetector and had given a positive UV detection response.

Acknowledgments

This work was supported by Universiti Sains Malaysia Research University Grant Scheme 1001/PFIZIK/8011005. The first author is grateful for the MyPhD scholarship awarded by Ministry of Education Malaysia.

References

[1] Monroy E, Omnes F and Calle F 2003 *Semicond. Sci. Technol.* 18 R33
[2] Bintsis T, Litopoulou-Tzanetaki E and Robinson R K 2000 *J. Sci. Food Agric.* 80 637
[3] Sang L, Liao M and Sumiya M 2013 *Sensors* 13 10482
[4] Gallagher R P and Lee T K 2006 *Prog. Biophys. Mol. Biol.* 92 119
[5] Lee W J and Hon M H 2011 *Appl. Phys. Lett.* 99 2011
[6] Liu H-Y and Liu G-J 2017 *IEEE Trans. Electron Devices* 64 1108
[7] Huang H, Xie Y, Yang W, Zhang F, Cai J and Wu Z 2011 *IEEE Electron Device Lett.* 32 530
[8] Fujishima A and Honda K 1972 *Nature* 238 37
[9] Frank S N and Bard A J 1977 *J. Am. Chem. Soc.* 99 303
[10] Vinodgopal K, Hotchandani S and Kamat P V 1993 *J. Phys. Chem.* 97 9040
[11] O'Regan B and Gratzel M 1991 *Nature* 353 737
[12] Chen X and Mao S S 2007 *Chem. Rev.* 107 2891
[13] Ge M, Cao C, Huang J, Li S, Chen Z, Zhang K-Q, Al-deyab S S and Lai Y 2016 *J. Mater. Chem. A* 4 6772
[14] Stamate M, Vascan I, Lazar I, Lazar G, Caraman I and Caraman M 2005 *J. Optoelectron. Adv. Mater.* 7 771
[15] Xie Y, Huang H, Yang W and Wu Z 2011 *J. Appl. Phys.* 109 23114
[16] Sasani Ghamsari M and Bahramian A R 2008 *Mater. Lett.* 62 361
[17] Liu S, Chen X and Chen X 2006 *Chinese J. Catal.* 27 697
[18] Kuo D-H and Shueh C-N 2003 *Chem. Vap. Depos.* 9 265
[19] Dunnill C W, Kafizas A and Parkin I P 2012 *Chem. Vap. Depos.* 18 89
[20] Dunnill C W H, Aiken Z A, Pratten J, Wilson M, Morgan D J and Parkin I P 2009 *J. Photochem. Photobiol. A Chem.* 207 244
[21] Swanepoel R 1983 *J. Phys. E.* 16 1214
[22] Xie Y, Wei L, Wei G, Li Q, Wang D, Chen Y, Yan S, Liu G, Mei L and Jiao J 2013 *Nanoscale Res. Lett.* 8 188
[23] Tsai T Y, Chang S J, Weng W Y, Wang S H, Chiu C J, Hsu C L and Hsueh T J 2012 *IEEE Photonics Technol. Lett.* 24 1584
[24] Sarkar A, Singh A K, Khan G G, Sarkar D and Mandal K 2014 *RSC Adv.* 4 55629
[25] Patel D B, Chauhan K R, Park S H and Kim J 2017 *Mater. Sci. Semicond. Process.* 64 137
[26] Liu H Y, Lin W H, Sun W C, Wei S Y and Yu S M 2017 *Mater. Sci. Semicond. Process.* 57 90

IOP Conf. Series: Journal of Physics: Conf. Series **1083** (2018) 012026 doi:10.1088/1742-6596/1083/1/012026

Impedance Study on NiO-$BaCe_{0.54}Zr_{0.36}Y_{0.1}O_{2.95}$ Composite Anode for Proton-Conducting Fuel Cell.

S M Senari[1], N Osman[2] and A M M Jani[3]

[1] Faculty of Applied Sciences, Universiti Teknologi MARA, 40450 Shah Alam, Selangor, Malaysia
[2] Faculty of Applied Sciences, Universiti Teknologi MARA, 02600 Arau, Perlis, Malaysia
[3] Faculty of Applied Sciences, Universiti Teknologi MARA, 35400 Tapah, Perak, Malaysia

fisha@perlis.uitm.edu.my

Abstract. A composite anode of NiO-$BaCe_{0.54}Zr_{0.36}Y_{0.1}O_{2.95}$ (NiO-BCZY) has been proposed as a suitable anode material for the application in a proton-conducting fuel cell (PCFC). A single anode supported cell with the structure of NiO-BCZY|BCZY|$La_{0.6}Sr_{0.4}Co_{0.2}Fe_{0.8}O_{3-\delta}$ (LSCF) was fabricated by a dry-pressing assisted with spin-coating technique. The cell was sintered in air at 1400°C for 3 hours with a heating rate of 10°C/min. Complex impedance measurements were conducted in H_2:N_2 (10%:90%) as the fuel and air as the oxidant after the cell was reduced in H_2:N_2 (10%:90%) for 7 hours. The typical impedance pattern of the single cell was resolved using an equivalent circuit consisting two parallel RQ elements connecting in series. The cell showed an overall resistance of 17.34 to 69.53.Ω cm^2 at a temperature ranging from 550 and 700°C. The SEM cross-sectional view of the single cell after impedance measurement showed a good adhesion between the three layers with characteristic dense electrolyte layer and a unique porous electrode. The experimental results indicated that NiO-$BaCe_{0.54}Zr_{0.36}Y_{0.1}O_{2.95}$ is a suitable candidate for anode material for the PCFC application.

1. Introduction

During past several years, solid oxide fuel cell (SOFC) becomes one of the clean energy generation systems that give a great impact to the research world. It can directly convert the chemical energy from fuel such as hydrogen and ethanol to the electrical energy effectively. Besides, it is one type of fuel cell which consists of a solid electrolyte and offers high energy conversion with minimum emissions, simple and environmental safety. The net conversion efficiencies reached as high as 70% had been recognized in the SOFC/gas turbine hybrid system which offers efficient electricity generation [1]. Unfortunately, when it comes to the commercial application, there are still some limitations such as expensive system, high operating temperature, poor stability and easily sulfur-poisoned in the longtime run [2,3]. Nowadays, the urgent demand for board commercialization focused on reducing the working temperature to intermediate/lower temperatures which can benefit both fabrication and stability.

Proton conducting fuel cell (PCFC) has a unique property of proton conductor from the anode to the cathode, while producing water at the cathode side which avoids fuel dilution problems compared to the conventional oxygen-ion conducting electrolytes [4]. This factor extends practical applications and leads to an important increase in overall cell efficiency. In addition, PCFC potentially lowering the cost and enhancing the durability and reliability of SOFC systems due to the lower activation

Content from this work may be used under the terms of the Creative Commons Attribution 3.0 licence. Any further distribution of this work must maintain attribution to the author(s) and the title of the work, journal citation and DOI.
Published under licence by IOP Publishing Ltd

energy and relatively high proton conductivity at intermediate temperatures attainable from proton conducting oxides enables reduced operating temperature PCFC.

Among the different types of solid-state proton conductors, the most extensively studied materials are simple perovskite structured ABO_3-based proton conductors. A stable and effective perovskite type barium cerate with co-doping of yttrium and zirconium (BCZY) was developed by Zuo et al. Its exhibited both adequate proton conductivity as well as sufficient chemical and thermal stability over a wide range of conditions relevant to fuel cell operation [5]. $BaCe_{0.54}Zr_{0.36}Y_{0.1}O_{2.95}$ has been proven for good electrolyte and showed the best performance among other composition of BCZY [6] and the most promising anode for PCFCs is Ni-BCZY [7]. Nickel is still the best choice for PCFC anode due to its high electrical conductivity, high catalytic activities and low cost. In this work, the combination of Ni and $BaCe_{0.54}Zr_{0.36}Y_{0.1}O_{2.95}$ based material to form composite anode of NiO-BCZY is expected to enhance the fuel oxidation at the PCFC anode. In all cases, a thin electrolyte layer is needed to reduce the resistance losses and to yield improved cell performance at reduced operating temperatures. So, in order to get a thin layer of electrolyte, anode-supported cells fabricated by a dry-pressing assisted spin coating technique have been chosen. In this study, the microstructure and electrochemical performance of the cell were examined.

2. Experimental

2.1 Material Preparation and Powder Synthesis.

Precursors used for the electrolyte, anode and cathode were metal nitrate salts. Powder samples of $BaCe_{0.54}Zr_{0.36}Y_{0.1}O_{2.95}$ (BCZY) compound were prepared by a sol-gel method as reported elsewhere [8]. Pure BCZY compound was finally obtained after calcination at 1100°C for 10 h in air. A stoichiometric amounts of $Ba(NO_3)_2$ (99%, ACROS), $Ce(NO_3)_3.6H_2O$ (99.5%, ACROS), $Zr(NO_3)_2O.xH_2O$ (99.5%, ACROS), $Y(NO_3)_3.5H_2O$ (99.9%, Aldrich) and $Ni(NO_3)_2.6H_2O$ (99%, ACROS) were used as starting materials for anode preparation. The nitrate salts were dissolved in deionized water. As a complexing agent, citric acid was then added to the solution with molar ratios of soluble metal ions: citric acid of 1:1.5. Ammonium hydroxide, NH_4OH (25%, HmbG) was added to the solution to adjust the pH value around 7. The resulting solution was slowly evaporated on a hot plate at 120°C. During the process, the browning gas (known as NOx) was released and a dark brown gel obtained. The gel was dried at 325°C in a furnace to yield black flakes. Then the sample was calcined at 1100°C with heating a rate of 10°C min^{-1} for 6 h to produce NiO-BCZY powders. LSCF cathode powder was synthesized by a sol-gel method as well [9].

2.2 Fabrication of Single Cell.

A button cell with a configuration of porous NiO-BCZY anode supported |BCZY electrolyte| LSCF cathode was fabricated by a dry-pressing assisted spin-coating technique. The NiO-BCZY powder was formed into disc pellet with a diameter of 13mm at a pressure 74 MPa. Then loose BCZY powder was uniformly distributed onto the anode substrate, co-pressed at 221 MPa and co-sintered at two-step sintering profile to obtain a dense BCZY electrolyte layer. Fine LSCF powder was mixed thoroughly together with an ethylcellulose-terpineol binder, to prepare the cathode slurry. The cathode slurry was then spin-coated on the BCZY electrolyte layer and was sintered at 950°C for 2 h in an air to form a tri-layer cell NiO-BCZY|BCZY|LSCF. Finally, Pt paste was painted onto the cathode layer as a current collector.

2.3 Characterization and Electrochemical Measurement.

The microstructure of the anode-supported single cell was investigated by a benchtop scanning electron microscope. The electrochemical impedance of the single cell was measured from 500-700°C at 50°C intervals by a lab-designed electrochemical system, with H_2:N_2 (10%:90%) as the fuel and stagnant air as the oxidant after the cell were reduced in H_2:N_2 (10%:90%) for 7 hours. The electrochemical impedance spectra were measured by ZIVE SP2 Electrochemical Workstation (ZIVELAB WonATech) connected to a personal computer. ZIVE® Smart Manager™ software was used for data acquisitions and analysis. The impedance spectra were obtained in the frequency range of 1 MHz to 0.1 Hz with ten steps per decade and signal amplitude of 10mV. The impedance spectra obtained were analyzed using ZMAN software.

3. Results and Discussion

Figure 1 presents the typical scanning electron microscopy (SEM) images of (a) cross-section view of the single cell and (b) higher magnification image of an interface of anode and electrolyte. Good adhesion is obtained between the three layers of cathode, electrolyte and anode. A dense electrolyte is observed in Figure 1(d), indicating that dry-pressing is as an effective way to fabricate dense electrolyte layer. Both the anode and cathode have the porous microstructure and no delaminations are found at the interfaces between electrodes and electrolyte.

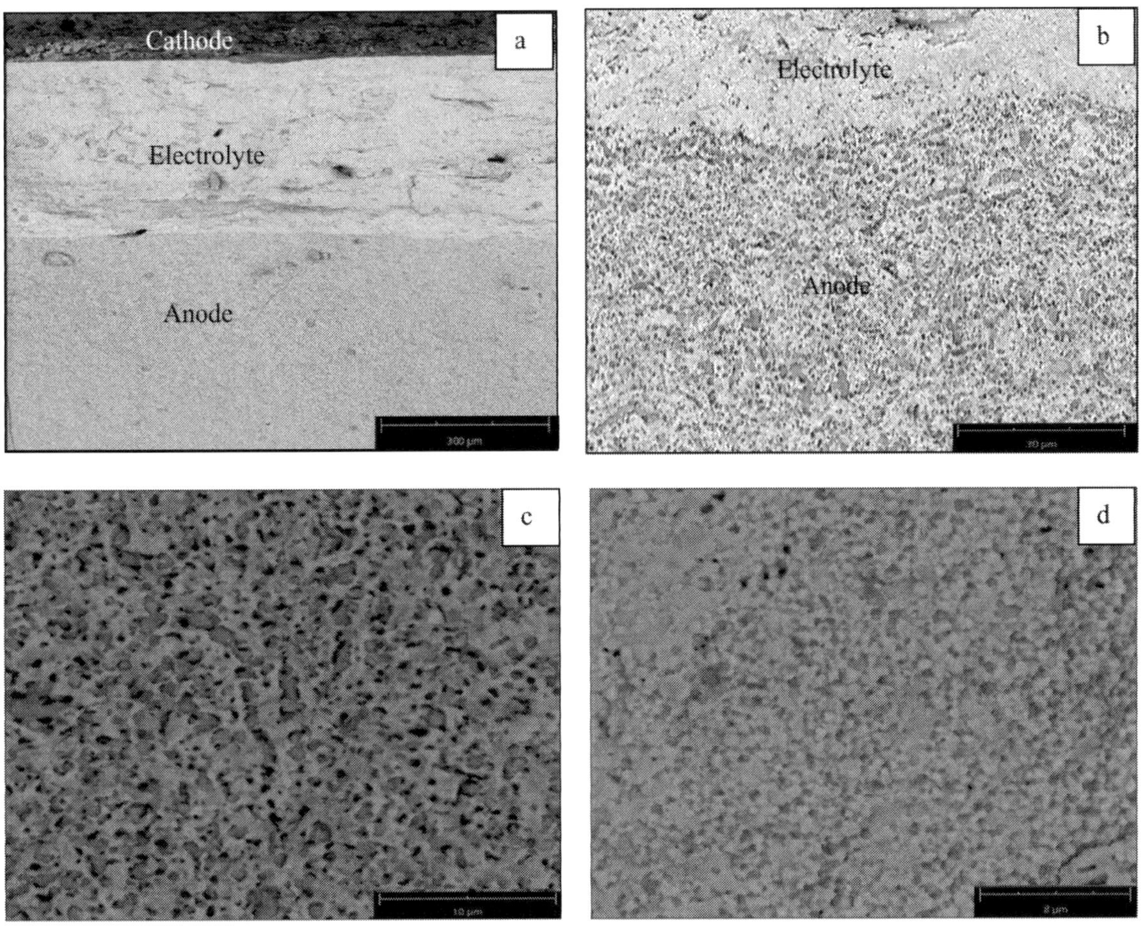

Figure 1. SEM image of (a) cross-section view of single cell, (b) higher magnification image of interface of anode and electrolyte, (c) anode surface and (d) electrolyte surface.

Electrochemical impedance analysis was performed for the anode-supported single cell by the lab-designed electrochemical system. Figure 2 shows the impedance spectra of the anode-supported single cell and the equivalent circuit (inset) used to separate all the responses. The typical impedance pattern of the anode-supported single cell was resolved using an equivalent circuit consisting two parallel RQ elements connecting in series. The intercept with the real axis at high frequencies represents the resistance of electrolyte (ohmic resistance), whereas the difference between the high-frequency and low-frequency intercepts with the real axis represents the sum of the electrode polarization resistances of the cell. The total cell resistances which represent the polarization and ohmic resistances as a function of the temperature were shown in Figure 3. Both the polarization and ohmic resistances decrease as the temperature increase, while the polarization resistances decrease more rapidly, typically from 56.00 Ω cm^2 at 550°C to 7.77 Ω cm^2 at 700°C. The values of ohmic resistances for the anode-supported single cell at 550, 600, 650 and 700°C were 13.53, 12.01, 10.95 and 9.57 Ωcm^2 respectively. It should be noted that the values of polarization resistances and ohmic resistances are higher than values in previously reported work [10,11]. The high ohmic resistance might be contributed from the different fabrication technique as well as the thickness of the electrolyte layer. The thicker electrolyte layer gives a higher ohmic resistance. These results of fuel cell tests could be enhanced by further investigation on the thickness of electrolyte and the fabrication technique. Anyhow, these results confirmed the functionality of proton conducting cermet anodes NiO–BCZY.

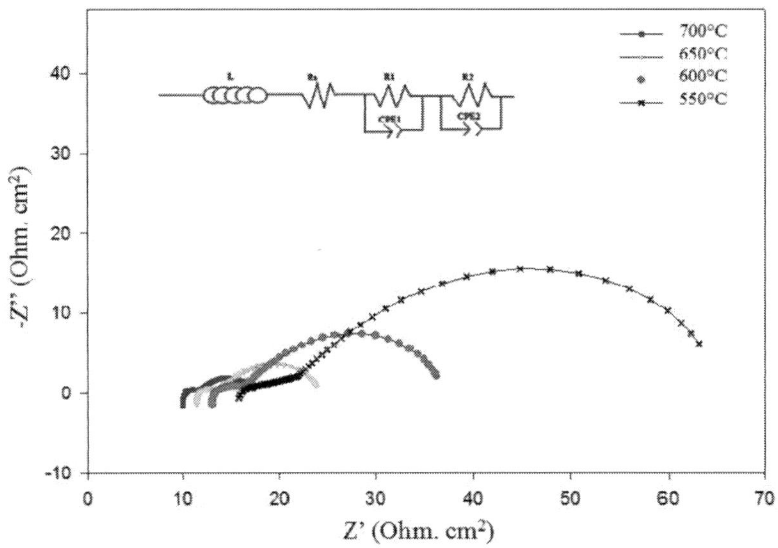

Figure 2. Impedance spectra of the anode-supported single cell under open circuit condition at 550-700°C.

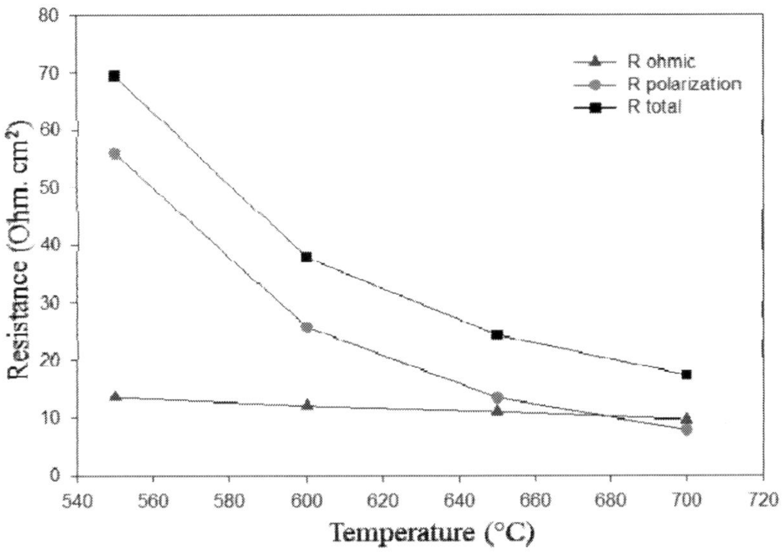

Figure 3. The total cell resistances, polarization resistances, and ohmic resistances (Rohm) obtained from the impedance spectra at different temperatures.

4. Conclusion

In this study, NiO-BCZY composite was fabricated and applied as an anode in PCFC with BCZY as the electrolyte. With a structure of NiO-BCZY|BCZY|LSCF, an excellent adhesion was obtained between the three layers with individual dense electrolyte layer and a unique porous electrode. The typical impedance pattern of the single cell was resolved using an equivalent circuit consisting two parallel RQ elements connecting in series. An increase in operating temperature from 550 to 700°C, results in the polarization resistance of the single cell decreasing from 56.00 to 7.77 Ωcm^2. The results indicate the NiO-BCZY composite to be a promising candidate anode for PCFC.

Acknowledgements

The authors would like to thank the Minister of Higher Education for the FRGS 1/2017 funding and Universiti Teknologi MARA for the facilities support.

References

[1] Shao Z P and Haile S M 2004 A High-Performance Cathode for The Next Generation of Solid Oxide Fuel Cells, *Nat.* 431 170-3
[2] Lee J G, Park J H and Shul Y G 2014 Tailoring Gadolinium-Doped Ceria-Based Solid Oxide Fuel Cells to Achieve 2 W cm^{-2} at 550 °C, *Nat. Commun.* 5 4045
[3] Steele B C H and Heinzel A 2001 Materials for Fuel-Cell Technologies, *Nature* 414 345–352
[4] Wang S, Zhang L, Yang Z, Fang S, Brinkman K and Chen F 2012 Two-step Co-Sintering Method to Fabricate Anode-Supported $Ba_3Ca_{1.18}Nb_{1.82}O_{9-\delta}$ Proton-Conducting Solid Oxide Fuel Cell, *J. Power Sources* 215 221-6
[5] Zuo C, Lee T H, Dorris S E, Balachandran U, and Liu M 2006 Composite Ni-$Ba(Zr_{0.1}Ce_{0.7}Y_{0.2})O_3$ Membrane for Hydrogen Separation. *J. Power Sources* 159(2) 1291–5
[6] Othman N W, Osman N, Ramli A and Safian D S 2016 Microstructure and Electrical Characterization of $Ba(Ce,Zr)O_3$ Solid Solution prepared by a TSS Method *J.Solid St Sci. & Technol. Lett.* 17(1) 61-6

[7] Lim D K, Kim J H, Chavan A U, Lee T R, Yoo Y S and Song S J 2016 Performance of Proton-conducting Ceramic-electrolyte Fuel Cell with BZCY40 electrolyte and BSCF5582 cathode *Ceram. Int.* 42 (3) 3776–3785

[8] Abdullah N A, Osman N, Hasan S and Hassan O H 2012 Chelating Agents Role on Thermal Characteristics and Phase Formation of Modified Cerate-Zirconate via Sol-gel Synthesis Route, *Int. J. Electrochem. Sci.* 7 9401–9

[9] Ismail I, Osman N and Md Jani A.M 2016 Tailoring The Microstructure of $La_{0.6}Sr_{0.4}Co_{0.2}Fe_{0.8}O_{3-\alpha}$ Cathode Material: The Role of Dispersing Agent. *J. Sol-Gel Sci. Technol.* 1-8

[10] Liu M, Gao J, Liu X and Meng G 2011. High Performance of Anode Supported $BaZr_{0.1}Ce_{0.7}Y_{0.2}O_{3-\delta}$ (BCZY) Electrolyte Cell for IT-SOFC. *Int. J. Hydrog. Energy.* 36 13741-5

[11] Yuan R H, He W, Zhang C, Ni M and Leung M K H 2017 Cobalt Free $SrFe_{0.95}Nb_{0.05}O_{3-\delta}$ Cathode Material for Proton-Conducting Solid Oxide Fuel Cells with $BaZr_{0.1}Ce_{0.7}Y_{0.2}O_{3-\delta}$ Electrolyte *Mater. Lett.* 200 75-8

IOP Conf. Series: Journal of Physics: Conf. Series **1083** (2018) 012027 doi:10.1088/1742-6596/1083/1/012027

Electrical Conductivity of Y^{3+}-doped $Ba(Ce,Zr)O_3$ in Wet Nitrogen Atmosphere

N S M Affandi[1], N Osman[2] and O H Hassan[3]

[1]Faculty of Applied Sciences Universiti Teknologi MARA, 40450 Shah Alam, Selangor, Malaysia
[2]Physics Department, Faculty of Applied Sciences Universiti Teknologi MARA Perlis, 02600 Arau, Perlis, Malaysia
[3]Faculty of Art & Design Universiti Teknologi MARA, 40450 Shah Alam, Selangor, Malaysia

fisha@perlis.uitm.edu.my

Abstract. Proton conducting ceramics such as barium cerate-zirconate oxide, $Ba(Ce,Zr)O_3$ is one of promising candidates to be used as an electrolyte material in intermediate temperature solid oxide fuel cell (IT-SOFC). In this work, single-phase of $BaCe_{0.54}Zr_{0.36}Y_{0.1}O_{2.95}$ (BCZY) powder was successfully prepared using a sol-gel method. The powder was pressed into a pellet with diameter, d=13 mm and thickness, t= ~1–2 mm using an uniaxial pressing technique with pressure of 369 MPa for 5 minutes. The pellet with relative density of 92% was obtained after underwent two-step sintering (TSS) method at T_1=1500°C and T_2=1450°C with holding time 1 minute and 12 hours respectively. Impedance data of the sample with grain in sub-micron size (as measured by SEM) was collected using an impedance spectroscopy (IS) in wet nitrogen atmosphere at temperature ranging from 500 to 800°C. All responses in the impedance spectrum were separated and fitted accordingly to an equivalent circuit based on a brick-layer model (BLM) with equivalent circuit of $R_1|Q_1–R_2|Q2$. Analysis of the impedance spectrum showed the sub-micron grain resulted in high grain boundary resistance with conductivity of 10^{-4} S cm^{-1} at 500 °C to 700 °C. The conductivity of the sample was thermally activated and obeyed Arrhenius behavior with activation energy (E_a) of 0.76 eV.

1. Introduction

Proton conductor fuel cell (PCFC) is a device that converts chemical energy into electrical energy which hold benefit over oxide-ion conductor fuel cell in terms of operating temperatures. It has been widely studied due to its benefits in intermediate temperature SOFC (IT-SOFC). Among all of the components in a single cell, electrolyte materials have been extensively investigated owing to its major contribution to the cell performance. Currently, $BaCeO_3$ and $BaZrO_3$ are the most investigated pair which is considered as the state of the art in proton-conducting oxide ceramic materials. $BaCeO_3$ alone had shown many advantages in term of conductivity, however, it lack in terms of stability. It has relatively low chemical stability in water and CO_2 atmospheres. Whereas $BaZrO_3$ shows good chemical stability but low total conductivity [1]. A work by Baral et al. [2] revealed that in CO_2 or wet operating conditions, barium cerates are unstable but conversely for barium zirconates which show good stability. Yet, it lack in terms of conductivity due to very high grain boundary resistance.

Content from this work may be used under the terms of the Creative Commons Attribution 3.0 licence. Any further distribution of this work must maintain attribution to the author(s) and the title of the work, journal citation and DOI.
Published under licence by IOP Publishing Ltd

Hence, researchers focused to one type of ceramic material in perovskite-oxide family, Y-doped $Ba(Ce,Zr)O_3$ which has been proposed as a good candidate for electrolyte component in intermediate temperature proton conductor fuel cell (PCFC). The material shows promising results at intermediate operating temperatures based on electrochemical and microstructure analysis [3-5]. Among all composition of $BaCe_{1-x-y}Zr_xY_yO_{3-\delta}$, the $Ba(Ce_{0.6}Zr_{0.4})_{0.9}Y_{0.1}O_{3-\delta}$ (BCZY) was chosen in this work due to its composition almost similar to that reported by J. F. Basbus et al [6]. They investigated the conductivity of $BaCe_{0.4}Zr_{0.4}Y_{0.1}$ which was $1x10^{-3}$ Scm^{-1}at temperature of 300 °C in wet 20% O_2/Ar atmosphere. However, no details impedance analysis was performed under wet N_2 condition on the mentioned BCZY composition, which would be the focus of this paper.

2. Experimental

Raw materials of $Ba(NO_3)_2$ (99%,ACROS), $Ce(NO3)_3$ (99.5%,ACROS), $Zr(NO_3)_2O.xH_2O$ (99.5%,ACROS) and $Y(NO_3)_{3.5}H_2O$ (99.9%,ACROS) were used to prepare BCZY powders. The starting materials of BCZY powder are weighed in stoichiometric amount accordingly and synthesised using a sol-gel process. A detailed preparation of the single-phase BCZY powder has been reported elsewhere [7]. The powder was pressed into a pellet of 13 mm in diameter and thickness about 1–2 mm using uniaxial pressing technique at 369 MPa for 5 minutes. The pellet was subjected to two-step sintering (TSS) process at 1500 °C for 1 minute and 1450 °C for 12 hours with respective heating rates of 5 $°Cmin^{-1}$ and 30 $°Cmin^{-1}$.

Geometrical method was used to measure density for each of the sintered pellet. The grain morphology of the pellet was observed using Benchtop SEM (Phenom XL) with 9000x magnification at 15kV. Pt paste was manually painted onto both side of the pellet and heated at 950°C for 2 hours to get a good contact between the Pt metal and BCZY electrolyte. Electrical measurement of the sample was completed using AC impedance spectroscopy in the frequency range between 42 Hz-1 MHz with an amplitude voltage of 500 mV. Impedance data was collected in wet N_2 atmosphere by using HIOKI 3532-5O LCR HiTESTER impedance analyser. Prior to the measurement, the pellet was stabilized for 24 hours in wet N_2 flow at 700 °C before going up to 800 °C and gradually step down for 50 °C for each temperature measurement. The impedance spectra obtained were analysed to extract the possible electrical responses produced by the sample using ZMAN software.

3. Result and Discussion

Figure 1 shows the micrographs of (a) polished surface area and (b) cross-section for BCZY pellet. The SEM image confirmed the dense microstructure of pellet at density of 92% and the grain was a mixture between small and large grains with average grain size of 274 nm as measured by ImageJ software. TSS method as reported by N. W. Othman et al. [8] proved to help the sample to achieve high density with small grain size. Furthermore, according to Abdullah S. S. C et al. [9] grain growth was controllable while the density of the pellet can be increased by executing TSS method. Therefore, it is important to attain high density which subsequently hardness is greatly dependent on the density of the pellet.

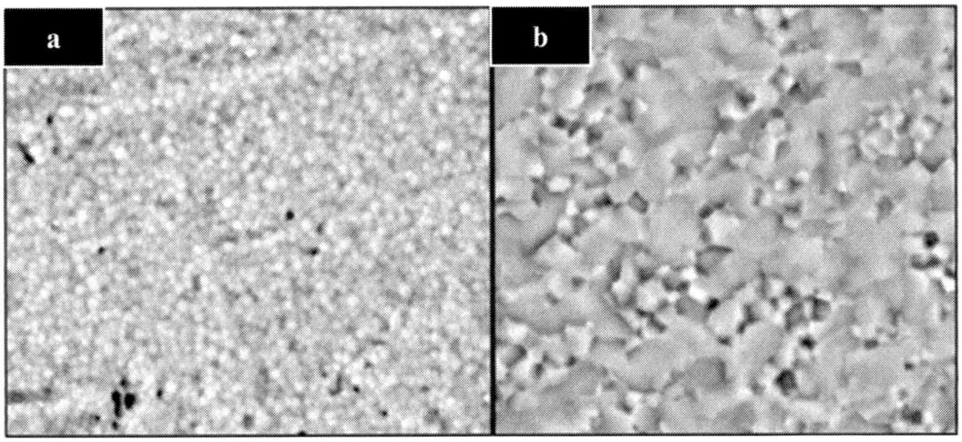

Figure 1. SEM micrographs of a) surface area of a polished $Ba(Ce_{0.6}Zr_{0.4})_{0.9}Y_{0.1}O_{3-\delta}$ pellet and b) cross section of a fractured BCZY pellet after TSS method.

Comparative Nyquist plot impedance data of the sample at 500 °C, 600 °C and 700 °C in wet N_2 atmospheres is shown in Figure 2. For all temperatures, there two arcs were observed. The first arc can be attributed to grain boundary (R_2) resistance in mid-frequency range while the second arc represent the interfacial resistance (R_3) in low-frequency range. The intercept value at Z' axis was taken as the grain resistance (R_1). The obtained results were aligned from those previously reported [10].

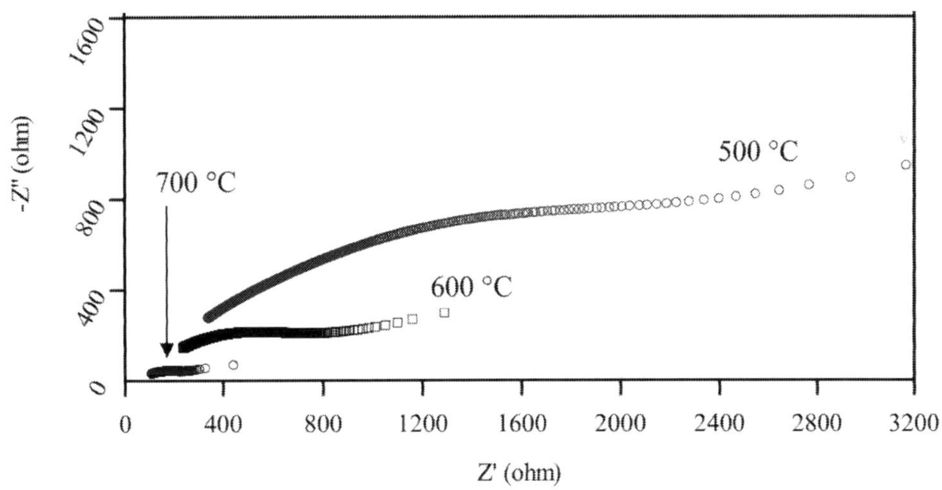

Figure 2. Typical Nyquist plot for $Ba(Ce_{0.6}Zr_{0.4})_{0.9}Y_{0.1}O_{3-\delta}$ pellet at different measurement temperatures

Equivalent circuit fitting was done to stimulate the resistance, capacitive, and inductive element components associated with the ac impedance responses [11]. The impedance spectrum circuit comprise of resistance with two parallel pair of resistance-constant phase element connected in series as shown in Figure 3. From the equivalent circuit and fitting procedures, the capacitance values can be calculated using equation 1;

$$C=Y^{\frac{1}{n}} \times R^{\frac{1}{n}-1} . \tag{1}$$

Where C is the capacitance value (F), Y and n refers to the parameter associated with the constant phase element (CPE) while R represents as the resistance [12].

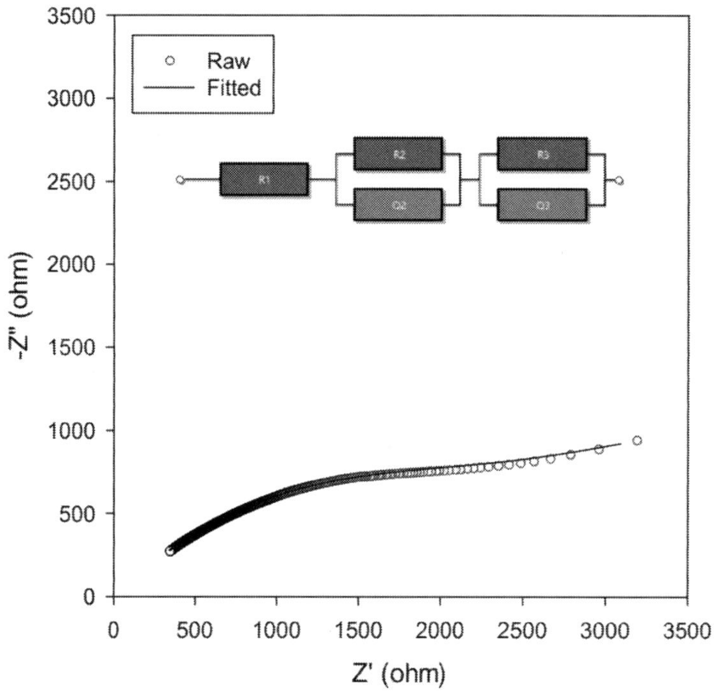

Figure 3. Impedance spectrum of experimental and simulation data were well fitted. (Insert is an equivalent circuit representing the responses in the spectrum of BCZY at 500 °C)

Table 1 shows the geometrical capacitance values for the total resistance response of $Ba(Ce_{0.6}Zr_{0.4})_{0.9}Y_{0.1}O_{3-\delta}$ at 500, 600 and 700 °C. The obtained capacitance values correlate with the usual range of grain boundary in order of 10^{-9} to 10^{-10} Farad at mid-frequency region [13].

Table 1. Capacitance range calculated from the values obtained from fitting procedure

Temperature (°C)	Capacitance, (F)
500	1.82×10^{-9}
600	2.06×10^{-9}
700	1.05×10^{-9}

The overall conductivity was calculated since the grain response was no longer possible at temperature above 300 °C due to incapability of impedance technique to measure the response as stated by Barison S. et al. [14]. The obtained conductivity was found to be 1.28×10^{-4} Scm^{-1} at 500 °C, 5.58×10^{-4} Scm^{-1} at 600 °C and 1.18×10^{-4} Scm^{-1} at 700 °C. The sample exhibit low conductivity values which can be explained by the small grain size owned by the BCZY sample, considering the grain boundary area were large and hence contributes to high total resistivity of the sample.

The Arrhenius graph of the total conductivity were plotted as shown in Figure 4. Activation energy, E_a that is the combination of energy of defect formation and energy of defect migration can be observed from the slop of Arrhenius plot [15] and was calculated to be 0.77 eV. Comparing to typical value of E_a for similar materials (0.78 eV) [16], that may be concluded it was due the occupied $Zr4^+$ and $Ce4^+$ ions in the perovskite B site has difference in electronegativity site and a higher repulsive interactions between Ce-H than Zr-H.

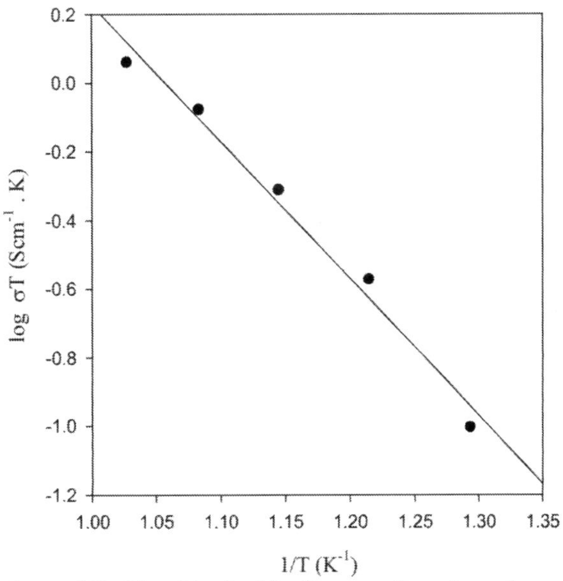

Figure 4. The σT values of $Ba(Ce_{0.6}Zr_{0.4})_{0.9}Y_{0.1}O_{3-\delta}$ as a function of temperature in wet nitrogen atmosphere.

4. Conclusion

The $Ba(Ce_{0.6}Zr_{0.4})_{0.9}Y_{0.1}O_{3-\delta}$ pellet with density of 92% and its grain in the sub-micron size was prepared via TSS method. Analysis of the impedance spectrum showed the sub-micron grain resulted in high grain boundary resistance. In wet N_2 atmosphere, the conductivity of the sample was found to be in the range of 10^{-4} S cm^{-1} at 500 °C to 700 °C with activation energy (E_a) of 0.76 eV.

Acknowledgement

This study is financially supported by the Ministry of Education (MoE) for the Grant Trans Displinary Research Grant 2016 (TRGS-2). Authors also thank the University Teknologi MARA for facilities and supports.

References

[1] Fabbri E, Epifanio A D, Bartolomeo E D, Licoccia S and Traversa E 2008 179 558–64.
[2] Baral A K, Choi S K, Byung K L and Jong-Ho L *J. Renew. Sustain. Energy* 1–9.
[3] Sawant P, Varma S, Wani B N and Bharadwaj S R 2012 *Int. J. Hydrog. Energy* 37(4) 3848–56.
[4] Nasani N, Ramasamy D, Mikhalev S, Kovalevsky A V and Fagg D P 2015 *J. Power Sources* 278 582–9.
[5] Hakim M, Joo J H, Yoo C Y, Kim B K and Yu J H 2015 *J. Eur. Ceram. Soc.* 35(6) 1855–63.
[6] Basbus J F, Arce M D, Prado F D, Caneiro A and Mogni L V 2016 *J. Power Sources* 329 262-7.
[7] Othman N W, Ramli A, Osman N, Abdullah S S C and Rahman M E A 2015 819(3) 129–33.
[8] Othman N W, Ramli A, Osman N, Abdullah S S C and Rahman M E A 2015 819(3) 129–33.
[9] Abdullah S S C, Abdullah R and Osman N 2014 962–6.
[10] Baral A K, Choi S K, Byung K L and Jong-Ho L 2014 *J. Renew. Sustain. Energy* 1–9.
[11] Li J, Luo J, Chuang K T and Sanger A R 2008 *Electrochimica Acta* 53 3701–7.
[12] Ismail I, Osman N and Jani A M M 2017*(Malaysia: iMIT-SIC Vol 2)* pp 177- 8
[13] Nasani N, Dias P A N, Saraiva J A and Fagg D P 2013 *Int. J. Hydrog. Energy* 38(20) 8461–70.
[14] Mortalo C, Barison S and Battagliarin M 2008 *J. Mater. Chem.* 5120–5128.
[15] Baral A K, Choi S K, Byung K L and Jong-Ho L 2014 *Mater Renew Sustain Energy* 1–9.
[16] Sawant P, Varma S, Wani B N and Bharadwaj S R 2012 37(4) 3848–56.

IOP Conf. Series: Journal of Physics: Conf. Series **1083** (2018) 012028 doi:10.1088/1742-6596/1083/1/012028

Effect of Calcium Concentration on the Microstructure and Electrical Properties of Ca-Doped PLSZT (Pb-La-Sr-Zr-Ti) System

F Yan Chin [1,a*], **T A B Tuan Nur Izzah** [1,b] and **M Hasmaliza** [1,c]

[1]School of Materials and Minerals Resources Engineering, Universiti Sains Malaysia, Engineering Campus, 14300 Nibong Tebal, Penang

[a]yanchin_fun@hotmail.com, [b]tnizzah@yahoo.com, [c]hasmaliza@usm.my

Abstract. Lanthanum and strontium modified Lead Zirconate Titanate (PLSZT) was doped with Ca^{2+} to improve the dielectric properties. Ca-doped PLSZT was prepared using high planetary ball mill, palletized and sintered at 1200 °C for 3 hours with heating rate 5 °C per minute. XRD results showed that all Ca-doped PLSZT samples consist of tetragonal and rhombohedral phase whereas SEM results showed that the grain size of the samples was decreased with an increase in calcium concentration but began to decrease at 8 mol% of calcium. The density was observed to decrease and the dielectric constant increased with the increase in concentration of calcium.

1. Introduction

The development of PZT materials is well established in electronic applications where numerous dopants have been added on the PZT systems to optimize their piezoelectric properties. In the last few decades, the synthesis of PZT modified with various dopants including soft isovalent doping had been extensively investigated for various applications [1,2]. Lead modified by isovalent cations such as Ba^{2+}, Sr^{2+} and Ca^{2+} shows undisturbed crystal symmetry which is responsible for the strong ferroelectric behavior of the samples [3].

Kalem et al. [4] have reported that the addition of Sr and La as co-dopants into PZT ceramics is able to stabilize the tetragonal phase. The incorporation of Sr^{2+} and La^{3+} into the PZT lattice has intense implications on the dielectric and piezoelectric properties of PZT ceramics. Bahanurddin et al. have doped PZT with Sr^{2+} and La^{3+} via high energy planetary ball mill process [5, 6]. They found that the evaporation of PbO during calcination and sintering processes can be avoided by virtue of its ability to skip the calcinations step. However, the piezoelectric properties were not mentioned by them. Meanwhile, Kour et al. [7] had studied the Ca^{2+} doped PZT and claimed that the dielectric constant increases with the increase in calcium concentration due to the increase in electronic polarization.

Therefore, calcium substituted PLSZT (Pb-La-Sr-Zr-Ti) for various dielectric applications have been studied. In this research, calcium was doped to "A" site of $Pb_{0.93}La_{0.02}Sr_{0.05}(Zr_{0.52}Ti_{0.48})O_3$ (PLSZT) to enhance its electrical properties. The influence of calcium concentration on the microstructure and dielectric properties of Ca-doped PLSZT were also studied.

Content from this work may be used under the terms of the Creative Commons Attribution 3.0 licence. Any further distribution of this work must maintain attribution to the author(s) and the title of the work, journal citation and DOI.
Published under licence by IOP Publishing Ltd

2. Materials and Method

The Ca-doped PLSZT (PCLSZT) ceramic was fabricated in accordance with the molecular formula $Pb_{0.93-x}Ca_xLa_{0.02}Sr_{0.05}(Zr_{0.52}Ti_{0.48})O_3$, with x = 0.00, 0.02, 0.04, 0.06, 0.08 and 0.10. The starting raw materials included lead (II) oxide (PbO), zirconium oxide (ZrO_2), titanium oxide (TiO_2), strontium carbonate ($SrCO_3$), lanthanum oxide (La_2O_3), and calcium carbonate ($CaCO_3$) were weighed with excess 10 wt.% PbO to present the loss of lead oxide. The PCLSZT powder was synthesized by high planetary ball mill method [5]. After milling process, the mixed powder was ground using agate mortar and pressed into pellet at 200 MPa. The green pellets were then sintered at 1200 °C for 3 hours with 5 °C/min heating rate. Phase analysis, surface morphology analysis of pellet samples were carried out by X-Ray Diffraction (XRD) (Bruker AXS D8 Advance) and Scanning Electron Microscopy (SEM) (Zeiss Supra 35VP), respectively. The bulk density of sintered pellets was calculated using Archimedes' Principle. The samples were painted with silver layer to make a conductive layer prior to measure the dielectric properties using LCR meter (RF Impedance/Material Analyzer 4291B Hewlett Packard).

3. Results and Discussion

3.1 X-Ray Diffraction (XRD) analysis

The X-ray diffraction pattern of undoped and Ca-doped PLSZT samples (PLSZT, PCLSZT2, PCLSZT4, PCLSZT6, PCLSZT8 and PCLSZT10) has been illustrated in Figure 1. XRD patterns indicated that the diffraction peaks of all samples were matched with pure perovskite structure. Diffraction pattern of the samples can be assigned to tetragonal related structure (ICDD card No. 98-007-6141) and rhombohedral (ICDD card No. 98-000-8531). There is no evidence of impurity peak as shown in the diffractrograms.

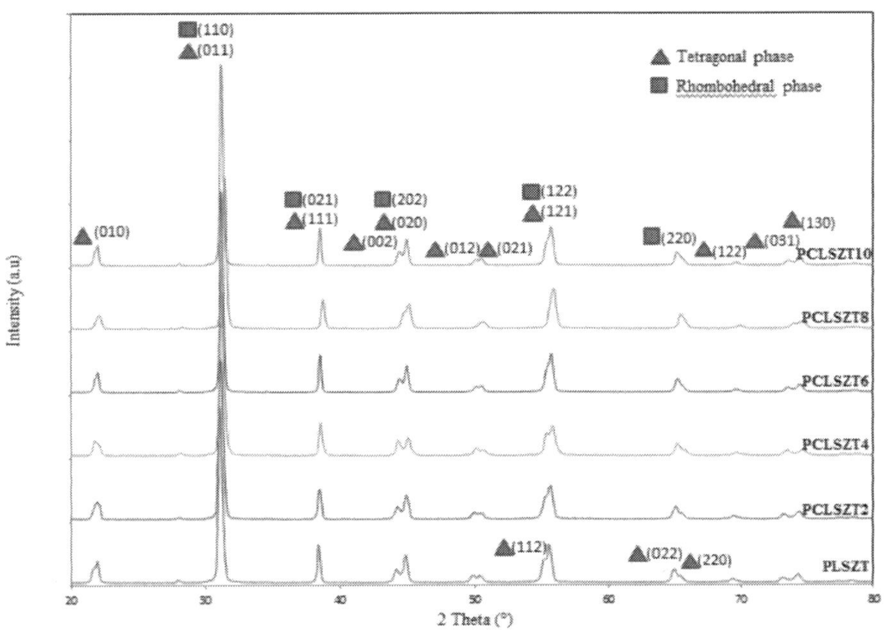

Figure 1. XRD pattern of sintered Ca-doped PLSZT pellets

By comparing the maximum intensity peak for all samples, samples with Ca dopants (PCLSZT2, PCLSZT4, PCLSZT6, PCLSZT8 and PCLSZT10) have slightly higher diffraction angle than the sample PLSZT. The peak shift towards the higher angle by increasing calcium concentration was due to the existence of lattice strain in the samples which are mainly due to the difference in ionic size of Ca^{2+} and Pb^{2+}. When Ca^{2+} was added, it occupied and replaced the sites of Pb^{2+}. The difference in ionic size of between Ca^{2+} and Pb^{2+} causes a significant strain to the lattice; hence stress was induced in the samples. Thus, this implies the incorporation of Ca^{2+} as a dopant in the PLSZT lattice [8].

3.2 Scanning Electron Microscopy (SEM) analysis

Figure 2 shows the SEM micrograph of the surface for all samples. Generally, no remarkable pores were observed with uniform grain size distribution. The grain size decreased as calcium concentration of the sample increased up to 6 mol % Ca. However, the grain size increased when 8 mol % Ca (PCLSZT8).

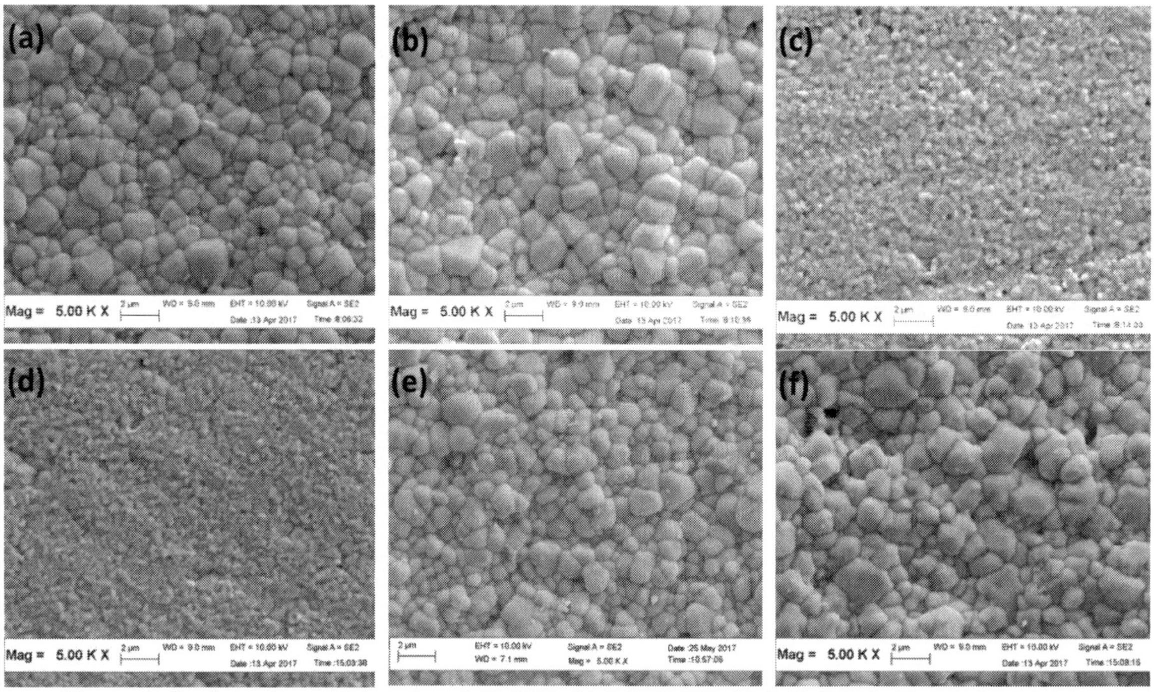

Figure 2. Surface view of (a) undoped PLSZT, (b) PCLSZT2, (c) PCLSZT4, (d) PCLSZT6, (e) PCLSZT8, and (f) PCLSZT10 samples at 5K magnification

The decrease in grain size as calcium concentration is increased corresponds to the diffusion of calcium due to ionic size mismatch and increase of lattice strain. The similar trend was observed by Kour et al. [7]. When the calcium was doped into PZT it replaced the lead, the decrease in the bond length gives rise to stress in the structure. The stress was increased with the increase in calcium concentration which obstructs the nucleation of the crystallite. As a result, the grain size of the samples was reduced. However, the increase in grain size began after 8 mol% of calcium was doped in PLSZT. The calcium content seems to promote the grain size enlargement. The enlargement of grain size was due to the increasing in mass transport at grain boundary and calcium dopant seems to induce that effect. In the study by Silva et al. [8], PZT ceramics were doped with calcium and the results showed that the grain size was enhanced with higher amount of calcium doped in PZT.

3.3 Density Analysis

Table 1 shows the bulk density for all samples. The density of Ca-doped PLSZT samples was lower than undoped PLSZT sample and the density had decreased slightly with increasing calcium substitution. This was due to the fact Ca^{2+} was doped into Pb^{2+} site of ceramic where Ca was lighter than Pb. Such results were also observed by Chu and Chen [9] in their study in which they showed similar trend where the density of ceramic reduced as the amount of Ca dopants increased.

Table 1. Bulk density for PLSZT doped with various calcium concentration

Sample	Bulk density (g/cm^3)
PLSZT	9.01
PCLSZT2	8.84
PCLSZT4	8.84
PCLSZT6	8.78
PCLSZT8	8.82
PCLSZT10	8.82

3.4 Dielectric Properties

The dielectric properties of PLSZT and Ca-doped PLSZT were measured in the frequency range from 1 kHz to 1 MHz at room temperature. Figure 3 shows the variation of dielectric constant, ε_r of all the samples at different frequencies measured at room temperature. It was observed that dielectric constant increases with calcium concentration. This phenomenon was due to two factors: (i) lattice distortion increased with the increase in calcium concentration due to ionic radius mismatch and (ii) local distortion in Pb site. The electronic polarization was affected leading to increase in the dielectric constant. This has been proved in previous study of effect of calcium modified PZT on electrical properties [7]. The decrease in dielectric constant with increasing frequency was due to the fact that polarization does not occur instantaneously with the application of electric field because of inertia. The delay response towards the alternating electric field leads to a decrease in dielectric constant. At low frequencies, all the polarizations have contributed. However, those with large relaxation times cease to respond and hence causing a decrease in dielectric constant at higher frequencies [10].

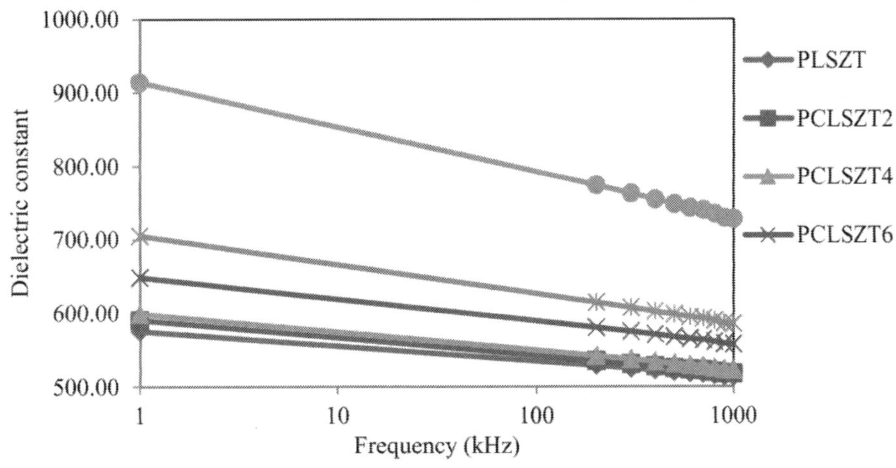

Figure 3. The variation of dielectric constant, k of all the samples at different frequency

Figure 4 shows the variation of dielectric loss, tan δ of all the samples at different frequencies measured at room temperature.

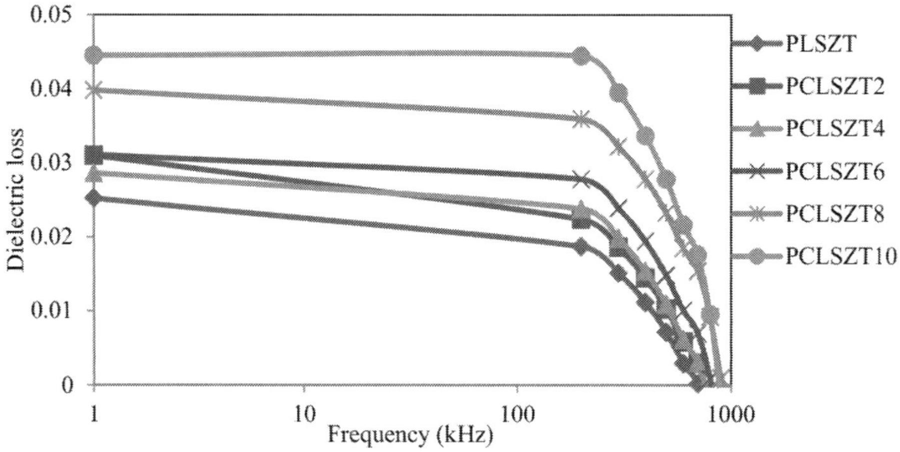

Figure 4. The variation of dielectric loss, tan δ of all the samples at different frequency

These results shows that the dielectric loss was enhanced with an increase in calcium concentration. It is due the fact that Ca creates Pb site defect in lattice and increases anisotropy in the material. Besides, size mismatch creates stress in the lattice site leading to anisotropy. Hence, the internal stress was not strong enough to switch the polarization back after the electric field was removed [7]. This phenomenon was in agreement with results found by Siddique et. al [11]

4. Conclusion

The effect of various calcium concentrations on the microstructure and dielectric properties of Ca-doped PLSZT (Pb-La-Sr-Zr-Ti) synthesized via high planetary ball mill reaction method were investigated. XRD results showed that pure PZT perovskite phase with tetragonal and rhombohedral structure were observed in all Ca-doped PLSZT (Pb-La-Sr-Zr-Ti) samples. The grain size decreased with addition of calcium concentration from 1.59 μm to 1.32 μm but began to enlarge again at 8 mol% calcium was added. The density of the sample also decreased with increasing calcium concentration. The dielectric constant and dielectric loss increased with the increase in the frequency as well as by increase in calcium concentration.

Acknowledgement

The authors would like to express their gratitude to Universiti Sains Malaysia and Ministry of Higher Education for funding this project through Fundamental Research Grants Scheme (FRGS 6071282).

References

[1] Panda, P. & Sahoo, B. 2015. *Ferroelectrics*, 474, 128-143.
[2] Andrich, E., *Electron. Appl,* 1965. 26(21): p. 123-44.
[3] Haertling, G.H., 1999. *J. Am.Cer. Soc.*, 82(4): p. 797-818.
[4] Kalem, V., İ. Çam, and M. Timuçin, 2011. *Cer. Int.*, 37(4): p. 1265-1275.
[5] Bahanurddin, N.F.K., et al. 2015. *Adv. Mat. Research*. Trans Tech Publ. p. 172 - 176
[6] Kainz, T., Bitschnau, B., Mautner, F. & Reichmann, K. 2016. *J. Eur. Cer. Soc.,* 36, 507-514.
[7] Kour, P., et al., 2014. *Solid State Comm.*, 190: p. 33-39.
[8] Silva, M.S., et al., 2015. *Mat. Sc. Forum*, 805.
[9] Chu, S.-Y. and C.-H. Chen, 2001. *Sensors and Actuators A: Physical*, 89(3): p. 210-214.
[10] Chopra, S., et al., 2003. *Solid State Comm.*, 127(4): p. 299-304.
[11] Siddiqui, M., Mohamed, J. J. & Ahmad, Z. A. 2017. *Cer. Int.,* 43, 2644-2649.

The Effect of Sr-substitution on the Electrical and Structural Properties in Colossal Magnetoresistive $La_{1-x}Sr_xMnO_3$ with $x = 0.05, 0.12, 0.18, 0.30$ and 0.40

Nor Azah Nik-Jaafar[1*], R. Abd-Shukor[2]

[1]Pusat PERMATApintar Negara®, Universiti Kebangsaan Malaysia, 43600 UKM Bangi, Selangor, Malaysia
[2]School of Applied Physics, Universiti Kebangsaan Malaysia, 43600 UKM Bangi, Selangor, Malaysia

[*]norazah_nj@ukm.edu.my

Abstract. The effect of Sr-substitution at the La-site in colossal magnetoresistive $La_{1-x}Sr_xMnO_3$ (x = 0.05, 0.12, 0.18, 0.30 and 0.40) on the electrical transport and structural property has been studied. Investigations were conducted through X-ray diffraction (XRD) analysis, temperature-dependent electrical resistance measurements and scanning electron microscope (SEM). Between x = 0.05 to 0.30, XRD analysis showed that at room temperature, the composition has orthorhombic structure. At x = 0.40 composition, the structure is rhombohedral. Electrical resistance measurements (30-290 K) showed that the material was insulating between x = 0.05 to 0.12, whereas increased substitution from x = 0.18 to 0.40 transformed the electrical property from insulating to metallic. The transformation in the transport property along with the conspicuous change in structure with increase Sr-substitution x provide a strong indication of a large electron-lattice coupling in this material.

1. Introduction

The novel properties of $Ln_{1-x}A_xMnO_3$ (Ln = La, Nd, Pr; A = Sr, Ca, Ba, Pb) remain a subject of research interest after more than two decades since the colossal magnetoresistance effect (CMR) was discovered in this material. At the time upon its discovery, CMR effect attracted wide attention due to its potential technological applications, therefore spurred intense research activity on this manganese oxide of perovskite structure.

Equally interesting is the rich structure-properties phase diagram of $La_{1-x}A_xMnO_3$ as a function of composition ($0.0 \leq x \leq 1.0$) and temperature. At both x = 0.0 and 1.0, $La_{1-x}A_xMnO_3$ is an antiferromagnetic (AFM) insulator at low temperatures. The insulating nature of the parent compound (x = 0.0) as well as the anisotropic magnetic interaction is particularly due to the Jahn-Teller distortion around Mn^{3+} ions. When the compound is progressively hole-doped (substitution of divalent cation at the La-site), the proportion of Mn^{4+} ions increases, which in turn decreases the orthorhombic distortion, leading to more cubic structures. The composition becomes rhombohedral or pseudocubic at certain value of x. In $La_{1-x}Sr_xMnO_3$, substitution of Sr transforms the compound from AFM insulator to ferromagnetic (FM) insulator at $x \approx 0.1$, and to ferromagnetic metal at $x \approx 0.16$ [1,2]. At $\sim 0.2 < x < 0.5$ a transition from a high temperature paramagnetic insulator (PMI) to a low temperature ferromagnetic metallic (FMM) phase occurs. It is in this composition range CMR effect the strongest [3]. The magnitude of decrease in resistivity upon the application of magnetic field is the highest in the region

Content from this work may be used under the terms of the Creative Commons Attribution 3.0 licence. Any further distribution of this work must maintain attribution to the author(s) and the title of the work, journal citation and DOI.
Published under licence by IOP Publishing Ltd

of Curie temperature T_C, which is accompanied by an insulator-metal transition at T_{IM} [4,5]. This simultaneous observation of itinerant electron and ferromagnetism is explained by Zener's double exchange mechanism [6]. Millis et al. argued that in addition of double exchange physics, there must be a strong electron-phonon interaction which arise from the Jahn-Teller splitting of the outer Mn d level [7]. This suggests the crucial role of the lattice and electron-lattice coupling. Therefore, in understanding the novel transport properties of perovskite manganates, it is imperative to consider the overall lattice symmetry as well as the local lattice distortions.

In this work, the effect of Sr-substitution on the structural and electrical properties of $La_{1-x}Sr_xMnO_3$ at doping compositions $x = 0.05, 0.12, 0.18, 0.30$ and 0.40 have been investigated.

2. Experimental Details

Samples of $La_{1-x}Sr_xMnO_3$ ($x = 0.05, 0.12, 0.18, 0.30$ and 0.40) were prepared by conventional solid state reaction method. Starting materials of La_2O_3, $SrCO_3$ and MnO_2 with purity of ≥99.9% were used. The mixed powders were grounded thoroughly and were calcined at 900°C for a period of 24 hours with intermediate grinding. The calcined materials were pressed into pellets of 13 mm in diameter and ~1.5 mm in thickness. This was followed by sintering in air at 1300°C for 12 hours and left to furnace-cooled.

The temperature-dependent electrical resistance measurements were carried out using the d.c. four-point probe method with silver paint contacts. Samples' crystallization were characterized by X-ray powder diffraction (XRD) technique using Bruker AXS D8 Advance Diffractometer with Cu-Kα radiation. The microstructure of the samples was examined using a Philips XL 30 Scanning Electron Microscope (SEM).

3. Results and Discussion

Figure 1 shows the XRD diffraction pattern of $La_{1-x}Sr_xMnO_3$ with $x = 0.05, 0.12, 0.18, 0.30$ and 0.40. In the composition range between $x = 0.05$ to $x = 0.30$ the material crystallizes in orthorhombic structure. Increased doping at $x = 0.40$ change the crystallization to rhombohedral structure with space group *R*-3*c*. Table 1 shows the lattice constants of the samples.

Table 1. Lattice constants of $La_{1-x}Sr_xMnO_3$ with $x = 0.05, 0.12, 0.18, 0.30$ and 0.40.

Sample *x*	Lattice parameters [Å]			Structure (Space group)	Volume [Å]^3
	a	*b*	*c*		
0.05	5.5368	5.5539	7.7506	Orthorhombic (*Pbnm*)	238.34
0.12	5.5093	5.5430	7.7944	Orthorhombic (*Pcmn*)	238.03
0.18	5.5373	5.5023	7.7853	Orthorhombic (*Pbnm*)	237.20
0.30	5.4512	7.7386	5.4981	Orthorhombic (*Pnma*)	231.94
0.40	5.4926	5.4926	13.361	Rhombohedral (*R*-3*c*)	403.08

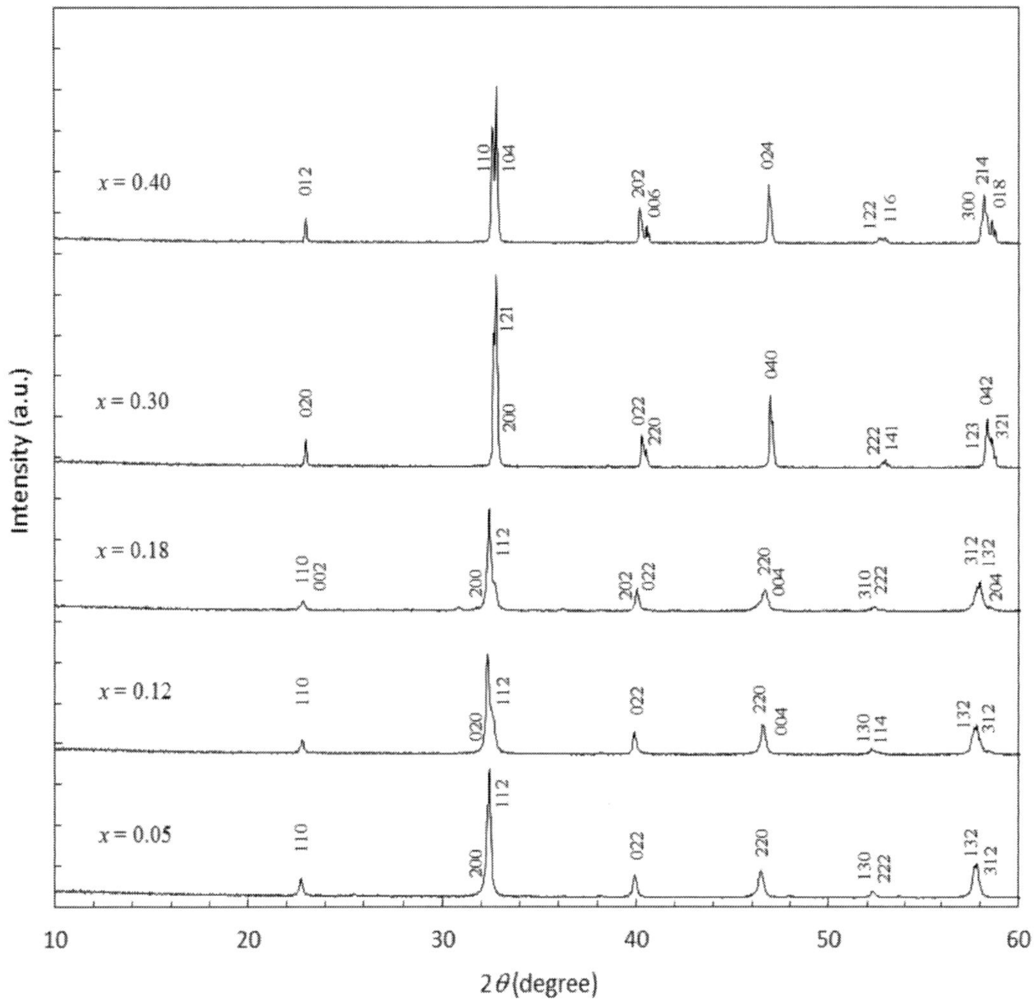

Figure 1. XRD patterns of $La_{1-x}Sr_xMnO_3$ with $x = 0.05$, 0.12, 0.18, 0.30 and 0.40.

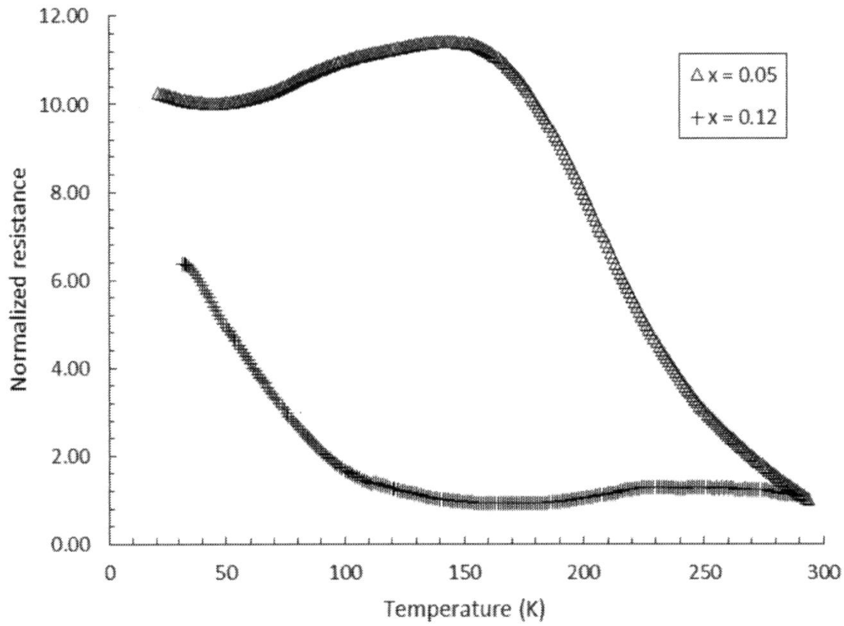

Figure 2. Normalized resistance versus temperature of $La_{1-x}Sr_xMnO_3$ for $x = 0.05$ and 0.12.

Figure 2 shows the results of temperature-dependent resistance measurement (30–300 K) for $x = 0.05$ and 0.12 samples. As temperature decreased, the $x = 0.05$ sample at first showed an insulating behavior until it reached a peak in resistance at 151 K. Below 151 K the resistance was decreasing with temperature characteristics of a metallic state although the resistivity remains at ~10 folds higher than that at room temperature. Sample $x = 0.12$ is an insulator over the whole measurement range.

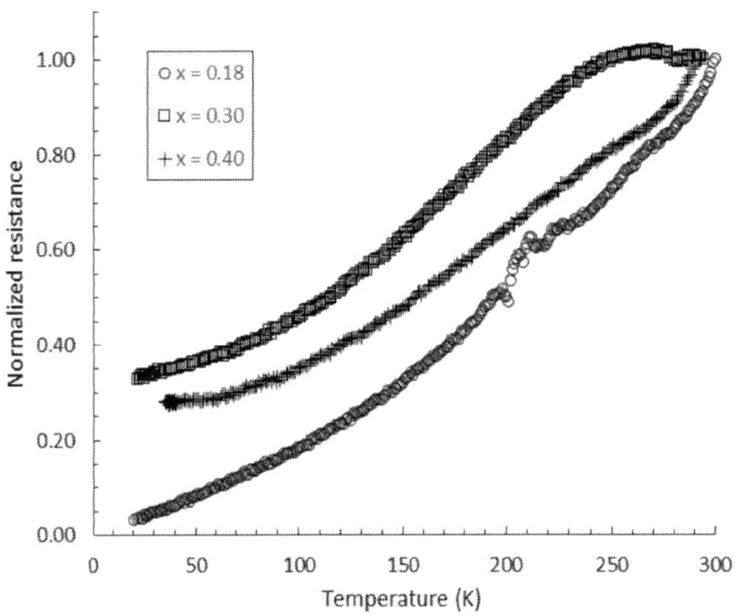

Figure 3. Normalized resistance versus temperature of $La_{1-x}Sr_xMnO_3$ for $x = 0.18$, 0.30 and 0.40.

Figure 3 shows the results of temperature-dependent resistance measurement (30–300 K) for x = 0.18, 0.30 and 0.40. At these compositions, the materials showed metallic resistivity, while the resistance variation with temperature suggests of a peak in resistivity at ~270 K for x = 0.30. It can be concluded that within composition range between x = 0.18 to 0.40, the material had low temperature metallic conductivity.

Figure 4(a) and 4(b) show the SEM micrograph at 3000× magnification for x = 0.05 and 0.30 respectively. The average grain size for x = 0.15 is ~5 μm, and the average grain size for x = 0.30 was 1-2 μm. Based on the SEM micrographs of all the compositions, it was shown in general that sample's grain size decreased with the increment in x.

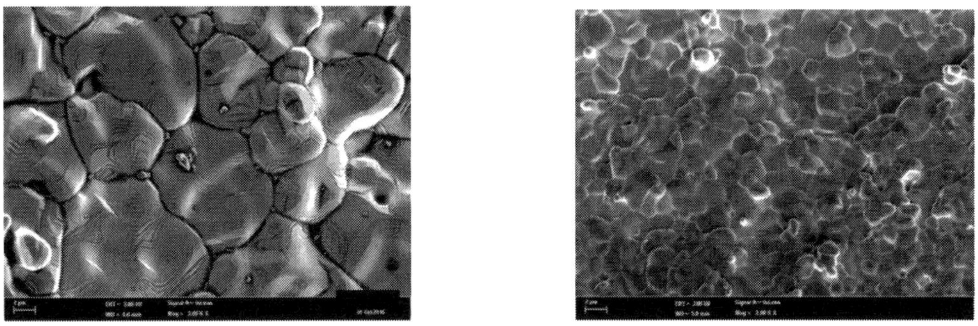

(a) (b)

Figure 4. Scanning electron micrograph of $La_{1-x}Sr_xMnO_3$ for (a) x = 0.05, and (b) x = 0.30.

At the amount of hole doping which corresponds to x = 0.18, 0.30 and 0.40, the resistivity of the compound was metallic below 300 K. By inferring the temperature-composition phase diagram of $La_{1-x}Sr_xMnO_3$ [1], it can be safely assumed that the insular-metal transition preceded the metallic phase at somewhere above 300 K, which is outside our measurement range. Hence, at x = 0.18 the low temperature transport property in $La_{1-x}Sr_xMnO_3$ was transformed from insulating to metallic. Previous works reported that the transformation from high temperature paramagnetic insulator into low temperature ferromagnetic metal was manifested at composition $x > 0.15$. The insulating resistivity was shown by the x = 0.05 and 0.12 compositions in this study were as expected as the insulating nature of the composition below $x \sim 0.15$ has been well established [4].

Between x = 0.05 to x = 0.30 composition, the structure of the sample at room temperature was orthorhombic. At x = 0.40, the composition had rhombohedral structure. Previous study on $La_{1-x}Sr_xMnO_3$ in the form of melt-grown crystals [1], structural transition from rhombohedral ($R\bar{3}c$) to orthorhombic (*Pbnm*) has been reported to occur for x = 0.15, 0.175 and 0.2 compositions at temperatures near 300 K. This structural transition accounted for the resistivity anomalies was shown by these compositions at those temperatures, whereby for the x = 0.175 composition this anomaly was exhibited at ~200 K, that was below the room temperature. Thus, this implies of the possibility that for the x = 0.18 composition in our study, the rhombohedral to orthorhombic transition may have taken place at above the room temperature. The same argument also applies to the x = 0.3 composition. The x = 0.3 to 0.4 composition has been shown in that study to crystallize in the rhombohedral structure over the whole temperature range below 500 K.

4. Conclusion

In conclusion, the electrical and structural properties of $La_{1-x}Sr_xMnO_3$ for x = 0.05, 0.12, 0.18, 0.30 and 0.40 in the temperature range below 300 K has been investigated. The electrical resistivity was insulating for x = 0.05 and 0.12, and metallic for x = 0.18, 0.30 and 0.40. The room temperature structural formation of all the compositions was orthorhombic, except for $x = 0.40$ which was rhombohedral. Sr-substitution at the La-site in $La_{1-x}Sr_xMnO_3$ was tantamount to hole doping, which increased Mn^{4+} proportion, therefore decreased the Jahn-Teller distortion. This effects the transformation from the insulating to metallic resistivity as observed at $x = 0.18$. The structural transformation from orthorhombic to rhombohedral at $x = 0.40$ was attributed to the decrease in the Jahn-Teller distortion around Mn^{3+} as La was progressively substituted with Sr [4]. The transformation of the transport property along with the conspicuous change in structure with increase Sr-substitution provided substantial evidence of a strong electron-lattice coupling in this manganese oxide of perovskite structure.

Acknowledgement
This work has been supported by the Ministry of Higher Education, Malaysia under research grant ERGS/1/2012/STG02/CUCMS/03/2. Nor Azah Nik-Jaafar would like to express her gratitude to Cyberjaya University College of Medical Sciences (CUCMS) for facilitating this research work throughout the whole duration of this project.

References

[1] Urushibara A, Moritomo Y, Arima T, Asamitsu A, Kido G and Tokura Y 1995 Insulator-metal transition and giant magnetoresistance in $La_{1-x}Sr_xMnO_3$ *Phys. Rev.* B 51 14103-14109
[2] Moritomo Y, Asamitsu A and Tokura Y 1997 Enhanced electron-lattice coupling in $La_{1-x}Sr_xMnO_3$ near the metal-insulator phase boundary *Phy. Rev.* B 56 12190-12195
[3] Bishop A R and Röder H 1997 Theory of colossal magnetoresistance *Current Opinion in Solid State and Materials Science* 2 244-251
[4] Rao C N R and Raychaudhuri A K 1998 Colossal magnetoresistance, charge ordering and other novel properties of manganates and related materials *Colossal Magnetoresistance, Charge Ordering and Related Properties of Manganese Oxides* ed Rao C N R and Raveau B (Singapore: World Scientific Publishing Co. Pte. Ltd.) pp 1-42
[5] Kusters R M, Singelton J, Keen D A, McGreevy R and Hayes W 1989 Magnetoresistance measurements on the magnetic semiconductor $Nd_{0.5}Pb_{0.5}MnO_3$ *Physica* B 155 362-365
[6] Zener C 1951 Interaction between the d-shells in the transition metals. II. ferromagnetic compounds of manganese with perovskite structure *Phys. Rev.* 82 403-405
[7] Millis A J, Littlewood P B and Shraiman B I 1995 Double exchange alone does not explain the resistivity of $La_{1-x}Sr_xMnO_3$ *Phy. Rev. Lett.* 74 5144-5147

External Quantum Efficiency of Electrochemically Prepared and Annealed Directly Stacked SLG/GZO/Cu_2O/CuO Photoactive Layer

Pei Loon Khoo[1, a *], Kenta Satou[1, b], Kota Nozaki [1, c], and Masanobu Izaki [1, d]

[1]Graduate School of Engineering, Toyohashi University of Technology, Toyohashi, Aichi 441-8580, Japan

[a]k1415108@edu.tut.ac.jp, [b]sato@tf.me.tut.ac.jp, [c]k163167@tf.me.tut.ac.jp, [d]m-izaki@me.tut.ac.jp

Abstract. A strategy to enhance the copper-oxides-based photovoltaic's performance is to introduce p-type semiconductor layers with different bandgap energies into the photoactive layer to extend the absorption bands and to increase the quantum efficiency. In this research, the strategy has been realized by a directly stacked Cu_2O/CuO photoactive layer that utilizes both copper oxides as light-absorbing layers. The directly stacked Cu_2O/CuO photoactive layers were fabricated with a novel facile method of rapid thermal annealing (RTA) of the Cu_2O samples (prepared by electrodeposition) in air at temperatures of 473-673 K. EQE performance of directly stacked Cu_2O/CuO shows widened absorption span at wavelengths above 630 nm under biased voltage was attributed to the light conversion from not only the Cu_2O layer but also the CuO layer. Widened absorption edge from 630 to 850 nm can be confirmed further from UV-vis characterization as a result of the formation of the directly stacked layers. Although the XRD patterns show insignificant changes between different annealing temperatures, FE-SEM observation indicates the formations of the CuO layer on the Cu_2O layer at approximate thicknesses of 20~150 nm. In spite of the EQE values are still low, further improvements can be implemented. However, this research highlights that both Cu_2O and CuO layers act as light-absorbing layers. This is proving the validity of the directly stacked Cu_2O/CuO photoactive layer in increasing the rate of electron-hole generation through expanded wavelength absorption, which provides insight into future designs of high-performance photoactive layers.

1. Introduction

Cuprous oxide (Cu_2O) and cupric oxide (CuO) are n-type semiconductors with bandgap energy of approximately 2.1 eV and 1.4 eV. These semiconductors have been receiving increasing attention as photoactive layers in photovoltaic devices and photocathode in photo-electrochemical water splitting, due to both materials have many advantages of being terrestrial abundant, low-cost, and non-toxic. An electrochemically-prepared ZnO/Cu_2O with a power conversion efficiency (PCE) of 1.28% have been reported by Izaki *et al.*[1] Since then, Cu_2O-based photovoltaics fabricated with various methods also have been reported, with improvements in the PCE. [2–5] However, there is a theoretical limit of the PCE of a photovoltaic, which depends on the bandgap energy of the light-absorbing photoactive layer. For the Cu_2O layer, it is approximately around 20%. [6,7] To break conventional PCE limit, the combination of light-absorbing layers with different bandgap energies is necessary. Multijunction

Content from this work may be used under the terms of the Creative Commons Attribution 3.0 licence. Any further distribution of this work must maintain attribution to the author(s) and the title of the work, journal citation and DOI.
Published under licence by IOP Publishing Ltd

photovoltaic is an example which makes use of multiple sub-cell containing different bandgap energies connected in series.[8,9] Quantum dots photovoltaics utilizes tunable bandgap energies to achieve high PCE's.[10–12] For copper oxide photoactive layer, a strategy to incorporate both Cu_2O and CuO as light absorbing layers containing different bandgap energies is viable through rapid thermal annealing (RTA).

It has been reported previously by our group, Cu_2O sample prepared by electrochemically and annealing shows improvement of electrical properties.[13,14] However, a facile method which involves annealing the sample in air at certain temperature was discovered to contribute to formation directly-stacked CuO layers on the Cu_2O while retaining lattice relationship.[15] In order to design future highly efficient copper oxides photovoltaic or photo-electrochemical devices, it is necessary to demonstrate the photoactivity of the directly-stacked Cu_2O/CuO layer.

2. Experimental Section

2.1 Electrodeposition of GZO/Cu_2O.

Conductive gallium-doped zinc oxide (GZO)-coated soda lime glass substrate was rinsed with acetone and then with deionized water in ultrasonic bath. The Cu_2O layer was then electrodeposited potentiostatically on the GZO substrate at –0.4 V referenced to an Ag/AgCl electrode in saturated KCl at an electric charge of 1.5 C cm^{-2} with a potentiostat (Hokuto Denko, HAL-3001) at 313 K. The potentiostat was connected to a Coulomb meter (Hokuto Denko, HF-301). The aqueous solution used contained 0.4 mol L^{-1} copper (II) acetate monohydrate ($Cu(CH_3COO)2 \cdot H_2O$, Nacalai Tesque, Inc.) and 3 mol L^{-1} lactic acid (Kanto Chemical, Co., Inc.). The pH was adjusted with NaOH to 12.5. The aqueous solution was prepared using deionized water purified by an Elix-UV system (Millipore).

2.2 Fabrication of directly-stacked GZO/Cu_2O/CuO.

After the Cu_2O layers were electrodeposited, it was rinsed thoroughly with deionized water and dried in vacuum for 30 minutes before annealing with rapid thermal annealing furnace (RTA, MILA 5000) in air. Annealing temperature was set at 473, 523, 573, and 673 K. Annealing time was 60 minutes.

2.3 Characterization of Cu_2O/CuO directly-stacked photoactive layer.

Surface and cross-sectional morphology were observed with field emission scanning electron microscope (FE-SEM, Hitachi High Technology SU8000). Structural characterization was carried out with x-ray diffraction (XRD, Rigaku RINT 2500) operated at 20 kV and 10 mA using Cu Kα radiation. Optical absorption spectra was recorded using a UV-Vis-NIR spectrophotometer with an integrating sphere (Hitachi High-Technology UV-4100). The external quantum efficiency (EQE) measurement for SLG/GZO/Cu_2O/CuO/Au devices was made by using a monochromator (Bunko Keiki, SM-250KB). The active area of the samples was defined by light-masking at 4 mm^2. A biased voltage of -0.01 was applied to annealed samples at temperature of 523, 573, and 673 K. The schematic structure and schematic band diagram of the device used for the EQE measurement are shown in Figure 1. It is predicted that the band of CuO shifted upwards under biased voltage, enabling charge collection and the expansion of absorption edge originating from the CuO layer at 523 K.

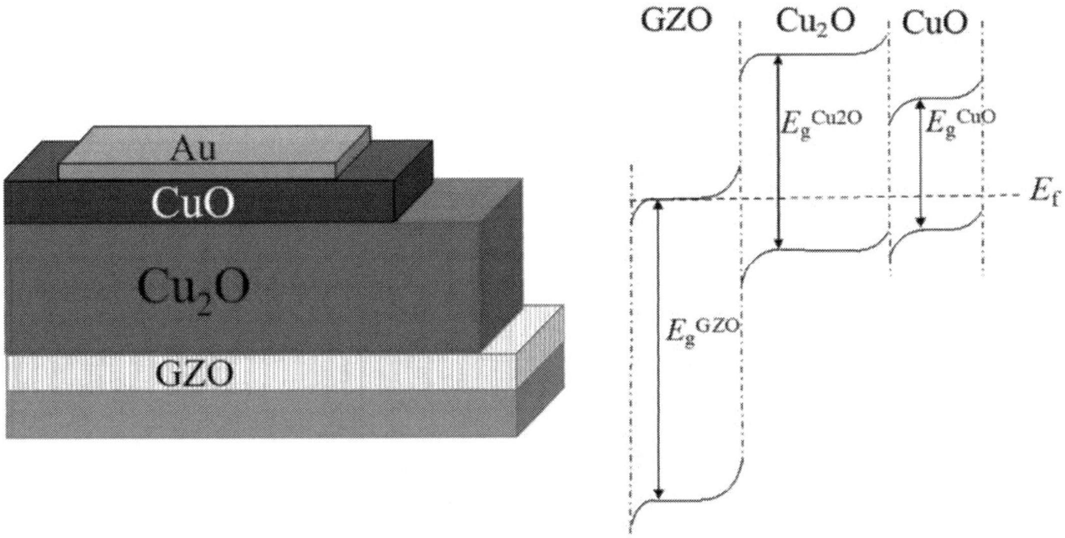

Figure 1: Schematic structure (left) and schematic band alignment (right) diagrams of the directly stacked SLG/GZO/Cu_2O/CuO.

3. Results and Discussions

Figure 2 shows the surface and cross-sectional FE-SEM images of the as-deposited sample and samples annealed in air. Continuous well-faced grains can be observed from the surface of the as-deposited GZO/Cu_2O sample. Three-sided pyramidal grains can be observed, which correspond to the corner of a cubic. The cross-sectional image in Figure 2(f) shows aggregates of columnar Cu_2O grains at an approximate thickness of 2 μm. The approximate thickness of the underlying GZO substrate is 50 nm. The well-faced grains at the surface transforms gradually with the increase of annealing temperature where the boundary of the grains diminish, especially at the temperature of 673 K. Cross-sectional observation reveals a distinguished CuO layer being formed at the surface starting at 523 K which grew into a thickness of approximately 150 nm at 673 K. At the same time, nanopores start to form in the Cu_2O lattice at 523 K (Figure 2(h)) and large voids can be observed at 673 K in multiple places including both in the CuO layer and at the heterojunction of GZO/Cu_2O. It is deduced that the outward diffusion of ions which occur during oxidation,[16] and the change of crystal structure from cubic of Cu_2O to monoclinic CuO caused the formation of nanopores and voids.

Figure 3 shows the XRD patterns of the as-deposited sample and samples annealed at 473, 523, 573, and 673 K. The highest peak at 34.3° corresponds to the (0001) plane of the GZO substrate, followed by the second highest peak at 36.4°, which corresponds to the (111) plane of the Cu_2O. The preferred orientation of the <111> Cu_2O is in good agreement with the FE-SEM image where the corner of a cubic structure can be observed at the surface. The XRD patterns of the samples are generally similar except for the sample annealed at 673 K. A small peak at 38.7° is observed, which is corresponds to the (111) plane of the CuO. The low intensity was due to the thinness of the CuO, which only formed at the surface, as observed in the FE-SEM images.

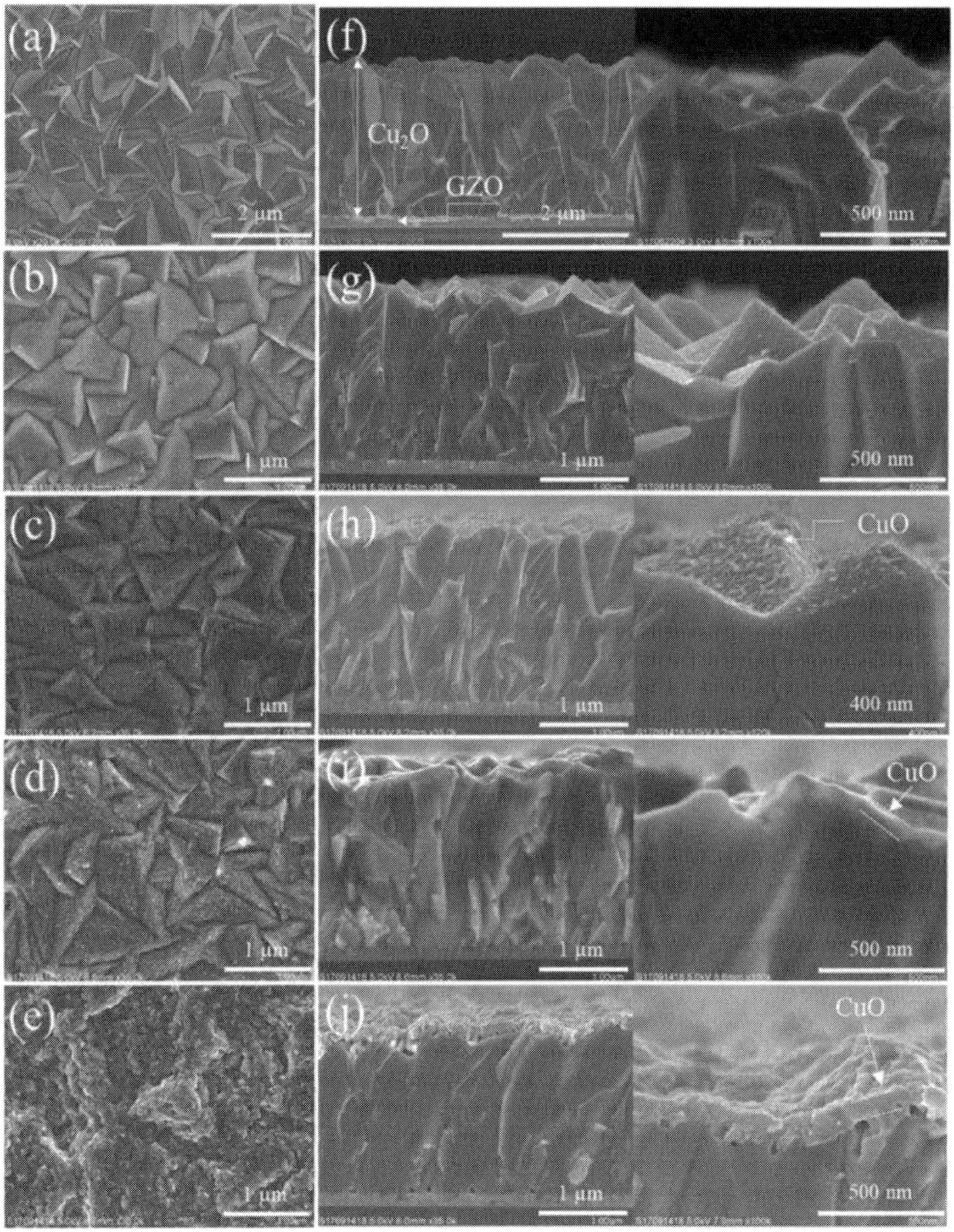

Figure 2: FE-SEM images of surface and cross-sectional images: (a)(f) as-deposited, annealed at (b)(g) 473, (c)(h) 523, (d)(i) 573, (e)(j) 673 K. Images on the right are magnified cross-sectional images near the surface.

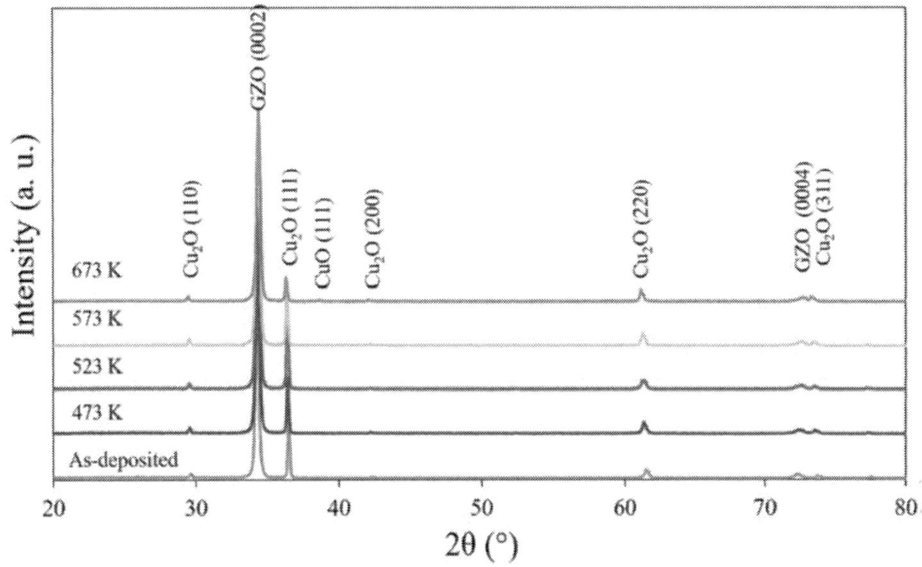

Figure 3. X-ray diffraction patterns of the as-deposited GZO/Cu_2O sample and samples annealed at 473, 523, 573, and 673 K in air

Figure 4 shows the absorbance of layers as-deposited, and annealed at 473, 523, 573, and 673 K. The absorption edge characteristic to the Cu_2O at around 650 nm can be seen for all samples except for 673 K, which shows higher absorbance above 600 nm and having an absorption edge which extends towards 850 nm. The color of the samples turned from reddish brown from as-depo to dark brown after annealing at 673 K, and the increase in absorbance can be attributed to the formation of CuO at the surface. The thickness of the Cu_2O layers for all samples could be observed to be approximately 2 μm before annealing and the changes were insignificant after annealing.

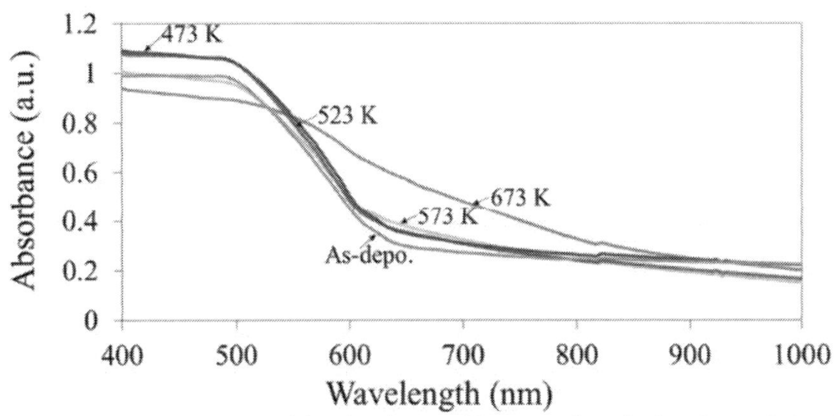

Figure 4. UV-Vis measurement of the absorbance of layers as-deposited, and annealed at 473, 523, 573, and 673 K.

External quantum efficiency (EQE) is an important evaluation that shows the ratio of the number of charge carriers collected by the photovoltaic to the number of photons in the photoactive layer. Figure 5 shows the EQE of samples: as-deposited, annealed at 473, 523, 573, and 673 K. For the as-deposited sample and sample annealed at 473 K, both display charge collection originated from Cu_2O layers which shows the characteristic absorption edge at around 650 nm and peaks at 44% and 43% at around 400 nm, corresponding to the UV-Vis results in Figure 3. Starting at the annealing temperature of 523 K, the EQE deteriorates significantly, which is speculated to be attributed to the formation of nanopores and voids that act as scattering defects that greatly hinder carrier charge mobility. However, with applied biased voltage, distinct EQE characteristics can be observed for the 523 K sample. This is significant due to the observed absorption edge of Cu_2O at 650 nm and the absorption edge of CuO which extends towards higher wavelength at approximately 850 nm. This shows that despite the presence of defects and the thinness of the CuO, carrier charge was generated and collected from the CuO layer under illumination, showing a 1.3% of EQE. The expansion of absorption demonstrates that with careful fabrication, the directly-stacked Cu_2O/CuO photoactive layers can act in tandem as light-absorbing layers.

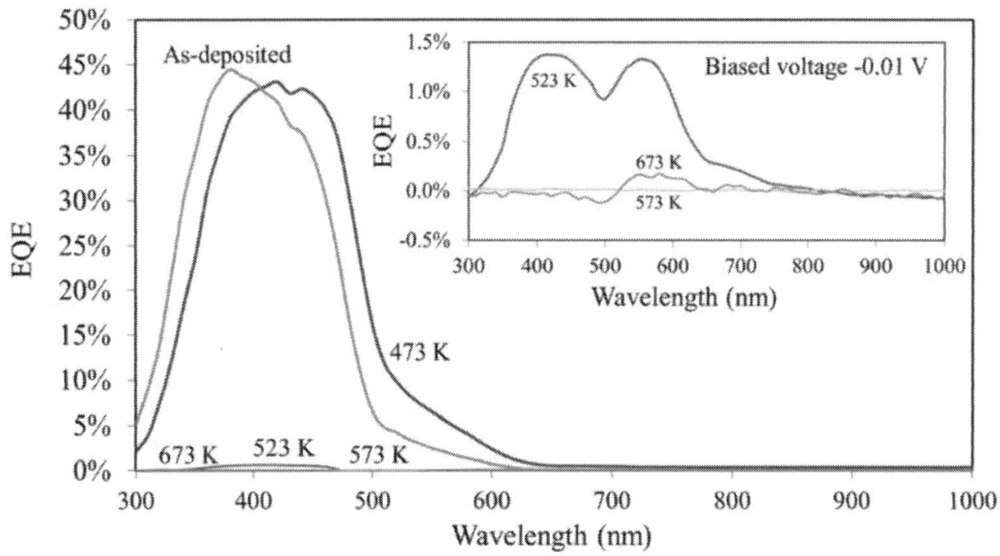

Figure 5. External quantum efficiency spectra of samples: as-deposited, and annealed at 473, 523, 573, and 673 K. Shown inset is spectra of samples annealed at 523, 573 and 673 K under biased voltage.

4. Conclusions

Directly-stacked Cu_2O/CuO layer has been fabricated with a facile method. Electrodeposition of Cu_2O layer on GZO substrate was carried out and annealed in 473, 523, 573, and 673 K in air for 60 minutes. Formation of Cu_2O/CuO layer was confirmed by FE-SEM, XRD and UV-vis measurements. At 673 K, the result of EQE shows that both the Cu_2O and CuO layers act as light-absorbing layers, and an expansion in the absorption width was achieved. Constraining the formation of defects by providing sacrificial Cu for oxidation, changing the thickness of Cu_2O to facilitate carrier movements from CuO, and further exploring the optimum oxidation temperature are the several improvements of the directly-stacked Cu_2O/CuO photoactive layer that can be implemented in future.

Acknowledgements
This work was supported by a Grant-in-Aid for Scientific Research (25281062) from the Japan Society for the Promotion of Science.

References

[1] M. Izaki, T. Shinagawa, K.-T. Mizuno, Y. Ida, M. Inaba, A. Tasaka, Electrochemically constructed p-Cu2O/n-ZnO heterojunction diode for photovoltaic device, J. Phys. D. Appl. Phys. (2007). doi:10.1088/0022-3727/40/11/010.

[2] Y.S. Lee, D. Chua, R.E. Brandt, S.C. Siah, J. V. Li, J.P. Mailoa, S.W. Lee, R.G. Gordon, T. Buonassisi, Atomic layer deposited gallium oxide buffer layer enables 1.2 v open-circuit voltage in cuprous oxide solar cells, Adv. Mater. 26 (2014) 4704–4710. doi:10.1002/adma.201401054.

[3] A.T. Marin, D. Muñoz-Rojas, D.C. Iza, T. Gershon, K.P. Musselman, J.L. MacManus-Driscoll, Novel Atmospheric Growth Technique to Improve Both Light Absorption and Charge Collection in ZnO/Cu 2 O Thin Film Solar Cells, Adv. Funct. Mater. (2013). doi:10.1002/adfm.201203243.

[4] T. Minami, H. Tanaka, T. Shimakawa, T. Miyata, H. Sato, High-efficiency oxide heterojunction solar cells using Cu2O sheets, Japanese J. Appl. Physics, Part 2 Lett. (2004). doi:10.1143/JJAP.43.L917.

[5] R. Wick, S.D. Tilley, Photovoltaic and Photoelectrochemical Solar Energy Conversion with Cu_2O, J. Phys. Chem. C. 119 (2015) 26243–26257. doi:10.1021/acs.jpcc.5b08397.

[6] S. Rühle, Tabulated values of the Shockley-Queisser limit for single junction solar cells, Sol. Energy. (2016). doi:10.1016/j.solener.2016.02.015.

[7] W. Shockley, H.J. Queisser, Detailed balance limit of efficiency of p-n junction solar cells, J. Appl. Phys. 32 (1961) 510–519. doi:10.1063/1.1736034.

[8] L. El Chaar, L.A. Lamont, N. El Zein, Review of photovoltaic technologies, Renew. Sustain. Energy Rev. 15 (2011) 2165–2175. doi:10.1016/j.rser.2011.01.004.

[9] K. Tanabe, A review of ultrahigh efficiency III-V semiconductor compound solar cells: Multijunction tandem, lower dimensional, photonic up/down conversion and plasmonic nanometallic structures, Energies. 2 (2009) 504–530. doi:10.3390/en20300504.

[10] E. Sargent, Colloidal quantum dot solar cells, Nat. Photonics. 6 (2012) 133–135. doi:10.1021/ja063166u.

[11] C.-H.M. Chuang, P.R. Brown, V. Bulović, M.G. Bawendi, Improved performance and stability in quantum dot solar cells through band alignment engineering, Nat. Mater. 13 (2014) 796–801. doi:10.1038/nmat3984.

[12] M. Graetzel, R.A.J. Janssen, D.B. Mitzi, E.H. Sargent, Materials interface engineering for solution-processed photovoltaics, Nature. 488 (2012) 304–312. doi:10.1038/nature11476.

[13] T. Shinagawa, M. Onoda, B.M. Fariza, J. Sasano, M. Izaki, Annealing effects and photoelectric properties of single-oriented Cu2O films electrodeposited on Au(111)/Si(100) substrates, in: J. Mater. Chem. A, 2013: p. 9182. doi:10.1039/c3ta11116d.

[14] P.L. Khoo, Y. Kikkawa, T. Shinagawa, M. Izaki, Effects on external quantum efficiency of electrochemically constructed n-ZnO/p-Cu_2O photovoltaic device by annealing, AIP Conf. Proc. 1865 (2017). doi:10.1063/1.4993370.

[15] M. Izaki, K. Fukuzawa, Photovoltaic layer and the preparation process, published patent application 2017–54917, March 16 2017

[16] G.J. Liu, Y.F. Zhu, K. Mimura, M. Isshiki, Effect of Alloying Element of S, Se or Te on Oxidation Resistance of Cu at 573-873 K, Open Corros. J. 3 (2010) 92–98.

Effect of Different Stoichiometry on the Electrochemical Behaviour of $LiCo_xNi_{1-x}O_2$ Cathode Materials

K Elong[1,3*], N Kamarulzaman[1,2] and R Rusdi[1]

[1] Centre for Nanomaterials Research, Institute of Science, Universiti Teknologi MARA, 40450 Shah Alam, Selangor, Malaysia
[2] School of Physics and Materials, Faculty of Applied Sciences, Universiti Teknologi MARA, 40450 Shah Alam, Selangor, Malaysia
[3] School of Chemistry and Environment, Faculty of Applied Sciences, Universiti Teknologi MARA, 40450 Shah Alam, Selangor, Malaysia

kelimah0907@salam.uitm.edu.my

Abstract. Pure, single phase, layered $LiCo_xNi_{1-x}O_2$ materials with good cation ordering are not easy to synthesize. Here, solid solutions of $LiCo_xNi_{1-x}O_2$ (x = 0.1,0.2 and 0.3) were synthesized using a self-propagating combustion route and characterized. All the materials were observed to be phase pure and that cobalt has been successfully substituted in the crystal lattice. Their electrochemical properties were investigated by a series of charge-discharge cycling in the voltage range of 3.0 to 4.3 V. It was found that some of the stoichiometries exhibit specific capacities comparable or better than that of $LiCoO_2$.

1. Introduction

Many research groups have investigated various cathode materials with layered and spinel structure. Layered materials from lithium transition metal oxides possess high theoretical specific capacities compared to that of the spinels. $LiCoO_2$ and $LiNiO_2$ are two layered materials that have potential application as cathode materials for Li-ion batteries. $LiCoO_2$ is a common commercial cathode material that has a high operating voltage and stable in structure but less environmentally benign and expensive production cost, that leads to the search for materials with less cobalt [1].

$LiNiO_2$ has the advantage of being cheaper, but it is more difficult to synthesize due to ionic radius of Li^+ (0.76 Å) is quite similar with Ni^{2+} (0.69 Å) [2]. Therefore, cation mixing between these two ions is more likely especially in the high content of nickel [3].

The weaknesses of $LiCoO_2$ and $LiNiO_2$ were overcome by incorporating $LiCoO_2$ and $LiNiO_2$ phases into $LiCo_xNi_{1-x}O_2$ compositions. This is because the presence of cobalt will be stabilizes the structure, thus favouring good reversibility of the intercalation and deintercalation reactions [4]. In this studies, three materials of $LiCo_xNi_{1-x}O_2$ (x = 0.1, 0.2 and 0.3) was prepared by a self-propagating combustion method.

2. Experimental

The $LiCo_xNi_{1-x}O_2$ (x = 0.1, 0.2 and 0.3) materials were synthesized using self-propagating combustion method. The annealed samples were then studied by X-Ray diffraction (Pan Analytical X'Pert Pro System) using Cu Kα radiation to identify the crystalline phases and purity of the materials. The

Content from this work may be used under the terms of the Creative Commons Attribution 3.0 licence. Any further distribution of this work must maintain attribution to the author(s) and the title of the work, journal citation and DOI.
Published under licence by IOP Publishing Ltd

morphology and particle size of the samples were examined by Field Emission Scanning Electron Microscope (JEOL JSM-7600F).

The cathode was prepared using 80 wt% active material, 10 wt% binder and 10 wt% graphite. The electrolyte was 1.0 M $LiPF_6$ in 1:1 volume ratio of ethylene carbonate (EC) and dimethylcarbonate (DMC). Celguard 2400 microporous polyethylene film was used as the separator and the anode was lithium metal. The cells were assembled in an Ar-filled glove box (Unilab Mbraun).

The cells were cycled using a constant current of 1 mA at a voltage range between 3.0 and 4.3 V. The charge-discharge performance of the cell was characterized using a battery tester, the Wonatech (WBCS 3000) electrochemical testing equipment.

3. Results and discussions

The XRD patterns for $LiCo_xNi_{1-x}O_2$ (x = 0.1 - 0.3) compounds are shown in Figure 1. Phase pure $LiCo_xNi_{1-x}O_2$ (x = 0.1 - 0.3) were obtained with no impurity peaks detected in the XRD patterns. This means that the Ni^+ ions have been successfully substituted in the $LiCoO_2$ structure. It can be clearly observed that all of the fingerprint peaks *viz.* (003), (101), (006), (012), (104), (018) and (110) are easily identifiable in all of the XRD patterns. All the diffraction peaks can be indexed with α –$NaFeO_2$ type structure based on the hexagonal crystal system (*R*-3*m* space group). They are isostructural with $LiNiO_2$ and $LiCoO_2$ phases as compared with the XRD patterns in the ICDD database. The XRD patterns clearly show that the materials were single phase and the sharp peaks in the spectra indicated a high degree of crystallinity of the materials.

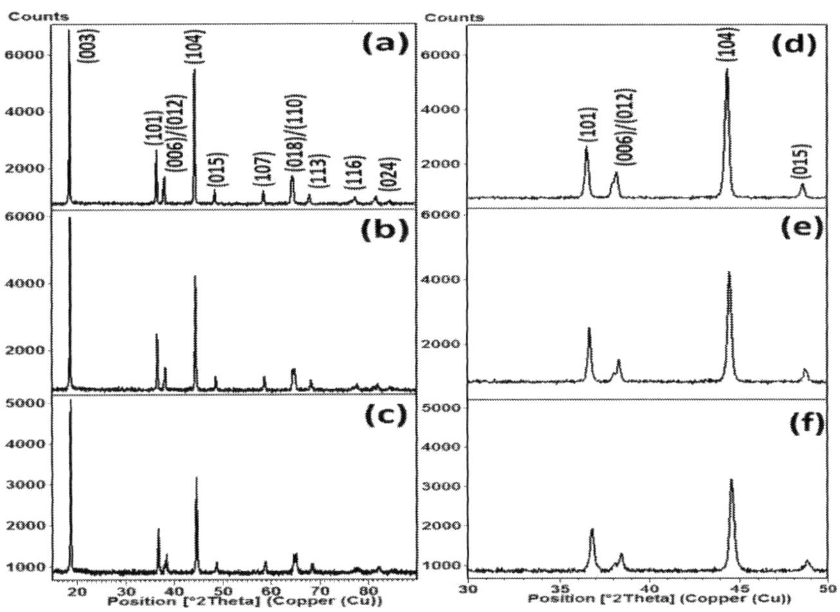

Figure 1. XRD patterns of **(a)** $LiCo_{0.9}Ni_{0.1}O_2$ **(b)** $LiCo_{0.8}Ni_{0.2}O_2$ **(c)** $LiCo_{0.7}Ni_{0.3}O_2$ cathode materials and enlargement of the XRD patterns at 30° and 50° for **(d)** $LiCo_{0.9}Ni_{0.1}O_2$ **(e)** $LiCo_{0.8}Ni_{0.2}O_2$ and **(f)** $LiCo_{0.7}Ni_{0.3}O_2$

The SEM images of $LiCo_xNi_{1-x}O_2$ (x = 0.1 - 0.3) are displayed in Figure 2. The images showed that all samples were well-formed polycrystalline materials. The $LiCo_{0.9}Ni_{0.1}O_2$ crystal size seemed larger than the $LiCo_{0.8}Ni_{0.2}O_2$ and $LiCo_{0.7}Ni_{0.3}O_2$ materials. It was also observed that as nickel content in the samples increased, the crystal size seemed to decrease. All the materials had mostly rounded polyhedral type crystals. $LiCo_{0.9}Ni_{0.1}O_2$ had crystallite size of about 50 – 300 nm which were bigger than those of $LiCo_{0.8}Ni_{0.2}O_2$ and $LiCo_{0.9}Ni_{0.1}O_2$ that had crystallite size of about 50 - 200 nm and 50 – 100 nm respectively. Some of the crystals seemed to aggregate slightly into larger particles.

Figure 2. SEM image of **(a)** $LiCo_{0.9}Ni_{0.1}O_2$ **(b)** $LiCo_{0.8}Ni_{0.2}O_2$ **(c)** $LiCo_{0.7}Ni_{0.3}O_2$ cathode materials with magnification of 50000x

The 1st cycle charge-discharge curves of the materials are shown in Figure 3. Figure 4 shows the specific capacities of the cells up to 30 cycles.

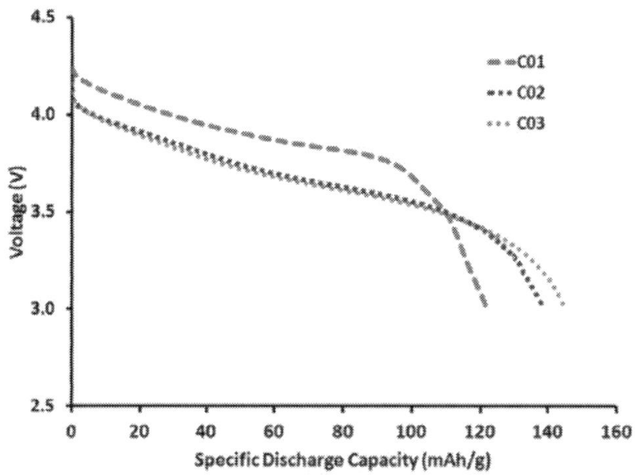

Figure 3. Initial specific discharge capacities of **(a)** $LiCo_{0.9}Ni_{0.1}O_2$ **(b)** $LiCo_{0.8}Ni_{0.2}O_2$ **(c)** $LiCo_{0.7}Ni_{0.3}O_2$ cathode materials

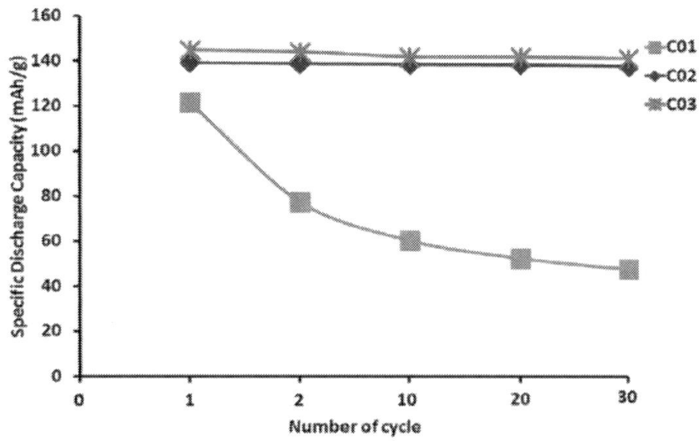

Figure 4. Cycling data of **(a)** $LiCo_{0.9}Ni_{0.1}O_2$ **(b)** $LiCo_{0.8}Ni_{0.2}O_2$ **(c)** $LiCo_{0.7}Ni_{0.3}O_2$ cathode materials

The electrochemical behavior of $LiCo_xNi_{1-x}O_2$ (x = 0.1 - 0.3) was summarized in Table 1.

Table 1. Discharge capacity of $LiCo_xNi_{1-x}O_2$ (x = 0.1 - 0.3) annealed at 700 °C for 48 h

Material 700 °C,48h	1st discharge capacity (mAh/g)	30th discharge capacity (mAh/g)	Capacity loss after 30th cycles (%)
$LiCo_{0.9}Ni_{0.1}O_2$	122	48	60.65
$LiCo_{0.8}Ni_{0.2}O_2$	140	138	1.43
$LiCo_{0.7}Ni_{0.3}O_2$	145	141	2.76

The material $LiCo_{0.7}Ni_{0.3}O_2$ has the best specific capacity of 145 mAh/g in the first cycle. It shows the highest initial discharge capacity compared to $LiCo_{0.8}Ni_{0.2}O_2$ and $LiCo_{0.1}Ni_{0.9}O_2$ with the capacity of 140 mAh/g and 122 mAh/g respectively. It also noted that $LiCo_{0.8}Ni_{0.2}O_2$ and $LiCo_{0.7}Ni_{0.3}O_2$ materials shows very good performance over 30 cycles with a capacity fade of only 1.43% and 2.76% which can be explained that these materials have potential in application as a cathode material for lithium-ion batteries. The discharge capacities of all the materials seem to converge at about the same value at the 30th cycle even though $LiCo_{0.7}Ni_{0.3}O_2$ has better initial capacities. $LiCo_{0.7}Ni_{0.3}O_2$ shows better performance due to the smaller crystallite size. This is because small crystallite size provides high surface area to volume ratio resulting in produce good electrochemical performance [5].

4. Conclusion

Nickel substitution in $LiCoO_2$ synthesized using the self propagating combustion method has been successfully done in $LiCo_xNi_{1-x}O_2$ (x = 0.1, 0.2, and 0.3). $LiCo_{0.7}Ni_{0.3}O_2$ material, annealed at 700 °C for 48 h, showed good promise for application in Li-ion batteries exhibiting an initial discharge capacity of 145 mAh/g. Different stoichiometry of the nickel substitution in $LiCoO_2$ effect the electrochemical behaviour of the cathode materials.

Acknowledgement

The authors would like to thank Institute of Research Management & Innovation (IRMI) for the grant 600-IRMI/MyRA 5/3/LESTARI (0073/2016) and Institute of Science, UiTM Shah Alam, Malaysia for financially funding this project.

References

[1] D.S. Adipranoto, T. Ishigaki, A. Hoshikawa, K. Iwase, M. Yonemura, K. Mori, T.Kamiyama, Y. Morii. M. Hayashi 2014 *Solid State Ionics* 262 92
[2] Y. Ding, R. Wang, L. Wang, K. Cheng, Z. Zhao 2017 *Energy Procedia* 105 2941
[3] K. Elong, N. Kamarulzaman, R. Rusdi, N. Badar, M.H. Jaafar 2013 *ISRN Condensed Matter Physics* 568191
[4] H.Rim, J. Song, D.R. Mumm 2014 *Ceramics Int.* 40 3511
[5] K. Elong, N. Kamrulzaman 2014 *Advanced Materials Research* Vol. 895 400
[6] J.M. Wang, J.P. Hu, C.Y. Ouyang, S.Q. Shi, M.S. Lei 2011 *Solid State Commun.* 151 234
[7] Md. M. Mallick, S. Vitta 2017 *Mater. Chem. Phys.* Volume 198 266
[8] M.H. Jaafar, N. Kamarulzaman, K.Elong, N.S. Mohamed, R. Rusdi 2013 *Inter. J. Electrochem. Sci.* 8 2254
[9] M. Song, I. Kwon, S. Shim,J.H. Song 2010 *Ceramics Int.* 36 1225
[10] A. Azahidi, R.Rusdi, N.A.M. Mokhtar, K.Elong 2017 *AIP Conference Proceedings* 1877 040005

IOP Conf. Series: Journal of Physics: Conf. Series **1083** (2018) 012032 doi:10.1088/1742-6596/1083/1/012032

Characterization of Electroless Ni-P and Ni-P/Al_2O_3 Composite Coating on Al6061 Alloy

C C Chai and A S Anasyida

Structural Materials Niche Area Group, School of Materials and Mineral Resources Engineering, Engineering Campus, Universiti Sains Malaysia, 14300 Nibong Tebal, Penang, Malaysia

Abstract. Aluminum and its alloy have a light density with good properties of stiffness and corrosion resistance. Although it has a good property, aluminum alloy still vulnerable to wear and corroded. A possible solution to protect the aluminum product from wear and corrosion is by applying coating such as nickel-phosphorus (Ni-P) coating onto its surface. Co-deposition of alumina (Al_2O_3) particles into Ni-P coating is expected to improve the wear resistance of the Ni-P coating layer. This study investigated the Ni-P and Ni-P/Al_2O_3 composite coating on Al 6061 alloy via electroless coating by varying the agitation speed and amount of Al_2O_3. The characterization of coatings includes morphology, coating thickness, hardness and wear properties were investigated by using field emission scanning electron microscopy (FESEM), optical microscope, Vicker microhardness and pin-on-disc tribometer. The agitation improved the adherence of alumina particles to the substrate. The increment amount of alumina in Ni-P/Al_2O_3 composite coating showed a decrement in percentage of nickel and phosphorus present in the coating but improved the hardness of the coating. The wear rate and coefficient of friction of the Ni-P/Al_2O_3 composite coating were lower than Ni-P coating.

1. Introduction

Electroless nickel plating on aluminum and many of its alloys is applied with a view to improve the surface hardness, corrosion protection and wear resistance. Aluminium and its alloy are considered a difficult substrate to be coated by autocatalytic nickel deposition. This is due to the other metals (e.g.) nickel does not adhere onto new formed aluminum oxide film. In order to create a suitable (oxide-free) aluminum surface for nickel deposition and adherent nickel coating, different pre-treatments for aluminum were developed. Most of these pre-treatments involve chemical dissolution of the aluminum oxide film and formation of an immersion deposit of another metal such as zinc and tin. Zinc immersion remained commonly used due to the fact that it can be adjusted to a large variety of aluminum alloys. The pre-treatment process involved many procedure and process parameter depending on aluminum alloy used.

The requirement towards uniformity of metal coatings and their characteristics such as hardness, resistance against corrosion, wear hardness, coefficient of friction etc., are constantly increasing. The properties of Ni-P can be improved by co-deposition with hard particles (e.g. SiC, WC, Al_2O_3) and dry lubricant (e.g. PTFE, MoS_2, graphite). The major challenges with the co-deposition of particles are the achievement of a high level of co-deposition, and the agglomeration of particles suspended in the electrolytes. Currently, research on electroless Ni-P coating is mostly focused on the effect of

Content from this work may be used under the terms of the Creative Commons Attribution 3.0 licence. Any further distribution of this work must maintain attribution to the author(s) and the title of the work, journal citation and DOI.
Published under licence by IOP Publishing Ltd

IOP Conf. Series: Journal of Physics: Conf. Series **1083** (2018) 012032 doi:10.1088/1742-6596/1083/1/012032

surfactant, pH, or reducing agents toward electroless Ni-P coating [1-4]. However, the agitation (stirring speed) during the coating process also one of the factors that might affect the properties of Ni-P coating. In this project, effect of agitation toward electroless Ni-P coating was studied. Researchers have reported that the addition of SiC improved [5] tribological behavior, while the others indicate worsening of wear behavior after introduction of SiC particles in Ni-P coating. However, Karthikeyan & Vijayaraghavan [6] have found that the hardness and wear resistance of steel substrate have improved by incorporation of Al_2O_3 in electroless nickel. Aluminium and alumina particles were selected as aluminum has higher strength and alumina particles can improve the wear resistance of Ni-P coating. In this study, Al_2O_3 particles with different amount were co-deposited in electroless coating and its surface morphology were characterized. Hardness and wear resistance for both Ni-P coatings and Ni-P/Al_2O_3 were compared.

2. Experimental Procedure

Al 6061 alloy with a size of 40 mm x 20 mm x 1 mm was used as the substrate. Alumina (99.98%) having particle sizes in the range 0.33 µm to 13.35 µm supplied by SkySpring Nanomaterials was used as the reinforcement material. The average alumina particles size was 4.02 µm The aluminum sheets were ground using SiC abrasive papers with grit of 400, 600, 800, 1200 and 2000 and followed by polishing using diamond suspension of 0.6 and 0.1 µm. The polished specimens were immersed in an ultrasonic bath containing acetone to ensure oil and dirt-free surface. Then the samples were immersed in a mixture of sodium carbonate and sodium phosphate dodecahydrate. After that, the samples were immersed in a mixture of sulphuric acid, phosphoric acid and propionic acid for 2 minutes at 65°C. For each step, the samples were washed and rinsed with distilled water. In order for chemical activation of the surface, the zincating pre-treatment was executed by dipping and stirring of samples into zincating bath for 1 min. The deposition of a uniform Zn layer at the Al surface requires double process of zincating process within a short time to avoid the formation of rough surfaces. Between the two zincating processes, Al substrates were dipped in 50% volume of nitric acid for 1 minute.

After zincating process, the Al substrates immediately immersed in the electroless nickel bath solution that was pre-heated at 89°C containing 8 g/L Al_2O_3 with different agitation speed. The agitation of electroless nickel coating process was carried out at different stirring speed 200, 250 and 300 rpm. Next, to produce Ni-P/Al_2O_3 composite coatings, different amount of Al_2O_3 particles (4 g/L, 6 g/L, 8 g/L) were added to the bath and kept in suspension by magnetic stirring and air bubbling. The composition of the composite electroless nickel solution is shown in Table 1. The zincate layer on the surface might become less activated if the substrate exposed longer time before immersed in electroless bath. Therefore, the nickel deposition might not occur on the substrate. Thus, electroless nickel process was prepared beforehand. The surface morphology and surface characterization of NiP/Al_2O_3 coatings were done by FESEM (LEO SUPRA 35VP) equipped with EDX spectrum analysis. The microhardness of the coatings was measured using Vicker microhardness Tester. The wear resistance of the coatings was determined by Ducon TR-20 pin-on-disc tester at sliding speed of 0.5 m/s, load of 3 kg and sliding distance of 1000 m. The specific wear rate is calculated based on Equation 1.1 (Karthikeyan & Ramamoorthy 2014).

$$Specific\ Wear\ Rate = \frac{W}{DL}\, kg\, N^{-1} m^{-1} \qquad \text{(Eq. 1.1)}$$

Where, W = mass loss of sample, D = sliding distance and L = load.

Table 1. Composition of electroless nickel solution

Function	Chemical used	Composition
Source of Nickel	Nickel (II) sulphate heptahydrate	0.1 mol
Reducing Agent	Sodium hypophosphite	0.35 mol
Complex Agent	Glycine	0.09 mol
	Lactic acid	0.40 mol
	Propionic acid	0.02 mol
Stabilizer	Copper (II) sulphate	4 ppm
	Lead (II) acetate trihydrate	2 ppm
Surfactant	Cetyl trimetylammonium bromide (CTAB)	0.3 g
Alumina	Al_2O_3	4, 6, 8 g/L

3. Results and Discussion

3.1 Determining agitation for the electrolysis process

Figure 1 shows the morphology of Ni-P/Al_2O_3 composite coating with different agitations. These coatings produced different size of nodules with similar shape. Similar observation has been reported by Sen et al. [7], which different size of nodules of Ni-CeO_2 is formed with different agitation. Figure 2 shows the cross-section images of Ni-P with alumina particles coating layer at different agitation speed. The coating layers for 200 rpm, 250 rpm and 300 rpm had almost the same thickness which was about 18 μm. The changes in the agitation of electroless nickel coating process did not show a significant influence on the thickness of coating layers. The samples were further analyzed using EDX. The EDX analysis in Figure 3 showed the increasing of nickel content by decreasing agitation. The lower amount of nickel content was detected in a coating that used agitation of 300 rpm. About 58.24 wt% of nickel were found in the coating with agitation of 300 rpm, while 82.25 wt% and 85.03 wt% of nickel were detected in coating with agitation of 250 rpm and 200 rpm, respectively. Bulasara et al. [8] also found that the higher agitation resulted in decreasing amount of nickel deposited in the coating. In addition, higher agitation would result in the decrease of deposition rate as nickel does not have enough time to be deposited onto the surface. Moonir-Vaghefi & Saatchi [9] also reported that higher agitation would cause the decrease of coating hardness. Based on the result, it can be concluded that the agitation that applied in an electroless nickel coating process not only maintain the temperature and improving the stability of the electroless nickel bath solution, but also influences the nickel deposited in the Ni-P with alumina particles coating layer. Thus, agitation for 200 rpm was used as it has the highest amount of nickel deposited.

3.2 Hardness

The hardness of Ni-P/Al_2O_3 composite coating was measured using Vicker micro hardness indenter. Seven measurements were done on the surface of the coating layer and the average value of microhardness of each sample was calculated and plotted in Figure 4. From the figure, the hardness of composite coating was increased to 90.2 HV with the addition of 4 g/L of alumina particles. Then, the hardness of composite coating showed an increasing trend with higher composition of alumina particles. The high hardness achieved with the addition of 8 g/L which recorded a hardness value of 95

HV. The improvement of the hardness of Ni-P/Al_2O_3 composite coating compared with binary Ni-P coating is due to alumina particles which were added in the coating process. As the alumina is one of the hard particles that used for wear resistance [10], this results in the improvement of the hardness of the coating layer. Alumina hard particles act as a barrier to plastic deformation of Ni–P matrix under the load [11]. As the higher amount of alumina particles added during the coating process, it produces a higher composition of alumina particles in coating layer. This result shows the similar trends with electroless Ni-P/WC composite coating that coated on steel [12]. They found that the hardness of Ni-P/WC composite coating on steel increases with an increment of WC composition. Meanwhile, in this research, increases of alumina particle composition used from 4 to 8 g/L during electroless nickel coating resulted in the increment of hardness of Ni-P/Al_2O_3 composite coating. Thus, sample with the highest hardness (8 g/L alumina) was used to evaluate the wear behaviour of Ni-P and Ni-P/Al_2O_3.

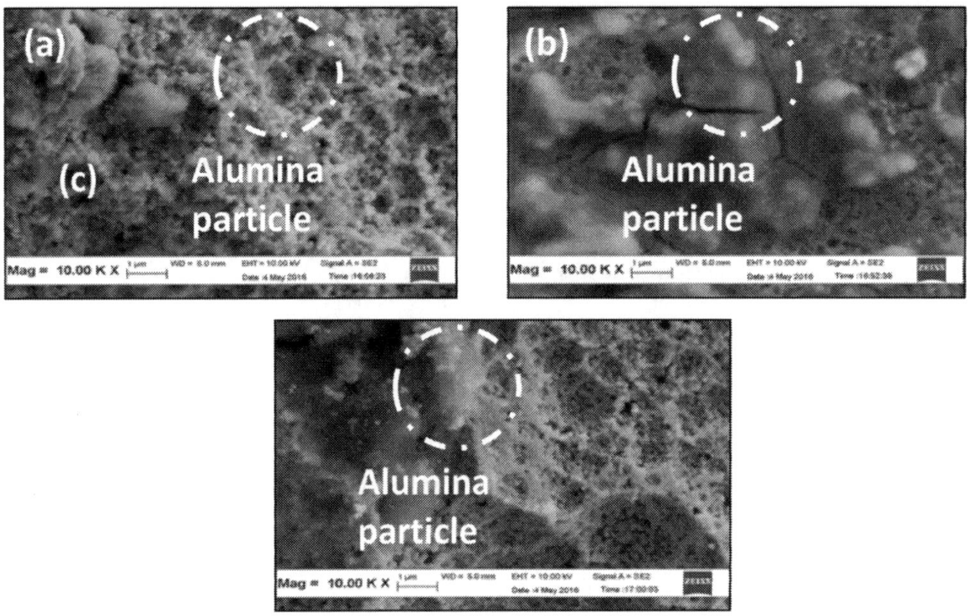

Figure 1. SEM morphology of Ni-P/Al_2O_3 coating layer with agitation of (a) 200 rpm, (b) 250 rpm, and (c) 300 rpm

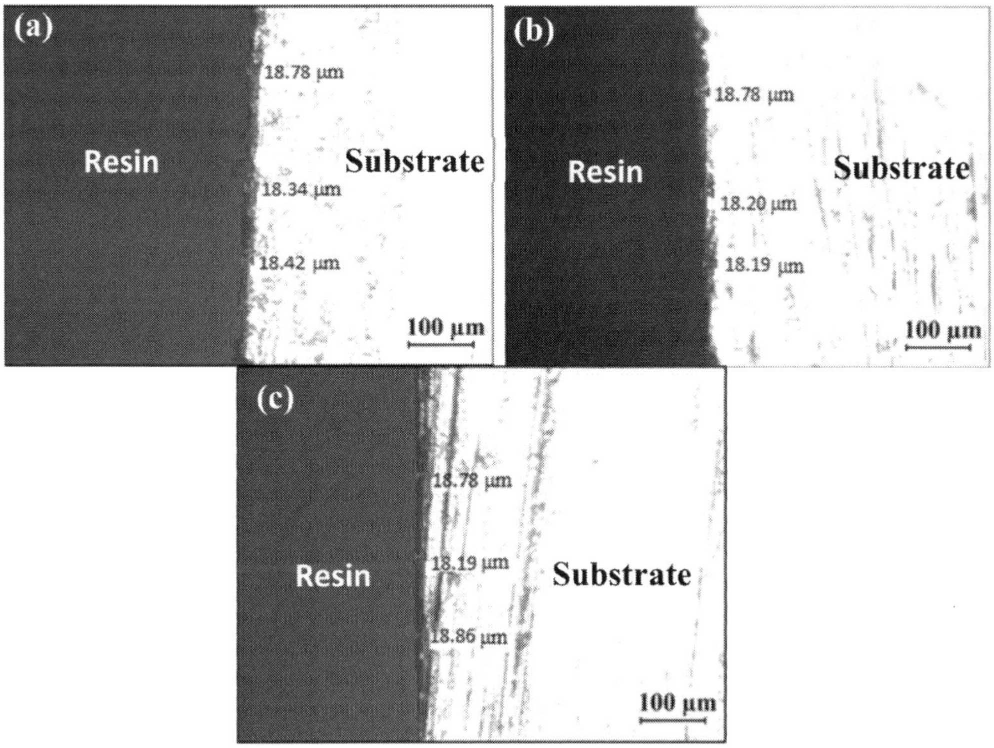

Figure 2. Thickness of Ni-P/Al_2O_3 coating with agitation of (a) 200 rpm, (b) 250 rpm and (c) 300 rpm

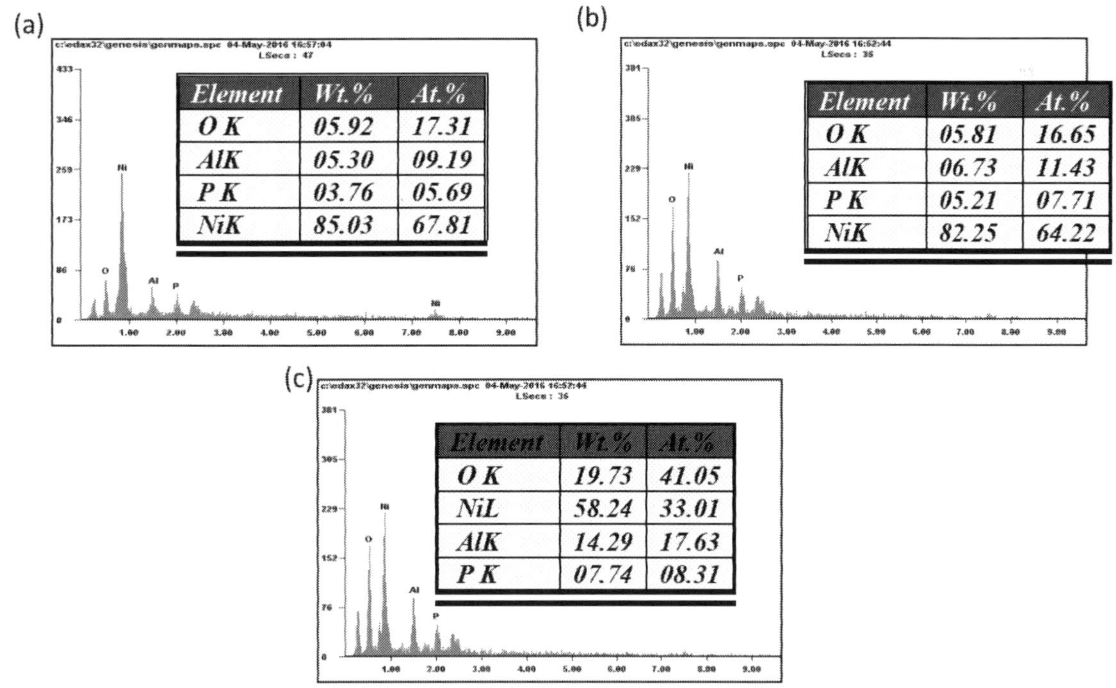

Figure 3. EDX analysis of Ni-P/Al_2O_3 coating with different agitation (a) 200 rpm, (b) 250 rpm, (c) 300 rpm

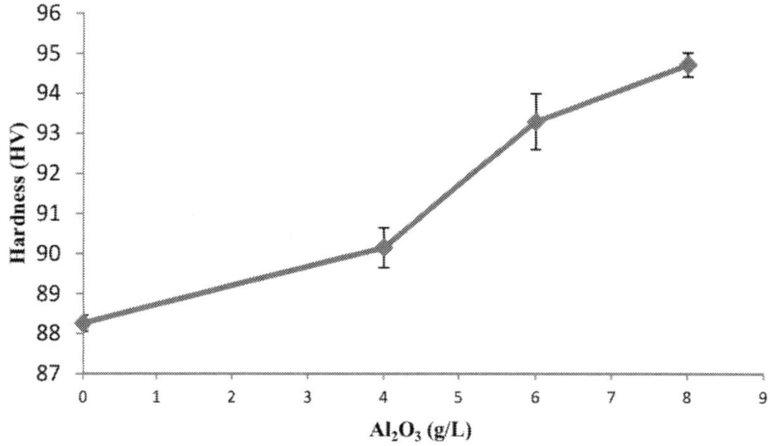

Figure 4. Average coating hardness of Ni-P/Al_2O_3 composite coating with different alumina particles concentration

3.3 Wear behavior

Wear test was performed to evaluate the Ni-P and Ni-P/Al_2O_3 composite coating performance on Al 6061 aluminium substrate at sliding speed of 0.5 m/s, sliding distance of 1000 m and at an applied load of 30 N. Figure 5 shows the wear rate of Ni-P/Al_2O_3 composite coating which was lower than Ni-P coating. This was due to the alumina particles acted as abrasive for the wearing toward the Ni–P coating. Alirezaei et al. [11] showed that the existence of hard alumina particles became a barrier to plastic deformation of Ni–P matrix under the load. Thus, the Ni-P/Al_2O_3 composite coating has lower wear rate. Devaraju et al. [13] have found that during wear process, the Al_2O_3 particles were released on the wear surface. Mahmoud et al. [14] and Tjong et al. [15] also found that both Al_2O_3 and boron nitride (BN) particles enhances the wear resistance.

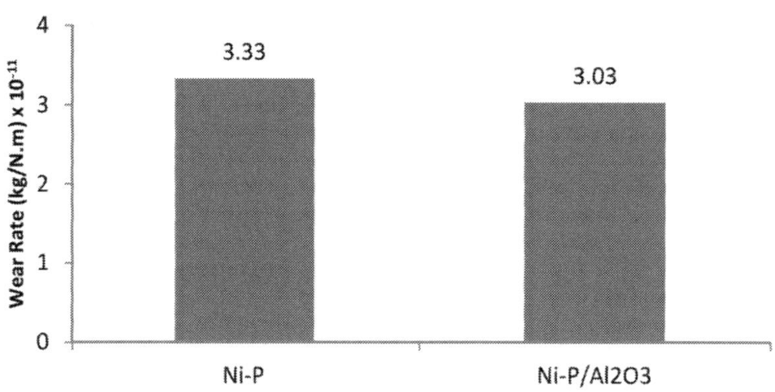

Figure 5. Wear rate of Ni-P coating and Ni-P/Al_2O_3 composite coating

4. Conclusion

The Ni-P/Al_2O_3 composite coating was successfully developed on the surface of Al 6061 alloy. The morphology of Ni-P/Al_2O_3 composite coating depends on the amount of alumina particle deposited.

The hardness of Ni-P/Al_2O_3 composite coating was improved with the increasing of alumina particles. The highest hardness of Ni-P/Al_2O_3 composite coating was 95 HV with containing 8 g/L of alumina particles. The wear rate of the Ni-P/Al_2O_3 composite coating was lower than Ni-P coating.

Acknowledgements
The authors gratefully thank Universiti Sains Malaysia for providing the facilities to complete the project.

References

[1] Afroukhteh, S, Dehghanian C and Emamy M 2012 Progress in Natural Science: Materials International 22 p 480–487.
[2] Chuang Y C, Chung S T, Chiu, S Y, Li C Y and Tsai W T 2013 Thin Solid Films 529 p 322–326.
[3] Dinakaran R, Arumugam A, Elansezhian R, 2013, Advance in Tribology 2013 p 304–308.
[4] Karthikeyan S and Ramamoorthy B 2014 Applied Surface Science 307 p 654–660.
[5] Malfatti C F, Veit H M, Santos C B, Metzner M, Hololeczek H and Bonino J P 2009 Tribology Letters, 36 p 165–173.
[6] Karthikeyan S and Vijayaraghavan L 2016 International Journal of Materials, Mechanics and Manufacturing 4 p 106-110.
[7] Sen R, Das S and Das, K 2011 Surface and Coatings Technology 205 p 3847–3855.
[8] Bulsara V K, Thakuria H, Uppaluri R and Purkait M K 2011 Journal of Materials Processing Technology 211 p 1488–1499.
[9] Moonir-Vaghefi S M and Saatchi A 1997 Metal finishing p 102–106.
[10] Sahoo P and Das S K 2011 Materials and Design 32 1760–1775.
[11] Alirezaei S, Monirvaghefi S M, Salehi M, Saatchi A 2004 Surface and Coatings Technology 184 p 170–175.
[12] Hamid Z A, El Badry S A and Aal A A 2007 Surface and Coatings Technology 201 p 5948–5953.
[13] Devaraju A, Kumar A and Kumaraswamy A 2013 Materials and Design 51 p 331–341.
[14] Mahmoud E R I, Takashi M, Shibayanagi T, Ikeuchi K 2010 Wear 268 p 1111–1121.
[15] Tjong S C, Lau K C and Wu S Q 1999 Metallurgical and Materials Transactions A 30 p 2551–2555.

IOP Conf. Series: Journal of Physics: Conf. Series **1083** (2018) 012033 doi:10.1088/1742-6596/1083/1/012033

Electrical Properties of Li Substituted $(Tl_{0.85}Cr_{0.15})Sr_{2-x}Li_xCaCu_2O_7$ Superconductor

J. Nur-Akasyah[1], K. Muhammad Aizat[1], Nurul Raihan Mohd Suib[1,2] and R. Abd-Shukor[1]

[1]School of Applied Physics, Universiti Kebangsaan Malaysia, 43600 Bangi, Selangor, Malaysia,
[2]Faculty of Applied Sciences, Universiti Teknologi MARA, 40450 Shah Alam, Selangor, Malaysia

ras@ukm.edu.my

Abstract. The effects of Li substitution for Sr in $(Tl_{0.85}Cr_{0.15})Sr_{2-x}Li_xCaCu_2O_7$ (Tl-1212) with x = 0 - 0.10 were investigated. Tl-1212 superconductor pellets were prepared via solid-state reaction and precursor method. The samples were characterized by x-ray diffraction and four-point probe methods. The XRD patterns showed that all samples formed mixed phases of Tl-1212 and $Ca_{0.3}Sr_{0.7}CuO_2$ (CSCO). The temperature-dependent resistance measurements showed metallic-behavior for all samples with onset transition temperature, $T_{c\ onset}$ between 96 K and 105 K and zero resistance temperature, $T_{c\ zero}$ between 85 K and 94 K. The transition-width, ΔT_c of all samples was in the range of 4 K to 11 K. The x = 0.02 and 0.07 sample showed increase in $T_{c\ onset}$ to 105 K and 102 K, respectively. This work showed that Li improved the transition temperature for x = 0.02 and 0.07 samples although the dominant phase was not Tl-1212.

1. Introduction

The discovery of high-temperature superconductivity in the TlBaCaCuO system by Sheng and Hermann in 1988 [1], have induced many scientists and researchers to investigate a new high temperature record. Since then, many new superconducting phases have been synthesized. Based on a research by Nakajima et. al, Tl-based 1212 type phases such as $TlBa_2CaCu_2O_7$ and $TlSr_2CaCu_2O_7$ have similar structures as YBCO-123 (also rare earth-123 i.e Nd and Gd) [2]. Some work reported that $TlSr_2CaCu_2O_7$ becomes superconducting at about 80 K [3]. Another reported that $TlSr_2RCu_2O_7$ (R = rare earth) becomes superconducting at 90 K [4]. Following this discovery, a novel Cr-substituted $TlSr_2CaCu_2O_7$ with T_c up to 110 K has been recorded [5]. In 2001, a study about the effect of transition metals substituted on $(Tl_{0.85}Cr_{0.15})Sr_2CaCu_2O_7$ system has recorded lower T_c reading with the substitution of Zn and Mn [6] and rare-earth elements [7] on $(Tl_{0.85}Cr_{0.15})Sr_2CaCu_2O_7$ had been reported. The co-precipitation method has been used to prepare ultra-fine $(Tl_{0.85}Cr_{0.15})Sr_2CaCu_2O_7$ powders [8].

Most of the efforts to improve T_c of the Tl-1212 phase were by addition of trivalent ions, especially the rare-earth elements [4,9]. Rare-earths elements have also been used to improve the $Bi_{1.6}Pb_{0.4}Ba_2Ca_2Cu_3O_{10}$ superconductor [10]. Substitution of single and monovalent alkali metals such as lithium into Tl-based high-temperature superconductors is interesting to study. Li is a light element that may give rise to high-frequency phonons if incorporated into the crystal lattice. The effective ionic size with 6-fold coordination of Li^{+1} (0.76 Å) is close to Cr^{+3} (0.62 Å), Cu^{+2} (0.73 Å), Tl^{+3} (0.885 Å) and

Content from this work may be used under the terms of the Creative Commons Attribution 3.0 licence. Any further distribution of this work must maintain attribution to the author(s) and the title of the work, journal citation and DOI.
Published under licence by IOP Publishing Ltd

Sr^{+2} (1.18 Å) [11]. Tl exists in both +3 and +1 oxidation states in the Tl-based high-temperature superconductors. Substitution of Li into $Tl_2Ba_2Ca_2Cu_3O_{10}$ shows T_c increases up to 117 K [12]. Most rare-earth elements with trivalent ions (R^{+3}) are effective in stabilizing the Tl-1212 phase.

The objective of this study was to investigate the effect of Li substitution on Sr sites in $(Tl_{0.85}Cr_{0.15})Sr_{2-x}Li_xCaCu_2O_7$ for $x = 0$ - 0.10. The X-ray diffraction method was used to determine the phase and the DC electrical resistance versus temperature measurements was performed to determine T_c.

This work showed that $Ca_{0.3}Sr_{0.7}CuO_2$ was presented as a major or minor phase in all samples. The monovalent Li^{+1} improved the transition temperature in some of the samples as this alkali metal was substituted into Tl-1212.

2. Experimental details

Samples with nominal compositions $(Tl_{0.85}Cr_{0.15})Sr_{2-x}Li_xCaCu_2O_7$ ($x = 0$, 0.02, 0.05, 0.07 and 0.10) were prepared by using the solid-state reaction method. Appropriate amounts of high purity (99.9+%) powders of Tl_2O_3, Cr_2O_3, $SrCO_3$ (or SrO), CaO (or $CaCO_3$), CuO and Li_2CO_3 from Sigma Aldrich were mixed and ground in an agate mortar. The precursor powders were then calcined in air at about 800°C for 48 h with several intermittent grindings to obtain homogenous powders. Appropriate amounts of Tl_2O_3 and Cr_2O_3 were then added to the precursor and completely mixed. The powders were then pressed into pellets with 13 mm diameter and 2 mm thickness under a load of 7 tons using a hydraulic press. The pellets were then placed in a ceramic boat and heated in a tube furnace at 1000°C in flowing oxygen for 4 min followed by furnace cooling to room temperature.

The powder X-ray diffraction method was used to determine the resultant phases. A Bruker model D8 Advance diffractometer with CuK_α source was used. The volume fraction of the phases were estimated by assuming that the amounts of Tl-1212 and the $Ca_{0.3}Sr_{0.7}CuO_2$ (CSCO) phases are proportional to the strongest diffraction line of each phase, i.e. the (103) reflection of the Tl-1212 phase and the (041) reflection of the CSCO phase [13,14]. The lattice parameters were calculated by employing at least 10 diffraction peaks. The percentage of the volume fraction for Tl-1212 and CSCO phase has been calculated by using the formula:

$$\text{Tl-1212 \%} = \frac{\sum I_{Tl-1212}}{\sum I_{CSCO} + \sum I_{Tl-1212}} \times 100\%$$

$$\text{CSCO \%} = \frac{\sum I_{CSCO}}{\sum I_{CSCO} + \sum I_{Tl-1212}} \times 100\%$$

The DC electrical resistance versus temperature measurements was carried out using the four-point probe method with silver paste contacts in conjunction with a closed cycle refrigerator from CTI Cryogenics Model 22 and temperature controller from Lake Shore Model 330. A constant current source between 1 and 100 mA was used throughout the measurements.

3. Results and discussion

The XRD patterns (Figure 1) shows the $x = 0$ and 0.07 substituted samples exhibited the Tl-1212 (space group P4/mm) as the major phase while CSCO (space group Cmcm (63)) exhibited as the major phase in the $x = 0.02$, 0.05 and 0.10 substituted samples (Table 1). CSCO phase was reported to be a minor phase in some high-temperature superconductors [14]. Our results were due to the calcination temperature (800°C) of the precursor as the melting point of Li_2CO_3 was 723°C. A study reported that 900°C calcination temperature of $(Tl_{0.85}Cr_{0.15})Sr_2CaCu_2O_7$ superconductor resulted in a mixed Tl-1212 and Tl-1201 phase [6]. This indicated that Li did not enhance the formation of the Tl-1212 phase. It can be seen from the data in Table 1 that the lattice parameters of all samples did not vary systematically from the non-substituted sample. For example, the $x = 0$ sample showed $a = 3.8202$ Å and $c = 12.0187$ Å and $x = 0.07$ sample showed $a = 3.8173$ Å and $c = 12.0187$ Å for Tl-1212 phase.

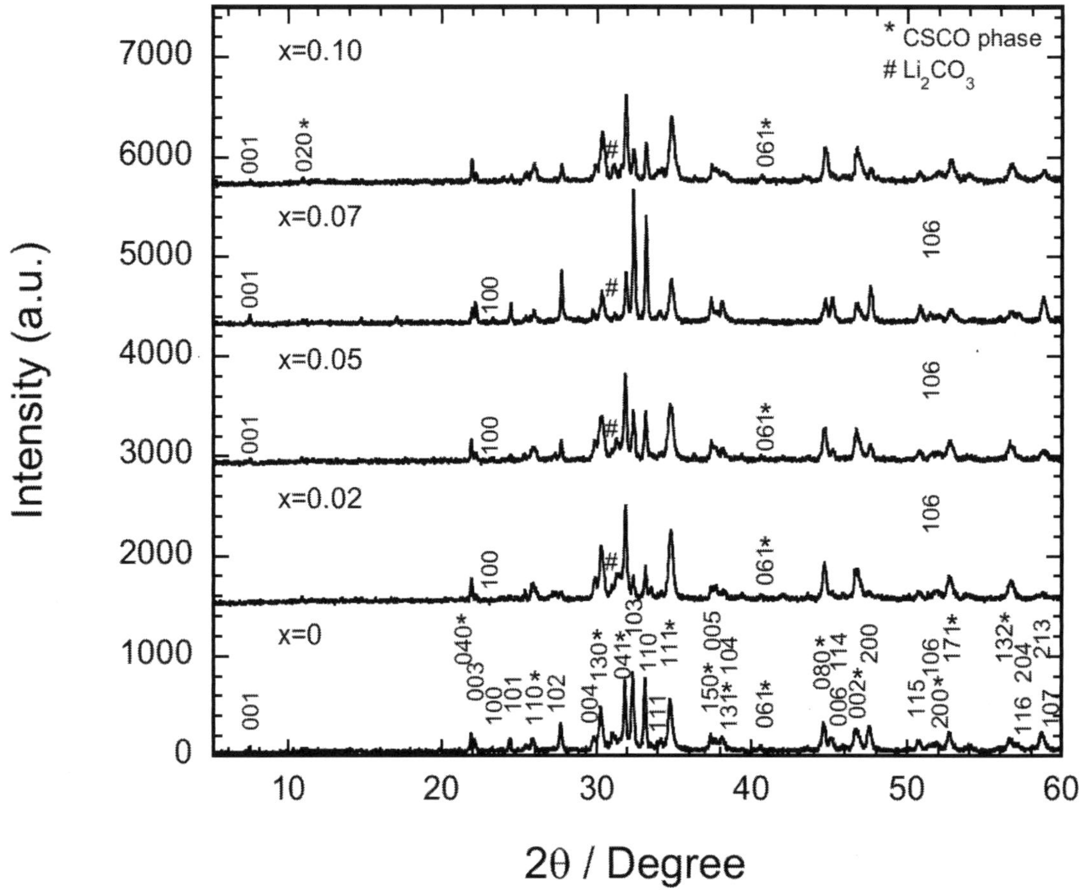

Figure 1. XRD patterns of $(Tl_{0.85}Cr_{0.15})Sr_{2-x}Li_xCaCu_2O_7$ for x = 0-0.10. * indicate CSCO and # indicate Li_2CO_3

The signature peak for Li_2CO_3 is at 2θ = 31.8° (which does not overlap with the Tl-1212 and CSCO peaks). From Table 1, Li substituted sample with x = 0.02 showed the highest amount of $Ca_{0.3}Sr_{0.7}CuO_2$ phase. The lattice parameter for CSCO phase (x = 0.02) Li substituted sample was a = 3.5193 Å, b = 16.2180 Å and c = 3.8944 Å. Li+1 was not substituted into the Tl-1212 crystal lattice. In the YBa2Cu3O7 system, Li was substituted into the Cu site [15].

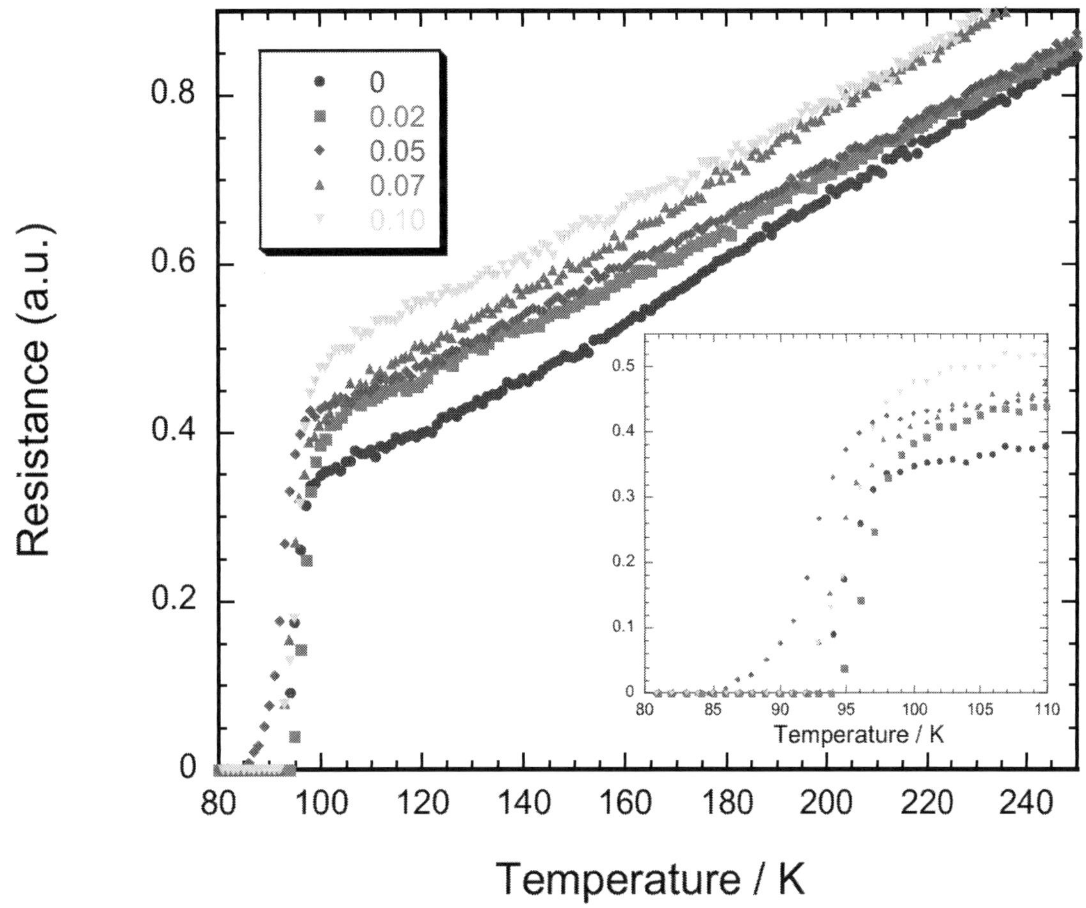

Figure 2. Electrical resistance versus temperature of $(Tl_{0.85}Cr_{0.15})Sr_{2-x}Li_xCaCu_2O_7$ for x = 0-0.10

The resistance versus temperature curves shows that the transition temperature was slightly increased as Li was substituted (Figure 2). The normal state resistance versus temperature curves showed metallic behavior for all samples. The onset transition temperature ($T_{c\ onset}$) of the samples was between 96 K and 105 K. The zero resistance temperature ($T_{c\ zero}$) was between 85 K and 94 K. Sample with x = 0.05 showed the lowest temperature reading for both $T_{c\ onset}$ and $T_{c\ zero}$ (96 K and 85 K). The transition-width (ΔT_c) of all samples is in the range of 4 K to 11 K (Table 1). Sample with x = 0.02 showed the highest increase in $T_{c\ onset}$ and $T_{c\ zero}$ although the major phase was CSCO. This indicated that the optimum Li composition for the $(Tl_{0.85}Cr_{0.15})Sr_{2-x}Li_xCaCu_2O_7$ superconductor was at x = 0.02 even though the Tl-1212 type phase was not the dominant phase in the sample. Anomalies in the electrical resistance and diamagnetic signals were reported at various temperatures in CSCO [16] but not observed in our measurements.

Table 1 shows $T_{c\ zero}$, $T_{c\ onset}$ ΔT_c, the percentage of the volume fraction $V_{Tl\text{-}1212}$ and the lattice parameter for Tl-1212 phase, and V_{CSCO} of the samples.

Table 1. $T_{c\ zero}$, $T_{c\ onset}$, ΔT_c, the percentage of the volume fraction $V_{Tl\text{-}1212}$ and the lattice parameter for Tl-1212 phase, and V_{CSCO} for $(Tl_{0.85}Cr_{0.15})Sr_{2-x}Li_xCaCu_2O_7$ ($x = 0 - 0.10$) samples

x	0	0.02	0.05	0.07	0.10
$T_{c\ zero}$ [K]	93	94	85	91	92
$T_{c\ onset}$ [K]	94	105	96	102	99
ΔT_c [K]	4	11	11	11	7
$V_{Tl\text{-}1212}$ / %	51	23	30	73	30
a (Å)	3.8202	3.8173	3.8187	3.8173	3.8173
c (Å)	12.0187	11.9975	12.0116	12.0187	11.9928
V_{CSCO} / %	49	77	70	27	70
a (Å)	3.5241	3.5193	3.5241	3.5186	3.5172
b (Å)	16.2339	16.2180	16.2339	16.2149	16.2180
c (Å)	3.8906	3.8944	3.8867	3.8852	3.8852

4. Conclusion

The effect of Li substitution on $(Tl_{0.85}Cr_{0.15})Sr_{2-x}Li_xCaCu_2O_7$ system was investigated. These results showed that Li was not able to stabilize the Tl-1212 phase. X-ray diffraction patterns showed that all samples formed mixed phases of Tl-1212 and $Ca_{0.3}Sr_{0.7}CuO_2$. Substitution of Li in $(Tl_{0.85}Cr_{0.15})Sr_{2-x}Li_xCaCu_2O_7$ superconductor enhanced the transition temperature in $x = 0.02$ and $x = 0.07$ sample.

Acknowledgements

This research was supported by Universiti Kebangsaan Malaysia under grant number AP-2015-006 and the Ministry of Higher Education of Malaysia under grant no. FRGS/1/2017/STG02/UKM/01/1.

References

[1] Z.Z. Sheng, A.M. Hermann, Bulk superconductivity at 120 K in the Tl–Ca/Ba–Cu–O system, Nature 332 (1988) 138–139.
[2] S. Nakajima, M. Kikuchi, Y. Syono, N. Kobayashi, Y. Muto, Superconductivity of high-Tc $TlBa_2Ca_{1-x}RE_xCu_2O_7$ (RE = Nd, Gd or Y) system, Physica C 168 (1990) 57–62.
[3] Z.Z. Sheng, A.M. Hermann, D.C. Vier, S. Schultz, S.B. Oseroff, D.J. George, R.M. Hazen, Superconductivity in the Tl–Sr–Ca–Cu–O system, Phys. Rev. B 38 (1988) 7074–7076.
[4] Z.Z. Sheng, Y. Xin, D.X. Gu, J. M. Meason, J. Bennett, D. Ford, D.O. Pederson, Semiconducting $TlSr_2RCu_2O_7$ (R = rare earth) and its superconducting derivatives, Z. Phys. B. 84 (1991) 349-352.
[5] Z. Z. Sheng, D.X. Gu, Y. Xin, D.O. Pederson, L.W. Finger, C.G. Hadidiacos, R.M. Hazen, A new 1212-type phase: Cr-substituted $TlSr_2CaCu_2O_7$ with T_c up to about 110 K, Mod. Phys. Lett. B 5 (1991) 635–642.
[6] R. Abd-Shukor, S.V. Lo, Effect of Zn and Mn substitutions on the formation and superconductivity of Tl-1212 type phase $(Tl_{0.85}Cr_{0.15})Sr_2CaCu_2O_7$, J. of Mat. Sci. Lett. 20 (2001) 159-161.
[7] R. Abd-Shukor, Nor Azah Nik Jaafar, Formation and superconductivity of Pr and Nd-substituted Tl-1212 phase $(Tl_{0.85}Cr_{0.15})Sr_2CaCu_2O_7$, J. of Mat. Sci.: Mater. in Elect. 10(9) (1999) 677-681.
[8] Imad Hamadneh, Yeong Wai Kuan, Lee Teik Hui & R. Abd-Shukor, Formation of $Tl_{0.85}Cr_{0.15}Sr_2CaCu_2O_7$ superconductor from ultrafine co-precipitated powders, Mater. Lett. 60 (2006) 734-736.
[9] L. R. Sung, J. F. Wang, W. H. Lee, Preparation and magnetic relaxation of Th-doped $TlSr_2CaCu_2O_7$ superconductors, Physica C 317 (2001) 317-323.
[10] H. Bagiah, S.A. Halim, S.K. Chen, K.P. Lim, M.M. Awang Kechik, Effects of rare earth nanoparticles (M = Sm_2O_3, Ho_2O_3, Nd_2O_3) addition on the microstructure and superconducting transition of $Bi_{1.6}Pb_{0.4}Sr_2Ca_2Cu_3O_{10+\delta}$ Ceramics, Sains Malaysiana 45(4) (2016) 643–651.
[11] R.D. Shannon, Revised Effective ionic radii and systematic studies of interatomic distances in

halides and chaleogenides, Acta Cryst. A32 (1976) 751-767.

[12] T.S. Kayed, H. Ozkan, N.M. Gasanly, Effect of lithium doping on the properties of Tl-based superconductors, Supercond. Sci. and Technol. 14 (2001) 738-740.

[13] F. Letouzé, M. Hervieu, C. Martin, A. Maignan, B. Raveau, Local ordering of the chromate groups in the "1212" superconductor $Tl_{0.9}(Cr_4)_{0.2}Sr_2Ca_{0.9}Cu_2O_{6.2-\delta}$, Physica C 296 (1998) 195–204.

[14] S. T. Misture, C. Park, R. L. Synder, B. Jobst, B. Seebacher, Powder-diffraction data for several solid solutions with the compositions $(Ca_xSr_{1-x})CuO_2$ and $(Ca_xSr_{1-x})_2CuO_3$, Powder Diffr. 10(4) (1995) 296-299.

[15] V. Antal, D. Volochova, V. Kavecansky, J Kovac, P. Diko, Influence of annealing in oxygen and argon on the superconducting properties of Li-doped YBCO single-grain bulks, Physica C 241 (2017) 22-29.

[16] S. Oh, J.N. Eckstein, Infinite-layer $(Ca_{1-x}Sr_xCuO_2)$ film growth by molecular beam epitaxy and effect of hole doping, Thin Solid Films 483 (2005) 301–305.

Effect of Different UV Light Intensity on Porous Silicon Fabricated by Using Alternating Current Photo-Assisted Electrochemical Etching (ACPEC) Technique

Siti Nurfarhana Sohimee[1, a *], Zainuriah Hassan[1, b], Naser Mahmoud Ahmed[2, c], Lim Way Foong[1, d], Quah Hock Jin[1, e]

[1]Institute of Nano Optoelectronics Research and Technology (INOR), Universiti Sains Malaysia, 11800 USM, Penang, Malaysia
[2]School of Physics, Universiti Sains Malaysia, 11800 USM, Penang, Malaysia

[a*]farhanasohimee@gmail.com, [b]zai@usm.my, [c]naser@usm.my, [d]wayfoong317@yahoo.com.sg, [e]jinquah1st@hotmail.com

Abstract. Alternating current (sine-wave a.c (50Hz)) photo-assisted electrochemical etching (ACPEC) process was used to produce the formation of porous silicon with different ultraviolet (UV) light intensity. The study aims to investigate the effect of different UV light illumination on the properties of porous silicon. The surface of n-type silicon (111) was selectively etched in the HF and ethanol solution with a ratio (5:20) for 30 minutes under different UV lamp intensity; 10%, 30% and 60%. The samples were characterized by using atomic force microscopy (AFM) and X-ray diffraction (XRD). The AFM measurements revealed porous silicon with 30% of UV light intensity has the highest surface roughness value with a pyramidal shape. XRD analysis showed FWHM value of etched samples increased as the crystallite size of the etched sample decreased.

1. Introduction

Porous silicon was accidentally discovered by Arthur Uhlir Jr. and Ingeborg Uhlir at Bell Laboratories in 1956 when they were trying to develop an electrochemical method to machine silicon wafers for use in microelectronic circuits [1]. Porous silicon can be defined as a nanostructured material, which are formed when the crystalline silicon wafers are etched electrochemically in electrolyte solution [2,3]. One of the most important characteristics of the porous silicon layers is it very large and reactive internal surface so that this internal surface would play an important role in those specific properties of porous silicon layers, which make this material from bulk one [4]. The easy and cheap fabrication method, associated with the good optical properties, for example increases the field of potential applications of porous silicon in integrated optics and photonics [4,5].

Electrochemical anodization is one of the process to produce porous silicon. Silicon does not dissolve in HF except if a current flows through the interface of the crystal and the solution, thus electrochemical etching occurs [3,6]. Therefore, by adjusting the formation parameters, porous silicon with a large range of porosities and morphologies can be produced, which can be tuned for several applications such as optoelectronic devices.

Content from this work may be used under the terms of the Creative Commons Attribution 3.0 licence. Any further distribution of this work must maintain attribution to the author(s) and the title of the work, journal citation and DOI.
Published under licence by IOP Publishing Ltd

The important parameters in the formation of porous silicon are electrolyte concentration, current density, etching duration, and illumination of ultra-violet (UV) light. UV light illumination is used to assist in the generation of electron-hole pairs, where etching proceeds through the oxidation and consequently, dissolution of the semiconductor surface [6]. Holes play an imperative role in converting the surface atom into a higher oxidation state [6,7]. The supply of holes available at the surface to participate in the oxidation reaction is greatly enhanced by the absorption of incident radiation, resulting in significantly enhanced etch rate [6].

The parameters mentioned above plays an important role in the formation of porous silicon because it can affect the surface morphology of porous silicon such as porosity and pore sizes. Porosity is defined as the fraction of void within the porous silicon layer and also can be determined easily by weight measurement [8]. The porosity of porous silicon can be range from 4% for macroporous layers to 95% for mesoporous layers and it can be increased by increasing current density, decreasing HF concentration, and thicker silicon layer [8,9]. The pore size of porous silicon can be divided into three categories, which are macroporous, mesoporous, and microporous. The pore width of microporous are less than 2 nm, while for mesoporous lies between 2 to 50 nm, and macroporous pore width larger than 50 nm.

Porous silicon can be achieved by the standard direct current (DC) and alternating current (AC). El-Bahar and Nemirovsky [10] reported on the formation of porous silicon under alternating current (AC) conditions instead of applying direct current (DC) for electrochemical process which has some advantages. The main advantages are (a) the alternating current formed porous silicon and exhibit higher mechanical stability, (b) the alternating current process can be performed without depositing backside contact.

In this work, we have fabricated porous silicon by using alternating current photo-assisted electrochemical etching (ACPEC) process at different intensity of the UV lamp.

2. Experimental Process

Single crystal wafer of n-type (111) Si were cut into square with area 1.0 cm^2. The silicon wafer was cleaned by Radio Corporation of America (RCA) method before the formation of porous silicon. This step had been done carefully to make sure all the samples were cleaned. Then, all the cleaned samples were used to produce porous silicon by using alternating current photo-assisted electrochemical etching technique.

The silicon wafer was fixed on a metal plate and sealed through O-ring, thus only the front side of the silicon wafer is contacted with the solution (HF:ethanol) in the Teflon cell. Two screws are used to clamp the Teflon cell tightly. Then, the silicon wafer was immersed in the mixture of HF and ethanol with ratio (5:20). The aim of this addition is helping to remove the H2 gas bubbles from the sample surface that was formed during the anodization process. The silicon wafer acts as the anode while the cathode is made of platinum wire. The constant current density of 30 mA/cm^2 was used to produce porous silicon under constant etching duration of 30 minutes. The samples were etched under different UV light intensity, 10%, 30%, and 60% of UV light sources LQ-400. After the etching process, all the samples were rinsed with deionized (DI) water. Figure 1 shows the set-up of etching method for porous silicon formation.

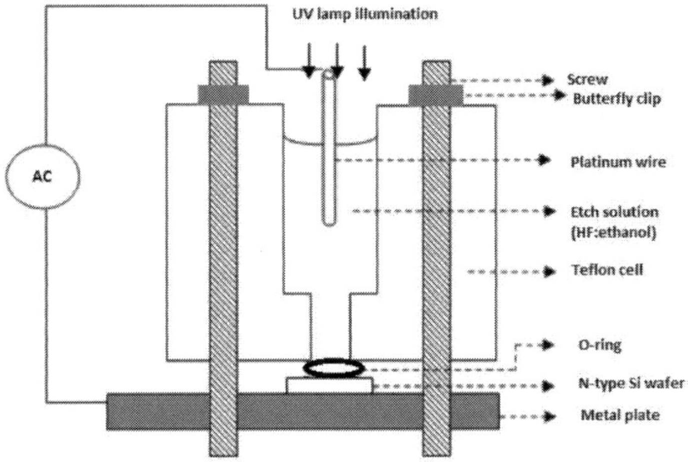

Figure 1. Schematic diagram of alternating current photo-assisted electrochemical etching (ACPEC) process.

The surface morphology of the porous silicon was characterized by atomic force microscopy (AFM) (Model:Dimension EDGE, BRUKER). NanoScope Analysis software was used to analyze the surface roughness and the average pore depth of the samples within the scan area of 5 x 5 μm. In addition, X-ray diffraction (XRD) (PANalytical X'Pert HighScore) was applied for the investigation of structural characteristics of porous silicon.

3. Results and discussion

The surface morphology of the etched sample with illumination from UV light shows the formation of porous structure. The illumination helps in generating the electro-hole pairs in the process of pore formation. Therefore, changes on the surface morphology can be observed by the presence of pores and etching of grain boundaries.

Figure 2 shows the AFM measurement of the porous silicon layer formed on the n-type Si sample at different intensity of UV light illumination starting from 10%, 30%, and 60% of UV light intensity.

NanoScope Analysis software was used to measure the surface roughness in root mean square (RMS) value and the result was tabulated in Table 1.

Based on Figures 2(a) and Figures 2(b), the etched surface with pyramidal shape was distributed over the surface. The pyramidal shape indicated the increase in the surface roughness, which is attributed to the etching parameters.

From Table 1, the RMS roughness value increase in the first two of the etched samples. The surface roughness of the sample etched by 30% of UV light illumination increased compared with that sample etched by 10% of UV light illumination, in which the etching occurs at the grain boundaries. The surface morphology of the sample etched by 30% of UV light illumination became relatively rougher as the pore size increased and more etching activity occurs at the grain boundaries. According to Khulathuraan et al. [11], the increased value in surface roughness is attributed to better formed pores of the silicon wafers.

Nevertheless, the RMS roughness value decreased as the intensity of the UV light illumination increased up to 60% due to reduction in pore size and insufficient etching activity occurs at the grain boundaries. This means that the electron-hole pairs that were generated were inadequate. The insufficient etching activity causes almost no formation of pores at 60% of UV light illumination. Thus, the RMS value decreased significantly.

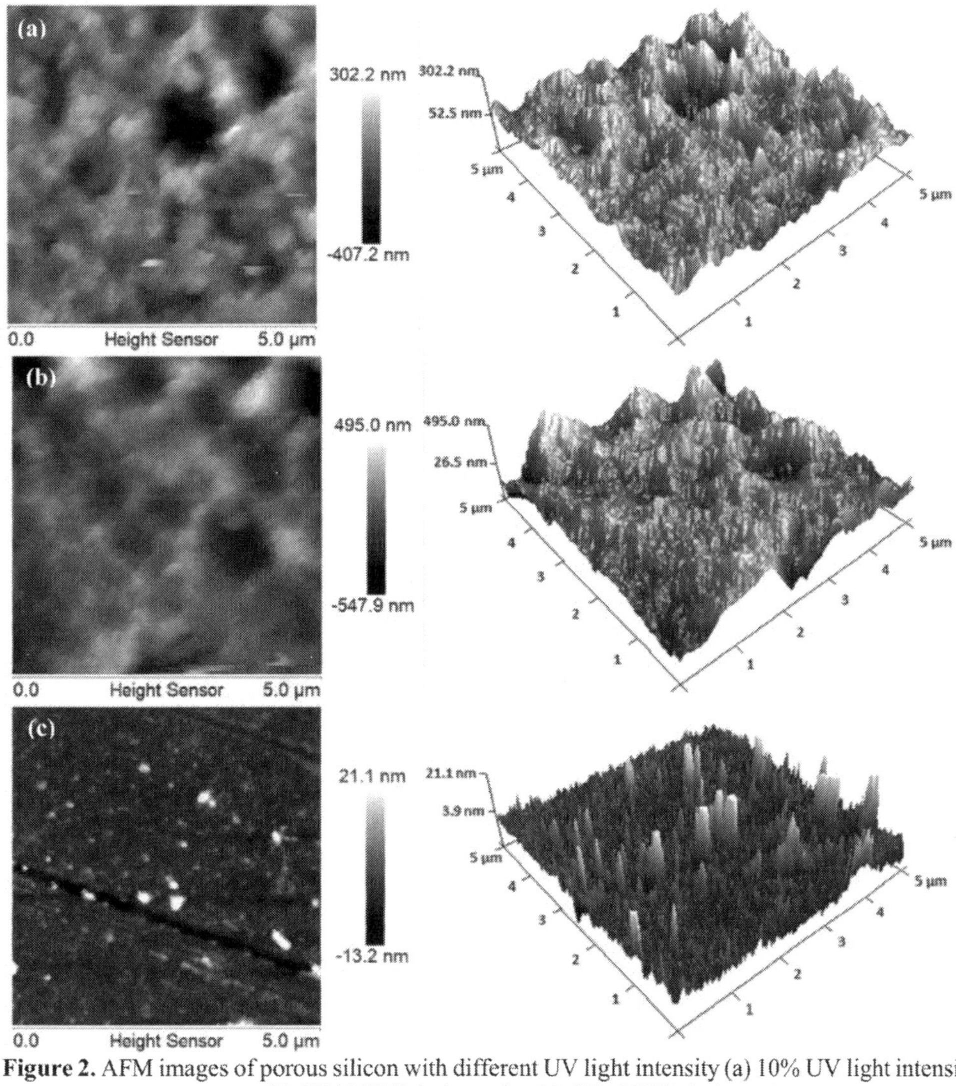

Figure 2. AFM images of porous silicon with different UV light intensity (a) 10% UV light intensity (b) 30% UV light intensity (c) 60% UV light intensity.

Table 1. RMS surface roughness and estimated average pore depth of etched samples at different intensity of UV illumination.

Sample	RMS Surface Roughness (nm)	Estimated average pore depth (nm)
10% UV light	92.80	226.61
30% UV light	128.00	321.72
60% UV light	3.71	35.84

X-ray diffraction (XRD) measurement was carried out to determine the crystalline properties of the samples. Figure 3 presents the 2θ XRD patterns for porous silicon etched at different amount of light intensity. It can be seen that porous silicon with 60% of light intensity have higher peak intensity compared to the porous silicon with 30% and 10% of light intensity. From the Figure 3, it can be observed that, at higher intensity, the peak was shifted slightly to lower angles. According to Radzali et al. [12], the shifting of peak occurred because of tensile strain experienced by the porous silicon after etching process.

Table 2 shows the summary of peak position 2θ, full-width half maximum (FWHM), and crystallite values of porous silicon. The crystallite size value can be determined by using the DebyeScherrer equation [11]:

$$D = \frac{K\lambda}{\beta \cos\theta} \qquad (1)$$

where D is the average crystallite, K is the Scherrer constant($K = 0.9$,) λ is the X-ray wavelength (λ = 1.5406 Å), β is the FWHM in radians, and θ is the diffraction angle calculated in radians. From the Table 2, porous silicon with 10% of UV light intensity was observed to have a smaller FWHM with a value of 0.072. Thus, the average crystallite size of the porous silicon with 10% of UV light intensity was observed to be higher compared to the other samples.

Table 2 Summary of 2θ peak position, FWHM, and crystallite size of porous silicon.

Sample	2θ (°)	FWHM (°)	Crystallite Size (nm)
10% UV light	28.5997	0.0720	113.87
30% UV light	28.7202	0.0960	85.42
60% UV light	28.5638	0.1378	59.49

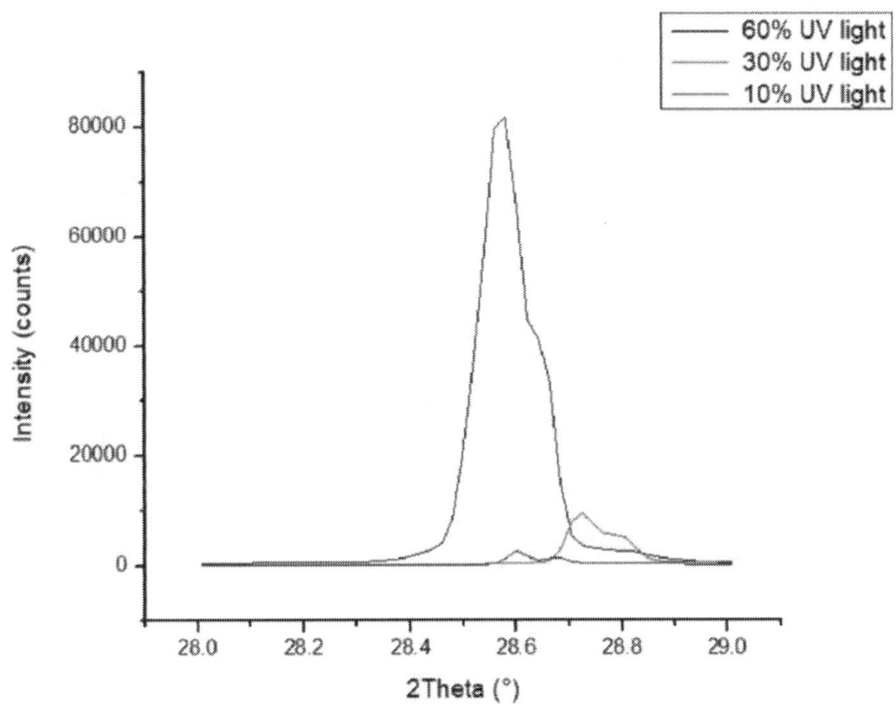

Figure 3. XRD patterns of porous silicon with different UV lght intensity.

4. Conclusion

In summary, alternating current photo-assisted electrochemical etching (ACPEC) technique was proven to be an effective technique to produce porous silicon. The surface of the porous silicon was investigated by AFM and XRD measurements. Amongst the sample, porous silicon with 30% of UV light intensity showed the highest surface roughness value with pyramidal shapes. From the XRD measurements, the FWHM value of etched samples increased as the UV light intensity increased. The FWHM value of etched samples increased as the crystallite size of the etched samples decreased. We strongly believe that further refinements and improvements of alternating current photo-assisted electrochemical etching technique will produce high porosity and uniformly distributed pores on the surface layer.

Acknowledgements

The authors acknowledge support from LRGS grant (203/CINOR/6720013) and Universiti Sains Malaysia.

References

[1] A. Uhlir, "Electrolytic Shaping of Germanium and Silicon," *Bell Syst. Tech. J.*, vol. 35, no. 2, pp. 333–347, 1956.
[2] M. Ramesh and H. S. Nagaraja, "The effect of etching time on structural properties of Porous silicon at the room temperature," *Mater. Today Proc.*, vol. 3, no. 6, pp. 2085–2090, 2016.

IOP Conf. Series: Journal of Physics: Conf. Series **1083** (2018) 012034 doi:10.1088/1742-6596/1083/1/012034

[3] Z. Thaira and A. Noor, "Impact of the Variability in the Current Density on the Porous Silicon Characteristi cs," vol. 50, pp. 488–493, 2014.

[4] H. A. Hadi, R. A. Ismail, and N. F. Habubi, "Fabrication and Characterization of Porous Silicon Layer Prepared by Photo-Electrochemical Etching in CH₃OH:HF Solution," *Int. Lett. Chem. Phys. Astron.*, vol. 8, pp. 29–36, 2013.

[5] O. Bisi, S. Ossicini, and L. Pavesi, "Porous silicon: a quantum sponge structure for silicon based optoelectronics," *Surf. Sci. Rep.*, vol. 38, pp. 1–126, 2000.

[6] A. Mahmood *et al.*, "A Novel AC technique for high quality porous GaN," *Int. J. Electrochem. Sci.*, vol. 8, no. 4, pp. 5801–5809, 2013.

[7] A. P. Vajpeyi, S. Tripathy, S. J. Chua, and E. A. Fitzgerald, "Investigation of optical properties of nanoporous GaN films," *Phys. E Low-Dimensional Syst. Nanostructures*, vol. 28, no. 2, pp. 141–149, 2005.

[8] "Porous_silicon @ en.wikipedia.org." [Online]. Available: https://en.wikipedia.org/wiki/Porous_silicon.

[9] R. Radzali, M. Z. Zakariah, A. Mahmood, A. F. A. Rahim, Z. Hassan, and Y. Yusof, "The effect of ecthing duration on structural properties of porous Si fabricated by a new twosteps alternating current photo-assisted electrochemical etching (ACPEC) technique for MSM photodetector," *AIP Conf. Proc.*, vol. 1875, 2017.

[10] A. and Y. N. El-Bahar, "A technique to form a porous silicon layer with no backside contact by alternating current electrochemical process."

[11] K. Kulathuraan, K. Mohanraj, and B. Natarajan, "Structural, optical and electrical characterization of nanostructured porous silicon: Effect of current density," *Spectrochim. Acta - Part A Mol. Biomol. Spectrosc.*, vol. 152, pp. 51–57, 2016.

[12] R. Radzali, Z. Hassan, N. Zainal, and F. K. Yam, "Structural and optical characteristics of porous InAlGaN prepared by photoelectrochemical etching," *J. Alloys Compd.*, vol. 622, pp. 565–571, 2015.

IOP Conf. Series: Journal of Physics: Conf. Series **1083** (2018) 012035 doi:10.1088/1742-6596/1083/1/012035

Effect of Medium on Interaction Forces between Atomic Force Microscopy (AFM) Tip and Gold Nanoparticle

Nurul Sabihah Zakaria[1, a *] **and Azlan Abdul Aziz**[2,3 b]

[1]School of Distance Education, Universiti Sains Malaysia (USM), Malaysia
[2] School of Physics, Universiti Sains Malaysia (USM), Malaysia
[3] Institute for Research in Molecular Medicine (INFORMM), USM Malaysia

[a]nurulsabihah@usm.my, [b]lan@usm.my

Abstract. Atomic Force Microscopy (AFM) has become powerful tool not only to study the surface topography but also the interaction forces between the tip and the sample. In this works, the measurement of the forces was determined by using AFM imaging in air and in liquid environment. Then, the force-distance (FD) curve was analysed to determine the interaction forces between the tip and the samples. The forces was computed from the pull-off forces of FD curve. When the FD curve are required in air, a thin layer of water (liquid contaminant layer) adsorbed on the sample surface exerts a meniscus or capillary force. Instead, when working in a liquid, the capillary force is eliminated and other forces becomes relevant. The force present is the van der Waals force and electric double layer force. The net force between the tip and the samples is the vector sum of several forces. In air, the dominant component of the adhesion force between the tip and the sample is believed to be the capillary force while in liquid, the dominant components present is the van der Waals force. The adhesion force ranging from 10 to 100 nN have been reported for imaging in air and the adhesion force of 5 to 104 nN observed in our studies lies well within this range. Meanwhile, in liquid, the adhesion force obtained from this work is in the range of 0.1- 40 nN and it is lies within the range reported previously which is as low as 100 pN. Detecting forces is an important step toward developing new analytical and biomedical devices such as biosensors.

1. Introduction

Atomic force microscopy (AFM) method is part of a scanning probe microscopy (SPM) technique used for surface characterization. It uses a sharp tip for scanning over a sample surface with the motion controlled precisely by a piezoelectric actuator to obtain information about the properties of the sample, including its topography image and surface interaction. AFM has been widely used in studying biomaterial characterization since it can be operated in air, vacuum, various gases and liquid environment. AFM visualises images by "feeling" the structure of a sample surface with a sharp probe rather than "looking" into the surface (Morris *et al.*, 2010).

In recent years, AFM has increasingly been used in nanotechnology especially for the visualization of biomolecules and force measurements. Originally AFM was used to image the topography of surfaces, but by modifying the tip it is possible to measure other quantities and also to perform various types of spectroscopy and analysis. AFM is more than a surface-imaging tool in that force measurements can be used to probe physical properties of molecules, such as molecular interaction. The main advantage of AFM in biology as compared to other methods is that it usually does not require specific

Content from this work may be used under the terms of the Creative Commons Attribution 3.0 licence. Any further distribution of this work must maintain attribution to the author(s) and the title of the work, journal citation and DOI.
Published under licence by IOP Publishing Ltd

sample preparation and allows measuring in liquid environment. The reason for choosing liquid media instead of air is not only because it is the natural physiological media for biological objects, but also due to the fact that all the interaction forces including unwanted ones are an order of magnitude smaller than in air allowing, for instance, to raise the resolution and to diminish image distortion. Although, it should be noted that measuring in liquids is much more complicated than imaging in air.

In nanobiotechnology, AFM can indicate the interaction of nanoparticles with biomolecules such as protein, DNA and others by measuring the force between the AFM tip and the sample surface via force spectroscopy (FS) measurement. Force spectroscopy or also known as force distance measurement is a method to measure interaction force occurring from the AFM tip to the surface of the samples (Neuman & Nagy, 2008). Force response is detected through beam deflection by a laser beam when a molecule of interest which is anchored at the end of the AFM tip to the surface of the sample is stretched (Ros *et al.,* 2004). Force spectroscopy can be analogized as a fishing rod, to pick up molecules at a sample surface and then to stretch them. It sounds easy but it is not certain whether it is a single molecule or a mass of molecules attached to the tip. Hence, to overcome this limitation, functionalization of the AFM tip to specifically attach to a point of interest in a sample surface must be introduced. The choice of functionalization approach of nanoparticles-biomolecules to the cantilever tip is of importance in order to obtain information on the interaction of nanoparticles with biomolecules.

Gold nanoparticles (AuNP) have been found to be one of the useful tools in biomedical applications because of its unique properties of being inert and less toxic, small size and large surface area to volume ratio, high reactivity to living cells, stability in high temperatures, ease of detection and others (Tiwari *et al.,* 2011). It can be synthesized into different shapes and sizes due to its ability to react and agglomerate with other nanoparticles in the surroundings. Due to its nano-size, it flows easily into various cells in the human body. Conjugation of nanoparticles to biomolecules generates hybrid materials that can be used to let the nanoparticles interact specifically with biological systems. Nanoparticle-biomolecule conjugates bring together the unique properties and functionality of both materials. If the aim is to use AuNP in biomedicine as diagnostic and therapeutic agents in cells or tissues, then, it is necessary to rightly choose the targeting component such as a monoclonal antibody (mAb), and the strategy to attach it on the surface of the particle (Cao-Milán & Liz-Marzán, 2014). Antibody conjugated gold nanoparticle probes allow for the study and visualization of cellular and molecular processes.

Naturally, AFM can perform imaging in all environment including in air, liquid and gases. For bioconjugated samples, AFM imaging in air is able to clearly track and illustrate the performance of attaching the biomolecules to the nanoparticle surface for the ratification of the conjugation while in liquid, the AFM provides information with high spatial and temporal resolution related to the real-time changes of shape or structure of bioconjugates (Tessmer *et al.,* 2013). Hence, AFM helps to understand the structural and mechanical properties of biomolecules. In summary, AFM known as a powerful tool for the structural investigation of biomolecular conjugation to AuNP surface and analysis of bio conjugation-based nanostructures. Understanding the bioconjugation and the physical interactions of bioconjugation on the surface of gold nanoparticles opens novel opportunities for their use in biomedicine.

2. Methodology

2.1. Mica surface functionalization

Mica surfaces were freshly cleaved just prior to experiment. The mica substrate should be changed from negative charge to positive charge in order to allow the negatively charged gold nanoparticle particles bound tightly to the surface of the mica. The mica surface was promoted by absorption method using 150-300 kDa of 0.01% Poly-L-Lysine (PLL). All the chemicals were obtained from Sigma Aldrich. The solutions were all prepared shortly before experiments. The PLL solution is prepared by ratio dilution of 1:10 with deionized water (Grobelny *et al.,* 2009). 50μl of dilute PLL solution was dropped on freshly cleaved mica using a micropipette. The solution was allowed to air dry completely under room temperature. Then, functionalized mica was rinsed with deionized water and dried under room temperature.

2.2. Deposition of AuNP conjugated goat anti-human (AuNP_IgA)
The 40 nm AuNP conjugated goat anti-human IgA of 0.02% sodium azide and 1% BSA was received from NANOBRI INFORMM, USM. 50 µl of AuNP_IgA was applied on functionalized mica using micropipette. Then, the substrate was air dried at room temperature. The mica was then rinsed gently with deionized water to remove loosely attached particles. Finally, the mica was air dried under room temperature.

2.3. AFM tip functionalization
The AFM tip was silanized in a 0.1% (v/v) toluene solution of 3-aminopropyl-triethoxysilane (APTES) for 15 minutes to coat it with a self-assembled monolayer of amines. Following the treatment with the amino silane, the tips were directly incubated for half an hour in a solution of 10 mg/ml O,O'-bis[2-(N-succinimidyl-succinylamino)ethyl] polyethylene glycol (bis(NHS)-PEG) in anhydrous chloroform containing 5µl/ml triethylamine as catalyst, and then rinsed using deionized water. Frequent washing step will remove the molecules of PEG that are not tightly bound to the AFM tip. The AFM tip is functionalized to control the place of attachment to the molecule of interest that lie between the tip and the substrate for further compressing and stretching by force spectroscopy measurements. Occasionally, a molecule will be picked up by the tip and stretched, as the sample and the tip are pulled apart. The PEG linker offers better control for the AFM tip to 'fish' the complementary molecules on the sample surface and to avoid more than one molecule binding to the tip at the same time, onwards, measuring the adhesion force upon the retraction of the tip.

2.4. AFM imaging
The FD measurement was carried out using triangular shaped OMCL-TR800PSA tip made of silicon nitride. The cantilever was 100 µm long, 13.4 µm wide and 0.8 µm thick with a spring constant of 0.57 N/m. The tip was 15 nm radius with a resonant frequency of 73 kHz. The AFM imaging was carried out in non-contact mode of operation in both ambient air and liquid environment. AFM imaging in liquid environment is the same manner as the AFM imaging in air. Phosphate Buffer Saline (PBS) with pH 7.4 was used in liquid environment. Only the cantilever's deflection against the Z scanner's position was measured. Acquisition of the FD curves in some places has been carried out to better repeatability and uniformity in all cases and at all time. In this study, the FD curve was obtained in a series of measurements taken at the same point, as well as at different points on the sample surface. Each time the tip interacts with the sample, a FD curve is captured. After AFM force measurement (obtaining force-vs-distance curves), the pull-off force data were directly extracted by the instrument-equipped software (XEI software).

3. Results and Discussion
Table 1 shows the pull-off force value for FD curve measurement for conjugation of 40 nm AuNP with goat anti-human (AuNP_IgA) on 150-300 molecular weight of 0.01% Poly-L-Lysine (PLL) functionalized mica using nonfunctionalized and functionalized OMCL tip imaging in air and liquid environment. The pull-off force represents the value of retrace data (when the Z scanner retracts) which means the tip is moved away from the sample. The magnitude of the pull-off force was higher when imaging in air environment. This is due to the meniscus force of liquid contaminant layer (water) that formed from the humidity condensation between the tip and the sample surface which acted against the pull-off during imaging in air environment. Basically, at low humidity condition, water on the tip as well as on the surface of the sample will accumulate at the point of contact to generate a meniscus.

These meniscus force is often referring as the capillary force between the tip and the sample surface. By placing the tip into contact with the substrate, capillary forces move the water on the tip and sample surface to the point of contact, and a meniscus forms at the periphery of the tip. In here, the magnitude of the pull-off force is a measure of the attractive attraction between the tip and the sample surface combined with the capillary force. The pull-off force is significantly influenced by the water in order to obtain the accurate values of the tip-sample interaction (Rozhok *et al.*, 2004).

The pull-off force for Table 1 is significantly higher about 3 folds after functionalized the tip. The meniscus force still occurs even though the tip coated with amine monolayer is used during imaging. Regardless, even with a functionalized tip, liquid contaminant layer (water) will move near the point of contact between the tip and sample surface and form a meniscus. During contact, some chemical bonds from the functionalization of the tip or adhesive bonds may result in non-conservative force.

Apparently, the attractive forces between the functionalized tip and water in the meniscus are weak enough to contribute to the value of the pull-off force (Rozhok *et al.,* 2004). Providing a stable hydrophobic surface via tip functionalization is one example to decrease the formation of liquid contaminant layer (water) on the surface, thereby decreasing the capillary forces (Chen & Soh, 2008). Besides, when obtaining the pull-off force in water, due to the coupling of the adhesion force and the double layer force, the magnitude of pull-off force becomes lower compared in air environment. Hence, the electric double layer on the surface of a samples in water can be influenced by the ionic strength of the liquid.

Table 1. Pull-off force for conjugation of 40 nm AuNP with goat anti-human (AuNP_IgA) on 150-300 molecular weight of 0.01% Poly-L-Lysine (PLL) functionalized mica using nonfunctionalized and functionalized OMCL tip imaging in air and liquid environment.

AFM tip	Pull-Off (nN)	
	Air	Liquid
Nonfunctionalized tip	30.47	0.904
Functionalized tip	103.7	37.10

Figure 1 shows force distance (FD) curve for non-functionalized OMCL tip imaging in (A) air and (B) liquid for conjugation of 40nm AuNP with goat anti-human (AuNP_IgA) deposited on glass slide functionalized with 150-300 molecular weight of 0.01 % Poly-L-Lysine (PLL). The x-axis of the FD curve represents the distance between the Z scanner and sample in the Z direction while y-axis of this curve represents the force exerted on the cantilever. The curve displays the approach (trace) curve in red colour and retract (retrace) curve in blue colour. The red line of the FD curve was obtained when the Z scanner extended towards the sample and the blue line was obtained when the Z scanner retracted from the sample surface. Therefore, the Z scanner would decline and incline as the tip snapped-in (red line) and pulled-off (blue line) at one point of the sample. At this point, there were changes on both, the cantilever deflection and force value.

Figure 1 and Fig. 2 (A) explaine that the detachment process is composed of single unbinding event between the tip and sample surface. It indicates that only one interaction between AFM tip and the sample is pulled apart as the cantilever retract from the sample surface. Meanwhile, Fig 1. and Fig. 2 (B) explaine that the detachment process is composed of several pull-offs. It indicates that more than one or several interactions between AFM tip and the sample is pulled apart as the cantilever retract from the sample surface. The tip and sample seem to have been separated from each other. However, there are a small number of AuNP and antibody molecules remains bound to the bonds that has been pulled tightly until they broke at a certain value. In the consideration that the molecules possibly situated at particular position on the tip as well as the sample, they probably have been separated at distinct distances. Therefore, the FD curve may have presented multiple detachments. Multiple interaction needs to be minimized when the purpose of the work is to study the individual interaction between the tip and the sample.

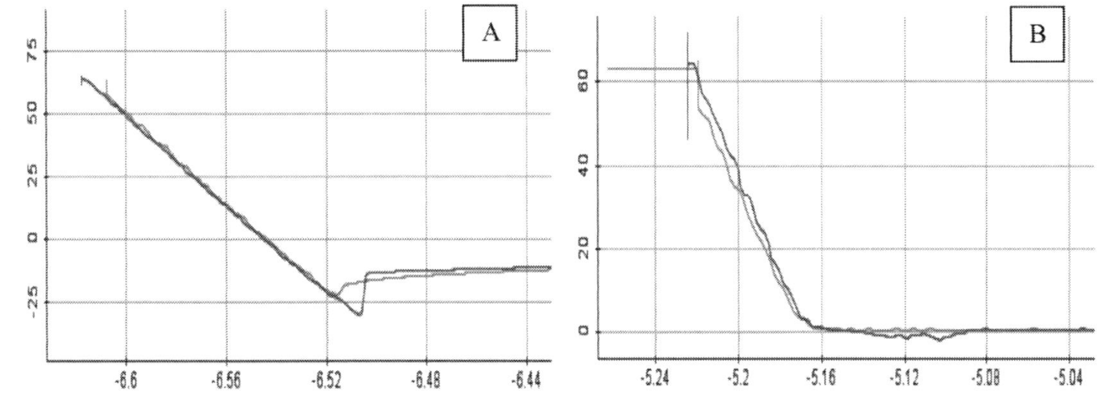

Figure 1. Force Distance (FD) curve for non-functionalized OMCL tip imaging in A) air and B) liquid for conjugation of 40nm AuNP with goat anti-human (AuNP_IgA) deposited on glass slide functionalized with 150-300 molecular weight of 0.01 % Poly-L-Lysine (PLL)

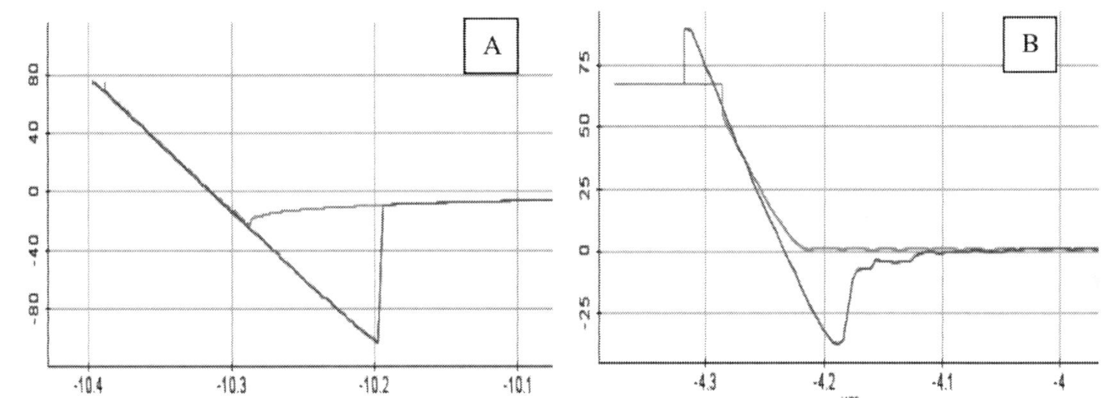

Figure 2. Force Distance (FD) curve for functionalized OMCL tip imaging in A) air and B) liquid for conjugation of 40nm AuNP with goat anti-human (AuNP_IgA) deposited on mica functionalized with 150-300 molecular weight of 0.01 % Poly-L-Lysine (PLL)

In this study, the 40 nm spherical shape of AuNP was used since it is a noble metal (exhibits a high chemical stability), easily to functionalize and has extraordinary capability to conjugate with various biomolecules (agglomeration induces a shift of the maximum absorption towards higher wavelength that reflects a color changes when the biomolecules are present in the sample) as well as the spherical AuNP was less toxicity (depends on the type of cells used) compared to the nanorod AuNP (Murphy *et al.,* 2008). Phosphate buffer saline (PBS) is used as a medium in liquid imaging because the pH value of PBS is set to be within the range of 7 to 7.6 as pH of most of biological materials fall between pH 7 to 7.6, but often set to be 7.4 as the pH of the blood nears 7.4. Furthermore, PBS has the osmolarity and ion concentration that matches those of the fluid inside the cells and cause isotonic and non-toxic to most cells.

4. Conclusion

The net force between the tip and the samples is the vector sum of several forces. The pull-off force in air was correlated to both capillary and electrostatic forces. In air, the dominant component of the adhesion force between the tip and the sample is believed to be the capillary force. Meanwhile, the pull-off force in liquid correlated to the Van der Waals force and double layer force. However, in liquid, the dominant component present between the tip and sample is believed to be the van der Waals force. When imaging is operated in ambient air environment, van der Waals forces which consistently appear at interfaces, are usually overshadowed by the emergence of stronger capillary forces resulting from the condensation of water vapour at the point of contact that occurs between the tip and the surface of the sample. Hence, the influence regarding capillary forces can be averted by conducting imaging in liquid environment which removes the capillary force which holds the tip to the sample surface that produces in ambient air environment from contaminant layer on a sample surface, mostly absorbed water, which determines the magnitude of adhesive forces.

Acknowledgements
This study was supported by the Universiti Sains Malaysia through Research University Grant (RU-PRGS) (1001/PFIZIK/845010).

References

[1] R Cao-Milán and L M Liz-Marzán 2014 Gold nanoparticle conjugates: recent advances toward clinical applications *Expert opinion on drug delivery*, vol. 11, no. 5, pp. 741–52
[2] Chen S H and Soh A K 2008 The capillary force in micro- and nano- indentation with different indenter shapes *International Journal of Solids and Structures*, 45, pp. 3122–3137
[3] J Grobelny, F W DelRio, N Pradeep, D I Kim, V A Hackley and R F Cook 2009 NIST - NCL Joint Assay Protocol , PCC-6 Size Measurement of Nanoparticles Using Atomic Force Microscopy *National Institute of Standards and Technology*, vol. 21702
[4] V J Morris, A R Kirby and A P Gunning 2010 *Atomic Force Microscopy for Biologist* Second Edition (Imperial College Press)
[5] C J Murphy, A M Gole, J W Stone, P N Sisco, A M Alkilany, E C Goldsmith and S C. Baxter 2008 Gold Nanoparticles in Biology: Beyond Toxicity\rto Cellular Imaging *Accounts Of Chemical Research*, vol. 41, no. 12
[6] Neuman K C, and Nagy A 2008 Single-molecule force spectroscopy : optical tweezers, magnetic tweezers and atomic force microscopy *Nature Methods*, *5*(6), pp. 491–505
[7] R Ros, R Eckel, F Bartels, and A Sischka 2004 Single molecule force spectroscopy on ligand-DNA complexes: from molecular binding mechanisms to biosensor applications *Journal of biotechnology*, vol. 112, no. 1–2, pp. 5–12
[8] S Rozhok, P Sun, R Piner, M Lieberman and C A Mirkin 2004 AFM Study of Water Meniscus Formation between an AFM Tip and NaCl Substrate *The Journal of Physical Chemistry B*, no. 108, pp. 7814–7819
[9] I Tessmer, P Kaur, J Lin and H Wang 2013 Investigating bioconjugation by atomic force microscopy *Journal of nanobiotechnology*, vol. 11, no. 1, p. 25
[10] P M Tiwari, K Vig, V A Dennis and S R Singh 2011 Functionalized Gold Nanoparticles and Their Biomedical Applications *Nanomaterials*, vol. 1, no. 1, pp. 31–63

Hydrothermal Synthesis and Structural Properties of V_2O_5 Nanoflowers at Low Temperatures

N.M. Abd-Alghafour[1], Naser. M. Ahmed[2], Z. Hassan[3], Munirah Abdullah Almessiere[4]

[1]Iraqi Ministry of Education, Al-Anbar, Iraq
[1]University of Al-Anbar, College of Science, Al-Anbar, Iraq
[2]School of Physics, Universiti Sains Malaysia, 11800 USM, Penang, Malaysia
[3]Institute of Nano Optoelectronics Research and Technology (INOR), Universiti Sains Malaysia, 11800 USM, Penang, Malaysia
[4]Physics Department, College of Science, University of Dammam, Saudi Arabia

na2013bil@gmail.com

Abstract. Vanadium pentoxide nanoflowers were synthesized through a simple hydrothermal method at low temperatures. The structure was fabricated by using NH_4VO_3 and oxalic acid as precursors with 0.1 M solution concentration. The morphology and structural properties of the nanoflowers were characterized using FESEM, XRD, Raman, and UV-Visible spectroscopy. The results reveal an orthorhombic structure with preferred orientation along (001) plane of the prepared V_2O_5 nanoflowers. Raman peaks also expressed the same structural features. FESEM images showed the V_2O_5 nanoflower with diameters in the range of 60-80 nm and length in 600-800 nm. A red-shift was observed in the characteristic absorption peak of V_2O_5 nanoflowers, which are attributed to the decrease of the bandgap of the samples.

1. Introduction

V_2O_5 has been gained significant interest in the applied research to range of applications [1]. V_2O_5 is the most stable among all vanadium oxides and has high oxidation state [2, 3]. The outstanding properties such as a direct band gap in the visible-light region (E_g = 2.2 to 2.7 eV) [4], multi-valance, good chemical and thermal stability, excellent thermoelectric property make V_2O_5 nanostructure is a suitable material for solar cells [5], gas sensor [6], optical-electrical switches [7], chemical sensing [8], electrochromic devices [9], and optoelectronic devices [10]. Therefore, one dimensional (1D) nanostructures of V_2O_5 are considered to be more appropriate for the device applications as compared with its other forms. Among these, V_2O_5 nanoflowers are being considered for applications in optoelectronic and electrochemical devices [11]. V_2O_5 nanostructures have been prepared by different techniques including chemical vapor deposition [12], magnetron sputtering [13], sol-gel method [14], pulsed laser deposition [15], electron beam evaporation [16], electrospinning [17], spray pyrolysis [18], and hydrothermal synthesis [19]. Among the various deposition techniques for the preparation of vanadium oxide nanostructures, hydrothermal technique is a relatively simple, cost-effective and environmental friendly route to prepare metal oxide nanostructures at low temperature and short reaction time [20].

Many researchers in the field have been developed to obtain 1D-V_2O_5 nanostructures. Parida et al. [21] synthesized of the V_2O_5 nanoflowers by a simple and novel cost-effective low temperature

Content from this work may be used under the terms of the Creative Commons Attribution 3.0 licence. Any further distribution of this work must maintain attribution to the author(s) and the title of the work, journal citation and DOI.
Published under licence by IOP Publishing Ltd

method. The nanoflowers are shown to exhibit large nonlinear optical absorption leading to optical limiting. Optical limiting can be observed due to a variety of nonlinear optical processes such as reverse saturable absorption, multiphoton absorption, and thermal scattering. Huang et al. [22] synthesized three-dimensional V_2O_5 nanoflowers by a simple two-step process, with a water bath method and thermal treatment. Each Nanoflower with the size of 2-3 μm is assembled by about 2000 nanorods and the diameter of each nanorod is around 80 nm. After calcination, all the diffraction peaks of the samples are in agreement with the JCPDS Card No. 41–1426 orthorhombic V_2O_5. Mu et al. [19] synthesized of the V_2O_5 nanostructures including nanoflowers, nanoballs, nanowires and nanorods by using hydrothermal method. The morphology of the product depends on the types of solvent and acid. The electrochemical test results indicate that the rod-like structure leads to a significant improvement of storage capacity, electrochemical kinetics, and rate capability. Chu et al. [23] fabricated of the V_2O_5 nanorods by using a simple one-step hydrothermal method. After calcination of the hydrothermal samples, nanorods can be grown on substrate with the diameters in the range 80-100 nm and lengths of 1-10 μm, respectively. It was found that the synthesized V_2O_5 nanorod is favourable in the development of smart window application and secondary lithium batteries. In the present work, we synthesized of V_2O_5 nanoflowers by using hydrothermal growth at low temperatures. The effect of post-annealing treatment on the structural evolution and optical properties of V_2O_5 nanoflowers were investigated in this study.

2. Experiment procedure

V_2O_5 nanoflowers was grown on glass substrate by using a hydrothermal method. The substrates were cleaned with an adequate amount of acetone and ethanol. After cleaning, the substrates were rinsed with deionised water and dried at room temperature. Hydrothermal synthesis was carried out for V_2O_5:0.5850 g NH_4VO_3 dissolved in 50 ml deionized water and then 1.1206 g oxalic acid was added to the system to form an orange solution. Ammonium metavanadate, NH4VO3 (0.5850 g) was dissolved in deionised water (50 ml). Oxalic acid (1.1206 g) was added into the solution and immediately. The resulting solution was. stirred for 15 minutes before the addition of 50 ml deionized water. Then, HCl (10 drops) was slowly added into the resulting solution. The solution was placed in a Pyrex glass bottle with a substrate positioned at the bottom of the glass and heated at 95°C for 3.5 hours and then allowed to cool down at room temperature overnight. After rinsed extensively with deionized water and dried in ambient air, V_2O_5 nanoflowers was observed homogeneously deposited on the glass substrate. The samples were annealed at 300°C for 1 hour then 500°C for 2 hours to create a V_2O_5 layer. The complete experiment procedures of the V_2O_5 nanoflowers are schematically shown in Figure 1.

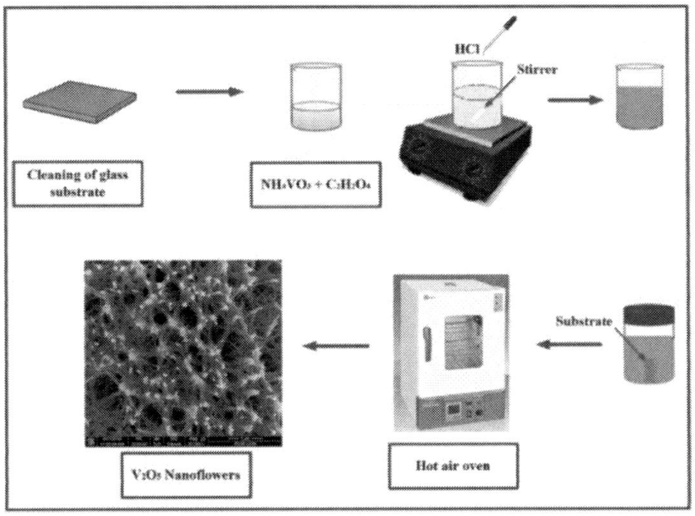

Figure 1: Schematic diagram of the experiment procedure used for the growth of the V_2O_5 nanoflowers.

The synthesized sample was investigated by using high resolution X-ray diffraction system (HR-XRD-X'Pert PRO MRD PW 3040 from PANalytical) operating with Cu-K_α radiation (λ =1.54056 Å) at 40 kV and 30 mA and field emission-scanning electron microscopy model FEI Nova NanoSEM 450. The optical properties were examined using a Raman spectrometer system (Horiba Jobin Yuon HR 800UV, Edi-son, NJ, USA) with Ar^+ as the excitation source operated at a wavelength of 514.55 nm (20 mW) and Shimadzu UV-vis 1800 spectrophotometer at wavelengths ranging from 300 to 1000 nm.

3. Results and discussion

3.1 FESEM observation

Figure 2(a-d) displays the morphology images of the prepared V_2O_5 nanoflowers at the pre- and post-annealing samples. Figure 2(a-b) shows that large area of the samples was covered with flower-like feature before annealing with diameter and length ranging from 50-70 nm and 150-250 nm, respectively. After annealing, it was easily observed that the V_2O_5 nanoflowers still have the similar structure with clearer flower-like feature., as is shown in Figure 2c. It was noted that the diameter and length increased with increasing temperatures. The diameter of the nanoflowers is around 60-80 nm and the length about 600-800 nm, as shown in Figure 3d. The diameter of the nanoflowers increases to about 80 nm, which is attributed to the release of H_2O and NH_3 during the thermal decomposition process of the pre-annealing samples [22].

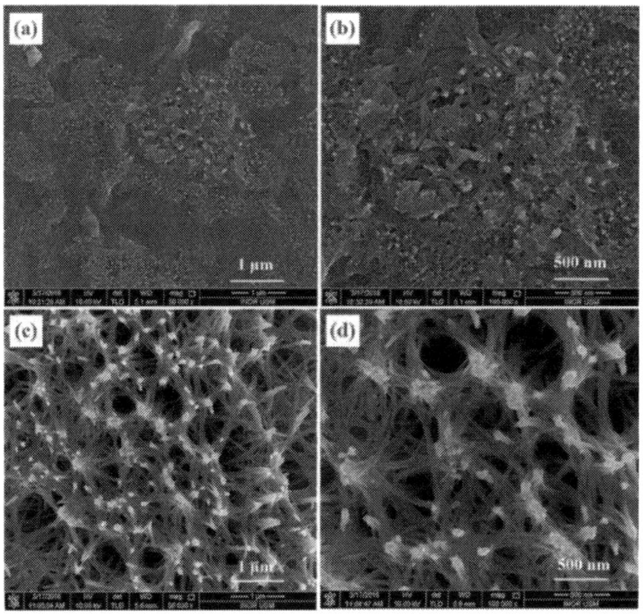

Figure 2. FESEM images of the V_2O_5 nanoflowers (a)-(b) pre-annealing, (c)-(d) post-annealing.

3.2 XRD analysis

The XRD diffraction patterns of the V_2O_5 nanoflowers synthesized onto glass substrates by hydrothermal method are shown in Figure 3. The figure shows the formation of diffraction peaks corresponding to V_2O_5 along different planes. A predominant peak was observed in the XRD diffraction pattern at 20.2°, which is related with the preferred orientation along (001) plane. Moreover, six reflection peaks were observed at 2θ values of 26.3°, 29°, 30.6°, 41.2°, 50.5° and 44.4° correspond (110), (400), (310), (002) and (501) reflections of the orthorhombic structure. All XRD peaks matched the orthorhombic phase of V_2O_5 (JCPDS card No. 00-041-1426). Additionally, the XRD patterns of the V_2O_5 nanoflowers are high in intensity and sharp due to the annealing temperature at 500°C for 2 hours, which is improved the crystalline quality of the V_2O_5 nanoflowers [22]. The XRD peaks show the formation of high crystallinity orthorhombic phase of V_2O_5, which is similar to the results reported by Qin et al. [24], Chen et al. [25], and Jin et al. [26]. These results reveal high crystallinity of the V_2O_5 nanoflowers as no peaks observed for impurities, which is observed no peaks of any other impurities.

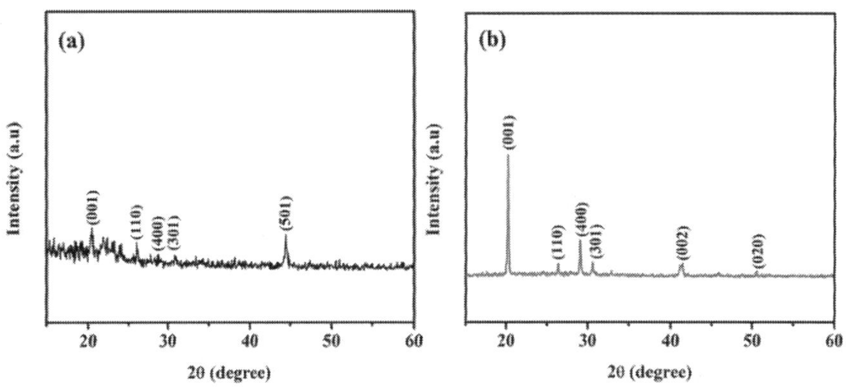

Figure 3. The XRD pattern formation and related phase matches of (a) V_2O_5 nanoflowers before annealing, (b) after annealing.

IOP Conf. Series: Journal of Physics: Conf. Series **1083** (2018) 012036 doi:10.1088/1742-6596/1083/1/012036

3.3 Raman spectra analysis

Raman spectra of V_2O_5 nanoflowers are shown in Figure 4. All Raman peaks match the orthorhombic structure [27, 28]. The Raman peaks of the V_2O_5 nanoflowers became observe at 143, 194, 281, 300, 405, 480, 523, 696, and 995 cm^{-1}, that can be assigned B_{3g}, A_g, B_{2g}, and B_{1g} modes, respectively [29, 30]. The intensity and positions of the Raman peaks match with the reference peaks of V_2O_5 material [31]. The high frequency peak at 995 cm^{-1} corresponds to the terminal oxygen (V=O) stretching mode, which is associated to the stretching mode related to the A_g symmetric vibrations of the shortest vanadium oxygen bond. The mode peak (V_2–O) at 696 cm^{-1} is associated to the corner-shared oxygens common to two pyramids. The Raman spectrum exhibited centred at 523 cm^{-1}, which was attributed to the triply coordinated oxygen (V3–O) stretching mode. The peaks positioned at 281, 405 cm^{-1} are assigned to the bending vibration of the V=O bonds. Furthermore, the peaks positions at 480 and 300 cm^{-1} are assigned to the bending vibrations of the bridging V–O–V and V3–O, respectively. The low frequency Raman scattering peaks at 143 and 194 cm^{-1} are strongly related with the layered structure.

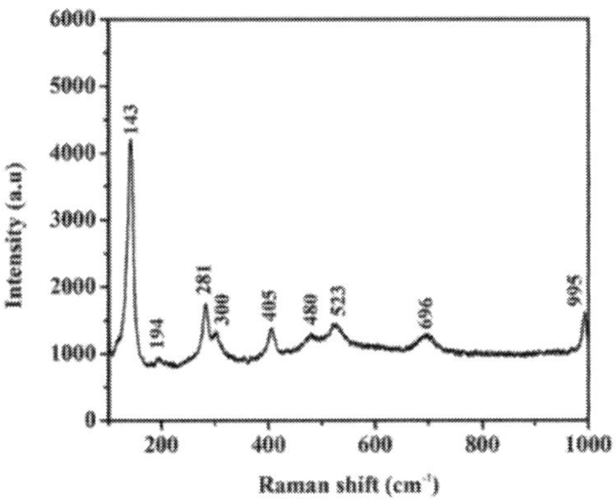

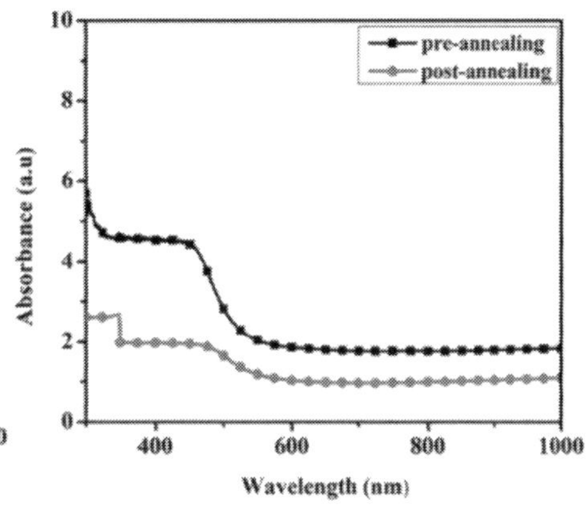

Figure 4. Raman spectra of the V_2O_5 nanoflowers. **Figure 5.** UV–visible absorption spectra.

3.4 Optical properties

The UV–visible diffuse absorption spectrum of prepared V_2O_5 nanoflowers is shown in Figure 5. It can be seen that the excitonic absorption peaks of V_2O_5 samples appear around 498.9 nm (E_g= 2.4 eV) and it shows a blue shift in comparison with the bulk V_2O_5 (E_g = 2.34 eV), which can be attributed to the quantum confinement effect due to electronic transitions from occupied 2p bands of oxygen to unoccupied 3d bands of vanadium oxide [9]. It is well known that the band maxima of the charge transfer transition of O → V^{n+} depends on the number of oxygen atoms surrounding the central vanadium ion. Hence, V^{5+} in tetrahedral coordination absorbs in the range 240-350 nm, in square pyramidal coordination at 350-450 nm and in octahedral coordination at 450-600 nm. In this spectrum a weak band at around 498.9 nm which can be attributed to lower energy charge transfer to V^{5+} species in octahedral coordination is shown. It can be observed clearly from the absorbance of the samples decreases slightly with increase in annealing. Additionally, a red-shift is observed in the characteristic absorption peak of V_2O_5 which is attributed to the decrease in the bandgap of the samples.

4. Conclusions

V_2O_5 nanoflowers with an average diameter of 80 nm were synthesized by a simple and cost-effective chemical hydrothermal method. The V_2O_5 nanoflowers annealed at 500°C exhibited an improved crystallinity and nanoflowers-like morphology. The XRD results indicated that the high crystallinity of the V_2O_5 product, which no peaks for impurities observed. The Raman spectra of the prepared V_2O_5 nanoflowers were symmetric with the Raman modes for a V_2O_5 phase. It can be clearly observed the absorbance of the samples decreases slightly with increase in annealing.

Acknowledgements
We gratefully acknowledge the support of the Universiti Sains Malaysia under RU Top Down grant (1001/CINOR/870019).

References

[1] Li B, Xu Y, Rong G, Jing M and Xie Y, 2006, Vanadium pentoxide nanobelts and nanorolls: from controllable synthesis to investigation of their electrochemical properties and photocatalytic activities, Nanotechnology, Vol. 17, No.10, Pp. 2560-2566.

[2] Luo Z, Wu Z, Xu X, Du M, Wang T and Jiang Y, 2010, Impact of substrate temperature on the microstructure, electrical and optical properties of sputtered nanoparticle V_2O_5 thin films Vacuum, Vol. 85, No.2, Pp. 145-50.

[3] Beke S, 2011, A review of the growth of V_2O_5 films from 1885 to 2010, Thin Solid Films, Vol. 519, No.6, Pp. 1761-71.

[4] Kim B H, Kim A, Oh S-Y, Bae S-S, Yun Y J and Yu H Y, 2008, Energy gap modulation in V_2O_5 nanowires by gas adsorption, Applied Physics Letters, Vol. 93, No.23, Pp. 233101-233103.

[5] Arbab E A and Mola G T, 2016, V_2O_5 thin film deposition for application in organic solar cells, Applied Physics A, Vol. 122, No.4, Pp. 405-412.

[6] Mane A, Suryawanshi M, Kim J and Moholkar A, 2017, Fast response of sprayed vanadium pentoxide (V_2O_5) nanorods towards nitrogen dioxide (NO_2) gas detection, Applied Surface Science, Vol. 403, Pp. 540-550

[7] Fisher B, Genossar J, Patlagan L, Chashka K and Reisner G, 2015, Electric-field-induced semiconductor–semiconductor transition in V_2O_5, Applied Physics A, Vol. 120, No.2, Pp. 435-442.

[8] Abd-Alghafour N, Ahmed N M, Hassan Z, Almessiere M A, Bououdina M and Al-Hardan N H, 2017, High sensitivity extended gate effect transistor based on V_2O_5 nanorods, Journal of Materials Science: Materials in Electronics, Vol. 28, No.2, Pp. 1364-1369.

[9] Talledo A and Granqvist C G, 1995, Electrochromic vanadium–pentoxide–based films: Structural, electrochemical, and optical properties, Journal of Applied Physics, Vol. 77, No.9, Pp. 4655-4666.

[10] Zhai T, Liu H, Li H, Fang X, Liao M, Li L, Zhou H, Koide Y, Bando Y and Golberg D, 2010, Centimeter-Long V_2O_5 Nanowires: From Synthesis to Field-Emission, Electrochemical, Electrical Transport, and Photoconductive Properties, Advanced Materials, Vol. 22, No.23, Pp. 2547-2552.

[11] Li J-M, Chang K-H and Hu C-C, 2010, A novel vanadium oxide deposit for the cathode of asymmetric lithium-ion supercapacitors, Electrochemistry communications, Vol. 12, No.12, Pp. 1800-1803.

[12] Nandakumar N K and Seebauer E G, 2011, Low temperature chemical vapor deposition of nanocrystalline V_2O_5 thin films, Thin Solid Films, Vol. 519, No.11, Pp. 3663-3668.

[13] Zou C, Yan X, Han J, Chen R and Gao W, 2009, Microstructures and optical properties of β-V_2O_5 nanorods prepared by magnetron sputtering, Journal of Physics D: Applied Physics, Vol. 42, No.14, Pp. 145402-145406.

[14] Raj D V, Ponpandian N, Mangalaraj D and Viswanathan C, 2013, Effect of annealing and electrochemical properties of sol–gel dip coated nanocrystalline V_2O_5 thin films, Materials Science in Semiconductor Processing, Vol. 16, No.2, Pp. 256-262.

[15] Beke S, Giorgio S, Kőrösi L, Nanai L and Marine W, 2008, Structural and optical properties of pulsed laser deposited V_2O_5 thin films, Thin Solid Films, Vol. 516, No.15, Pp. 4659-4664.

[16] Kumar A, Singh P, Kulkarni N and Kaur D, 2008, Structural and optical studies of nanocrystalline V_2O_5 thin films, Thin Solid Films, Vol. 516, No.6, Pp. 912-918.

[17] Ban C, Chernova N A and Whittingham M S, 2009, Electrospun nano-vanadium pentoxide cathode, Electrochemistry Communications, Vol. 11, No.3, Pp. 522-525.

[18] Abd-Alghafour N, Ahmed N M, Hassan Z, Mohammad S M, Bououdina M and Ali M, 2016, Characterization of V_2O_5 nanorods grown by spray pyrolysis technique, Journal of Materials Science: Materials in Electronics, Vol. 27, No.5, Pp. 4613-4621.
[19] Mu J, Wang J, Hao J, Cao P, Zhao S, Zeng W, Miao B and Xu S, 2015, Hydrothermal synthesis and electrochemical properties of V_2O_5 nanomaterials with different dimensions, Ceramics International, Vol. 41, No.10, Pp. 12626-12632.
[20] Vernardou D, Vlachou K, Spanakis E, Stratakis E, Katsarakis N, Kymakis E and Koudoumas E, 2009, Influence of solution chemistry on the properties of hydrothermally grown TiO_2 for advanced applications, Catalysis Today, Vol. 144, No.1, Pp. 172-176.
[21] Parida M R, Vijayan C, Rout C S, Sandeep C S, Philip R and Deshmukh P, 2010, Room temperature ferromagnetism and optical limiting in V_2O_5 nanoflowers synthesized by a novel method, The Journal of Physical Chemistry C, Vol. 115, No.1,Pp. 112-117.
[22] Huang J, Qiao X, Xu Z, Ouyang H and Li J, 2015, V_2O_5 nanoflowers assembled by nanorods as cathode material for lithium-ion batteries, Micro & Nano Letters, Vol. 10, No.12, Pp. 686-688.
[23] Chu J, Kong Z, Lu D, Zhang W, Wang X, Yu Y, Li S, Wang X, Xiong S and Ma J, 2016, Hydrothermal synthesis of vanadium oxide nanorods and their electrochromic performance, Materials Letters, Vol. 166, Pp. 179-182.
[24] Qin Y, Fan G, Liu K and Hu M, 2014, Vanadium pentoxide hierarchical structure networks for high performance ethanol gas sensor with dual working temperature characteristic, Sensors and Actuators B: Chemical, Vol. 190, Pp. 141-148
[25] Chen Y, Chen C, Chen W, Liu H and Zhu J, 2015, Influence of thermal-decomposition temperatures on structures and properties of V_2O_5 as cathode materials for lithium ion battery, Progress in Natural Science: Materials International, Pp. 1-6.
[26] Jin W, Yan S, An L, Chen W, Yang S, Zhao C and Dai Y, 2015, Enhancement of ethanol gas sensing response based on ordered V_2O_5 nanowire microyarns, Sensors and Actuators B: Chemical, Vol. 206, Pp. 284-290.
[27] Meng L-J, Silva R A, Cui H-N, Teixeira V, Dos Santos M and Xu Z, 2006, Optical and structural properties of vanadium pentoxide films prepared by dc reactive magnetron sputtering, Thin solid films, Vol. 515, No.1, Pp. 195-200.
[28] Lee S-H, Cheong H M, Je Seong M, Liu P, Tracy C E, Mascarenhas A, Pitts J R and Deb S K, 2002, Microstructure study of amorphous vanadium oxide thin films using raman spectroscopy, Journal of applied physics, Vol. 92, No.4, 1893-1897.
[29] Lee S-H, Cheong H M, Seong M J, Liu P, Tracy C E, Mascarenhas A, Pitts J R and Deb S K, 2003, Raman spectroscopic studies of amorphous vanadium oxide thin films, Solid State Ionics, Vol. 165, No.1,Pp. 111-116.
[30] Su Q, Lan W, Wang Y and Liu X, 2009, Structural characterization of β-V_2O_5 films prepared by DC reactive magnetron sputtering, Applied Surface Science, Vol. 255, No.7, Pp. 4177-4179.
[31] Julien C, Guesdon J, Gorenstein A, Khelfa A and Ivanov I, 1995, The influence of the substrate material on the growth of V_2O_5 flash-evaporated films, Applied surface science, Vol. 90, No.3, Pp. 389-391.

Seed/Catalyst-Free Growth of ZnO Nanoleaves on ZnO Seed Layer/Glass by Thermal Evaporation Method

F. H. Alsultany[1], Z. Hassan[2], N. M. Ahmed[1] and H. R. Abd[1]

[1]School of Physics, Universiti Sains Malaysia, 11800 USM, Penang, Malaysia
[2]Institute of Nano Optoelectronics Research and Technology (INOR), Universiti Sains Malaysia, 11800 USM, Penang, Malaysia

foratusm@gmail.com, zai@usm.my

Abstract. We report the seed/catalyst-free growth of 2D ZnO nanoleaves via thermal evaporation method. These nanoleaves were grown on (75 ± 5 nm) ZnO seed layer, which were deposited on glass substrates by radio frequency (RF) sputtering. Prior to synthesized ZnO nanoleaves, the sputtered ZnO seeds were annealed using the continuous wave (CW) CO_2 laser at 450 °C in air for 15 min. The morphology and the structure of ZnO nanoleaves were investigated by controlling the preparation temperature of each of the Zn powder and the glass substrate under O_2 and Ar gases. Studies were carried out on ZnO nanoleaves using X-ray diffraction, field emission scanning electron microscopy and UV–vis spectrophotometer. The results showed that the sample grow in the hexagonal wurtzite structure with preferentially oriented along (002) direction, good crystallinity and high transmittance. This study serves as the basis for further research on the growth of 2D ZnO nanostructures on a cost-effective glass substrate and at low temperature.

1. Introduction

Zinc oxide (ZnO) nanostructure is an II–VI compound semiconductor with a direct and wide energy band gap (3.37 eV), a native n-type and a high exciton binding energy. ZnO nanostructure is non-toxic, low cost, high thermal stability and high transmittance in the visible region [1]. ZnO has been extensively used in various fields, such as environmental science, optoelectronics, actuators and piezoelectric transducers, solar cells, and catalysts. Many chemical and physical methods are available and can be used to prepare ZnO nanostructures, such as atomic layer deposition, magnetron sputtering (RF), spray pyrolysis, sol-gel method, and thermal evaporation. Among these methods, thermal evaporation is considered to be one of the important methods employed to synthesize ZnO because of its unique features, namely, catalyst-free growth, simple, cost-effectiveness, and easily controlled growth parameters, such as growth rate, distribution on the substrate, and film thickness [2]. Thermal evaporation presents a wide range of nanostructures with various morphologies and with different shapes and sizes such as nanobelts [3], nanosheets [4], nanoscrews [5], nanonails [6], and nanoleaves [7]. The objectives of this study are to use a low cost substrate in a synthesis of the 2D ZnO nanoleaves and to determine the effects of growth temperatures to improve the structural, morphological, and optical properties of 2D ZnO nanoleaves on glass substrates.

Content from this work may be used under the terms of the Creative Commons Attribution 3.0 licence. Any further distribution of this work must maintain attribution to the author(s) and the title of the work, journal citation and DOI.
Published under licence by IOP Publishing Ltd

2. Experimental

ZnO nanoleaves were grown on ZnO seed layer coated glass substrate by thermal evaporation method in a horizontal two-zone tube furnace. The seed layers of ZnO were prepared by radio RF sputtering onto glass substrate, with a thickness of around 75±5 nm. Prior to growth nanoleaves, the seeds were annealed using continuous wave laser (CO_2 laser, tube 70cm, China) at temperatures of 450 °C in the air for 15 min. High purity metallic Zn powder as a first source material was loaded into a quartz boat. This boat with Zn powder was transferred into the center of the furnace using a quartz tube. The ZnO seed coated glass substrate was placed horizontally in the downstream direction of the Zn material. In the first zone, the Zn powder is gradually heated up from room temperature to 550 °C at a rate of 10 °C /min. Purity Ar gas was used during the growth process of nanoleaves as a carrier gas in reaction zone during thermal evaporation. Purity O_2 gas was employed as a second source material which was fed after the temperature reached 550 °C. The pumping of O_2 gas into the reaction zone continued for 90 min. White material was formed on the glass substrate after the evaporation process was completed. The quartz tube was cooled naturally to room temperature.

Morphological observations of ZnO nanoleaves were carried out using a field emission scanning electron microscopy (FESEM) model (FEI/Nova Nano SEM450). Structure analysis was studied using an X-ray diffraction (PANalytical X'Pert PRO) equipped with Cu-Kα radiation (λ= 1.5418 Å). The tube furnace system (model: F21100- 33), works at 1240 W, 230 V, 5.4 A, and frequency of 50/60 Hz Phase-1. The transmission curves were obtained using spectrophotometer type; a Varian, Cary system 5000 UV, Vis-NIR spectrophotometer was used to determine the optical transmission spectra.

3. Results and discussions

The morphologies of the ZnO nanoleaves grown on ZnO seeds/glass substrate via thermal evaporation of Zn powder in the presence of O2 gas was measured by FESEM. Figure 1 (a, b) shows high and low magnification images of the morphology of ZnO nanoleaves that were grown on 75 nm thick layer which appears as a large area, high-density and uniform growth of ZnO that resembles the tree leaves. Therefore, this structure can be called a nanoleaf structure. The diameter of most of these leaves ranges approximately 200 nm length and 50 nm width.

Figure 1. High and low magnification FESEM image of the catalyst-free growth of ZnO .nanoleaves on ZnO seed layers

Figure 2 shows the EDX spectra with Zn and O ratios of ZnO nanoleaves structure which was grown on 75 nm ZnO seed layer. This sample revealed good stoichiometry of the synthesized sample. The EDX showed that this sample contains 50.28 At.% Zn and 49.72 At.% O, as shown in the table inset on Figure 2.

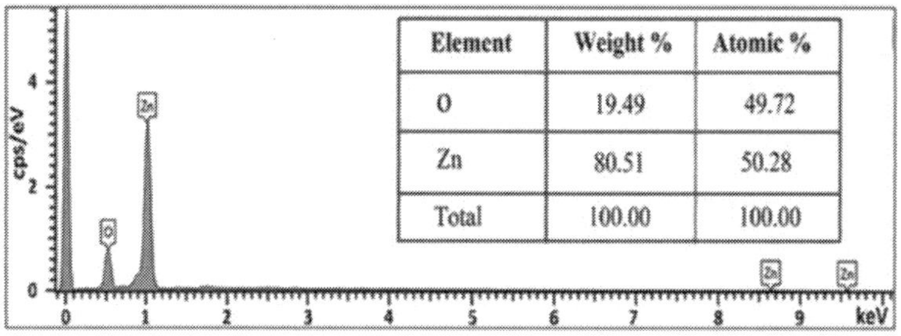

Element	Weight %	Atomic %
O	19.49	49.72
Zn	80.51	50.28
Total	100.00	100.00

Figure 2. EDX spectrum of the ZnO nanoleaves grown on ZnO seed layer. Top-right inset shows the table of sample composition ratio.

3.1. *Crystalline structurer*

Figure 3 shows the XRD pattern of the ZnO nanoleaves grown on ZnO/glass substrates. All peaks of the sample could be indexed to the hexagonal phase with a wurtzite structure of ZnO according to ICSD 01-089-1397. The high and sharp diffraction peak of (002) direction indicates ZnO leaves grow along the c-axis direction. An extensive XRD pattern of ZnO film appeared is extending from 31.67° to 72.21° in the 2θ region, because this is typical diffraction patterns for a glass substrate.

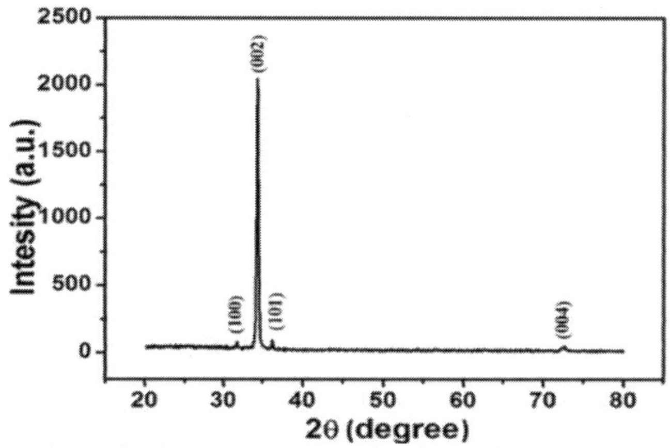

Figure 3. The XRD patterns of ZnO nanoleaves grown on ZnO/glass substrates.

3.2. *Optical properties*

Figure 4 (a) shows the UV–vis transmission of the ZnO nanoleaves on a glass substrate in a wavelength range of 200- 800 nm. The average transmittance is estimated to be 84.46% for ZnO nanoleaves film. The high optical transmittance of this sample in the visible light range indicates the high homogeneity and low surface roughness [8]. The Eg of the sample was derived by plotting the Tauc formula and taking the extrapolation of the linear portion from the plot of $(\alpha h\upsilon)^2$ versus the photon energy (hv), as shown in Figure 3 shows. 4 (b). The Eg of the ZnO nanoleaves is 3.27 eV which agrees with bulk value of ZnO (3. 37 eV) [9].

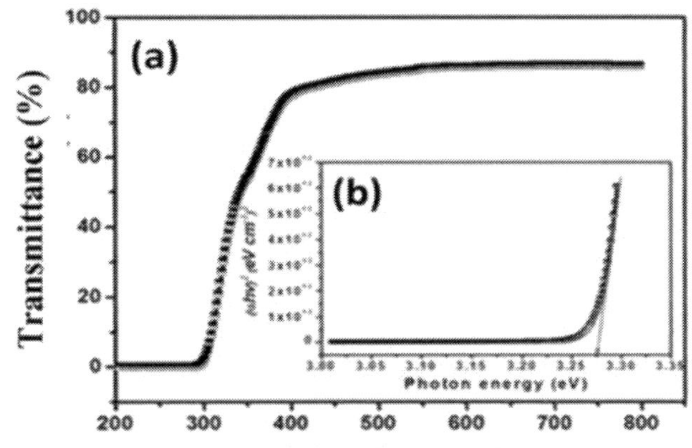

Fig. 4: (a) UV–vis transmission of the ZnO nanoleaves on glass substrate, (b) $(\alpha h\upsilon)^2$ as a function of photon energy ($h\upsilon$) for the sample.

4. Conclusion

In summary, the growth of 2D ZnO nanoleaves on ZnO/glass substrates was successfully carried out on glass substrate via thermal evaporation. It was found that the preparation of the nanoleaves by using growth temprature of 550 °C for Zn powder with the vertical position of the substrate under O_2 and Ar gases. XRD measurement has shown the formation of high crystal quality of the ZnO film along the c-axis direction. The average UV–vis transmission was estimated to be 84.46% and the absorption shoulder was in the UV region (300– 400 nm). white light, low color temperature, and a suitable rendering index was recorded by controlling the powder sintering temperature.

Acknowledgements

The authors acknowledge the support from LRGS grant (203/CINOR/6720013) and Universiti Sains Malaysia.

References

[1] Wang Y *et al* 2015 *Mater. Chem. Phys.* 266 153.
[2] Alsultany F H *et al* 2016 *Ceramics International.* 13144 42.
[3] Pan Z *et al* 2001 *Nanobelts of semiconducting oxides. Science* 1947 291.
[4] Park J *et al* 2006 *Curr. Appl. Phys.* 1020 6.
[5] Liao L *et al* 2005 *Appl. Phys. Lett.* 083106 86.
[6] Song X *et al* 2007 *J. Phys. Chem. Solids.* 1681 68.
[7] Zhai B. G *et al* 2015 *Applied Surface Science.* 842 324.
[8] Alsultany F H *et al* 2016 *Superlattices and Microstructures.* 79 92.
[9] Alsultany F H *et al* 2016 *Materials Research Bulletin.* 63 79.

IOP Conf. Series: Journal of Physics: Conf. Series **1083** (2018) 012038 doi:10.1088/1742-6596/1083/1/012038

Comparative Studies on The Transfer of Chemical Vapor Deposition Grown Graphene Using Either Electrochemical Delamination or Chemical Etching Method

Raed Abdalrheem*, F.K. Yam, Abdul Razak Ibrahim, K.P. Beh, Y.Z. Ng, F.H.A. Suhaimi, H.S. Lim, M.Z. Mat Jafri, Ammar A. Oglat

School of Physics, Universiti Sains Malaysia, 11800 Minden, Penang, Malaysia.

rsheyyab2017@gmail.com, mjafri@usm.my

Abstract. In this research, the electrochemical delamination (ECD) and chemical metal etching (CME) techniques were performed to transfer the chemical vapor deposition (CVD) grown graphene onto glass and SiO2 substrates. Principally, graphene was grown on a copper substrate by chemical vapor deposition (CVD) under specified conditions. Atomic force microscopy (AFM), Raman spectroscopy and X-ray Diffraction (XRD) were employed to investigate the morphological aspects and qualities of transferred graphene samples, while the UV-VIS spectroscopy was also applied to compare their transparencies. The Raman spectra analyses result revealed that Raman-bands positions of the transferred graphene could be shifted by the using various transfer methods or substrate. However, the thickness and the visible light transparencies of the transferred graphene samples were almost similar. The graphene sample transferred via ECD has portrayed lowest D to G-bands ratio (ID/IG) which indicates better graphene.

1. Introduction

The Chemical vapor deposition (CVD) is the most common method used for growing graphene. The method excels other methods due to large production quantities and high graphene quality. In addition, the CVD method can use various carbon precursors and metallic substrates [1]. Usually, the CVD grown graphene is deposited on metallic substrates such as copper and nickel. The fact that graphene is exploited in various applications, it must be transferred onto various substrates such as polymers [2], oxides [3], metallic [4] and others substrates [5]. Typically, transferring graphene onto the SiO2/Si wafer was widely common. This substrate is optimally suitable to study the graphene characters. However, transferring monolayered graphene of an atom thickness without protection layer is almost an impossible task. To overcome this obstacle, a spin coating of a polymeric layer such as PMMA is required. The PMMA layer onto the graphene/substrate eases graphene holding and ensure its protection [6]. Some techniques used for graphene include thermal release tape (TRT), free-floating and Bubble transfer [7]. Though, the efficient technique to transfer graphene without altering its properties is challenging. In CME transferring processes the common chemical metallic etchants solutions like iron (III) chloride (FeCl3), iron (III) nitrate (Fe(NO3)3) and ammonium persulfate (APS) are used to etch the metallic substrates. During the process, the chemicals induced metallic contaminations into graphene film [8]. However, graphene transfer by the ECD technique occurs when an electrical field is applied to graphene/substrate (cathode) and another metallic electrode (anode) in an electrolyte solution. It leads

Content from this work may be used under the terms of the Creative Commons Attribution 3.0 licence. Any further distribution of this work must maintain attribution to the author(s) and the title of the work, journal citation and DOI.
Published under licence by IOP Publishing Ltd

to the formation of the hydrogen bubbles between the graphene and the metallic substrate; this caused the detachment of PMMA/graphene film. It was reported that the ECD has fewer residues compared to CME. However, the PMMA residues involve defects which degrades graphene optical and electrical properties [9]. In this research, ECD (bubble transfer) and CME (free floating) techniques were used to transfer the deposited graphene/copper onto glass and SiO2/Si substrates. The qualities, defects intensities, and transparencies of transferred graphene samples were compared.

2. Materials and Methods

2.1 Atmospheric Pressure Chemical Vapor Deposition (APCVD) grown graphene:

Copper foil (Cu, thickness: 25 µm, purity: 99.9%, supplied by STREM) was used as a catalytic substrate to deposit graphene. The furnace temperature was ramped up to 1050°C. Cu foils loaded into the cold zone of furnace quartz tube. After that, the copper foils were purged under nitrogen (3000 sccm) for 5 min. Subsequently, the copper foils were annealed at 1050°C under the flowing gas mixture (H_2:N_2, 200:1000 sccm) for 10 min. The samples were subjected to a second purging process using nitrogen (1000 sccm) for 5 min. The graphene was synthesized by APCVD method where the methanol (purity: 99.8%) was used as a liquid precursor of carbon atoms. Thus, hydrogen (30 sccm) has flowed through a gas bubbler containing methanol. The N_2 flow rate has remained the same (1000 sccm) during both growth and cooling process. The graphene growth time was 30 min. The samples were allowed to cool down under N_2 environment to room temperature.

2.2 Transferring of Graphene Films onto Desired Substrates:

The deposited graphene on copper was transferred onto glass to measure the visible light transparency while the SiO_2/Si substrates (commercial wafer of 300nm oxide thickness on P-type Si (111)) were used for better characterizations. Initially, a layer of polymethyl methacrylate PMMA was spin-coated on graphene/copper foil (2000 rpm for 10 sec). Then, the PMMA coated sample was dried in ambient for 2 min at 175°C. The previous process is essential to hold and protect the graphene sample. Secondly, the PMMA/graphene layer was released from Cu substrate by using three chemical etching solutions, that were ferric chloride ($FeCL_3$), ferric nitride ($Fe(NO_3)_3$, 2M) and ammonium persulfate (APS,0.6M). The suspended PMMA/graphene film is transferred to DI water to eliminate the copper residues.
The sample is transferred onto insulating substrates such as glass or SiO_2/Si. The samples were dried at 150°C for 15 min to achieve more adhesion on the new substrate. Then, the acetone is used to dissolve the PMMA. Subsequently, the graphene/glass is dried for 5 min at 70 °C. Finally, an additional annealing ambient at 400 °C for 20min is required to eradicate the PMMA residues. In ECD, the Power supply of (3-4V) has been applied to the graphene/substrate (cathode) and Pt electrode (anode) inside (NaCl,0.5M) electrolyte solution. Then the sample rinses with distilled water and ethanol much time until release from the substrate. After that, the sample has been scooped by the desired substrate.

2.3 Graphene Characterization

Atomic force microscopy (AFM) was used to scan the morphological aspects of the transferred graphene samples on SiO2/Si and glass substrates. Raman spectroscopy (JOBIN-YVON HR 800) measurements were conducted on the graphene/Si-SiO_2 and graphene/glass. The wavelength of incident LAZER was 514.45 nm operated at 20 mW. Furthermore, the XRD (PANalytical X' Pert PRO MRD-PW3040) measurements were carried out to identify the graphene growth for the samples on SiO2/Si substrates. In addition, UV-Vis spectroscopy (Agilent-Cary 5000) was utilized to study the transparencies of graphene/glass samples.

The positions of Raman bands depend on the lattice vibrational frequency (V). The vibrational frequency (V) depends on the bonds force constant (k) and the mass of atoms (m) in the vibrated lattice as shown in equation. 1. The bonds force constant (k) can be changed by the variations of carbon atoms positions and the atoms bonds length. Consequently, when these features alter under any effects, the frequency of Raman bands will be changed. These effects include stress or strain in the graphene sheets, which causes changes in lattice spacing, resulting to a change in vibrational modes wavenumbers[10]. The graphene number of layers (n) which depending on its transparency (T%) can be calculated [11] using equation. 2:

$$\nu = (\frac{1}{2\pi})\sqrt{k/m} \quad (1)$$

$$n = \frac{100 - T\%}{2.3\%} \quad (2)$$

3. Results and Discussion

The samples were presented according to the chemical etching solutions that have been used in the transferring processes S-$FeCl_3$, S-$Fe(NO_3)_3$ and S-APS. The S-ECD is denoted for the sample which has been transferred by electrochemical delamination. Figure. (1-a) shows a ($10m\mu \times 10\mu m$) selected image of transferred graphene sample on SiO_2/Si substrate which scanned under (AFM). From the figure, the morphological layer can be observed which indicates the multi-layered graphene based on the root mean square (RMS) of surface roughness (~20 nm). Moreover, the coverage area of the transferred graphene showed similar colour in their appearance. Both the wrinkles and impurities were pointed out in the figure. Figure. (1-b) presents the XRD spectra of transferred graphene on SiO_2/Si substrate. The spectrum revealed that the synthesis of multi-layered graphene film of carbon peak was obtained at 26.4 degree.

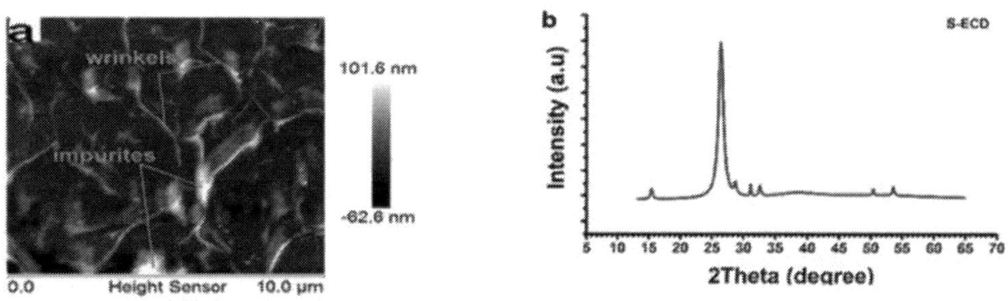

Figure 1. **(a)** AFM image of CVD grown graphene on SiO_2/Si substrate. **(b)** XRD-spectrum of graphene/SiO_2. the carbon peak appears at $2\theta = 26.4°$.

Raman analyses are an efficient tool for extracting the graphene thicknesses and qualities. Raman spectra of graphene showed the three notable bands which are considered as a fingerprint of the graphitic compounds, the D, G, and 2D-peaks were centered at 1350, 1580 and 2700 cm^{-1} respectively [12]. The 2D- to G-peak intensities ratios (I_{2D}/I_G) and the full width of half Maximum FWHM of 2D-band were used to obtain the number of graphene layers, while the (I_D/I_G) ratio indicates to graphene quality and defects/lattice disorder. Figure. (2-a,b) presents the Raman spectrum for the transferred graphene samples on SiO_2/Si and glass substrates. The Raman-bands (D, G, and 2D) have been shifted according to various graphene transfer techniques.

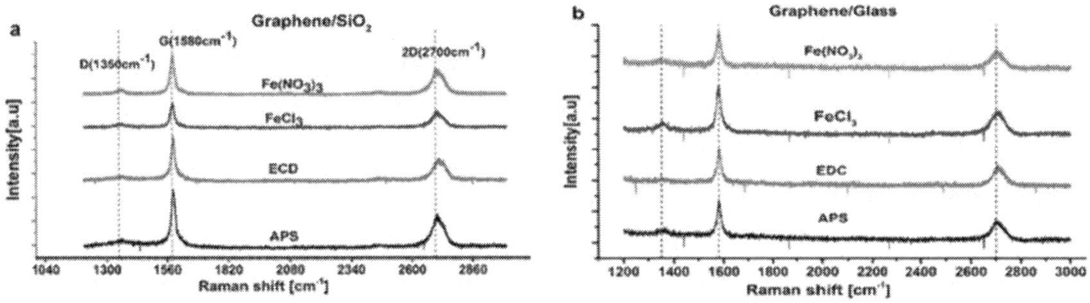

Figure. 2 : Raman spectra of transferred graphene samples using different transfer techniques. for (**a**) SiO_2/Si and (**b**) glass substrates.

The Lorentzian curve fit was carried out to obtain the shift of D, G and 2D-bands positions concerning their ideal positions (1350, 1580 and 2700 cm^{-1}) respectively. From Figure. (3-a, b) it can be noted that SiO_2 caused higher G and 2D-bands shifts than the glass substrates for all samples under study. In addition, the G-bands shifts of transferred graphene on glass substrates were very small. However, Figure. (1-f) Showed that the sample (S-ECD) has the highest 2D-bands upshift for different substrates, while the sample (S-$Fe(NO_3)_3$) has the lowest 2D and G-bands shifts. From Figure. (3-c), it can be observed that the D-band shift occurred by changing either the graphene transfer method or the substrate type. The upshift of D-band for the studied samples was higher than other Raman-bands (2D and G). briefly, Raman spectra analyses revealed that the Raman-bands of the transferred graphene could be shifted by using various transfer methods or different substrates. Furthermore, the unintentionally doping of graphene films are the potential reason for the 2D and G-bands shifts[13]. Whereby the D-bands shift attributed to various impurities types in various samples [14].

IOP Conf. Series: Journal of Physics: Conf. Series **1083** (2018) 012038 doi:10.1088/1742-6596/1083/1/012038

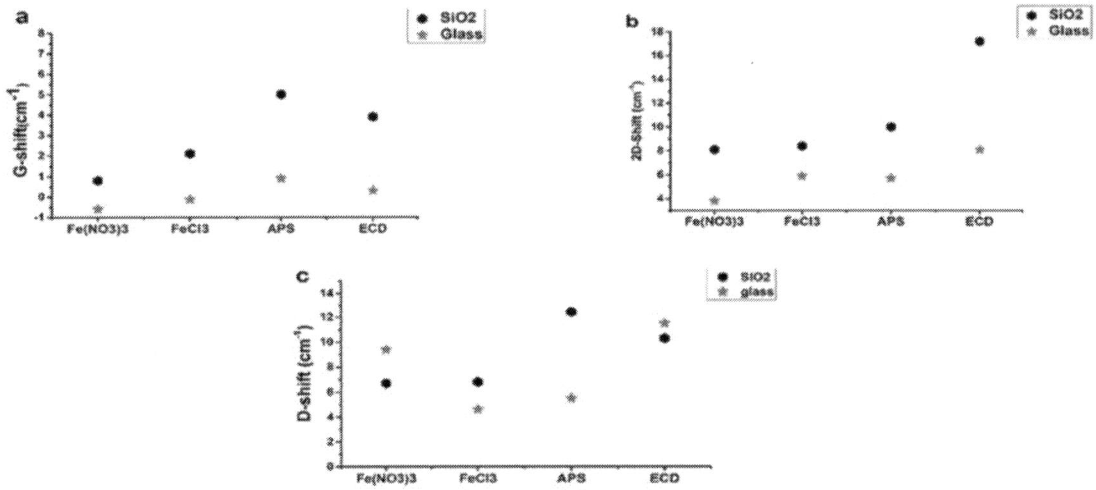

Figure. 3 : positions shifts of Raman-bands ; (**a**) G-bands (**b**) 2D-bands and (**c**)D-band for graphene/SiO2 and graphene/glass under various transfer techniques.

Figure. 4 displayed the primary features of the transferred graphene samples. It was observed that the transferred graphene samples on SiO_2 and glass substrates showed multi-layered graphene according to I_{2D}/I_G ratios (~0.5-0.6) as shown in Figure. (4-a). This range short indicates a similar number of layers which signifies that the graphene thickness was not affected by the changing of transfer methods. The transparencies values which appear in Figure. (4-b) has been employed to obtain the graphene number of layers. The visible light absorption of monolayered graphene is ~ 2.3%. The transparencies values were ranging between 76% to 80% as shown in Figure. (4-b). The convergence of transparencies values was noticed to reflects the convergence in I_{2D}/I_G ratios which resulted in obtaining the same number of graphene layers, starting from 9 to 10 layers. The slight differences in transparencies measurements could be due to the existence of various defect levels. Furthermore, the increasing of attenuation coefficient ($\propto$) caused by wrinkles could change the graphene transparency [15]. However, the quality of graphene and the defects levels in its structures can be verified based on the I_D/I_G ratios. According to Figure. (4-c), it can be noted that the (S-ECD) graphene sample has the lowest I_D/I_G ratio leading to better graphene quality and lower defect level. Furthermore, the SiO_2 substrates showed better graphene characters than glass. These results are consistent with other studies where the SiO_2 substrate was described as the best choice for graphene characterizations [16]. The crystallite size (L_a) and the average distance between defects points (L_D) have been calculated based on Tuinstra-Koenig empirical relation [17]. From Figure (4-d, e) it can be noted that the L_a and L_D of graphene on silicon oxide substrates were more significant than its values on glass substrates. Moreover, the S-ECD sample has the highest L_a and L_D comparing with other samples. These results are expected because the L_a and L_D are inversely proportional to I_D/I_G ratios.

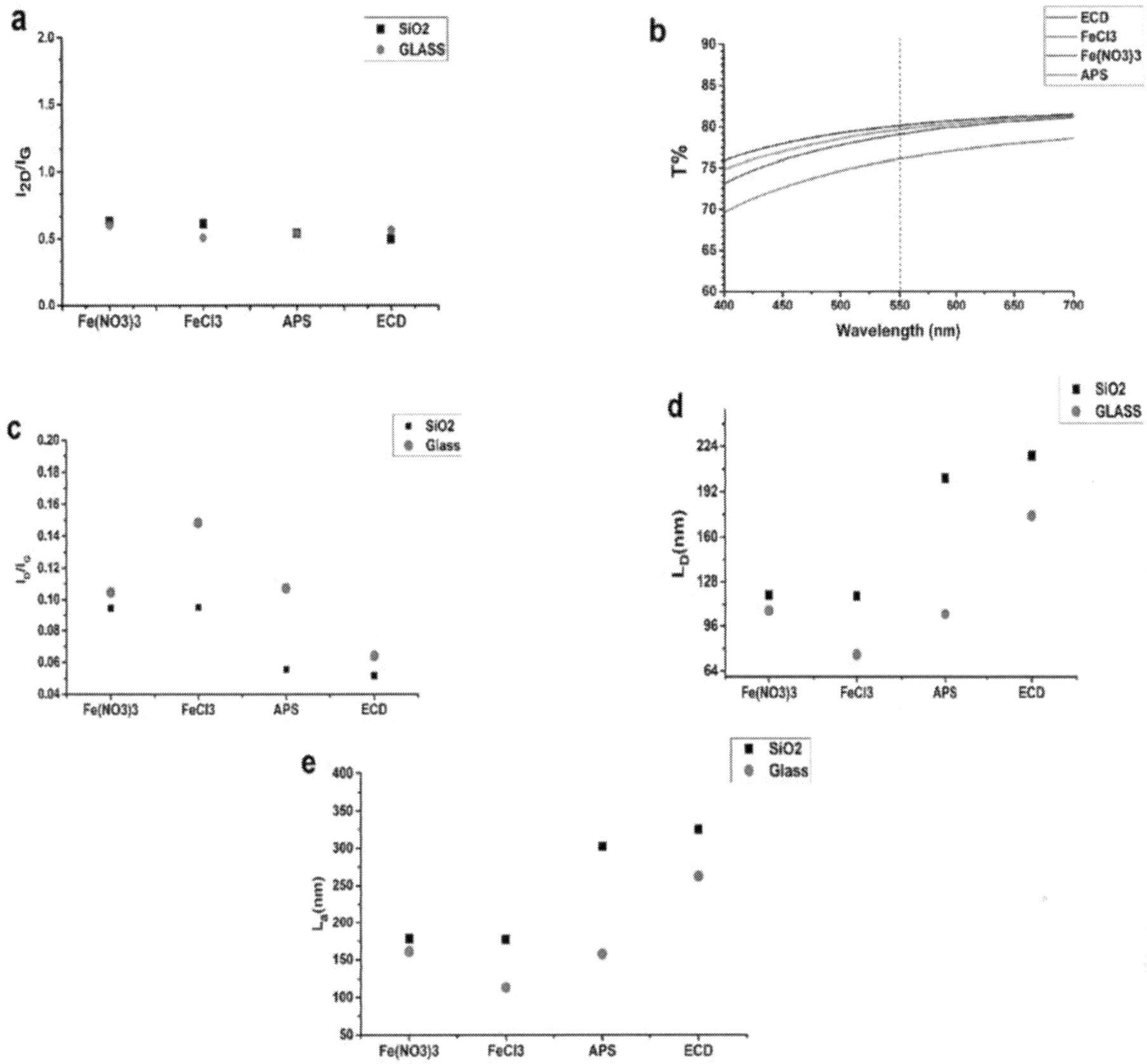

Figure. 4: the primary features of the transferred graphene samples on various substrates; (a)2D to G ratios (I_{2D}/I_G), (b) visible light transparencies (T%), (c) D to G ratios(ID/IG), (d) crystallite sizes (L_a), (e) average distance between defects points (L_D).

4. Conclusion

The ECD and CME have been performed to transfer the CVD grown graphene deposited on copper substrates to either glass and SiO_2 substrates. The transferred graphene samples were multi-layered (9-10 layer) according to Raman spectroscopy, XRD, AFM and UV-Vis spectroscopy. Raman spectra analyses revealed that the Raman-bands positions of the transferred graphene were shifted using various transfer methods or different substrates. The notable shifts in D-bands are attributed to various impurities types in the samples. The convergence of the transparencies values reflects the convergence in the I_{2D}/I_G ratios that gave almost the same graphene thickness for the samples. Comparatively, the SiO_2 substrate is better than glass for graphene characterizations. The graphene sample transferred via ECD has portrayed lowest D to G-bands ratio (I_D/I_G) which indicates better graphene quality.

Acknowledgments
This work is financially supported by the grant "Clad (Topological Insulators and Transition Metal Dichalcogenides) Planar Photonic Devices" (203/PFIZIK/6770002), which is under the LRGS grant "Graphene and Other 2-Dimensional Nanomaterial" (grant no.: LRGS(2015)/NGOD/UM/KPT) from Ministry of Higher Education (MOHE) of Malaysia. All supports from USM and various parties are gratefully acknowledged.

References

[1] Lavin-Lopez, M., L. Sanchez-Silva, J. Valverde, and A. Romero, *CVD-graphene growth on different polycrystalline transition metals.* AIMS MATERIALS SCIENCE, 2017. 4(1): p. 194-208.

[2] Chen, J.H., M. Ishigami, C. Jang, D.R. Hines, M.S. Fuhrer, and E.D. Williams, *Printed graphene circuits.* Advanced Materials, 2007. 19(21): p. 3623-3627.

[3] Novoselov, K.S., A.K. Geim, S.V. Morozov, D. Jiang, Y. Zhang, S.V. Dubonos, I.V. Grigorieva, and A.A. Firsov, *Electric field effect in atomically thin carbon films.* science, 2004. 306(5696): p. 666-669.

[4] Kim, J.-S., D.W. Kim, H.T. Jung, and J.W. Choi, *Controlled lithium dendrite growth by a synergistic effect of multilayered graphene coating and an electrolyte additive.* Chemistry of Materials, 2015. 27(8): p. 2780-2787.

[5] Gannett, W., W. Regan, K. Watanabe, T. Taniguchi, M. Crommie, and A. Zettl, *Boron nitride substrates for high mobility chemical vapor deposited graphene.* Applied Physics Letters, 2011. 98(24): p. 242105.

[6] Reina, A., H. Son, L. Jiao, B. Fan, M.S. Dresselhaus, Z. Liu, and J. Kong, *Transferring and identification of single-and few-layer graphene on arbitrary substrates.* The Journal of Physical Chemistry C, 2008. 112(46): p. 17741-17744.

[7] Leclercq, J. and P. Sveshtarov, *The Transfer of Graphene: A Review.* Bulgarian Journal of Physics, 2016. 43(2).

[8] Liang, X., B.A. Sperling, I. Calizo, G. Cheng, C.A. Hacker, Q. Zhang, Y. Obeng, K. Yan, H. Peng, and Q. Li, *Toward clean and crackless transfer of graphene.* ACS nano, 2011. 5(11): p. 9144-9153.

[9] Wong, C.H.A. and M. Pumera, *Electrochemical delamination and chemical etching of chemical vapor deposition graphene: contrasting properties.* The Journal of Physical Chemistry C, 2016. 120(8): p. 4682-4690.

[10] Liu, Q., B. Guo, Z. Rao, B. Zhang, and J.R. Gong, *Strong two-photon-induced fluorescence from photostable, biocompatible nitrogen-doped graphene quantum dots for cellular and deep-tissue imaging.* Nano letters, 2013. 13(6): p. 2436-2441.

[11] Bonaccorso, F., Z. Sun, T. Hasan, and A. Ferrari, *Graphene photonics and optoelectronics.* Nature photonics, 2010. 4(9): p. 611-622.

[12] Malard, L., M. Pimenta, G. Dresselhaus, and M. Dresselhaus, *Raman spectroscopy in graphene.* Physics Reports, 2009. 473(5): p. 51-87.

[13] Iqbal, M., A.K. Singh, M. Iqbal, and J. Eom, *Raman fingerprint of doping due to metal adsorbates on graphene.* Journal of Physics: Condensed Matter, 2012. 24(33): p. 335301.

[14] Ferrari, A.C., J. Meyer, V. Scardaci, C. Casiraghi, M. Lazzeri, F. Mauri, S. Piscanec, D. Jiang, K. Novoselov, and S. Roth, *Raman spectrum of graphene and graphene layers.* Physical review letters, 2006. 97(18): p. 187401.

[15] Li, X., Y. Zhu, W. Cai, M. Borysiak, B. Han, D. Chen, R.D. Piner, L. Colombo, and R.S. Ruoff, *Transfer of large-area graphene films for high-performance transparent conductive electrodes.* Nano letters, 2009. 9(12): p. 4359-4363.

[16] Wang, Y.Y., Z.H. Ni, T. Yu, Z.X. Shen, H.M. Wang, Y.H. Wu, W. Chen, and A.T. Shen Wee, *Raman studies of monolayer graphene: the substrate effect.* The Journal of Physical Chemistry C, 2008. 112(29): p. 10637-10640.

[17] Mallet-Ladeira, P., P. Puech, C. Toulouse, M. Cazayous, N. Ratel-Ramond, P. Weisbecker, G.L. Vignoles, and M. Monthioux, *A Raman study to obtain crystallite size of carbon materials: a better alternative to the Tuinstra–Koenig law*. Carbon, 2014. 80: p. 629-639.

IOP Conf. Series: Journal of Physics: Conf. Series **1083** (2018) 012039 doi:10.1088/1742-6596/1083/1/012039

Properties of Zinc Oxide Nanorods as Ethanol Sensor

Haslinda Abdul Hamid[1,2, a], Zainovia Lockman[1] and Khairunisak Abdul Razak[1,b]

[1]School of Materials and Mineral Resources Engineering, Engineering Campus, Universiti Sains Malaysia, 14300 Nibong Tebal, Penang, Malaysia
[2]Faculty of Applied Sciences, Universiti Teknologi MARA, 13500 Permatang Pauh, Penang, Malaysia

[a]hasli8366@ppinang.uitm.edu.my, [b]khairunisak@usm.my

Abstract. In this work, the zinc oxide (ZnO) nanorods were studied as a potential ethanol sensor for environmental monitoring. Nanorods were grown on indium tin oxide substrate (ZnO/ITO) by using the hydrothermal method with the aid of seeds. The morphology of the ZnO nanorods was characterized by scanning electron microscopy (SEM), energy dispersive X-ray (EDX) and X-ray diffractometer (XRD). The performance of the ZnO nanorods for the electrochemical detection of ethanol was investigated by using cyclic voltammetry (CV). The three electrodes CV system shows high catalytic activity towards the oxidation of ethanol in 1 M potassium hydroxide (KOH) solution. The ZnO/ITO sensor shows high sensitivity of 8.17 $\mu AmM^{-1}cm^{-2}$ and the detection limit of 5.2 mM. The linear range is from 7-25 mM with linear correlation coefficient of 0.9589 with good stability. Therefore, ZnO/ITO electrode has great potential as ethanol sensor.

1. Introduction

Volatile organic compounds (VOCs) have been identified as highly toxic or carcinogenic in nature and may have both short and long term impacts on human health as well as on the natural ecosystem. VOCs are defined as organic compounds that are present in the atmosphere as gases, but could be present in liquids or solids under normal conditions of temperature and pressure. Ethanol is colorless, flammable, slightly odor and volatile. Ethanol is commonly used as household chemical, car manufacturing, pharmacology, engine fuel and solvent [1]. Ethanol has an effect with the sick building syndrome (SBS), which is used to define symptoms of reduced of well-being or healthiness. Due to the widely usage and negative effects, early discovery and monitoring of these toxic and hazardous chemicals is thus mandatory for environmental safety purposes.

Various analytical techniques, such as liquid chromatography [2], infrared spectroscopy [3] and mass spectroscopy [4] have been used to detect ethanol concentration but there are non-portable and expensive. Due to high cost and complex process, it is necessary to develop high sensitivity, inexpensive, portable and fast response for ethanol detection. For example, chemical sensors using semiconductor metals have attracted widespread attention to play a noteworthy role for the detection and observing poisonous harmful chemicals. Among these materials, ZnO is promising material for this applications due to its wide and direct band gap, high exciton binding energy, biocompatibility, excellent electrical characteristics, non-toxic and good thermal stability [5]. The 1-D ZnO nanostructures, for example nanorods have been widely studied for sensing applications due to their

Content from this work may be used under the terms of the Creative Commons Attribution 3.0 licence. Any further distribution of this work must maintain attribution to the author(s) and the title of the work, journal citation and DOI.
Published under licence by IOP Publishing Ltd

high surface to volume ratio, crystallinity and charge confinement ability [6]. A number of methods have been reported in literature for the fabrication of 1-D ZnO nanostructures such as hydrothermal method [7], sonochemical growth [8] and chemical bath deposition [9]. Several works have been done to achieve high performance of ZnO such as doping [10-11] or in nano-composites ZnO [12] for their sensing properties. These works lead to complicated work and time consuming.

In this work, facile hydrothermal method was used to prepare ZnO nanorods. This method is relatively low-cost, ease handling and works at low temperature conditions. Moreover, ZnO nanorods were prepared in aqueous method without adding any organic solvents or surfactants. To the best of our knowledge, limited information can be obtained on using undoped or as-grown ZnO nanorods for the detection of ethanol in liquid phase using a simple and reliable CV technique which showed the redox reactions. The detection of ethanol is based on the response currents resulted from the ethanol oxidation in alkaline media and the results exhibit excellent electro-catalytic activity for ethanol detection. In addition, the sensitivity (S), limit of detection (LOD) and linear dynamic range (LDR) are also evaluated accordingly.

2. Experimental method

In this work, the ITO electrodes were cleaned by using a standard cleaning method. Ammonium hydroxide (NH_4OH), hydrogen peroxide (H_2O_2) and distilled water in a ratio of 1:4:20 were mixed and heated at 60 °C for 20 minutes. In order to remove water residues, ITO electrodes were immersed in 2-propanol and dried in an oven for few minutes. ZnO seed layer was prepared by sol-gel technique using zinc acetate dehydrate [Zn $(CH_3COO)_2$. $2H_2O$] in methanol. The mixture was stirred vigorously at 60°C for 20 minutes. Then, 15 mL mono-ethanolamine was added drop by drop into the solution under constant stirring for 2 hours at 60°C. The solution was aged for 24 hours. For deposition process, the ZnO seed layer was dropped on the ITO electrodes and dried in an oven at 150°C for 20 minutes. This process was repeated 3 times. The coated ITO electrodes were placed in the furnace at 500°C for 2 hours for annealing process. Hydrothermal process was used to grow ZnO nanorods. In this process, the ZnO nanorods were grown in aqueous solutions of zinc nitrate [Zn- $(NO_3)_2$] and hexamethylamine (HMT) in deionized water. The molar ratios of these solutions were 1:1. The electrodes were placed in a screw-capped bottle in a pre-heated oven at 80°C for 4 hours. The fabrication process is shown in figure 1. Electrochemical measurement was conducted using a three electrode system at room temperature where the ZnO/ITO acted as the working electrode. Besides, an Ag/AgCl was used as the reference electrode and platinum as the auxiliary electrode. Series of experimental works were performed to optimize the different parameters of electrochemical reaction, such as the effect of ethanol concentration, the effect of KOH, the effect of scan rate and also the stability of electrode in ethanol. The morphology and structure of ZnO nanorods were examined using SEM attached with EDX and XRD.

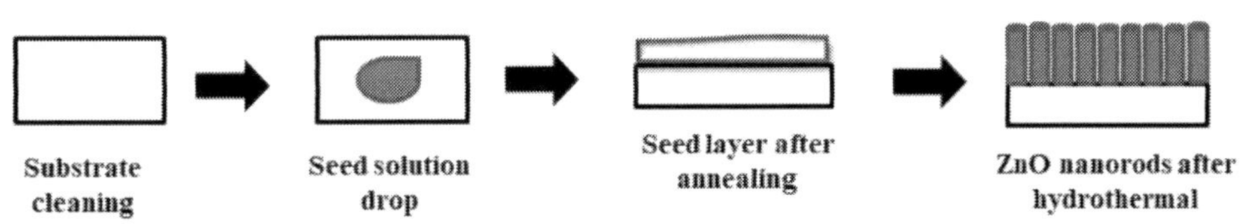

Fig. 1 Fabrication steps of ZnO nanorods electrode.

3. Results and discussion

The surface morphology of ZnO nanorods prepared seed is shown in Figure 2a. The result showed that the seed solution contains homogeneous nanoparticles and uniform in size with an average diameter of 60 nm. It is essential to have ZnO nanocrystal seeds layer during preliminary phase of hydrothermal reaction as it can expedite the well-arranged and vertical growth of ZnO nanorods on substrates. The reason can be ascribed to lower nucleation energy barrier and easier heterogeneous nucleation route due to the identical lattice structure of ZnO seed layer [11]. Moreover, uniform surface roughness of ZnO seed layer can provide more nucleation sites and limit the disorder movement of nucleation sites. Figure 2b shows the SEM image of ZnO nanorods. Nanorods are grown in very high density and well distributed. The diameters of ZnO nanorods are in average of 80 nm. Figure 2b (inset) shows the nanorods possess a hexagonal surface and grew evenly on the surface of the substrate.

(a) (b)

Figure 2: The SEM images of surface morphologies of (a) seed layer and (b) nanorods.

The produced ZnO nanorods was used as the working electrode by using the three electrodes system and measured by CV technique using ethanol as a target chemical. Figure 3 shows CV sensing response and calibration curves of the ZnO nanorods electrode in 1 M KOH solution with the presence and absence of 1 M ethanol. No obvious anodic peak was observed without ethanol, suggesting that there is no redox reaction occurs on the surface of the electrode. In solution with ethanol, a substantial increase in the current value with applied potential is clearly revealed after ethanol addition. The peak is wider compared to without ethanol, which can be affected both by adsorption routes and by electron transfer activities at the electrode surface.

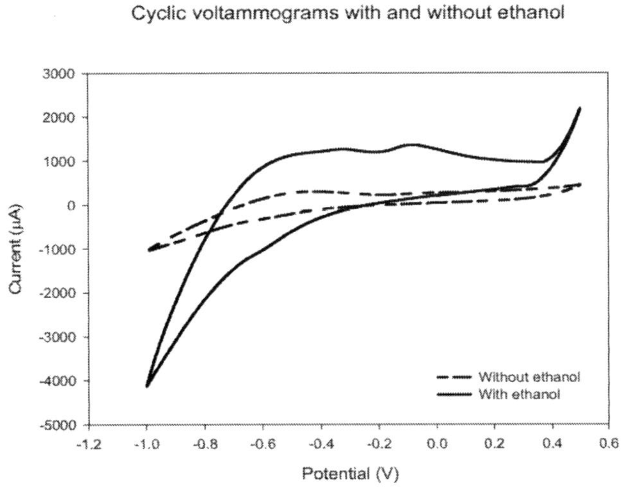

Figure 3: The CV analysis with and without ethanol.

Figure 4 shows the CV of the 1 M ethanol solution containing KOH solutions of various concentrations ranging from 0.1 M to 6 M. The peak current increases as the KOH concentration increases from 0.1 to 1M. However, further increase in the KOH concentration leads to a decrease in the peak current. This could be due to excessive hydroxide ions, (OH^-) from KOH could inhibit the adsorption of ethanol ions on the electrode surface. Furthermore, the higher the concentration of KOH in the solution, the higher the pH would be and can cause massive of hydroxide ions [12]. Figure 5 shows the CV of the ZnO nanorods in the ethanol solution for the concentration ranging from 7-25 mM. Meanwhile, inset Figure 5 shows the calibration curve with good linearity of current-concentration, which was plotted at -0.1V. The increase in the current with the concentration is probably because of the active sites of the surface with ethanol molecules. The sensitivity was calculated from the slope of the calibration plot is 8.17 $\mu AmM^{-1}cm^{-2}$. The linear dynamic range is 7-25 mM with square of the correlation coefficient r^2=0.9589 and the limit of detection is 5.2 mM.

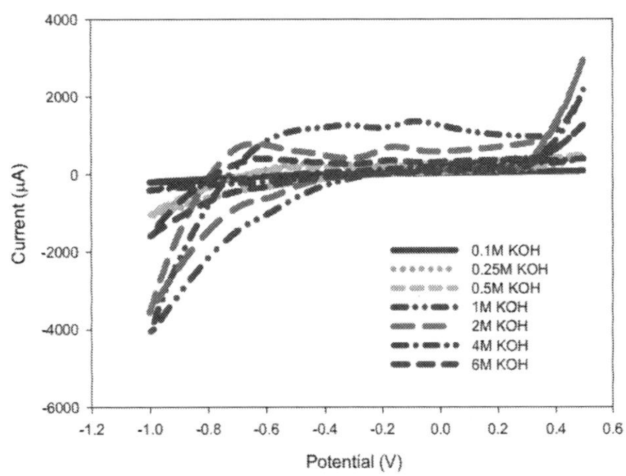

Figure 4: Variation of KOH concentration.

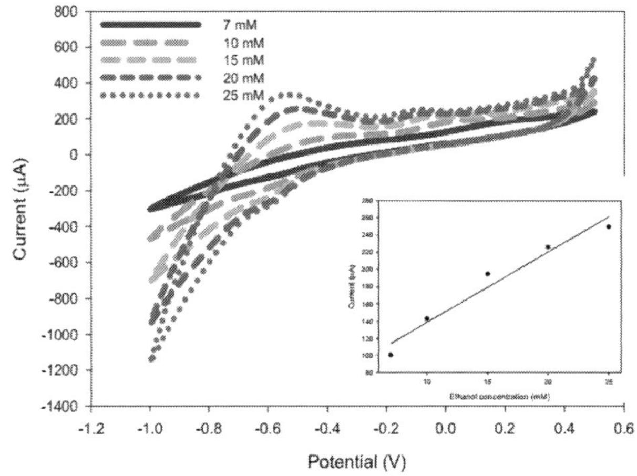

Figure 5: Variation of ethanol concentration.

The effect of scan rate was investigated in 1 M ethanol in 1 M KOH at scan rate from 10-50 mVs^{-1} as shown in Fig. 6. The CV results showed an increase of current when the scan rate was increased. The oxidation peak widen with increasing the scan rate. This behavior was due to the kinetic behavior of ions in this process [12]. Higher scan rate denotes faster reaction of electron transfer on the surface of electrode. The electrode adsorption happens for all species of ions in its chemical reaction. This is also evidence for the occurrence of higher current. However, at the low scan rate, the electrode reaction is limited due to the slower electron transfer and therefore, the electrode has sufficient time for its oxidation process and which also indicates the slower diffusion of ions during low scan rate. Therefore, ZnO nanorods can be used efficiently for a high performance of ethanol sensing.

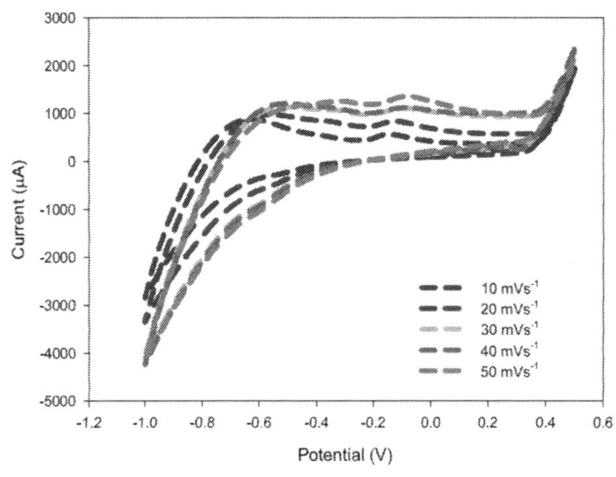

Figure 6: Typical cyclic voltammetry for 1M ethanol at various scan rates.

4. Conclusion

In this work, 1-D ZnO nanorods electrode was successfully fabricated for the ethanol sensor application by using simple hydrothermal method with the aided of seeds. The morphology of ZnO nanorods were confirmed by SEM and showed the nanorods were grown in hexagonal shape and in high density. The ZnO nanorods was utilized as an efficient electron facilitator for the fabrication of ethanol chemical sensor, which exhibited sensitivity of 8.17 $\mu AmM^{-1}cm^{-2}$ with detection limit of 5.2 mM. The produced ZnO nanorods could be competently used as an effective ethanol sensor for environmental sensing.

Acknowledgement

The authors appreciate the technical supports from the School of Materials and Mineral Resources, and Institute for Research in Molecular Medicine (INFORMM), Universiti Sains Malaysia. This research was funded by a grant 1001/Pbahan/870049. One of the authors acknowledges Universiti Teknologi MARA.

References

[1] D. G. Peters, C. M. McGuire, E. M. Pasciak, A. a. Peverly, L. M. Strawsine, E. R. Wagoner, and J. Tyler Barnes, "Electrochemical dehalogenation of organic pollutants," *J. Mex. Chem. Soc.*, vol. 58, no. 3, pp. 287–302, 2014.

[2] K. Honeychurch, "Review: The Application of Liquid Chromatography Electrochemical Detection for the Determination of Drugs of Abuse," *Separations*, vol. 3, p. 28, 2016.

[3] A. C. Silva, L. F. B. Lira Pontes, M. F. Pimentel, and M. J. C. Pontes, "Detection of adulteration in hydrated ethyl alcohol fuel using infrared spectroscopy and supervised pattern recognition methods," *Talanta*, vol. 93, pp. 129–134, 2012.

[4] K. Kar and W. Cheng, "Using Mass Spectrometry to Detect Ethanol and Acetaldehyde Emissions from a Direct Injection Spark Ignition Engine Operating on Ethanol/Gasoline Blends," pp. 1–14, 2011.

[5] I. Udom, M. K. Ram, E. K. Stefanakos, A. F. Hepp, and D. Y. Goswami, "One dimensional-ZnO nanostructures: Synthesis, properties and environmental applications," *Mater. Sci. Semicond. Process.*, vol. 16, pp. 2070–2083, 2013.

[6] K. Khun, Z. H. Ibupoto, C. O. Chey, J. Lu, O. Nur, and M. Willander, "Comparative study of ZnO nanorods and thin films for chemical and biosensing applications and the development of ZnO nanorods based potentiometric strontium ion sensor," *Appl. Surf. Sci.*, vol. 268, pp. 37–43, 2013.

[7] L. Chow, O. Lupan, G. Chai, H. Khallaf, L. K. Ono, B. Roldan Cuenya, I. M. Tiginyanu, V. V. Ursaki, V. Sontea, and a. Schulte, "Synthesis and characterization of Cu-doped ZnO one-dimensional structures for miniaturized sensor applications with faster response," *Sensors Actuators, A Phys.*, vol. 189, pp. 399–408, 2013.

[8] G. Bandekar, N. S. Rajurkar, I. S. Mulla, U. P. Mulik, D. P. Amalnerkar, and P. V. Adhyapak, "Synthesis, characterization and photocatalytic activity of PVP stabilized ZnO and modified ZnO nanostructures," *Appl. Nanosci.*, vol. 4, no. 2, pp. 199–208, 2013.

[9] S. H. Chiu and J. C. a Huang, "Chemical bath deposition of ZnO and Ni doped ZnO nanorod," *J. Non. Cryst. Solids*, vol. 358, no. 17, pp. 2453–2457, 2012.

[10] M. Faisal, S. B. Khan, M. M. Rahman, A. Jamal, A. M. Asiri, and M. M. Abdullah, "Smart chemical sensor and active photo-catalyst for environmental pollutants," *Chem. Eng. J.*, vol. 173, no. 1, pp. 178–184, 2011.

[11] E. Tavakolian, J. Tashkhourian, Z. Razmi, H. Kazemi, and M. Hosseini-Sarvari, "Ethanol electrooxidation at carbon paste electrode modified with Pd-ZnO nanoparticles," *Sensors Actuators, B Chem.*, vol. 230, pp. 87–93, 2016.

[12] Z. Yb and M. M. Rahman, "RSC Advances for environmental safety †," *RSC Adv.*, vol. 7, pp.

22627–22639, 2017.
[13] Z. Yang, L.-M. Li, Q. Wan, Q.-H. Liu, and T.-H. Wang, "High-performance ethanol sensing based on an aligned assembly of ZnO nanorods," *Sensors Actuators B Chem.*, vol. 135, no. 1, pp. 57–60, 2008.
[14] R. Devaraj, K. Venkatachalam, and P. M. Razad, "Role of seed layer: lattice orientation attachment and structural stabilization on the lateral growth of ZnO nanorods," *J. Mater. Sci. Mater. Electron.*, vol. 27, pp. 4011–4018, 2016.

Nanoparticle Optical Properties: Size Dependence of a Single Gold Spherical Nanoparticle

A.R. Shafiqa[a *], Azlan Abdul Aziz[b] and Baharak Mehrdel[c]

School of Physics, Universiti Sains Malaysia, 11800, Penang, Malaysia

[a]sitinurshafiqa95@gmail.com, [b]lan@usm.my, [c]baharak_mehrdel@yahoo.com

Abstract. The optical properties of a spherical gold nanoparticle have become our interest due to its remarkable characteristics when interacting with the light. In this work, the dependence of surface plasmon resonance on the size of the gold nanosphere has been studied with varying diameter from 20 to 100 nm. The optical spectra of the single isolated gold nanoparticles are simulated by employing the versatile MNPBEM toolbox. The wavelength corresponding to the maximum extinction, absorption as well as scattering redshifts (shift to longer wavelengths) were observed as the nanoparticle size increased. The rate of change of scattering (Δsca) and absorption (Δabs) relative to the extinction is calculated to correlate these two properties. For the nanoparticle with size less than 70 nm, Δabs is larger than Δsca, whereas for the diameter of nanoparticles above 70 nm, they show that Δsca is larger than Δabs. Due to this, it is found that absorption efficiency initially increases with nanoparticle size, which eventually reaches a maximum at a size of 70 nm and then begins to decrease with further increase in the nanoparticle diameter. The observed volcano trend of absorption efficiency is the result of electromagnetic retardation and radiative scattering. It can be concluded that the intermediate-size nanoparticle has the highest absorption efficiency with an optimal size of 70 nm for gold nanoparticles.

1. Introduction

Gold nanoparticles have unique optical properties that can be easily tuned, which are different from bulk gold. These intriguing characteristics provide a plethora of advantages in various fields particularly in the biomedical field as its optical properties can be applied in biosensing, drug delivery and diagnostic imaging [1,2]. One of the major attributes offered by nanoparticles is the interchange of light with electrons on its exterior. At a specific light wavelength, a phenomenon known as surface plasmon resonance occurs when the frequency of the incident photon is resonant with the collective electrons oscillations on the surface of the nanoparticle [3]. As the incident light with an electric field passed through the nanoparticle, there is a displacement of electrons, which located at the boundary of nanoparticles relative to the positive ionic lattice, leading to the net charge difference. In turn, it gives rise to Coulombic restoring force, which is responsible for restoring the system to its equilibrium state and results in the creation of dipolar oscillation of electron or also called as surface plasmon oscillation [4]. Mie theory postulated by Gustav Mie in 1908 has been used to explain the color variation of colloidal gold with particle size. Mie calculated the surface plasmon resonance by solving Maxwell's equation with appropriate boundary conditions for small sphere interacting with electromagnetic field [5]. When the size of the particles is much smaller than the wavelength of light ($2r << \lambda$), it is sufficient to consider only the dipole oscillation contributes to the interaction of the light-nanoparticle interaction (dipole approximation) and it is possible to employ the quasistatic approximation, means that the electric field

Content from this work may be used under the terms of the Creative Commons Attribution 3.0 licence. Any further distribution of this work must maintain attribution to the author(s) and the title of the work, journal citation and DOI.
Published under licence by IOP Publishing Ltd

of light is assumed to be constant [6-8]. Under this circumstance, the polarizability of the sphere, α is given by Clausius-Mossotti relation as:

$$\alpha = 3\varepsilon_o V(\frac{\varepsilon-\varepsilon_m}{\varepsilon+2\varepsilon_m}) \quad (1)$$

where ε_o is the permittivity of vacuum, ε_m is the dielectric constant of surrounding medium, ε is the dielectric function of the metallic nanoparticle and V is the volume of spherical nanoparticle. This equation indicates that the polarizability of the nanoparticle directly depends on the size of the nanoparticle as $V=\frac{4}{3}\pi r^3$.

Generally, the size of the nanoparticle affects the spatial distribution of the polarization charges over the surface as well as the separation of positive and negative charges. Thus, this effect influences the peak maxima, plasmon resonance wavelength and plasmon bandwidth as shown by many researchers both theoretically and experimentally. Based on their works, however, the size effect of gold nanoparticles is either studied up to 80 nm of diameter or emphasizing on the extinction spectra only [9-12]. In date, the contribution of the absorption and scattering properties to the total extinction of a single gold spherical nanoparticle has not yet been investigated in detail. Therefore, in this present work, the size dependence of the gold nanospherical on the total extinction, absorption and scattering have been highlighted.

2. Methods and Parameters

In order to simulate the size dependency of gold nanoparticles on the surface plasmon resonance, the MNPBEM toolbox was used, provided by Ulrich Hohenester, in which has been described elsewhere in details [13-15]. Therefore, the most concise of introductions was provided. The major purpose of the toolbox is to compute Maxwell's equations for dielectric nanoparticles with homogeneous dielectric properties, which are separated by sharp boundaries. This homogenous dielectric environment condition is advantageous as only the boundaries between different dielectric materials need to be discretized rather than the whole volume, eventually results in faster simulations as compared to DDA and FDTD method [16]. In this work, a spherical gold nanoparticle embedded in water of refractive index, n=1.33 [17] was investigated. The diameter of the gold nanosphere was increased by 10 units from 20 to 100 nm, while other optical characteristics were fixed. The data of dielectric constant for gold was taken from the experimental data of Johnson and Christy [18]. Full BEM simulation (BEMret) was chosen as the most appropriate solver as it solves the simulations based on the full Maxwell equations which works perfectly for all cases. Moreover, this solver also takes retardation effect into account which is significantly important for large nanoparticles (in the case of gold, diameter greater than about 20 nm) [19].

3. Results and Discussion

Figure 1 illustrates the optical efficiency as a function of the wavelength for different values of the nanoparticle diameter, with Figure 1a presenting the extinction efficiency, Figure 1b the scattering efficiency and Figure 1c the absorption efficiency. It is worth mentioning that all results in Figure 1 are presented in terms of efficiency instead of cross sectional area for clearer and better understanding. From Figure 1a, the dipole plasmon resonance of the gold nanoparticle gives a peak in extinction at about 522 nm for the 20 nm gold particle which is in good agreement with previous works [10].

By increasing the size of gold spherical nanoparticles, there is a remarkable redshift in the position of the wavelength corresponding to the maximum extinction efficiency. Additionally, this pattern can also be perceived in scattering efficiency spectrum as shown in Figure 1b. These expected results are related to the dependence of free electron contribution to the dielectric function which can be modified by changing the size of the particle. Electromagnetic retardation which results from the depolarization of the light field across the particle surface causes the Coulombic restoring force on the electron cloud becomes weaker. Hence, the frequency needed to initiate the surface plasmon oscillation is much lower (higher resonance wavelength). As depicted in Figure 1c, the size effect is strongly shown in the absorption efficiency, where two trends of intensity can be seen. As spherical nanoparticles get larger, the dipole resonance peaks redshifts from 521 to 543 nm. However, it is found that increasing the size of nanoparticles gives rise to an increase in the absorption efficiency, which eventually reaches a maximum and then begins to decrease with further increase in the nanoparticle diameter. For the sake of clarity, the plot of the absorption maxima against particle size is shown in Figure 2.

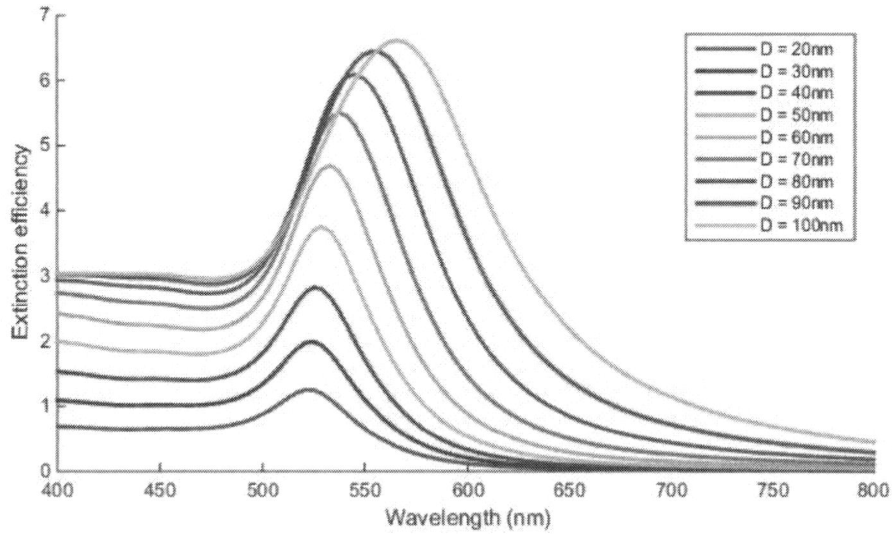

Figure 1a: The simulated extinction efficiency with different size as a function of wavelength.

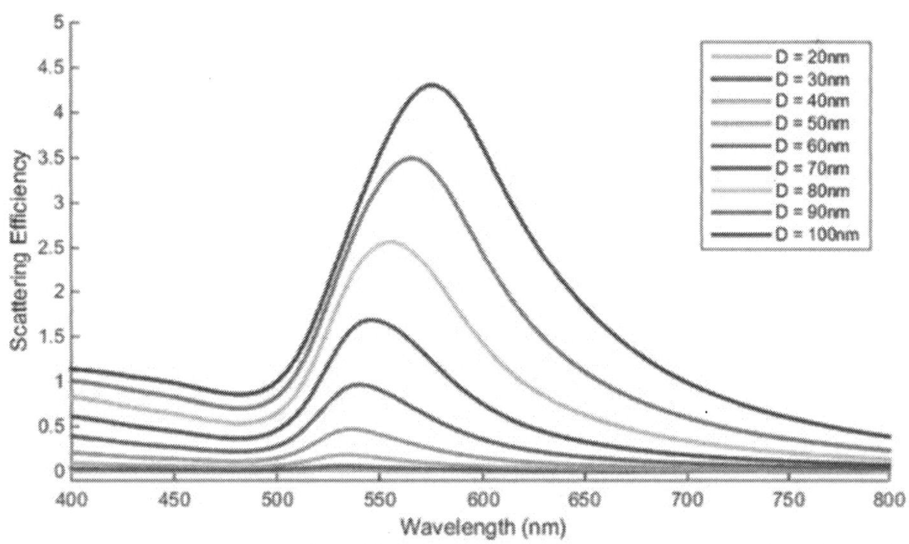

Figure 1b: The simulated scattering efficiency with different size as a function of wavelength.

The observed trend resembles a volcano shape, which is the result of effects 1) radiative scattering and 2) electromagnetic retardation. To correlate the optical properties, namely absorption and scattering, the rate of change of scattering (Δsca) and absorption (Δabs) relative to the extinction is calculated. It shows that when the diameter of nanoparticles below 70 nm, Δabs is larger than Δsca, whereas nanoparticles with size greater than 70 nm, they show that Δsca is larger than Δabs.

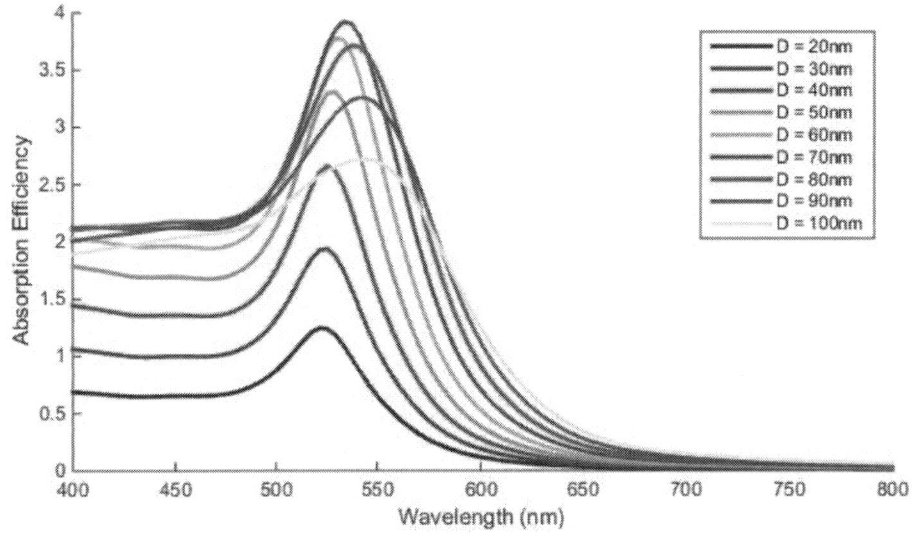

Figure 1c: The simulated absorption efficiency as a function of wavelength.

At small size, the nonradiative absorption dominates, which is due to the collision of electron with the nanoparticle surface [19]. The total damping rate in small nanoparticles can be written as

$$\gamma = \gamma_{bulk} + \frac{Av_f}{R} \tag{2}$$

This damping is included in the dielectric function of the metal as shown in Equation 3:

$$\varepsilon(\omega) = \varepsilon_{\infty} - \frac{{\omega_p}^2}{i\omega\gamma + \omega^2} \tag{3}$$

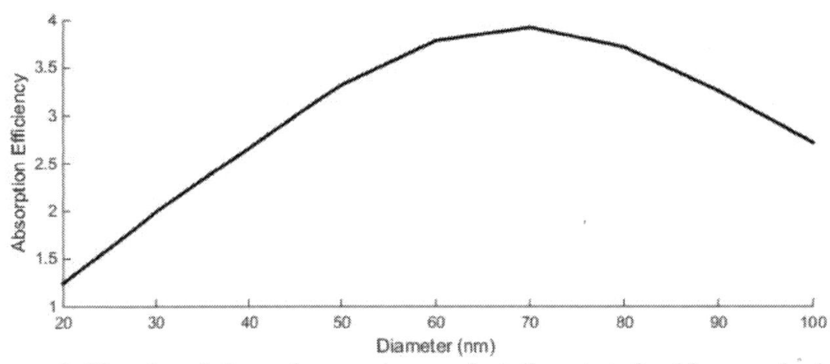

Figure 2: The plot of absorption maxima against diameter of gold nanospherical.

This contribution is valid if the size of the gold nanoparticle is smaller than its mean free path of electrons (for gold = 40 nm). At large size, the radiative scattering dominates in which its damping rate increases with increasing nanoparticle size as $(kR)^3$. There is more scattering as opposed to absorption. We can conclude that the nonradiative absorption dominates in small nanoparticle whereas the radiative scattering dominates in larger particle. Since the nanoparticle size approaches the wavelength of light, the electric field across the nanoparticle can no longer be considered as uniform. Thus, the incident light field is perturbed across the particle. This leads to electromagnetic retardation or dynamic depolarization in which causes damping for large particles. Theoretically, as the size of the nanoparticle is comparable to the wavelength of light, the nanoparticle is no longer homogeneously polarized by the light due to the electromagnetic retardation, results in the excitation of higher-order oscillation mode. The emergence of higher oscillation mode peak (quadrupole, octupole etc.) is always located at shorter wavelength relative to the dipolar one [7]. For a mode of order l, the general resonance condition is given as Equation 4

$$\varepsilon_r = -(\frac{l+1}{l})\varepsilon_m \tag{4}$$

where l=1 for dipole, l=2 for quadrupole and so on.

Figure 3 shows the dependence of a) extinction efficiency and b) ratio of scattering to the absorption on the nanoparticle size. It is shown that with an increasing nanosphere diameter, the extinction as well as the ratio of scattering to absorption (C_{sca}/C_{abs}) increase exponentially. From this figure, for 20 nm gold nanoparticle, total extinction is predominantly contributed by absorption. Gradually, scattering becomes important, but absorption still contributes significantly with diameter is increased until up to 70 nm in which scattering starts to be larger than absorption. Thus, for small nanoparticle, absorption dominates while scattering becomes significant for larger nanoparticle.

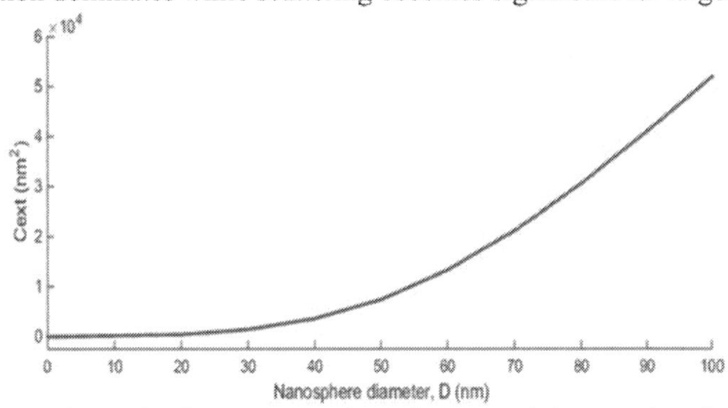

Figure 3a: The extinction cross-section of nanoparticles as a function of size.

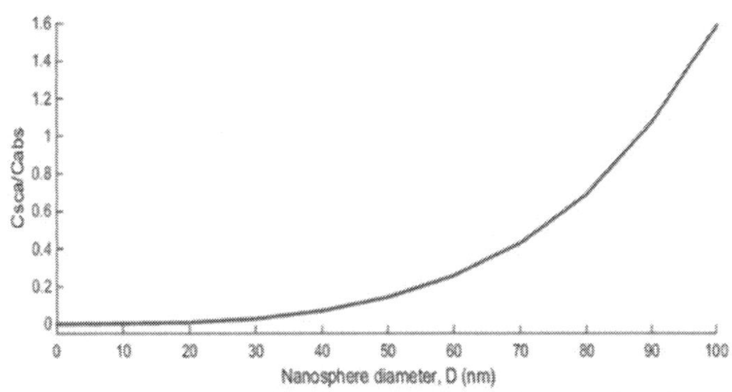

Figure 3b: The ratio of scattering to the absorption as a function of size.

4. Conclusion

This work demonstrated the size dependence of gold nanospheres on the dipole plasmon resonance by using MNPBEM toolbox. It was found that the dipole plasmon resonance wavelength of the gold spherical nanoparticle redshifted (shift to longer wavelengths) with increasing of the nanoparticle size. Nevertheless, the absorption spectrum showed a distinct feature that is not possessed by extinction and scattering spectra, in which a significant damping in intensity had been observed for nanoparticle larger than 70 nm. Consequently, the plotted spectrum for absorption efficiency resembled a volcano shape with the maximum intensity at 3.92 for 70 nm gold nanosphere. Therefore, it can be concluded that the intermediate-size nanoparticle has the highest absorption efficiency with an optimal size of 70 nm for an isolated gold spherical nanoparticle.

Acknowledgement
The authors are grateful for the financial support from Universiti Sains Malaysia, Grant No. 203/PFIZIK/6763003 under Transdisciplinary Research Grant Scheme.

References

[1] Cai W, Gao T, Hong H and Sun J 2008 *Nanotechnol Sci Appl.* 1 17-32
[2] Khan A, Rashid R, Murtaza G and Zahra A 2014 *Trop. J. Pharm. Res.* 7, 1169–77
[3] Louis C and Pluchery O 2012 *Gold Nanoparticles for Physics, Chemistry and Biology* (Université Pierre et Marie Curie, France)
[4] Huang X and El-Sayed M A 2010 *J. Adv. Res.* 1 13–28
[5] Hergert W and Wriedt T 2012 *The Mie Theory Basics and Application* Springer-Verlag Berlin Heidelberg
[6] Jain P K and El-sayed M A 2010 *Chemical Physics Letters* 487 153–164
[7] Rayford II C E, Schatz G and Shuford K 2005 *Nanoscape* 2 27-33
[8] Fan X, Zheng W and Singh D J 2014 *Light: Science & Applications* 3 e179
[9] Jain P K, Lee K S, El-Sayed I H and El-Sayed M A 2006 *The Journal of Physical Chemistry B* 110 7238–48
[10] Tangod V B, Raikar P, Mastiholi B M and Raikar U S 2015 *J. Shivaji Univ.* 41, 1-2
[11] Doak J, Gupta R K, Manivannan K, Ghosh K and Kahol P K 2010 *Physica E* 42 1605-09
[12] Near R D, Hayden S C, Hunter R E, Thackston D and El-sayed M A 2013 *The Journal of Physical Chemistry C* 117 23950-55
[13] Hohenester U and Trügler A 201 *Computer Physics Communications* 183(2) 370–381
[14] Waxenegger J, Trügler A and Hohenester U 2015 *Computer Physics Communications* 193 138–150
[15] Hohenester U 2014 *Computer Physics Communications 185*(3) 1177–87
[16] Myroshnychenko V, Rodríguez-Fernández J, Pastoriza-Santos I, Funston A M, Novo C, Mulvaney P, Liz-Marzán L M and García de Abajo F J 2008 *Chem. Soc. Rev.* 37 1792–1805
[17] Jain P K and El Sayed M A 2007 *Nano Letters* 7 2854-58
[18] Johnson P B and Christy R W 1972 *Phys. Rev. Letters* 11 541
[19] Link S and El Sayed M A 2000 *The Journal of Physical Chemistry B* 19 409-453

IOP Conf. Series: Journal of Physics: Conf. Series **1083** (2018) 012041 doi:10.1088/1742-6596/1083/1/012041

Simulation of Surface Plasmon Resonance on Different Size of a Single Gold Nanoparticle

Norsyahidah Md Saleh[1] and A. Abdul Aziz[1]

[1]School of Physics, Universiti Sains Malaysia, 11800 USM, Penang, Malaysia.

syahidahsaleh@gmail.com, lan@usm.my

Abstract. In this work, the surface plasmon resonance wavelength position is calculated using metal nanoparticle boundary element method (MNPBEM) toolbox from Matlab classes. The surface plasmon resonance wavelength position is determined by full Maxwell's equations. We present a simulation optical properties study of the dependence of the plasmon resonance spectrum of individual gold nanoparticles on the dielectric function and size change in the extinction efficiency spectrum. Previous studies have shown that the shift of the gold nanoparticle plasmon resonance wavelength increased when the size of gold nanoparticles was increased. However, the dielectric function of gold nanoparticle showed a different shifting when using a higher spectral range. In this work, the attention is focused on the size effects of gold nanoparticle on the position of the plasmon resonance wavelength. The aim of this work is to study the behaviour of the resonance position for different diameter size of gold nanoparticle by using boundary element method simulations. The gold nanoparticle diameter between 10 nm to 100 nm is used in determine the shifting of the surface plasmon resonance. The results in this work indicated that the position of the plasmon resonance wavelength for gold nanoparticle was redshift when the size of gold nanoparticle was increased. However, dielectric function measured by previous experiments show that the high spectral resolution gives the significant redshift to the extinction efficiency spectrum.

1. Introduction

Optical properties are closely related to the electrical and electronic properties of the material. There must be some other factors that can also be seen when dealing with optical properties. Optical properties usually can be referring to the interaction of electromagnetic radiation with matter. For instance, an electromagnetic wave of a single frequency entering a medium from a vacuum. This wave could be reflected, refracted or absorbed. There are only two possibilities phenomena that can be occur when there is the interaction of a medium with electromagnetic radiation which is scattering and absorption. When we consider a wider spectrum of frequencies, then some part of the spectrum would be absorbed and the other frequencies would be scattered [1].

Gold nanoparticles interaction with light is strongly dictated by their environment, size and physical dimensions. Oscillating electric fields of a light ray propagating near a colloidal nanoparticle interact with the free electrons causing a concerted oscillation of electron charge that is in resonance with the frequency of visible light. These resonant oscillations are known as surface plasmons. By collective conduction band electron oscillation in response to the electrical field of the light radiation, the surface plasmon absorption dominate the optical properties of metal nanostructures in the visible region [2]. Also for particles that consist of alkali, noble, and various other metals the surface plasmon

Content from this work may be used under the terms of the Creative Commons Attribution 3.0 licence. Any further distribution of this work must maintain attribution to the author(s) and the title of the work, journal citation and DOI.
Published under licence by IOP Publishing Ltd

show themselves as pronounced optical resonances in the visible or UV parts of the spectrum. The resonance frequency of the oscillation such as surface plasmon energy is determined the dielectric properties of the metal and the surrounding medium and by the particle shape and size [3].

For this work, MNPBEM Matlab toolbox is used to calculate full Maxwell's equation [4]. From this, we can predict the position of plasmon resonance. For this work, we choose to compare the dielectric function that are very common which is by Johnson and Christy with dielectric function of gold that the latest to be calculated which is by Olmon etc. Dielectric function of gold was measured by Johnson and Olmon are different since they approach different method in calculating the dielectric function of gold. For Johnson and Christy, they have calculated the dielectric function by using the index of refraction of films was measured by reflection and transmission from 0.50 to 6.5 eV (190 nm to 2.48 μm) in 1972 [5]. But for Olmon, in 2012 they performed precise spectroscopic ellipsometry measurements to determine the dielectric function across a broad spectral range (300 nm to 25 μm) 0.05-4.14 eV with high spectral resolution [6].

2. Modelling and simulations parameter

The work only focusing on the single particle model which have been calculated using MNPBEM toolbox (metallic nanoparticles using boundary element method) [4]. The size of the gold nanoparticles is ranged between 10-100nm. The particle is "comparticle" which is compound of particles in a dielectric environment and the shape of particle is describe by "trisphere" function which is load points on sphere from file and perform triangulation. The changing parameters for "trisphere" is diameter of the particles and the vertices. For this measurement, 144 vertices was used to calculate the interaction happened on the surface of the nanosphere. For the solver, the full Maxwell's equations was used since we want to see the shifting of the extinction peak. The quasistatic plane-wave excitation and cross section was the parameters for "plane-wave-stat" function and the direction of E-field polarisation was along x-axis and light propagation was along z-axis. The wavelength range was between 400nm to 700nm. To have a smoother peak of extinction cross section 300 points was set to be calculated by the boundary element method solver [4], [7].

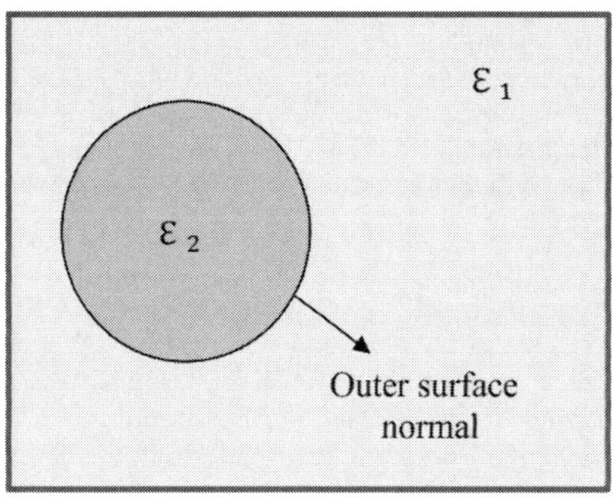

Figure 1 Single gold nanoparticle in dielectric environment [4].

Where, ε_1 is the dielectric constant of medium and ε_2 is the dielectric function of gold nanoparticles. In this work, we compare two different dielectric functions of gold which are by Johnson and Christy; and Olmon etc. Johnson and Christy is widely used nowadays but Olmon etc. came out with the recent calculation of dielectric function which have a broad spectral range from 300nm to 25 μm. (0.05-4.14 eV) with high spectral resolution in 2012. In 1972, Johnson and Christy measured the index of refraction of gold films with thicknesses 34 and 46 nm that were deposited onto fused-quartz substrates by reflection and transmission from 0.50 to 6.5 eV (190 nm to 2.48 μm) [6]. The dielectric function have a relation with refractive index given by:

IOP Conf. Series: Journal of Physics: Conf. Series **1083** (2018) 012041 doi:10.1088/1742-6596/1083/1/012041

$$\varepsilon(\omega)=n^2 \quad (1)$$

Where $\varepsilon(\omega)$ is dielectric function and n is refractive index of gold nanoparticles [8]. To compare those two dielectric functions stated above, the bulk refractive index of the surrounding media was changed between air (n=1.00) and water (n=1.33) and the sensitivity of the bulk refractive index was calculated. The sensitivity is defined as:

$$S=\Delta\lambda/\Delta n \quad (2)$$

Where $\Delta\lambda$ is the shift of the extinction peak and Δn is the refractive index change of the media [7]. From the comparison, we can determine which dielectric function is more significant for this kind of measurement. The signification was determined by which dielectric function have larger peak shift of extinction cross section.

3. Results and discussion

The normalised extinction cross sections (between air and water) simulated for two different dielectric functions of single-particles model with a diameter of 70nm are presented in Figure 2. Although, the peak wavelength between air for Johnson and air for Olmon is nearly the same. The peak shift (between air and water) of the two dielectric functions have 3nm in difference. From the previous calculation for Johnson and Palik, the difference was nearly the same [7]. From here, we can see that Olmon have longer peak shift (between are and water) compared to Johnson. Taking these observations into consideration, the Olmon dielectric function was used for further calculation.

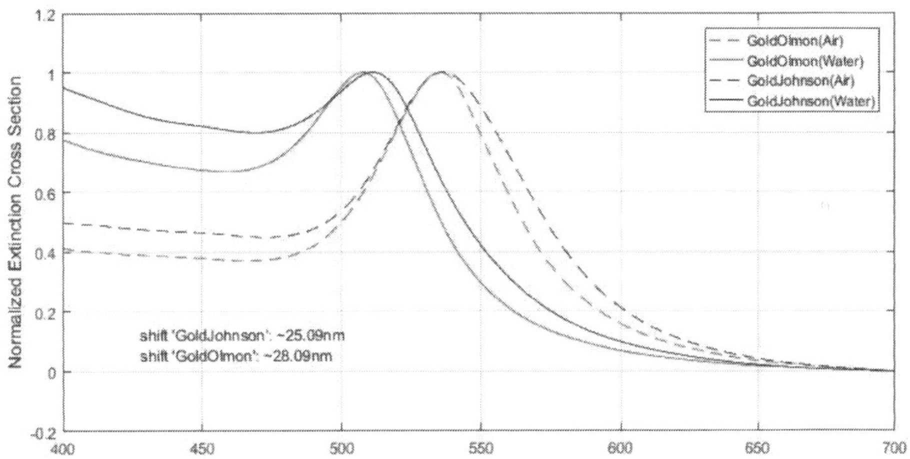

Figure 2 Normalized extinction spectra of single 70 nm particles in water/air. Comparison of the two available tabulated gold dielectric functions ('GoldJohnson', 'GoldOlmon').

Figure 3 shows the normalized extinction efficiency of gold nanoparticles with diameter change from 10 to 100 nm. From the graph, we can see that the peak of wavelength is redshifted when the diameter of gold nanoparticles was increased. This redshift occurs due to the perturbation of the density of the conduction electrons in the particles. Therefore the redshift of the peak wavelength is size dependence. The other reason is due to the electromagnetic retardation which results from the depolarization of the light field across the particle surface. The peak wavelength of the extinction efficiency against wavelength was the results of surface plasmon resonance to occur. Besides the redshift, the peak also broadening when the size of the gold nanoparticles increase. The broadening of the peaks occurs due to the dephasing of the coherent electron oscillation. The large bandwidth corresponds to rapid loss of the coherent electron motion [9].

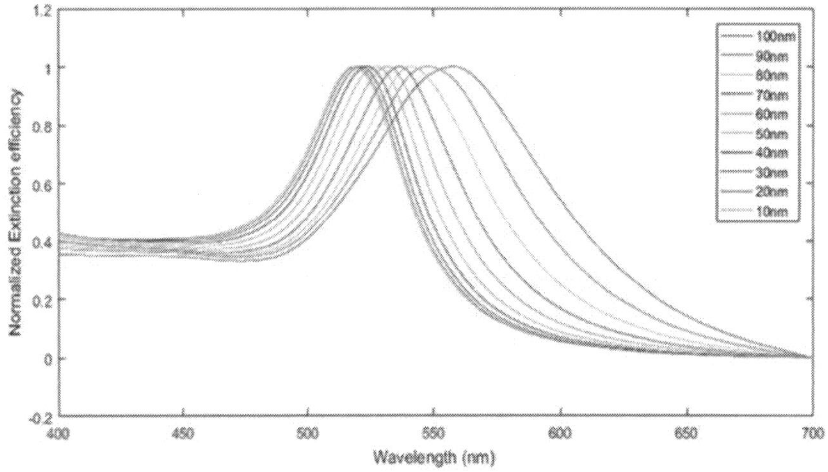

Figure 3 Normalized extinction efficiency of 10-100nm particle diameter in water (n=1.33).

The absolute peak shifts of the extinction efficiency (between air and water) simulated for different diameter of single gold nanoparticle are presented in Figure 4. From here, we can see that when we change the dielectric constant of surrounding medium, the peak shift (between air and water) was increased when the diameter of the gold nanoparticles increased. The regression value also agrees with the increment of the gold nanoparticles diameter, where the wavelength peak shift increase. Since surrounding medium is one of the parameter of the shifting of plasmon resonance due to the boundary conditions which include the electric potential inside and outside the gold nanoparticles. This is because dielectric function of gold nanoparticles is the parameter that depends on frequency as the surrounding medium is a constant and real parameter [10].

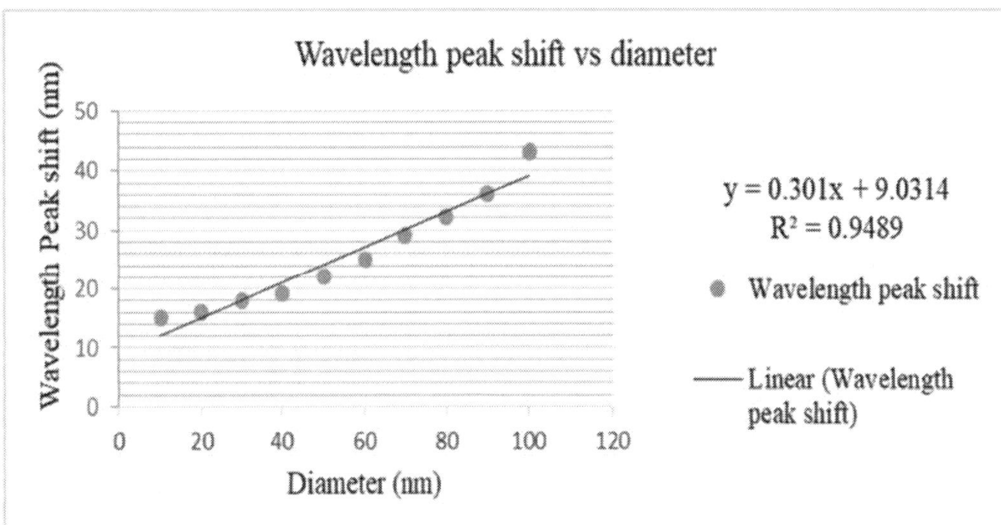

Figure 4 Peak shift (between air and water) against diameter of gold nanoparticles (10-100nm).

4. Conclusion

The plasmonic properties of single-spherical gold nanoparticle systems were investigated with MNPBEM Matlab toolbox. It was found that between Johnson and Olmon dielectric function of gold, Olmon have longer peak shift (between air and water) compared to Johnson. Olmon dielectric function have been used to further the measurement. The surface plasmon resonance appears in the visible or near-infrared range for gold nanoparticle [10]. From the simulation, we can state that the diameter between 10 nm to 100 nm were in the visible or near-infrared range. Therefore, 70 nm of gold nanoparticles was in the good size for further measurement since it peak shift (between air and water) and the broadening of the peak was optimum. Meanwhile, different size of gold nanoparticle was applicable for different applications based on its requirement. The examples of application such as biosensor, catalyst application, and therapeutic agents [1].

Acknowledgement
Authors would like to thank Universiti Sains Malaysia for funding this research work through its project Optical and Structural Properties of Metallic and Composites Nanoparticles in Liquid Media 203/PFIZIK/6763003).

References
[1] J. Z. Zhang, Optical Properties and Spectroscopy of Nanomaterials, Singapore: World Scientific Publishing Co. Pte. Ltd., 2009.
[2] S. Dhara et al., "'Spillout' effect in gold nanoclusters embedded in c-Al2O3(0001) matrix," Chem. Phys. Lett., vol. 399, no. 4–6, pp. 354–358, 2004.
[3] T. Klar, M. Perner, S. Grosse, G. von Plessen, W. Spirkl, and J. Feldmann, "Surface-Plasmon Resonances in Single Metallic Nanoparticles," Phys. Rev. Lett., vol. 80, no. 19, pp. 4249–4252, 1998.
[4] A. T. Ulrich Hohenester, "MNPBEM – A Matlab toolbox for the simulation of plasmonic nanoparticles," Computer Physics Communications, vol. 183, pp. 370-381, 2012.
[5] P. B. Johnson and R. W. Christy, "Optical Constants of the Noble Metals," Phys. Rev. B, vol. 6, no. 12, pp. 4370–4379, 1972.
[6] R. L. Olmon et al., "Optical dielectric function of gold," vol. 235147, pp. 1–9, 2012.
[7] A. Bonyár, "Simulation of the Refractive Index Sensitivity of Coupled Plasmonic Nanostructures Simulation of the refractive index sensitivity of coupled plasmonic nanostructures," Procedia Eng., vol. 168, no. January, pp. 962–965, 2017.
[8] M. Quinten, Optical properties of Nanoparticle System, Weinheim: Wiley-VCH Verlag & Co. KGaA, 2011.
[9] S. Link and M. a. El-Sayed, "Shape and size dependence of radiative, non-radiative and photothermal properties of gold nanocrystals," Inter. Rev. Phys. Chem., vol. 19, no. 3, pp. 409–453, 2000.
[10] O. P. Catherine Louis, Gold Nanoparticles for Physics, Chemistry and Biology, London: Imperial College Press, 2012.

The Sensitivity of Surface Plasmon Resonance Damping for Colloidal Silver Nanoparticles

B Mehrdel[1], A Abdul Aziz [1,2]

[1]School of physics, Universiti Sains Malaysia, 11800 USM, Penang, Malaysia
[2]Nano-Biotechnology Research and Innovation Lab (NanoBRI), Institute for Research in Molecular Medicine (INFORMM), Universiti Sains Malaysia, 11800 USM, Penang, Malaysia.

baharimehr@gmail.com

Abstract. The impact of environmental conditions (ionic strength and type of electrolytes) on the damping of surface plasmon resonance spectra of silver nanoparticles (AgNPs) colloids suspensions were investigated. This study is structured for analyzing the interaction of three different sized (20, 50, and 80 nm) silver nanoparticles, which electrostatically stabilized (sodium citrate) with non-adsorbent media (e.g. NaCl and NaOH). The absorption spectra were observed using a UV- visible / NIR spectrophotometer in the range of 300 – 1000 nm. Taking advantage of the sensitivity of surface plasmon resonance (SPR) on the refractive index of the surrounding medium on the silver nanoparticles, the role of ionic media's refractive index in changing the sensitivity of SPR on damping and plasmon resonance shift were investigated experimentally and compared with theory results thoroughly. The results of experiment revealed that the absorption spectra are blue shifted, due to the increase in perturbation of conduction electron density of AgNPs when the electron density of surrounding media increases and electrons transfer from Cl^- or OH^- ions to the NPs. The results of sensitivity of SPR damping corroborate the assumption that the damping of resonance strongly depend on type and concentration of ionic surrounding media (chemical interface damping). It is demonstrated that the sensitivity of SPR damping is depended on primary particle size.

1. Introduction

Silver colloids suspension exhibit a diversity of research interest for many decades as a result of their intriguing electronic and optical properties. The optical properties of silver nanoparticles (AgNPs) are strongly dependent on the particle size, shape, coating, and local dielectric of surrounding media [1-5]. The excitation of surface plasmon resonance (SPR) of AgNPs that is due to the coupling of incident wavelength to the collective oscillation mode of the conduction electrons and contribution of interband transition from the 4d to the 5sp band in the AgNPs are responsible for the generally observed increase in the optical absorption spectra [6]. The change in the absorption spectra is a function of varying the chemical and physical properties of the environment of AgNPs colloidal surface. Studies in the literature report, the analysis of the surrounding medium refractive index effect on SPR response of metal NPs [4, 5]. According to linear correlation between refractive index and concentration in non-absorbing media [7], the results are presented as a function of concentration.

Despite the concern regarding the environmental impacts of the AgNPs, there was lack of information available with regards to the effect of non-adsorbent ionic media (e.g. NaCl and NaOH)

Content from this work may be used under the terms of the Creative Commons Attribution 3.0 licence. Any further distribution of this work must maintain attribution to the author(s) and the title of the work, journal citation and DOI.
Published under licence by IOP Publishing Ltd

concentration sensitivity on coupling plasmon resonance by considering the NPs size effect. Therefore, the main objective of this study was to investigate SPR sensitivity (S_exp) of AgNPs suspensions on to the concentration of non-adsorbent ionic media. The numerical differentiation was utilized to explain the sensitivity of damping (nm^{-1}) of absorption efficiency for SPR on the basis of the medium refractive index and salt concentration of different sized AgNPs. The sensitivity of absorption of plasmon resonance from Mie theory was compared to experimental sensitivity.

2. Experimental Details

Spherical AgNPs (brand Cytodiagnostic) with three different size (e.g. 20, 50, and 80 nm) were investigated. All AgNPs were purchased from Sciencewerke Co. (Singapore) were supplied in 2mM citrate. Non-functionalized standard AgNPs were used. Size distribution and morphology of AgNPs suspension are analyzed using "Phillips CM12 with Docu Version 3.2 image analysis" transmission electron microscopy (TEM) and Dynamic Light Scattering (DLS) to verify the samples size. The hydrodynamic diameter (HDD) was measured using a zetasizer nonseries Malvern instrument with 633 nm laser source. Zeta potential of aqueous AgNPs were measured using the Smoluchwaski's approximation with the same instrument. The aqueous AgNPs colloidal absorption spectra were measured in the wavelength range of 300 – 1000 nm. The presence of the surface plasmon resonance (SPR) peak verified using a UV-visible / NIR spectrophotometer (Shimadzu UV/VIS/NIR UV3600). The HDD and SPR peak of AgNPs were examined as a function of salt solution concentration (0.01, 0.03, 0.04, 0.1 and 1 M) of suspended media. Sodium chloride (NaCl) and sodium hydroxide (NaOH) which are ionic compounds with a molar mass of 58.44 g/mol and 40 g/mol respectively (brand Sigma-Aldrich) were used as a function of salt solution concentration. Surface charge properties of three different AgNPs size were examined under same conditions. The AgNPs were analysed following the steps, (1) preparation of salt concentration (e.g. 0.01 M NaCl), (2) 30% volume fraction AgNPs suspensions were dispersed in 3 mL cuvette, (3) instantly, samples of AgNPs suspended in salt were analysed for SPR peak then particle size and zeta potential. The TEM image and size distribution of 20, 50 and 80 nm primary AgNPs are shown in Figure 1. Table 1, summarises the primary particles hydrodynamic size (d. nm), zeta potential (ζ. mV), FWHM (full width at half maximum) and the area of absorption spectra.

Table 1: AgNPs primary particle hydrodynamic size, zeta potential, FWHM and area.

AgNPs	DLS*		FWHM	Area
	(d. nm)	(ζ. mV)	(nm)	(100%)
20 nm	23.13	-35.3	63.46	56.76
50 nm	54.13	-55.6	66.71	66.45
80 nm	83.82	-59.9	181	159.3

*DLS (Dynamic Light Scattering)

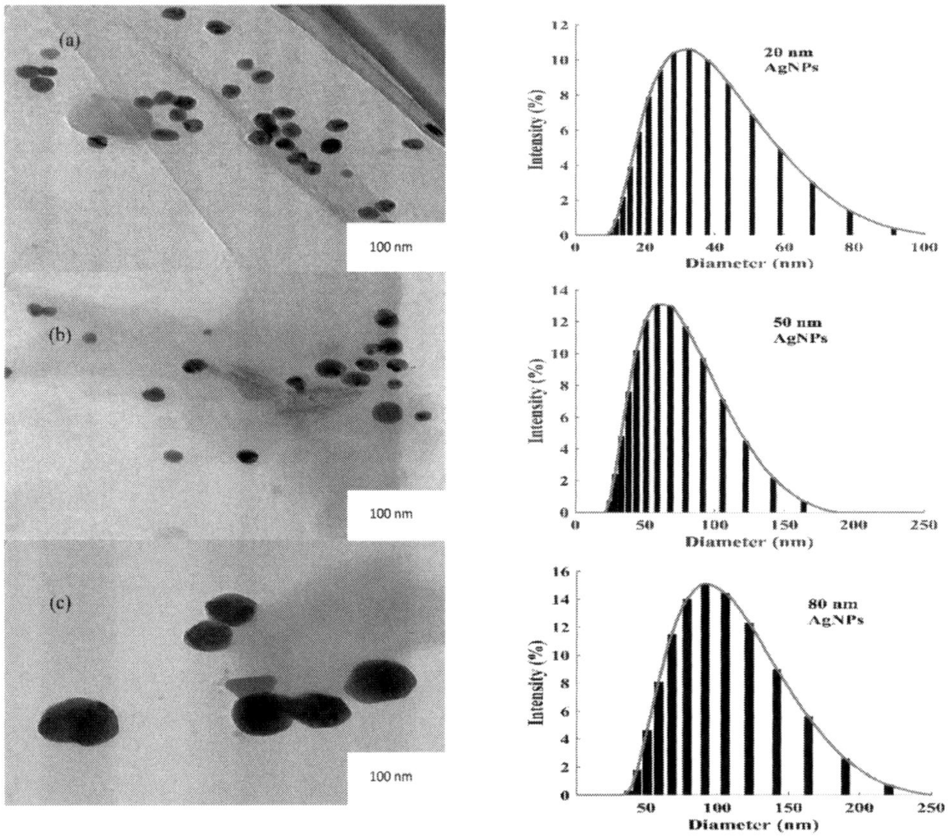

Figure 1: Size distribution histogram of (a) 20 nm, (b) 50 nm and (c) 80 nm AgNPs. TEM image indicates the physical structure of AgNPs used in this work.

3. Result and Discussion

The calculated absorption spectra of Mie theory for the size of NPs varied from 10 to 90 nm. Figure 2(a), shows the particle size dependence of the maximum value of plasmon resonance absorption efficiency as a function of salt concentration. Mie theory results showed the absorption was increased by increasing the particle size of 30 nm. In the case of AgNPs more than 30 nm the absorption started to damp, due to the particles larger than 30 nm scattered light more than absorption. Therefore, AgNPs are more suitable due to its higher scattering contribution to absorption as compared to gold nanoparticles with the same size [8]. However, the redshift of SPR was significant for the particles size as a function of surrounding media concentration from the Mie simulation (Figure 2(b)). This is due to the changes of refractive index of non-adsorbents surrounding media (e.g. NaCl).

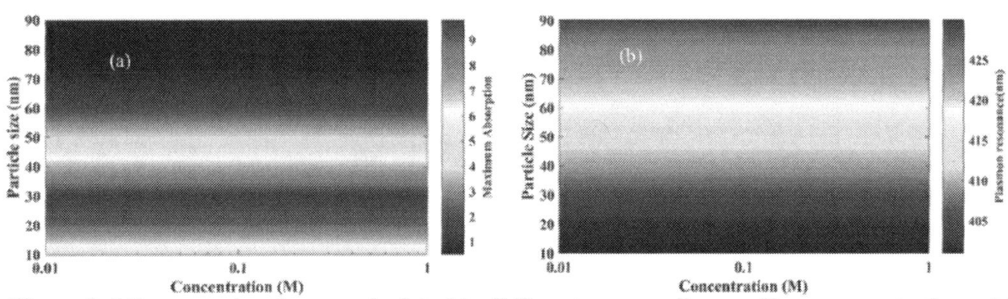

Figure 2: Mie scattering theory calculated in different surrounding medium concentrations. (a) particle size vs. maximum value of absorption. (b) particle size vs. surface plasmon resonance.

The Mie theory calculated the damping sensitivity (nm^{-1}) of absorption efficiency (Q_Abs) as a function of AgNPs size to explore the sensitivity of SPR to the surrounding medium refractive index change as shown in Figure 3. The sensitivity increased with AgNPs size and reached the maximum of the plateau for a diameter of 20 nm. Based on the results of Mie calculation for the sensitivity of damping in 0.04 M concentration (refractive index = 1.3334) of media (Figure 3), one may conclude that the particle size of AgNPs colloidal is the important factor to enhancement and damping of plasmon resonance spectra.

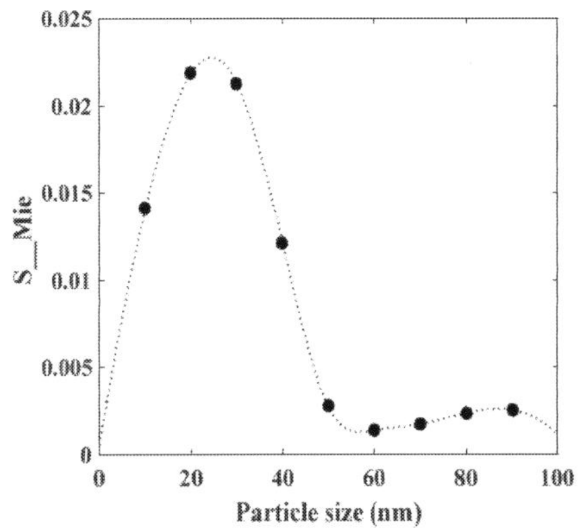

Figure 3: The change of Mie scattering theory sensitivity of damping (nm^{-1}) vs. particle size (nm).

The anionic component of the surrounding ionic medium had a significant impact on the stability of AgNPs colloidal suspension. Aggregation was induced by adding the small amount of an ionic solution (e.g. NaCl and NaOH) stronger than threshold concentration which depended on the type of ionic media. The anionic component of the surrounding media electrolyte (Cl^- vs OH^-) had a considerable impact on the measured zeta potential (surface charge) and the aggregation of the electrostatically stabilised AgNPs as presented in Table 2, for different sizes of NPs. The use of OH^- rather than Cl^-, as a surrounding medium of AgNPs resulted in higher measured zeta potential (mV) over the concentration range (0.01- 1 M) for three sizes of AgNPs. The increase in the measured values of zeta potential for citrate AgNPs suspended in NaOH was most likely the results of increasing OH^- concentration. Ag atoms at the surface structure of the NPs were unsaturated to the OH^- and H_2O molecule (nucleophilic), which can donate a pair of electrons. Therefore, increasing the OH^- concentration allowed the OH^- to effectively compete for the empty sites on the surface of AgNPs which produced a negative surface charge [9, 10]. However, adding the NaOH rather than NaCl resulted in a larger degree of aggregation at citrate AgNPs (Table 2). This indicated that how a minor change in the surrounding media charge could cause a significant change in the aggregated particle size and surface charge density. In general, AgNPs exhibiting high surface charge densities and surface plasmon resonance are extremely sensitive to electron charge density. The results of the experiment (Table 2) revealed that the absorption spectra were blue shifted (50 nm and 80 nm AgNPs suspensions), due to the increase in perturbation of conduction electron density of AgNPs when the electron density of surrounding media increased and electrons transfer from Cl^- or OH^- ions to the NPs [10].

Table 2: Effect of types of electrolyte on the hydrodynamic diameter (HDD) and SPR of AgNPs. (a) 20 nm, (b) 50 nm, and (c) 80 nm silver suspended in sodium citrate.

(a) 20 nm silver suspended in sodium citrate										
	NaCl					NaOH				
Molarity of media [a]	DLS [b]			Zeta potential [mV]	SPR peak [nm]	DLS [b]			Zeta potential [mV]	SPR peak [nm]
	HDD [nm]	PDI [c]	FWHM [d]			HDD [nm]	PDI [c]	FWHM [d]		
0.01	119.9	0.55	39	-24.2	401	36.01	0.57	40	-22.4	400
0.03	80.81	0.25	38	-9.8	401	28.32	0.76	42	-36.7	402
0.04	55.06	0.14	38	-24	402	34.78	0.51	44	-35.9	402
0.1	311.6	0.43	104	-29.3	391	384.6	0.28	82	-35.4	404
1	164	0.31	127	-20.7	412	382.1	0.25	281	-14.8	403
(b) 50 nm silver suspended in sodium citrate										
	NaCl					NaOH				
Molarity of media [a]	DLS [b]			Zeta potential [mV]	SPR peak [nm]	DLS [b]			Zeta potential [mV]	SPR peak [nm]
	HDD [nm]	PDI [c]	FWHM [d]			HDD [nm]	PDI [c]	FWHM [d]		
0.01	84.74	0.40	39	-24	410	84.94	0.26	41	-38.2	413
0.03	76.25	0.15	36	-18.1	410	95.79	0.20	40	-32.8	412
0.04	108.6	0.22	33	-30.6	408	99.62	0.26	42	-27.5	410
0.1	169.3	0.23	92	-25.2	399	220.4	0.26	71	-25.6	408
1	183.6	0.26	77	-10.2	405	277.3	0.29	101	-13.6	405.5
(c) 80 nm silver suspended in sodium citrate										
	NaCl					NaOH				
Molarity of media [a]	DLS [b]			Zeta potential [mV]	SPR peak [nm]	DLS [b]			Zeta potential [mV]	SPR peak [nm]
	HDD [nm]	PDI [c]	FWHM [d]			HDD [nm]	PDI [c]	FWHM [d]		
0.01	110.2	0.22	153	-34	472	84.06	0.17	153	-54.8	475
0.03	163.8	0.28	196	-28.9	463	211.4	0.39	204	-48.8	470
0.04	189.2	0.29	315	-20.8	448	217.9	0.26	227	-45	462
0.1	236.2	0.26	206	-20.4	455	237.9	0.28	238	-33.5	454
1	328.7	0.32	194	-21.2	486	305	0.26	249	-13.5	452

[a] Concentration of NaCl and NaOH. [b] Dynamic light scattering. [c] Polydispersity Index. [d] Full width half maximum (nm).

Depending on the results of sensitivity of SPR damping for NaCl (Figure 4), one may conclude that the surrounding media of citrate AgNPs colloidal was the important factor to the enhancement and damping of plasmon resonance spectra. This means that the type and concentration of surrounding media can affect the sensitivity of damping of plasmon resonance, which strongly depends on primary citrate AgNPs size (Figure 3).

The calculated and empirical results of the plasmon resonance sensitivity (S) are shown in Figure 4. The calculated results indicated the magnitude of S_Mie in the case of 20 nm was higher than larger particles (50 and 80 nm) (Figure 3). The S_Mie was in good agreement with the calculated maximum absorption which is varied with particle size. The magnitude of absorption for citrate AgNPs 20 nm was higher than citrate AgNPs of 50 and 80 nm (Figure 2(a)). Figure 4, shows that the experimental SPR sensitivity (S_exp) of citrate AgNPs suspended in NaCl solution depended on particle size. Therefore, the value of S_exp for citrate AgNPs 20 nm and 50 nm was almost the same, whereas in the case of 80 nm shows lower value. The main difference between S_exp of these three -citrate AgNPs was from the low concentration of salt until threshold concentration where the aggregation of AgNPs started.

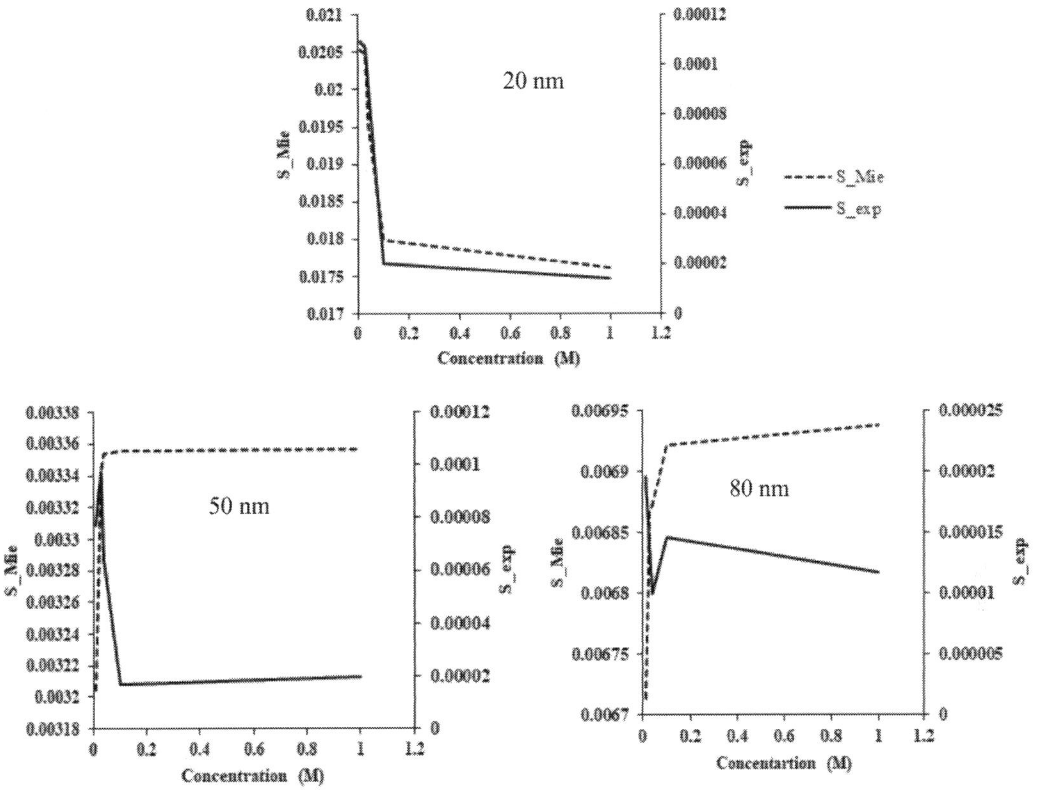

Figure 4: Calculated and empirical sensitivity of SPR damping for three different AgNPs size (20, 50 and 80 nm).

For confirmation and better understanding of S the additional experiments (using the same 20, 50 and 80 nm AgNPs colloidal) with different ionic media were performed. In this work, NaOH (sodium hydroxide) was deliberately chosen as the new aqueous solution medium. The AgNPs colloidal was suspended in different concentrations of NaOH solutions. For the clarity of purpose, the applied concentrations were the same value of previous experiment for NaCl. Compared trend of S_exp for AgNPs colloidal in NaOH and NaCl solutions are shown in Figure 5. The comparison between S_exp of the two media showed an insignificant difference at low concentration (0.01- 0.03 M) for 20 and 80 nm AgNPs. However, the most significant variation in S_exp was observed for 50 nm AgNPs on the low concentration (0.01- 0.03 M) of media (Figure 5).

In Figure 6, the S_exp of absorption spectrum of AgNPs in electrolyte matrix solutions (0.04 M) before NPs starting to aggregate is shown. The two ionic compounds (Cl^- and OH^-) were compared and showed that these two compounds very insignificantly different in value of sensitivity. However, in the case of 50 nm AgNPs the Cl^- showed higher sensitivity. The reason for this might be due to AgNPs being suspended in NaCl aggregated faster than NaOH (Table 2).

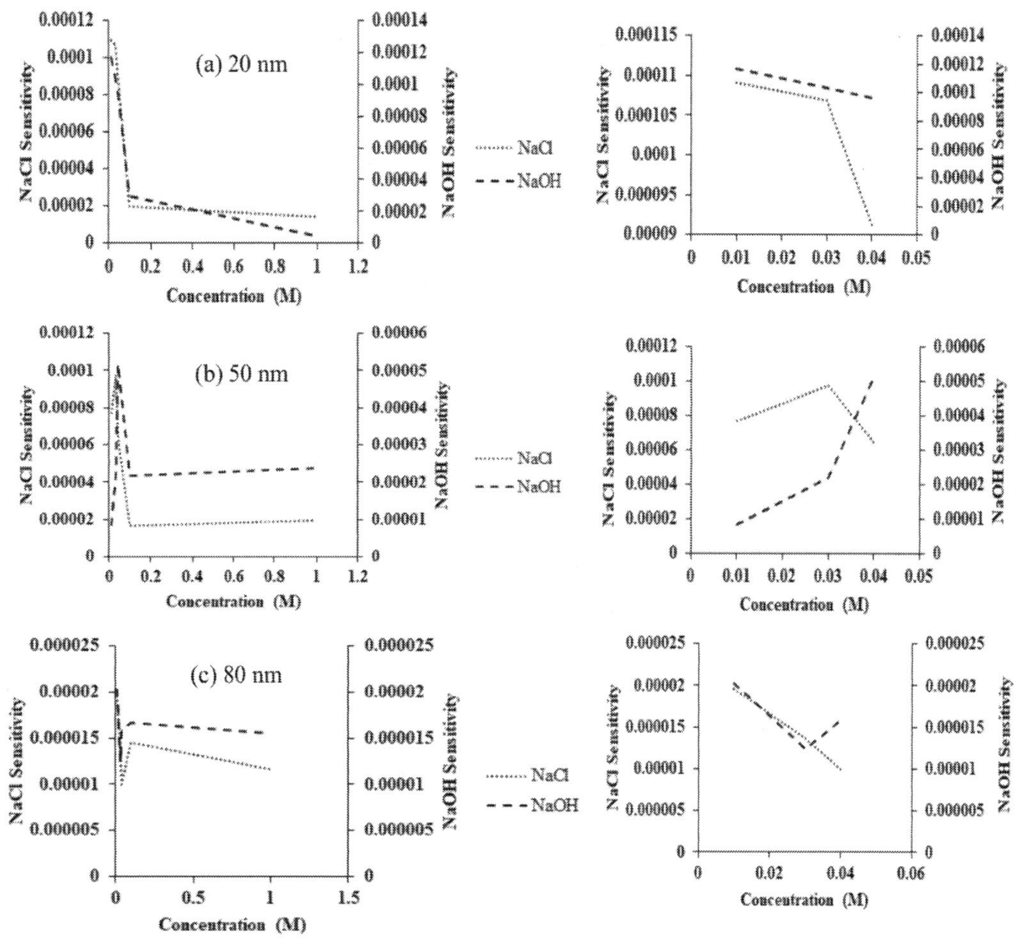

Figure 5: The left hand side shows comparison between NaCl and NaOH solutions for SPR sensitivity of damping for AgNPs and the right hand side shows the zoom at low concentration. (a) 20 nm, (b) 50 nm, and (c) 80 nm.

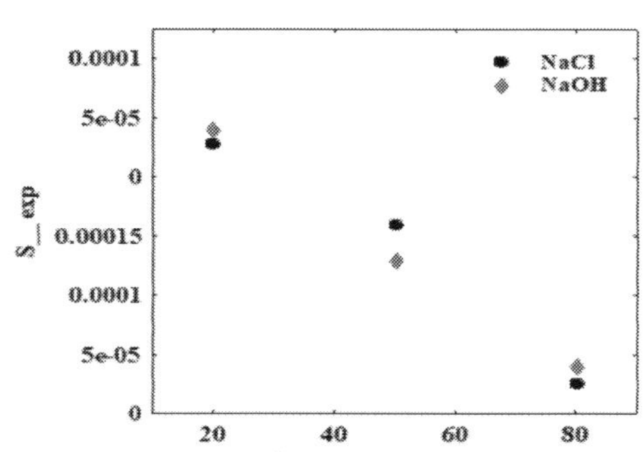

Figure 6: Comparison between sensitivity of damping (S_exp (nm^{-1})) for NaCl and NaOH (0.04 M) vs particle size (nm).

The finding of this study indicated that the SPR was very sensitive to the changes in the concentration and type of local surrounding medium of AgNPs in the presence of almost similar refractive index which changed gradually from 1.333 to 1.34 for different concentrations (0.01, 0.02, 0.03, 0.04, 0.1, 0.152, and 1 M) of monovalent electrolyte matrix solutions. When the AgNPs were suspended in the media, which is contained the Cl^- ions, the absorption spectra damp more than other for 20 and 80 nm AgNPs. Consistent with the results in Figure 6, the plasmon resonance is a function of concentration and type of ions contained in solution despite the dependence on refractive index of surrounding media and interparticle distance results from aggregation of AgNPs. Furthermore, the S strongly depended on primary particle size of NPs.

4. Conclusion

In conclusion, experiments demonstrated that the values of maximum absorption spectra for the AgNPs suspension based were sensitive to the concentration and type of surrounding medium. The OH^- was less effective than Cl^- ions, especially in the case of particles 20 and 80 nm. Overall, the calculated and observed sensitivity of damping was attributed to the size decency of AgNPs.

Acknowledgments

The authors would like to thank Universiti Sains Malaysia for funding this research work through its Transdisciplinary Research Grant Scheme (203.PFIZIK.6763003).

References

[1] Sotiriou G Aand Pratsinis S E 2011 *Current opinion in chemical engineering* 1 3-10
[2] Marambio-Jones Cand Hoek E M 2010 *J. Nanopart. Res* 12 1531-1551
[3] Loza K, Diendorf J, Sengstock C, Ruiz-Gonzalez L, Gonzalez-Calbet J, Vallet-Regi M, Köller Mand Epple M 2014 *J. Mater. Chem. B* 2 1634-1643
[4] Saison-Francioso O l, Lévêque G t, Boukherroub R, Szunerits Sand Akjouj A 2015 *J. Phys. Chem. C* 119 28551-28559
[5] Zhang X-F, Liu Z-G, Shen Wand Gurunathan S 2016 *Int. J. Mol. Sci* 17 1534
[6] Lee K-Sand El-Sayed M A 2006 *J. Phys. Chem. B* 110 19220-19225
[7] Glover Fand Goulden J 1963 *Nature* 200 1165-1166
[8] Quinten M 2010 *Optical properties of nanoparticle systems: Mie and beyond*, John Wiley & Sons.
[9] Badawy A M E, Luxton T P, Silva R G, Scheckel K G, Suidan M Tand Tolaymat T M 2010 *Environ. Sci. Technol.* 44 1260-1266
[10] Mulvaney P, Linnert Tand Henglein A 1991 *The Journal of Physical Chemistry* 95 7843-7846

IOP Conf. Series: Journal of Physics: Conf. Series **1083** (2018) 012043 doi:10.1088/1742-6596/1083/1/012043

Effect of pH on the Crystal Growth of ZnO Nanomaterials and Their Band gap Energies

M F Kasim[1, 3], N Kamarulzaman[1,2], R Rusdi[1,2] and A A Rahman[4]

[1]Centre for Nanomaterials Research, Institute of Science, Level 3 Block C, Old Engineering Building, Universiti Teknologi MARA, 40450 Shah Alam, Selangor, Malaysia
[2]School of Physics and Materials Studies, Faculty of Applied Sciences, Universiti Teknologi MARA, 40450 Shah Alam, Selangor, Malaysia
[3]School of Chemistry and Environment, Faculty of Applied Sciences, Universiti Teknologi MARA, 40450 Shah Alam, Selangor, Malaysia
[4]Centre for Science Foundation Studies, Univesity of Malaya, 50603 Kuala Lumpur, Malaysia

muhdfir@salam.uitm.edu.my

Abstract. Zinc oxide nanostructures with vary pH have been successfully synthesize using a sol-gel method. The nanostructured materials were annealed at low temperature of 300 °C for 3 hour. The effect of pH of the synthesis condition of ZnO on the structural and optical properties of the nanostructures were studied using X-ray Diffraction (XRD), Field Emission Scanning Electron Microscopy (FESEM) and UV-Vis spectrophotometer. XRD results revealed that only sample at very acidic condition was not pure with the presence of impurities. This implies that acidic condition require more energy in order to produce pure and single phase of ZnO nanostructures. FESEM results showed that at very acidic and basic condition, the crystallites tended to agglomerate to form bigger particles. The band gap energy of the ZnO nanostructures at pH 6 and 9 were 3.276 and 3.300 eV respectively. Interestingly, it was found that as the aspect ratio of nanorods increased, the band gap energy of the material also increased.

1. Introduction

Zinc oxide (ZnO) has received much attention from many researchers as promising candidate for semiconductor device applications [1-4]. This is due to their properties of wide and direct band gap semiconductor (3.37 eV) at room temperature [3, 5-9]. ZnO is being considered as a strong competitor to GaN, which also possess hexagonal structure but the advantage of ZnO over GaN is that ZnO has a large exciton binding energy (60 meV) that should favour light emission at room temperature. Therefore, ZnO has applications in ultraviolet (UV) optoelectronic devices such as UV light detectors, UV light emitting diodes, p-n junction devices and semiconductor lasers. ZnO also has been used in piezoelectronic transducers, gas sensors, transparent conductive films and bulk acoustic wave devices [5].

A wide variety of techniques have been used to synthesize ZnO nanostructures. This includes the hydrothermal methods [4], pulsed laser deposition (PLD) [5], combustion [10], and sol-gel method

Content from this work may be used under the terms of the Creative Commons Attribution 3.0 licence. Any further distribution of this work must maintain attribution to the author(s) and the title of the work, journal citation and DOI.
Published under licence by IOP Publishing Ltd

[1]. Different synthesis method will yield different type of morphology. ZnO is famous for having variety of morphology such as rod, spherical, needle, tube, flower and etc [2, 11]. Each morphology will have different properties that will suit a particular applications. So, it is very important to be able to control the synthesis route in order to obtain desired morphology. This work has focused on the effects of pH condition on the structural, morphology and also the optical properties of ZnO.

2. Experimental

The synthesis of ZnO nano powders were prepared by using a sol-gel method. The starting materials used was zinc acetate dehydrate (R & M chemicals, 99.5% purity). This material as first has been dissolved with absolute ethanol and then was stirred for about 1 hour. The pH of the material was adjusted to acidic condition using 1.0 M of nitric acid. Whereas for basic condition, the solution was added with 1.0 M of ammonium hydroxide. In this work, 3 different pH condition were studied which were pH 3, 6 and 9. After the desired pH of the solution was obtained, the material was slow dried and a white precursor was obtained. The samples were then grinded using an agate mortar to obtain fine powders of ZnO. Thermal studies were done by using a simultaneous thermogravimetric analyzer (STA), SETARAM SETSYS Evolution 175. Based on the TG/DSC, the precursors were annealed at 300 °C for 3 h. The annealed precursors were characterized by using X-ray diffraction (XRD) using the PANalytical X'pert Pro MPD diffractometer. The XRD measurements were done using a bragg-brentano optical configuration. The morphology and crystallite size of the materials were then examined by using field emission scanning electron microscopy (FESEM - JEOL JSM-7600F). The band gap energies of the materials were determined using a UV-Vis spectrophotometer, the Perkin Elmer Lambda 950 UV-Vis-NIR. The measurements were done in reflection mode and in ambient conditions.

3. Results and discussion

The XRD patterns for ZnO nanostructures annealed at 300 °C for 3 h are shown in Fig. 1. The diffraction patterns show that only ZnO pH 6 and 9 sample were in good agreement with the ICDD reference number 01-089-0511 indicating that the synthesized samples were single phase with hexagonal structure and space group P63mc. Whereas for ZnO pH 3 sample was impure due to the acidic condition (pH 3) consist of high concentration of H^+ ion that leads to the difficulty for precursor to form ZnO and more energy are needed in order to form pure ZnO phase. For the pH of 6 and 9, and due to the low concentration of H^+ and high concentration of OH^-, the formation of ZnO are much easier compared to pH 3 condition. The reaction between OH^- and Zn^+ plays an important role in the formation of ZnO nanostructures [12]. The narrow and high intensity peaks for pure samples showed that the synthesized ZnO samples were well crystalline.

IOP Conf. Series: Journal of Physics: Conf. Series **1083** (2018) 012043 doi:10.1088/1742-6596/1083/1/012043

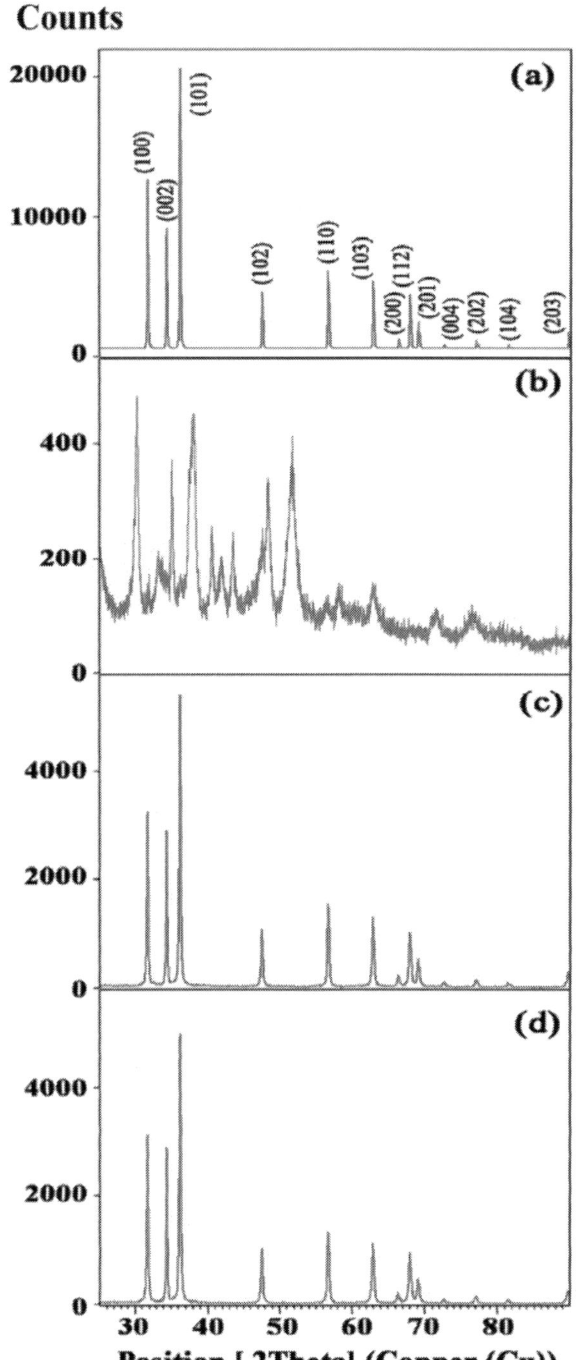

Figure 1: The XRD patterns of (a) ZnO ICDD 01-089-0511 (b) pH 3 (c) pH 6 and (d) pH 9 of ZnO annealed at 300 °C for 3 hour.

FESEM images of ZnO with varies pH are shown in Fig. 2. It can be seen that for impure ZnO pH 3 sample (Fig. 2 (a)) has cube like shape. Meanwhile, the morphology for pure ZnO pH 6 and 9 samples were short and long nano rods shape. It was found that the sample synthesize at very acidic medium cause the crystallites to become agglomerate and form bigger particles. For sample that

IOP Conf. Series: Journal of Physics: Conf. Series **1083** (2018) 012043 doi:10.1088/1742-6596/1083/1/012043

synthesize at basic medium shown in Fig. 2 (c), it was found that the sample possess very long and thin nano rods shape. These long and thin nano rods were stick together to form bigger rods. Unlike at very acidic and basic medium, for pH 6 it can be seen clearly that the crystallites were not agglomerate. The average dimension and aspect ratio for pure samples were obtained by measuring the dimension of 50 crystallites for each sample and listed in Table 1.

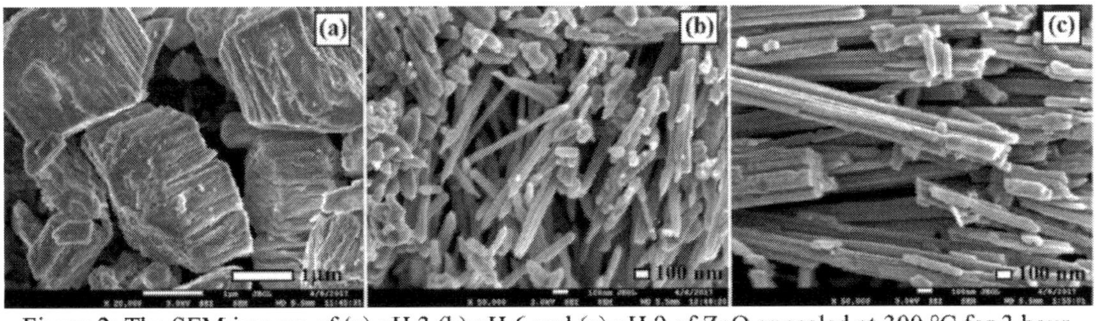

Figure 2: The SEM images of (a) pH 3 (b) pH 6 and (c) pH 9 of ZnO annealed at 300 °C for 3 hour.

Table 1. The average dimension and aspect ratio for ZnO pH 6 and 9 samples.

Sample	Morphology	Diameter (nm)	Length (nm)	Aspect ratio (Length/Diameter)
ZnO_pH 6	Short rods	82.50	899.15	10.90
ZnO_pH 9	Long nanorods	73.62	1652.39	22.44

Further studies on the light absorption properties of pure samples have been carried out using UV-Vis spectrophotometer. Fig. 3 (a) shows the UV-Vis reflectance curves of ZnO nano rods. The nano rods sample were measured at room temperature in the wavelength of 200-800 nm. It was found that the absorption edge of the ZnO pH 9 was slightly shifted to the left as compared to ZnO pH 6 sample. This indicates that the absorption wavelengths have decreased. Band gaps can be evaluated from the absorption edges. The band gap energies of ZnO materials were determined by using Tauc plots. Equation (1) below is used,

$$(\alpha h\upsilon)^2 = C' (h\upsilon - E_g) \qquad (1)$$

where α is the absorption coefficient of the ZnO at a certain value of wavelength λ, h is Planck's constant, C' is the constant, υ is the frequency of light, E_g is the band gap energy. The Tauc plot graph for ZnO pH 6 and 9 samples are shown in Fig. 3 (b). The determination of band gap energy using Tauc plot graph was explained by R. Rusdi et al. [2]. The band gap energy for ZnO pH 6 and 9 are 3.276 and 3.300 eV respectively. It was found that nano rods with higher aspect ratio have larger band gap energy. The dimension of the crystallites plays critical role in the band gap energy of the materials. The effects of dimension of crystallites on the band gap energy of materials were well studied in our previous work [1].

IOP Conf. Series: Journal of Physics: Conf. Series **1083** (2018) 012043 doi:10.1088/1742-6596/1083/1/012043

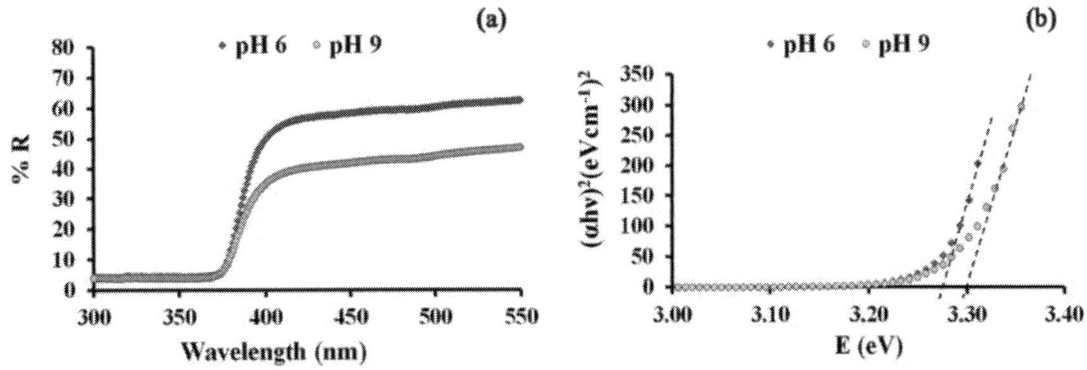

Figure 3: The results of (a) UV-Vis reflectance spectra and (b) Tauc plots for ZnO annealed at 300 °C for 3 hour.

4. Conclusion

The ZnO with morphology of long nano rods have successfully synthesized via sol-gel method at very low temperature. It was found that basic synthesis condition was more easy and suitable to obtaine ZnO nano rods compared to acidic condition. The morphology of the materials with vary pH revealed that pH affected the shape of the crystals. The very acidic and basic ZnO nanostructures tended to form agglomeration and then leaded to bigger particles. At very basic condition, it was found that ZnO had very long rods which were 1652.39 nm. It was also found that the ZnO nano rods with higher aspect ratio had larger band gap energy. In conclusion, by controlling the pH, the desired morphology of the crystals can be obtained.

Acknowledgement

The authors would like to thank the Universiti Teknologi MARA (UiTM) for the grants 600-IRMI/Dana KCM 5/3/LESTARI (103/2017) for funding this work and also Institute of Science, Universiti Teknologi MARA, Shah Alam, Malaysia for their support to this work.

References

[1] Kamarulzaman N, Kasim M F and Chayed N F 2016 *Results Phys.* 6 217-230
[2] Rusdi R, Rahman A A, Mohamed N S, Kamarudin N and Kamarulzaman N 2011 *Powder Technol.* 210 18-22
[3] Qi K, Cheng B, Yu J, Ho W 2017 *J. Alloys Compd.* 727 792-820
[4] Kumaresan N, Ramamurthi K, Ramesh Babu R, Senthuraman K and Moorthy Babu S 2017 *Appl. Surf. Sci.* 418 138–146
[5] Shan F K, Liu G X, Lee W J and B.C. Shin 2006 *J. Cryst. Growth* 291 328-333
[6] Hitkari G, Singh S and Pandey G 2017 *Nano-structures and Nano-objects* 12 1-9
[7] Sahoo K, Biswas A and Nayak J 2017 *Sens. Actuators, A* 267 99-105
[8] Du J, Zhang Y, Liu L, Qv M, Lv Y, Yin Y, Zhou Y, Cui M, Zhu Y and Zhang H 2017 *Chemosphere* 187 368-375
[9] Prajapati C S, Visser D, Anand S, Bhat N 2017 *Sens. Actuators, B* 252 764-772
[10] Karunakaran C, Rajeswari V and Gomathisankar P 2011*Superlattices Microstruct.* 50 234-241
[11] Eadi S B, Kim S and Jeong S W 2017 *J. Chem. Vol.* 2017
[12] Rayerfrancis A, Balaji Bhargav P, Ahmed N, Chandra B and Dhara S 2015 *Physica B* 457 96–102

Development of Gold Nanoparticles Modified Electrodes for the Detection of Heavy Metal Ions

Ilias Ait Tayeb[1] and Khairunisak Abdul Razak[1, 2]

[1]School of Materials and Mineral Resources Engineering, Universiti Sains Malaysia, Nibong Tebal, Penang 14300, Malaysia
[2]NanoBiotechnology Research and Innovation, Institute for Research in Molecular Medicine, Universiti Sains Malaysia, 11800 USM, Penang, Malaysia

khairunisak@usm.my

Abstract. Following the trend of developing mercury-free electrodes for environmental monitoring. Gold-based electrodes are proving to be excellent alternatives, due to their high sensitivity and selectivity especially in the monitoring of heavy metals. In this study, gold nanoparticles (AuNPs) modified screen printed carbon electrodes (SPCE) have been developed for the purpose of simultaneous detection of mercury and lead ions in water. AuNPs were synthesized using the seeding-growth method. Cyclic voltammetry (CV) was performed to investigate the electrochemical properties of the developed SPEs. AuNPs of average size of 50 nm showed optimal electroactivity. Square-wave Anodic stripping voltammetry (ASV) was used for the detection of the target heavy metals. Analysis were conducted with acetate buffer as the supporting electrolyte. Optimal concentration of acetate buffer was found at 0.2 M. The limit of detection (LOD) of Hg was 1.18 μg/L, while the LOD of Pb was 4.62 μg/L.

1. Introduction

Heavy metals are regarded among the most serious pollutants due to their non-biodegradability in the environment and high toxicity to human health. For example, Pb is known to interfere with calcium, therefore, hindering the release of neurotransmitters which can lead to major neurological or neurobehavioral complications especially in infants and young children. Hg and its respective compounds are known to have major effects on the central nervous system leading to various neurological complications [1, 2]. Moreover, these heavy metals tend to be persistent in the environment and bio-accumulate in human body through food consumption. Nevertheless, with the rapid increase of industrial activities and inappropriate disposal of wastewaters, heavy metal concentrations are bound to increase above their specific threshold set by regulatory organizations such as World Health Organization (WHO). Taking into consideration the hazardous effects of these heavy metals, their detection is therefore considered of huge importance and necessity.

Currently, spectroscopic techniques such as Inductive Coupled Plasma Mass Spectroscopy (ICP-MS) and Flame Atomic Absorption Spectroscopy (FAAS) are the commonly adopted approaches for the purpose of trace level heavy metals detection. However, the main drawbacks of these established methods come from their expensive instrumentation, their incapability for in-field analysis and some fall short for multi elemental analysis [3]. In contrast, electrochemical analysis is receiving extensive attention as substitutes for spectroscopic techniques. The techniques employed for the analysis are

Content from this work may be used under the terms of the Creative Commons Attribution 3.0 licence. Any further distribution of this work must maintain attribution to the author(s) and the title of the work, journal citation and DOI.
Published under licence by IOP Publishing Ltd

known to be suitable for the selective detection of metal species, therefore they are widely used also for *in- situ* applications. In addition, they require no or minimum sample alteration, thus dramatically reducing chances of contaminations by reagents or losses by adsorption on containers [4].

Many types of materials have been implemented for fabrication and developments of working electrodes in electrochemical sensing applications, such as mercury, bismuth, gold and carbon glassy. Additionally, with the progressive trend towards miniaturization of devices, screen-printed electrodes (SPEs) arose as alternatives to the conventional electrodes which make them well-suited for in-field analysis. SPEs are devices fabricated by printing layers of electroconductive and insulative inks on planar plastic or ceramic substrates. One of the main advantages of SPEs is that they are manufactured by mass production technology while maintaining high cost efficiency. In addition, they provide high reliability for electrochemical analysis with few usages, giving them their huge range of applications [5].

On the other hand, nanostructured materials have been investigated extensively for applications in electrochemistry. This is attributed to their large effective surface area and fast mass transport [6]. Several studies have investigated the effect of nanoparticles on enhancing the electrochemical performance of electrodes. For example, AuNPs were successfully utilized in the modification of ultramicroelectrode array for the detection of copper ions [7]. A comparison with the non-modified electrodes showed a 4 time increase of the output current after AuNPs modification. Additionally, a sensitive method for the simultaneous detection of Pb and Cu was investigated by developing AuNPs modified screen-printed gold electrode (SPGE). The limit of detection (LOD) was found to be at 2.2 ppb and 1.6 ppb for Pb and Cu, respectively.

In this study, a disposable AuNPs modified SPCE was developed for the simultaneous detection of mercury and lead. AuNPs were produced using the seeded-growth method. Cyclic voltammetry (CV) and square wave anodic stripping voltammetry (SWASV) were used in the electrochemical analysis. Systematic experiments were carried out for the concentration of supporting electrolyte and determination of LOD.

2. Experimental Method

AuNPs were synthesized using the seeding-growth method. This method consists of two stages: the seeding stage and the growth stage. At the seeding stage, trisodium citrate and gold chloride ($HAuCl_4.3H_2O$) was used to synthesize the AuNPs seeds. At the growth stage, reduction of gold chloride was achieved by adding hydroxylamine hydrochloride ($NH_2OH.HCl$) as the reducing agent. Colloidal AuNPs with different particle sizes were produced in this stage. This was achieved by adding 7 ml, 4 ml, and 2 ml of the seed solution to 50 ml of Milipore water. Then 1 mL of hydroxylamine amine ($NH_2OH.HCl$) was added into each solution. After the synthesis was completed, 0.02% of sodium azide was then added to the suspension. This was essential to prevent the colloid from coagulation. The characterization of the AuNPs was performed by using XRD and Zeta-sizer. The electrochemical performance of the AuNPs were tested using cyclic voltammetry (CV). The electrochemical properties of bare SPE was compared to the electrochemical properties of the AuNPs modified SPE. The CV analysis were performed in 0.1 mol/L of KCl solution (pH 7.0) containing 2.0×10^{-3} mol/L of potassium ferrocyanide, $[Fe(CN)6]^{3-}$. Additionally, the scanning range of the CV analysis was performed from -0.3 V to 0.6 V, starting at 0 V with the scan rate of 0.01 V/s and potential step of 0.01 V. ASV was conducted for the detection of Hg and Pb. The tests were conducted in solutions containing concentrations of Hg and Pb ranging from 5 μg/L – 25 μg/L. The AuNPs modified SPE were immersed, and the solution was stirred at 300 rpm.

3. Results and Discussion

3.1. Characterization

XRD patterns of AuNPs produced with varying volumes of seeds solution are shown in Figure 1. Five peaks were clearly observed at 32°, 38°, 64.5°, 78°, and 81.6° that can be assigned to the (111), (200),

(220), (311) and (222) planes that matched with the ICDD No. 089-3697. Thus, the XRD patterns show that AuNPs have FCC crystal structure and crystalline in nature.

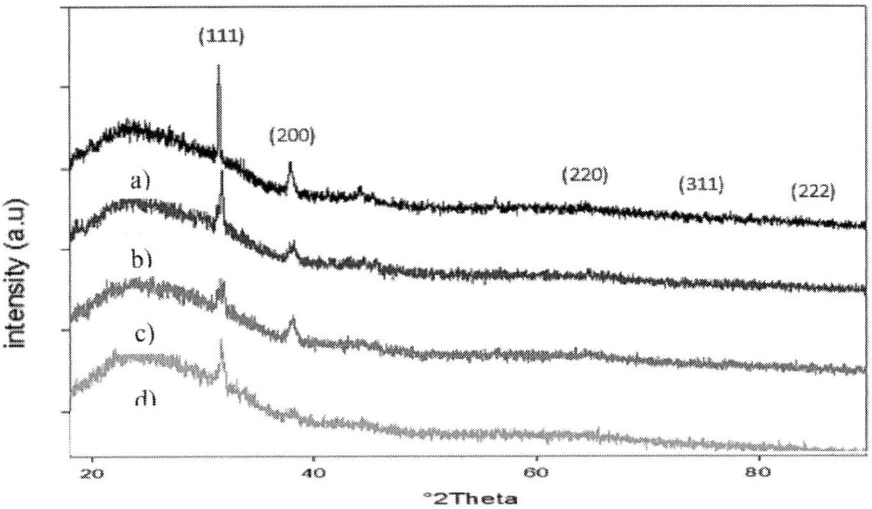

Figure 1. XRD patterns of AuNPs synthesized by tuning volumes of AuNPs seeds. (a) 2 ml, (b) 4 ml, (c) 7 ml, (d) seed.

After the completion of the seeding and growth stages, the average sizes of the produced particles were determined. Figure 2 (a-d) show the average sizes of the particles in the seed solution and the three different solutions when the volume of AuNPs seeds was varied. It is expected that average particle size would increase compared to the average size of the seed particles size if growth alone occur on the seed surface. The results were obtained from Zeta-sizer characterization of the produced solutions. The seed solution of AuNPs were effectively enlarged to AuNPs with average sizes of 28.39 nm, 46.47 nm, and 53.85 nm by varying volume of AuNPs seeds: 7 ml, 4 ml, and 2 ml respectively.

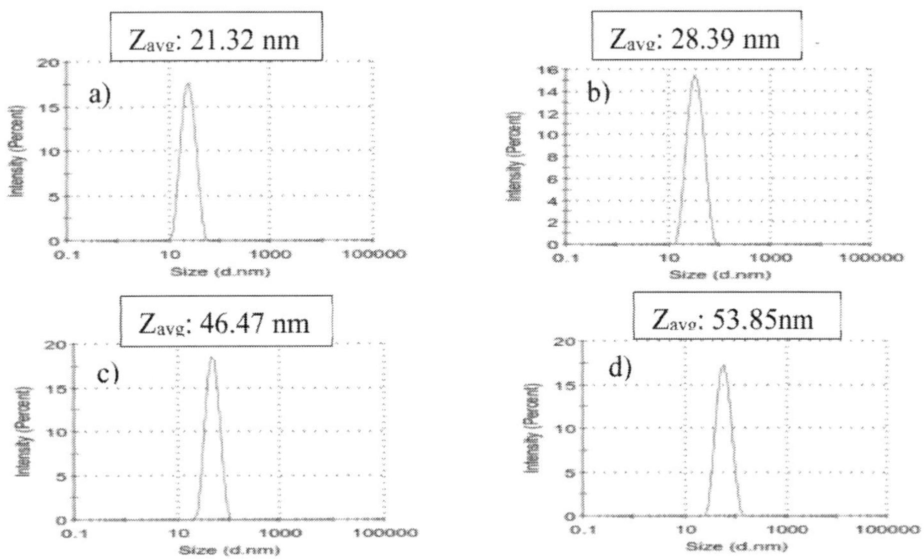

Figure 2. Size distribution of AuNPs obtained using Zeta- sizer analysis after seeding-growth process a) seed solution b) 7 ml of seed, c) 4 ml of seed d) 2 ml of seed.

3.2. Cyclic Voltammetry

The electrochemical behaviour of the bare SPE and AuNPs modified SPE was investigated using CV. SPEs were modified with AuNPs obtained from the above solutions. The CV analysis of AuNPs modified SPEs were performed in 0.1 mol/L of KCl solution (pH 7.0) containing 2.0×10^{-3} M of potassium ferrocyanide $[Fe(CN)6]^{3-}$. Additionally, the scanning range of the CV analysis was performed from -0.3 V to 0.6 V, starting at 0 V with the scan rate of 0.01 V/s and potential step of 0.01 V. These CV curves are shown in Figure 3. The peak current obtained for the bare SPE was 22.93 μA. As can be seen from Figure 3, the peak currents increased as the average size of the AuNPs increased. The increase in the peak currents for the AuNPs modified SPEs could be explained in terms of the enhanced electrocatalytic activity and larger active surface area of AuNPs, as it improves the mass-transport rate of electrons between the SPE and $[Fe(CN)6]^{3-}$ ions, therefore inducing faster electron transfer kinetics. In addition, the larger electrochemical surface-active area is not only beneficial for enhancing the sensor sensitivity, but also advantageous for lowering the detection limit.

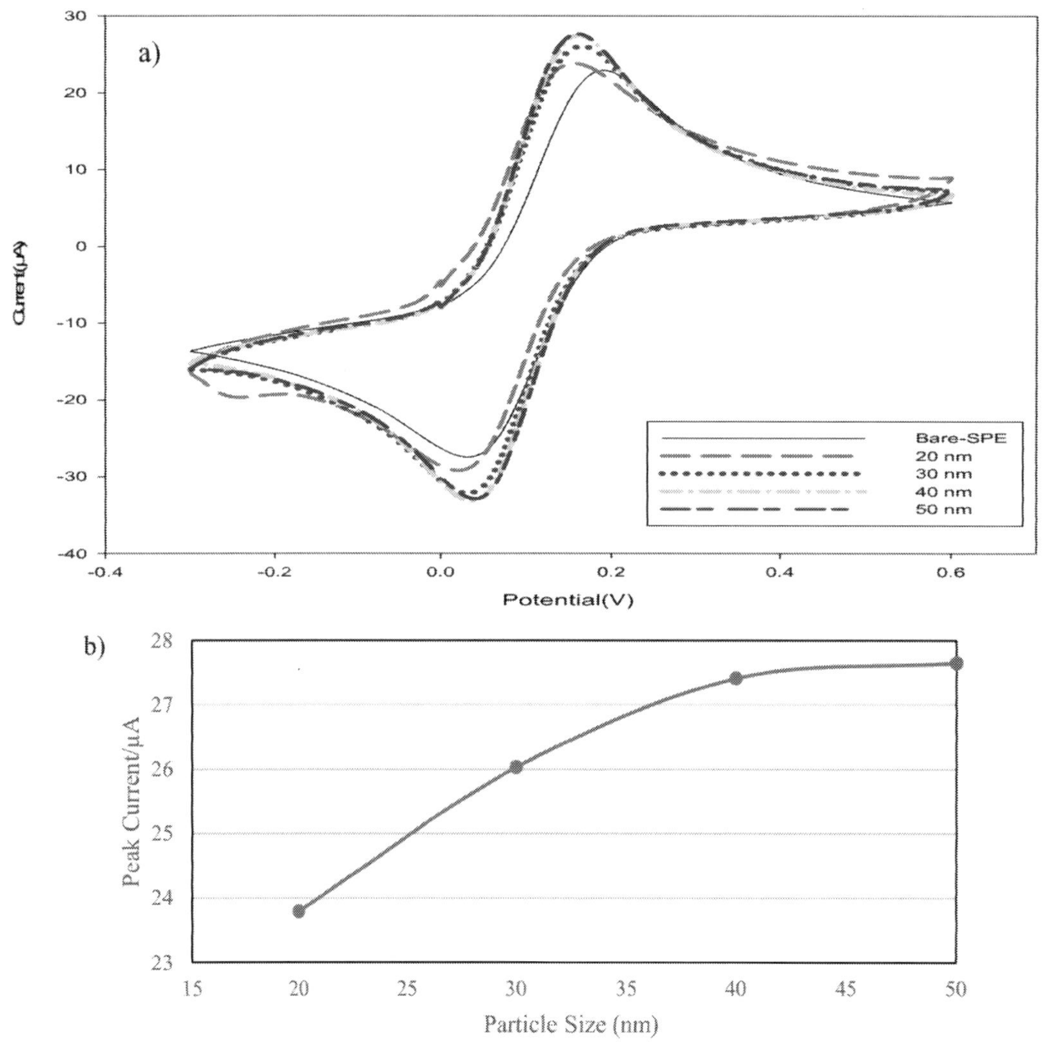

Figure 3. Cyclic Voltammograms of bare SPE and modified SPEs with different average sizes of AuNPs, b) Effect of average particle size of AuNPs on the electrochemical behavior of modified SPCE

3.3. Anodic Stripping Voltammetry

The characteristic stripping potentials of Hg and Pb were around 0.74V and 0.03 V respectively. Plot of the resulting stripping voltammograms versus scan potential is displayed in Figure 4a with varied concentration ranging from 5 μg/L – 25 μg/L. The obtained voltammograms showed an increase of the stripping currents of both Hg and Pb upon increasing the concentration of Hg and Pb. The stripping peak currents of Hg and Pb were extracted and a calibration curve of peak currents versus concentration is shown in Figure 4b. In the range of 5 ppb to 25 ppb, peak currents of Hg and Pb presented very good linearity with the increasing concentration. The calibration equation of Hg is $I_{Hg} = 0.2534C + 8.705$ with correlation coefficient of 0.9981 and sensitivity of 0.2534 μA/ppb. Whereas the calibration equation of Pb is $I_{Pb} = 0.376C - 0.54$ with correlation coefficient of 0.9851 and sensitivity of 0.376 μA/ppb. Consequently, the LOD of the developed analytical method was evaluated. The LOD values were found at 1.18 and 4.62 ppb for Hg and Pb, respectively.

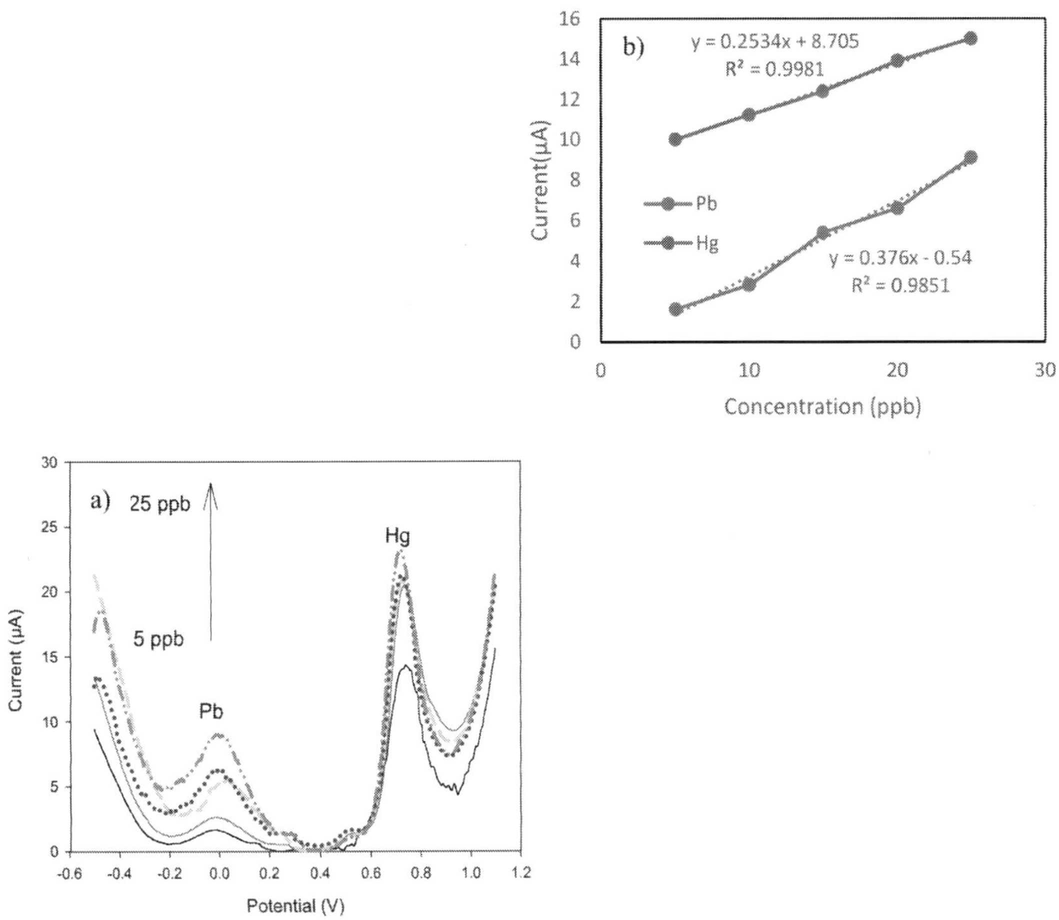

Figure 4. a) Stripping voltammograms of AuNPs modified SPCE with different concentration of Hg and Pb, b) Calibration plot for increasing concentration of Pb and Hg with their corresponding calibration equation

4. Conclusion

Gold nanoparticles were synthesized via seeding-growth method. The average particle sizes ranged from 20 nm to 50 nm. Characterization were conducted using XRD and Zeta-sizer analysis. SPCEs were modified by the produced AuNPs of different particle sizes. Cyclic voltammetry was performed to examine the electrochemical behavior of the modified SPCE. The produced AuNPs solution with 50 nm average size was found to have the best electrochemical performance. A volume of 5 μL of 50nm AuNPs was found to be optimal in modification of SPCE, due to its larger surface area. For the detection of the target heavy metals, Hg and Pb, ASV analysis were performed. The LOD of Hg was measured to be at 1.18 μg/L, while the LOD of Pb was measured to be at 4.62 μg/L.

Acknowledgment
We would like to acknowledge the support provided by the School of Materials and Mineral Resources and the Institute for Research in Molecular Medicine, Universiti Sains Malaysia.

References

[1] Goyer R A 1995 Nutrition and Metal Toxicity *Am J Clin Nutr.* 61 646s-50s.
[2] Langford N J and Ferner R E 1999 Toxicity of mercury *J Hum Hypertens.* 13 651-6
[3] Cui L, Wu J and Ju H 2015 Electrochemical sensing of heavy metal ions with inorganic, organic and bio-materials *Biosens Bioelectron.* 63 276-86
[4] Hayat A and Marty J L 2014 Aptamer based electrochemical sensors for emerging environmental pollutants *Front Chem.* 2
[5] Wan H, Sun Q, Li H, Sun F, Hu N and Wang P 2015 Screen-printed gold electrode with gold nanoparticles modification for simultaneous electrochemical determination of lead and copper *Sensor Actuat B-Chem.* 209 336-42
[6] Rassaei L, Nebel M, Rees N V, Compton R G, Schuhmann W and Marken F 2010 Discharge cavitation during microwave electrochemistry at micrometre-sized electrodes *Chem Commun.* 46 812-4
[7] Wang J, Bian C, Tong J, Sun J and Xia S 2013 Microsensor Chip Integrated with Gold Nanoparticles-Modified Ultramicroelectrode Array for Improved Electroanalytical Measurement of Copper Ions *Electroanal.* 25 1713-21

Electrical Properties of Nano Bi_2O_3 Added (Bi,Pb)Sr-Ca-Cu-O Superconductor

Nurul Raihan Mohd Suib[1,2], J. Nur-Akasyah [1], K. Muhammad Aizat[1], R. Abd-Shukor[1],

[1]School of Applied Physics, Universiti Kebangsaan Malaysia, 43600 UKM Bangi, Selangor, Malaysia,
[2]Fakulti Sains Gunaan, Universiti Teknologi MARA, 40450 Shah Alam, Selangor, Malaysia

ras@ukm.edu.my

Abstract. In this study, Bi_2O_3 nanoparticles were added into $(Bi_{1.6}Pb_{0.4})Sr_2Ca_2Cu_3O_{10}$ (BPSCCO) superconductor. Ultrafine BPSCCO powders were prepared by the co-precipitation method. Bi_2O_3 with weight percent (wt. %) 0, 0.04, 0.06, and 0.10 was added during the final heating stage. The samples were investigated by x-ray powder diffraction method. Critical current density, J_c was measured between 40 K and 77 K with the 1 μV/cm criterion. DC electrical resistivity was measured using the four point probe method. This result showed that the highest J_c was observed for sample with 0.06 wt. % Bi_2O_3. A gradual decrease of J_c and transition temperature T_c was observed for samples with greater than 0.06 wt. % Bi_2O_3.

1. Introduction

There are many studies done to improve the critical current density of (Bi,Pb)Sr-Ca-Cu-O (BPSCCO) superconductor. The low operating temperature required by low-temperature metallic superconductors (LTS), with transition temperatures below 23 K has limited their practical use in power application [1]. Existing HTS wires technology need to be improved for grid application. The application of superconductor in power grid can be performed at liquid nitrogen temperature [2, 3].

The critical current density, J_c can be determined from the current-voltage characteristics. In transport measurements, electric field criterion such as 1 μV/cm are often used to determine J_c [4]. Addition of other element in BPSCCO can act as impurity for pinning center to improve J_c and transition temperature T_c [5-8]. Among the Bi-based materials, $Bi_2Sr_2Ca_2Cu_3O_{10}$ (Bi-2223) phase, which is also called high T_c phase, is of great interest due to its higher critical temperature (~110 K), atmospheric stability and ability to be rolled into long wires or tapes [9].

In application of BSCCO phase superconductor, the major limitation is the intergrain weak links and flux pinning capability. Effective artificial flux pinning is introduced to enhance flux pinning capability. The flux pinning capability could be improved by microstructural control such as concentration of dislocations and planar faults, tiny secondary phase and irradiation damage zones [10]. From a previous study, HTSC tapes were prepared with the optimum amount 0.01 wt. % nano Bi_2O_3 addition. In the current study nanosize Bi_2O_3 were added into (Bi,Pb)Sr-Ca-Cu-O with 0 to 0.15 wt. % [11]. The objectives of this study were to investigate the effect of nano Bi_2O_3 on the transport properties of (Bi,Pb)Sr-Ca-Cu-O superconductor.

Content from this work may be used under the terms of the Creative Commons Attribution 3.0 licence. Any further distribution of this work must maintain attribution to the author(s) and the title of the work, journal citation and DOI.
Published under licence by IOP Publishing Ltd

2. Experimental Details

Bi(Pb)-Sr-Ca-Cu-O high-temperature superconductor was prepared via the co-precipitation method. The chemicals used in this method were bismuth acetate 99.99 %, strontium acetate, not mentioned percentage (Aldrich), lead acetate 99.5 %, calcium acetate 90 %, (Fluka) and copper acetate 98 % (Alfa Aesar). Oxalic acid, acetic acid, 2- propanol and deionized water were used as solutions in this synthesis. The acetate powders were diluted in acetic acid solution with a stoichiometric amount (solution A). The mixture was stirred for about 2 h at 80 oC to dilute the entire chemical to form dark blue colour solution. Oxalic acid (0.5M) was mixed to the water: isopropanol with ratio 1:1.5 to form solution B.

Solutions A and B were mixed at 0-3 °C for 30 min to form blue navy slurry. The precipitate was filter and dried at 80-100 °C for 8 h. The blue powder was calcined at 730 °C for 12 h. This calcination was to remove impurities from the powder. The powder was reground and calcined again at 845 °C for 24 h. This was followed by the addition of Bi_2O_3 nanopowder Aldrich 99.9 % (0, 0.04, 0.06 and 0.10 wt. %). The powder was pressed into pellets and sintered at 845 °C for 48 h in a furnace. DC electrical resistivity was measured using the four point probe technique in combination with CTI cryogenics closed-cycle refrigerator (Model 22). The transport critical current density J_c was measured using pellets cut into bar shape using by the four-point probe method with the 1 μV/cm criterion between 40 and 77 K. The XRD patterns of the sample were recorded using a Bruker XRD diffractometer with CuKα radiation.

3. Results and Discussion

Table 1 shows the onset transition temperature (Tc-onset), zero resistance temperature (Tc-zero), transition width ΔTc, critical current density Jc at 77 K and volume fraction of Bi-2212 and Bi 2223 phase. The volume fraction of Bi-2223 phase was calculated using

$$V_{2223} = \frac{\sum I2223}{\sum I2223+\sum I2212+\sum Iothers} \text{ x } 100\%$$

where $\sum$ *I*2223, $\sum$ *I*2212 and $\sum$ *I* others are the sum of intensities of Bi-2223, $Bi_2Sr_2CaCu_2O_8$ (Bi-2212) phase and another phase, respectively [12,13]. Volume fraction for Bi-2223 and Bi-2212 are shown in Table 1. Bi-2223 peaks decreased with Bi_2O_3 addition. The volume fraction of the highest Bi-2223 phase is the non-added sample. The low T_c phase of Bi-2212 was observed in the 0.10 wt. % sample.

Figure 1 shows the XRD patterns of the non-added sample, 0.04, 0.06, and 0.10 wt. % Bi_2O_3 added samples. Most of the peaks in both non-added and Bi_2O_3 added samples belong to the high T_c phase (Bi-2223) with a few peaks corresponding to the low T_c phase (Bi-2212).

The addition at 0.10 wt % sample showed the lowest T_c, which showed the intensity of high phase (Bi,Pb) -2223 (H) decreased with increase of the low phase of (Bi,Pb) -2212 (L). (Bi,Pb) -2212 (L) phase exists at 0.10 wt. % sample much more than other samples. For sample 0.06 wt. % XRD pattern (Bi,Pb) -2223 (H) was nearly same with 0.04 wt% except for (Bi,Pb) -2212 (L) peak 1115L which diminished.

Figure 2 shows the resistance versus temperature curve for all samples. The highest T_c achieved was at 0.04 wt. % (106 K), with optimum amount of addition during the final processing of BPSCCO samples can effectively improve the flux pinning and increased T_c as well [14]. Normally, the superconducting transition temperature decreases with the increase of impurities [15]. The sample with 0.10 wt. % showed the lowest T_c value with 103 K.

Table 1. Onset transition temperature ($T_{c\text{-onset}}$), zero resistance temperature ($T_{c\text{-offset}}$), transition width $\Delta\ T_c$ and critical current density J_c at 77 K

Bi_2O_3 (wt. %)	$T_{c\text{-onset}}$ (K)	$T_{c\text{-zero}}$ (K)	$\Delta\ T_c$ (K)	J_c (mA/cm^2)	Volume fraction Bi-2223	Bi-2212
0	112	105	7	982	89	11
0.04	113	106	7	1322	85	15
0.06	112	105	7	5047	82	18
0.10	111	103	8	3757	77	23

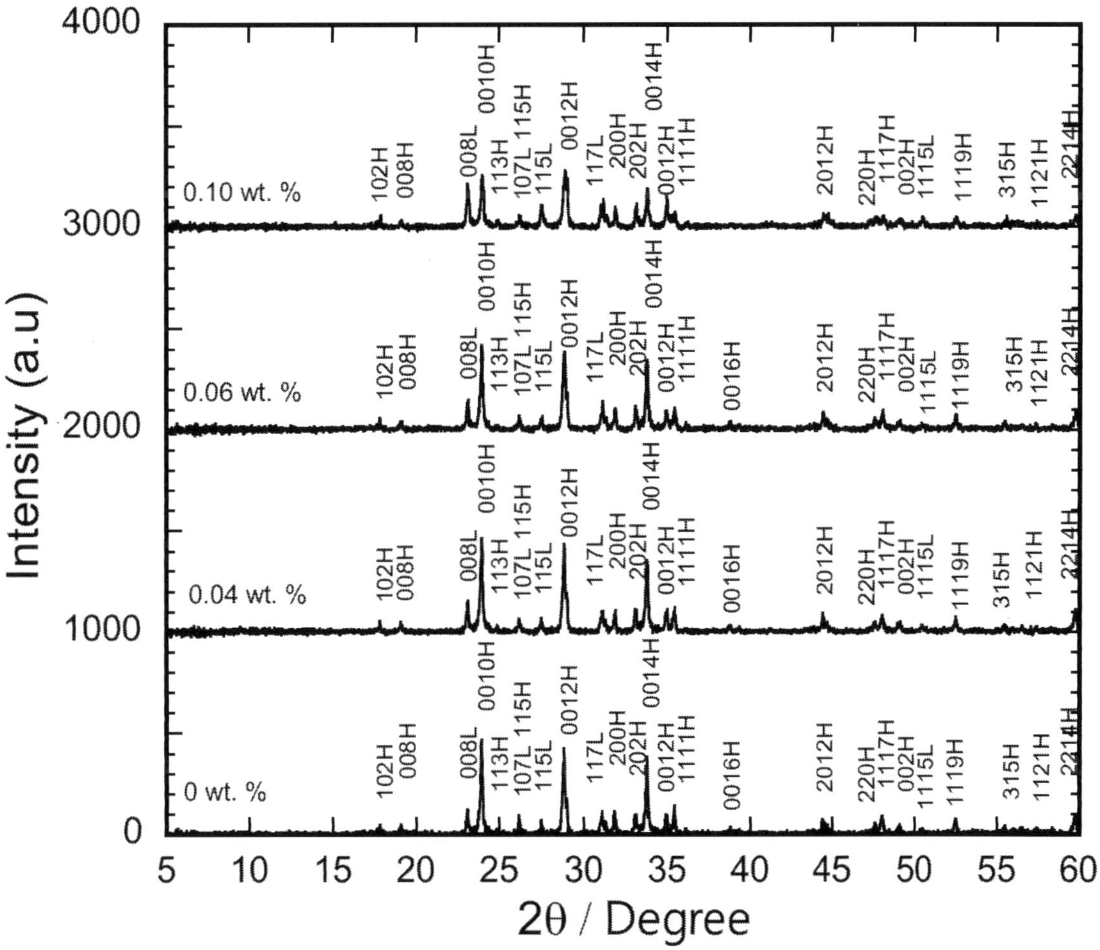

Figure 1. XRD patterns for $Bi_{1.6}Pb_{0.4}Sr_2Ca_2Cu_3O_{10}$ of the non-added sample and samples added with 0.04, 0.06 and 0.10 wt. % Bi_2O_3. (H) denotes the high- T_c phase (Bi-2223) and (L) denotes the low- T_c phase (Bi-2212)

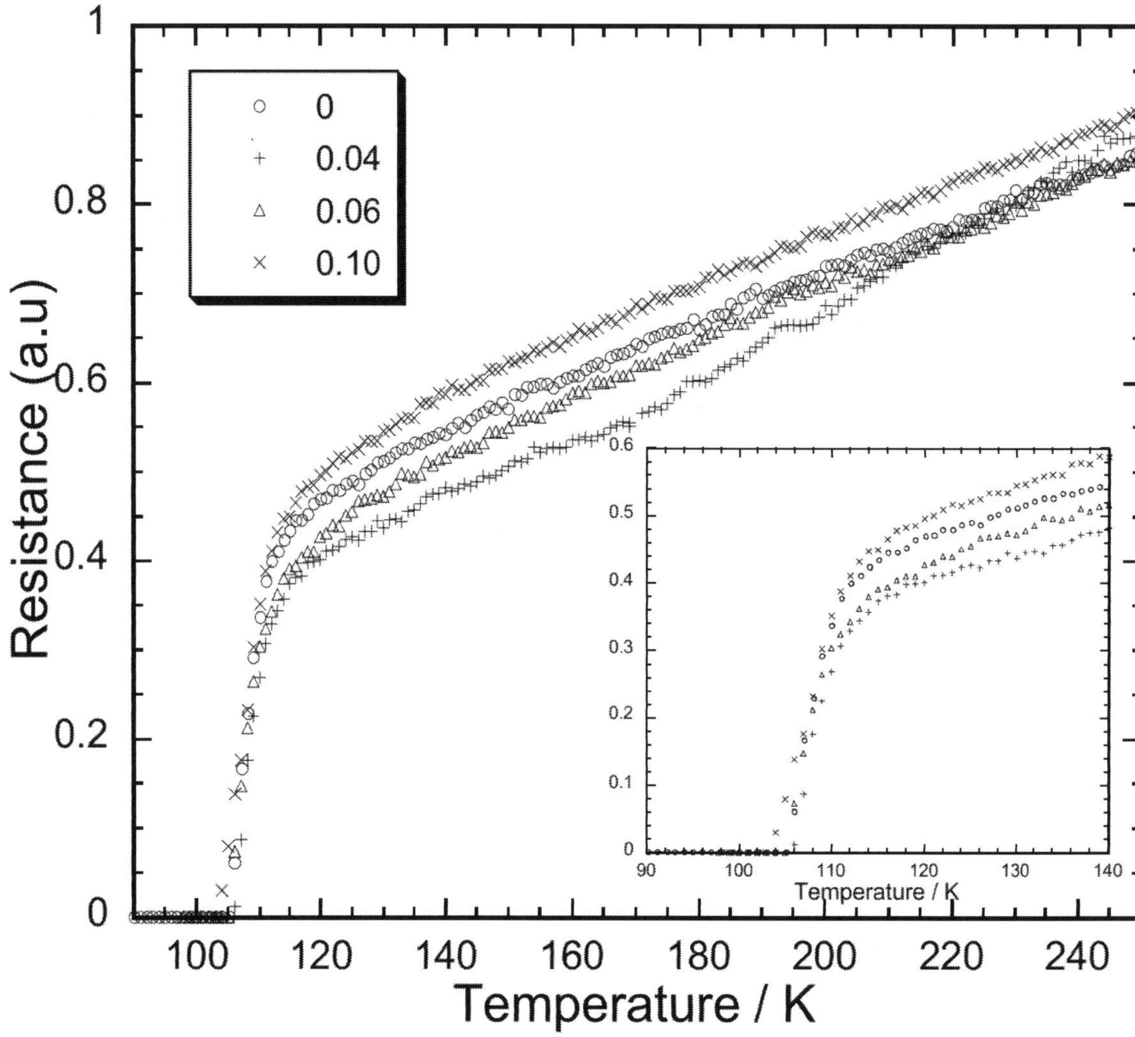

Figure 2. Resistance versus temperature curves for $Bi_{1.6}Pb_{0.4}Sr_2Ca_2Cu_3O_{10}$ of the non-added sample and samples added with 0.04, 0.06 and 0.10 wt % Bi_2O_3. The inset shows the curve near T_c

The critical current density, J_c of all samples were measured at 40 K, 50 K, 60 K, 70 K and 77 K are shown in Figure 3. The highest J_c was observed at 40 K for 0.06 wt. % with the highest value (7396 mA/cm^2). J_c at 77 K for this sample was 5237 mA/cm^2. The J_c of the non-added sample showed J_c at 40 K was 1833 mA/cm^2 and J_c at 77 K was 982 mA/cm^2. The J_c value of 0.10 wt. % sample at 40 K was 5237 mA/cm^2 and at 77 K was 3757 mA/cm^2. All samples with the addition of nano Bi_2O_3 showed a higher transport critical current density compared to the non-added sample. Overall, J_c decreased with increasing temperature and this due to the thermally activated flux creep [16].

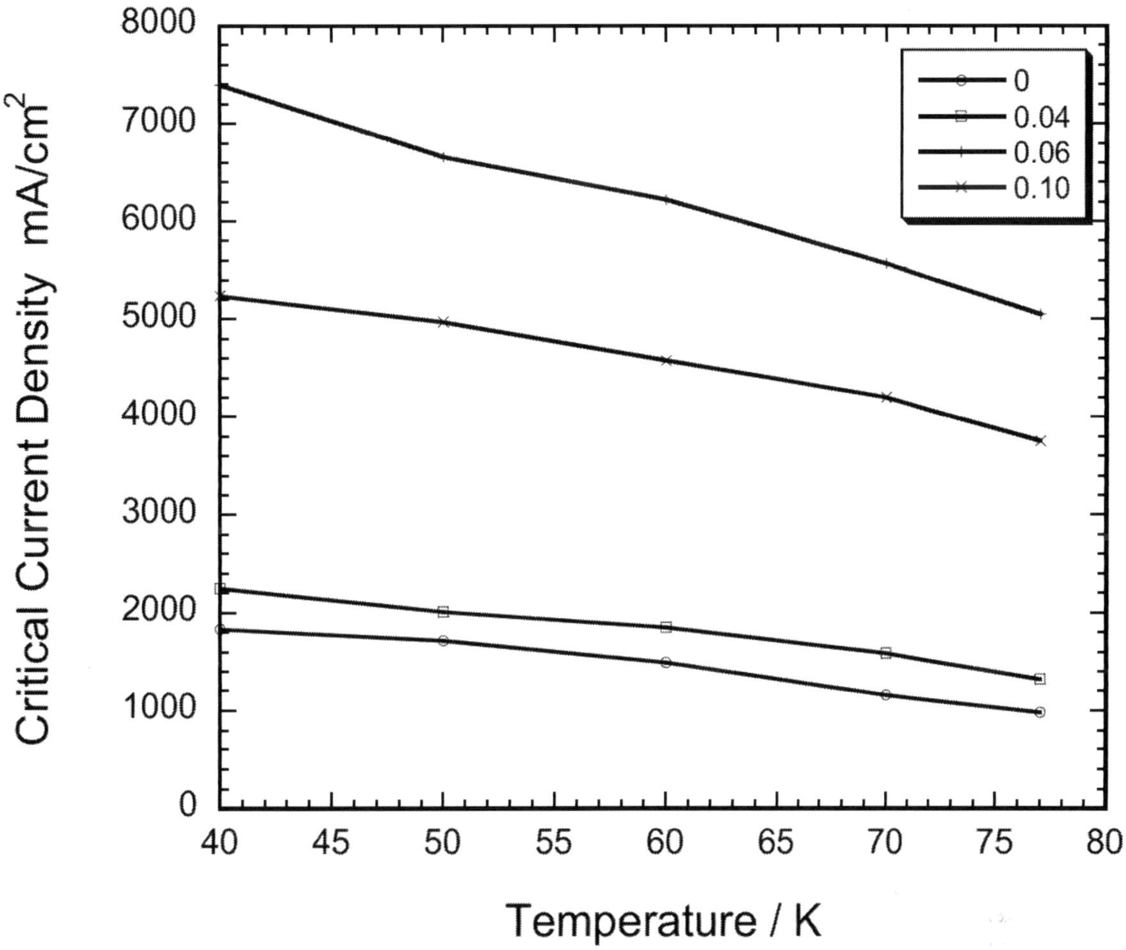

Figure 3. Critical current density, J_c of $Bi_{1.6}Pb_{0.4}Sr_2Ca_2Cu_3O_{10}$, of the non-added sample and samples added with 0.04, 0.06 and 0.10 wt. % Bi_2O_3 at temperature 40, 50, 60, 70 and 77 K

4. Conclusion

Nano Bi_2O_3 addition in (Bi,Pb)-2223 superconductor improved J_c for all added samples. An appropriate amount of Bi_2O_3 optimized the critical current density of (Bi,Pb)-2223. The highest J_c was observed in the 0.06 wt. % sample. XRD pattern showed the increase of (Bi,Pb)-2212 phase which lowered T_c, but did not affect J_c. The excess amount of Bi_2O_3 for 0.10 wt. % sample contributed to the low T_c.

Acknowledgments

This work was supported by Universiti Kebangsaan Malaysia under grant number AP-2015-006 and the Ministry of Higher Education of Malaysia under grant number FRGS/1/2017/STG02/ UKM/01/1.

References

[1] A.P. Malozemoff, Does the electric power grid need a room temperature superconductor?, Physica C: Superconductivity, 494 (2013) 1-4.
[2] K. Funaki, M. Iwakuma, K. Kajikawa, M. Hara, J. Suchiro, T. Ito, Y. Takata, T. Bohno, S.I. Nose, M. Konno, Y. Yagi, H. Maruyama, T. Ogata, S. Yoshida, K. Ohashi, H. Kimura, K. Tsutsumi, Development of a 22 kV/6.9 kV single-phase model for a 3 MVA HTS power

transformer, IEEE Transactions on Appiled Superconductivity, 11 (2001) 1578-1581.

[3] X. Ma, Q. Hao, G. Liu, H. Zheng, X. Xu, G. Jiao, S. Zhang, C. Li, P. Zhang, Influences of Pb content on the critical current of Bi-2223 multi-filamentary tapes, Materials Letters 162 (2016) 5-8.

[4] J.P. Rush, C.J. May-Miller, K.G.B. Palmer, N.A. Rutter, A.R. Dennis, Y.H. Shi, D.A. Cardwell, J.H. Durrell, Transport in Bulk Superconductors: A Practical Approach?, IEEE Transactions on Applied Superconductivity 26 (2016) 1-4.

[5] A. Agail, R. Abd-Shukor, Effect of Different Nanosized NiO Addition on Ag-Sheathed $(Bi_{1.6}Pb_{0.4})Sr_2Ca_2Cu_3O_{10}$ Superconductor Tapes, Journal of Superconductivity and Novel Magnetism 27 (2013) 1273-1277.

[6] R. Abd-Shukor, I. Kong, E.L. Lim, N.A. Mizan, H.A. Alwi, M.H. Jumali, W. Kong, Enhanced Critical Current Density of FeF_2 Added $YBa_2Cu_3O_{7-\delta}$, Journal of Superconductivity and Novel Magnetism 25 (2012) 957-960.

[7] R. Abd-Shukor, W. Kong, Magnetic field dependent critical current density of Bi–Sr–Ca–Cu–O superconductor in bulk and tape form with addition of Fe_3O_4 magnetic nanoparticles, Journal of Applied Physics 105 (2009) 07E311.

[8] A. Zelati, A. Amirabadizadeh, A. Kompany, H. Salamati, J. Sonier, Effect of Eu_2O_3 Nanoparticles Addition on Structural and Superconducting Properties of BSCCO, Journal of Superconductivity and Novel Magnetism 27 (2014) 1369-1379.

[9] M. Mubeen, M. Anis-ur-Rehman, Study of Superconducting Properties in Bismuth-Based Cerium Doped High-Tc Superconductors, Journal of Superconductivity and Novel Magnetism 26 (2012) 1123-1127.

[10] M. Roumié, S. Marhaba, R. Awad, M. Kork, I. Hassan, R. Mawassi, Effect of Fe_2O_3 Nano-Oxide Addition on the Superconducting Properties of the (Bi,Pb)-2223 Phase, Journal of Superconductivity and Novel Magnetism 27 (2013) 143-153.

[11] N.A.A. Yahya, A. Al-Sharabi, N.R.M. Suib, W.S. Chiu, R. Abd-Shukor, Enhanced transport critical current density of (Bi, Pb)-2223/Ag superconductor tapes added with nano-sized Bi_2O_3, Ceramics International 42 (2016) 18347-18351.

[12] T. Matsushita, A. Suzuki, T. Kishida, M. Okuda, H. Naito, The effect of Ag on the superconductivity of Bi_{2-x} Pb_x Sr_2 Ca_2 Cu_3 Oy superconductors prepared by an optimum thermal procedure, Superconductor Science and Technology 7 (1994) 222.

[13] I. Hamadneh, S.A. Halim, C.K. Lee, Characterization of $Bi_{1.6}Pb_{0.4}Sr_2Ca_2Cu_3O_y$ ceramic superconductor prepared via coprecipitation method at different sintering time, Journal of Materials Science 41 (2006) 5526-5530.

[14] M. Annabi, A. M'Churgui, I.Ben Azzouz, M. Zououi, M. Ben Salem, Addition of nanometer Al_2O_3 during final processing of (bi, Pb)-2223 superconductor, Physica C, 405 (2004) 25-33.

[15] P.K. Mukherjee, Effect of Non-magnetic Impurities on Iron-based Superconductors, Journal of Superconductivity and Novel Magnetism 29 (2015) 323-327.

[16] W. Kong and R. Abd-Shukor,Enhanced Electrical Transport Properties of Nano $NiFe_2O_4$ added $(Bi_{1.6}Pb_{0.4})Sr_2Ca_2Cu_3O_{10}$ Superconductor, J Supercond Nov Magn 23 (2009) 257–263.

IOP Conf. Series: Journal of Physics: Conf. Series **1083** (2018) 012046 doi:10.1088/1742-6596/1083/1/012046

Molecular Structure Investigation On Organic Chalcone Derivative of (*E*)-3-(4-Bromothiophen-2-1-(3-nitrophenyl)prop-2-en-1-one: A Combined Experimental and Theoretical Study

Dian Alwani Zainuri[1, a *], Suhana Arshad[1, b], Nuridayanti Che Khalib[1,c], Muhamad Fikri Zaini[1, d] and , Ibrahim Abdul Razak[1, e]

[1]X-ray Crystallography Unit, School of Physics, Universiti Sains Malaysia, 11800 USM, Penang, Malaysia
[1]X-ray Crystallography Unit, School of Physics, Universiti Sains Malaysia, 11800 USM, Penang, Malaysia.

[a]dianalwani14@gmail.com, [b]suhanaarshad@usm.my, [c]yantie_198@yahoo.com, [d]epy.pasoh1@gmail.com, [e]arazaki@usm.my

Abstract. A new chalcone derivative namely, (*E*)-3-(4-Bromothiophen-2-1-(3-nitrophenyl)prop-2-en-1-one was successfully synthesized using Claisen-Schmidt condensation reaction method. The compound crystallized in the orthorhombic, *Pnma* space group. The synthesized molecule was well characterized using spectroscopic techniques of Fourier Transform Infrared (FTIR), Nuclear Magnetic Resonance (NMR) and Ultraviolet-Visible (UV-Vis) absorption. Molecular and crystal structures were determined by single crystal X-ray diffraction method. The molecular geometry obtained from X-ray structure determination was further optimized by Density Functional Theory (DFT) using B3LYP/6-311++G(d,p) basis set in the ground state to calculate the vibrational wavenumbers and isotropic chemical shifts. TD-DFT/B3LYP method has been employed to calculate the electronic properties and frontier molecular orbital (FMO). The discrepancies between experimental and theoretical values are due to the fact that the molecule is calculated as an isolated molecule in the gaseous phase. The intermolecular interactions in the solid state contribute to the comformational changes of the molecular structure and the subsequently the crystal structure.

1. Introduction

Chalcone have gained significant attention in modern chemistry because of their wide range of applications [1]. Chalcone compound a group of open-chain flavonoid are very interesting due to their biological significances and pharmacological properties such as anti-fungal, anti-bacterial, anti-inflammator, anti-tuberculosis and anti-oxidant [2,3,4,5]. In addition, chalcones are excellent models for studying photo-induced electron transfer processes, which play a key role in different field such as polymer, photo, optic and laser physics [6,7]. A number of studies on chalcones have been carried out in the past few years [8,9,10]. Joseph et al. [11] has reported the synthesis of 3,4-Dimethoxy chalcone where this crystal has been characterized by single crystal x-ray diffraction, electronic and vibrational spectral studies. Density Functional Theory (DFT) has proved to be highly successful in describing the

Content from this work may be used under the terms of the Creative Commons Attribution 3.0 licence. Any further distribution of this work must maintain attribution to the author(s) and the title of the work, journal citation and DOI.
Published under licence by IOP Publishing Ltd

structural and electronic properties in a vast class of materials, ranging from atoms and molecules to simple crystals and complex extended system [10]. DFT also has become a common tool in first-principles calculations aimed at describing or predicting properties of molecular and condensed matter systems [10]. The present studies explain the single crystal x-ray structure and spectroscopic (FTIR, NMR and UV-vis) studies which were calculated using the DFT-B3LYP method at 6-311G++ (d,p) basis set and compared with experimental result. In order to study the intra- and intermolecular hydrogen bond interactions of the compound, single crystal X-Ray Diffraction (XRD) analysis has been employed. The geometry optimizations of the possible stable conformers were performed, and the corresponding relative energies were compared.

2. Methodology

2.1. Synthesis of compound

3-Nitroacetophenone (0.1 mol, 0.08257g) were dissolved with methanol (20 ml) about 10-15 mins. Then, 4-Bromo-2-thiophene (0.1 mol, 0.095g) were added and the solution were stir about another 10-15 mins. Then, the solution was dissolved in the presence of NaOH and was stir for another 4 hours until the precipitate form. After stirring for 4 hours, the reaction mixture was poured into cold water (50 ml) and stir for 5 mins to 10 mins. The precipitate solid was filtered, dried and recrystallized by acetone to get corresponding chalcone.

2.2. Spectroscopy analyses

The FTIR spectrum was recorded in the range of 4000-400 cm^{-1} with Perkin Elmer Spectrum GX Frontier Spectrophotometer using KBr pellet. The spectrum was recorded at room temperature, with 8 scan numbers and the spectral resolution of 4.0 cm^{-1}. 1H NMR and 13C NMR spectra were recorded in DMSO-d6 by BRUKER AVANCE 500 MHz spectrometer. The chemical shifts (δ) are reported in parts per million (ppm) downfield from internal tetramethysilane (TMS) internal reference. UV- Vis spectrum was recorded in acetonitrile solution using a UV-Visible Spectrophotometer Model Cary 5000 UV-VIS-NIR in the spectral region of 200-800 nm.

2.3. X-ray single crystal analysis

A single crystal of title chalcone with dimension 0.77 x 0.17 x 0.05mm^3 was used to collect the X-ray diffraction data on a Bruker APEX II duo CCD [12] diffractometer using MoKα radiation (λ= 0.71073 Å). The data collection was performed using APEX2 software [12] while the cell refinement and data reduction were performed using SAINT software [12]. The crystal structure was solved by Direct Method using the program SHELXTL [13] and refined by full-matrix least squares technique on F^2. SADABS was used to correct the data for absorption effects [12]. All geometrical calculations were carried out using the program PLATON [14].

2.4. Computational Methods

In this study, all the theoretical computations were performed by GAUSSIAN 09W software package [15]. The geometry optimization was carried out using the initial geometry generated from standard geometrical parameters of B3LYP methods, adopting 6-311++G (d,p) basis sets to characterize all stationary points as minima. B3LYP is chosen based on which ones give the best fit to a certain body of experimental data. This is proved by several literatures that used B3LYP as a functional for theoretical calculation of chalcones [10, 16]. B3LYP was published in 1993, it contains 3 empirical mixing parameters, which regulate the mixing of 5 components (including Hartree-Fock, LDA and Becke 88 exchange, VWN and LYP correlations) [17]. The optimized molecular structure of title compound is given in Fig. 1 (b). The molecular structure, fundamental vibrational frequencies and intensity of the vibrations bands are interpreted with the aid of structure optimizations and normal coordinate force field calculations based on density functional theory (DFT/B3LYP) methods combination with 6-311++G (d, p) basis set. The computed harmonic frequencies at B3LYP/6-311G(d,p) level of basis set were scaled by 1.0032 for frequencies less than 1700cm^{-1} and 1.0006 for higher frequencies. The time dependent density functional theory (TD-DFT) at 6-311++G (d, p) has been used to determine HOMO-LUMO energies, oscillator strengths, absorption wavelengths λ_{max} of the electronic transitions and compare with the experimental UV absorption spectra. ^{1}H NMR and ^{13}C NMR chemical shifts were calculated with gauge-independent atomic orbital (GIAO) approach by applying the DFT method with B3LYP/6-311++G (d, p) functional. The GIAO approach allows the computation of absolute chemical shielding due to the electronic environment of the individual nuclei. In order to obtain the calculated results we have transformed the absolute shielding returned by the program in chemical shifts by subtracting it from the absolute shielding of tetramethylsilane (TMS) and compared with the experimental data.

3. Result and discussion

3.1. Molecular Structure and Crystal Packing Analysis

The title compound ($C_{13}H_8BrNO_3S$) is crystallized in Orthorhombic space group *Pnma* and the molecular structure is shown in Fig. 1(a). The theoretical of bond lengths, bond angles and torsion angles correlate well with the experiment data. But some values slightly larger and smaller, because the deviations of theoretical values are performed in gaseous phase and experimental results are performed on solid phase of molecule. In order to investigate the agreement for bond length, bond angles and torsion angles, the root means square deviation values between experimental and calculated geometry parameter were calculated. The root means square deviation (RMSD) values of geometry parameters were calculated using the following expression.

$$\sqrt{\frac{1}{n}\sum_{i}^{n}(V_i^{cal} - V_n^{exp})^2} \quad (1)$$

Where n is the number of the experimental or calculated data. The RMSD value for bond lengths, bond angles and torsion angles, computed from DFT methods with the experimental value were found to be 0.019, 0.78 and 0.0 respectively. Any variation recorded between the calculated and experimental values is attributed to the intermolecular interactions. All non- hydrogen atoms were refined anisotropically. The hydrogen atoms bound to C atoms were positioned geometrically [C-H = 0.95 Å] with U_{iso} (H) = -1.2 U_{eq} (C).The molecular structure adopts *trans* configuration with respect to the C7=C8 double bond across the ethylenic bridge (O1/C5-C7). The compound is observed to be essentially planar. The torsion angles of the central C4-C5=C6-C7 are 180.0^0 for XRD and DFT. The discrepancies between experimental geometrical parameters and theoretical values are due to the fact that the molecule is calculated as an isolated molecule in the gaseous phase. All the hydrogen bond interactions are shown in Fig. 1 (c). In the crystal packing of (*E*)-3-(4-Bromothiophen-2-yl-1-(3-nitrophenyl) prop-2-en-1-one, the molecules are connected *via* intermolecular C1-H1A···O1^i and C9-H9A···O2^i hydrogen bonds. The interactions form $R_2^2(17)$ ring motifs [18]. These ring motifs further connected into one dimensional ladder along the *a*-axis. In addition, S···O short contacts are also observed in the crystal packing (Fig. 1c) and further link the molecules into infinite one dimensional chain along the *a*-axis where the short contact distance is 3.077 Å.

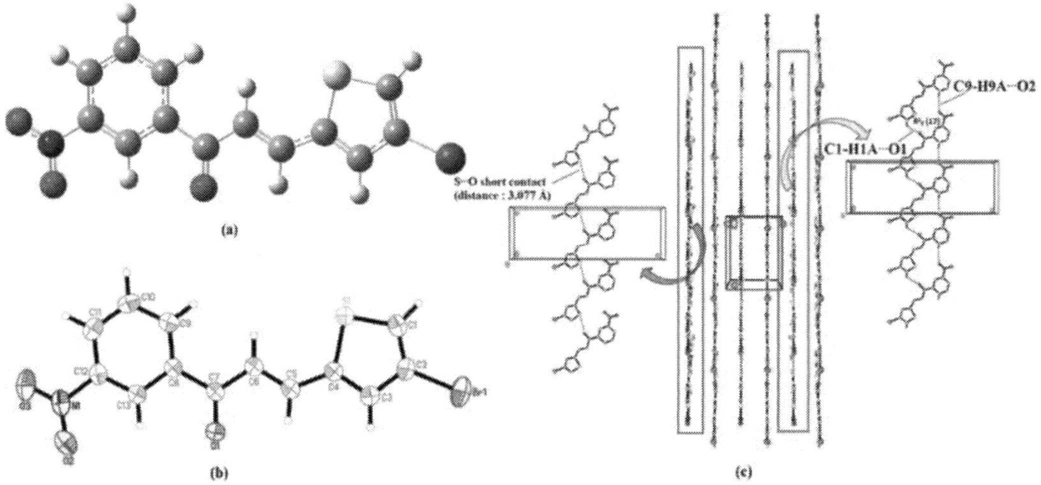

Figure 1. (a) The molecular structure of the compound ; (b) The optimized structure of the compound at DFT/B3LYP 6-311++G(d,p); (c) The crystal packing showing C-H···O and S···O short contacts interactions that link that molecules into an infinite chain along the viewed down the *a*-axis.

3.2. Fourier Transform Infrared Spectroscopy (FTIR) Analysis

The experimental and simulated FTIR spectrum using DFT methods with B3LYP/6-311++G (d, p) functions are shown in Fig. 2. The calculated harmonic wavelength are usually higher than the corresponding experimental quantites because of the combination of electron correlation effects and basis set deficiencies [19]. The summary of all bands was shown in Table 1.

The most characteristic bands in the spectra of nitro compounds are due to the NO_2 stretching vibrations, which are the two most useful group vibrations, not only because of their spectral positions but also for their strong intensities [20]. The NO_2 asymmetrical and symmetrical stretching vibrations are observed at 1580.46 cm^{-1} and 1344.14 cm^{-1}, respectively. Theoretically, this mode appears at 1617.7-1593.9 cm^{-1} for asymmetric stretching and 1376.4 cm^{-1} for symmetric stretching. Romani *et al.* [21] have assigned NO2 assymmetrical stretching range 1540-1584 cm^{-1}, and the symmetrical stretching at 1335 cm^{-1}. The hydrogen bonding has very insignificant effect on asymmetric stretching, it shows appreciable effect on the symmetric NO2 vibration [22], hence the deviation in these vibration wavenumbers from the experimental values is justified. The NO2 scissoring modes are assigned at 915.1-854.6 cm^{-1} theoretically. Karnan *et al.* [23] reported the NO2 scissoring modes designated to the band 835 cm^{-1}.The theoretical results also shows that the rocking modes are assigned at 321.15-313.00 cm^{-1}, wagging modes are assigned to the IR band at 16.7-11.9 cm^{-1}, while the twisting modes for chalcone are assigned at 39.3 cm^{-1}. The deformation vibrations of NO2 group (rocking, wagging, and twisting) contribute to several normal modes in the low frequency region [24, 25].

In this compound, the absorption bands corresponding to the C-H stretching vibrations are observed at 3191.89 cm^{-1}while the calculated DFT ν_{CH} are found at 3256.9-3153.4 cm^{-1}.The C-H stretching mode occurs above 3000 cm^{-1} and is typically exhibited as a multiplicity of weak to moderate bands compared with the aliphatic C-H stretch [26]. From the previous reported C-H stretching frequencies, the wavenumber appear in the range of 3125-3000 cm^{-1}[27].The band in the region 1382.9-1066.8 cm^{-1} and 1014.3-749.7 cm^{-1} are assigned to C–H in-plane and C–H out-of-plane bending vibration, respectively. The CH in-plane bending vibrations usually occur in the region of 1300–1000 cm−1 while the CH out-of-plane bending modes generally occur in the region of 1000–710 cm^{-1}[28, 29].In this region the bands are not affected significantly by nature of the substitutions. All the aromatic C-H stretching bands are found to be weak and this is due to the decrease in the inductive effect, which in turn is caused by the increase in the chain length of the substituent [30]. The wavenumber of the C=O stretch due to carbonyl group mainly depends of the bond strength, which in turn depends upon inductive, conjugative, steric effects and lone pair of electron on oxygen [31]. The carbonyl stretching C=O vibration is expected in the region of 1750-1680 cm^{-1}[31]. The observed C=O stretching vibration is assigned at 1660.01cm^{-1}. The position of the C=O vibration is very sensitive to the various factors like physical state, electronic effects by substituents and ring strains. This value indicates the formation of conjugated enone of chalcone. The calculations give values at 1724.1-1637.1 cm^{-1}. The assignments are also in good agreement with the literature value, 1659 cm^{-1}[5].

In this compound, the C=C stretching vibration is observed at 1603 cm^{-1} while calculated values are assigned at 1637.1-1617.7 cm^{-1}. The C=C stretching vibration shows good agreement between calculated and experimental data. The C=C stretching mode is expected in the region of 1667-1640 cm^{-1}[32]. According to Socrates [32] the C=C stretching mode is expected around 1600 cm^{-1} when conjugated with C=O group. Generally, characteristic C-Br stretching vibration is observed in the region 585.70-508.20 cm^{-1}. Theoretically, C-Br is calculated at 558.3-507.6 cm^{-1}. Krishnakumar *et al.* [33] have assigned the C-Br vibration in the frequency range 1300-550 cm^{-1}. The C-S stretching appears in the range of 809.69-703.86 cm^{-1} (Experimental) and 818.36 cm^{-1} (Theoretical). Kumar *et al.* reported C-S stretching vibrations observed at 722-710 cm^{-1} and calculated at 727-712 cm^{-1} [10].

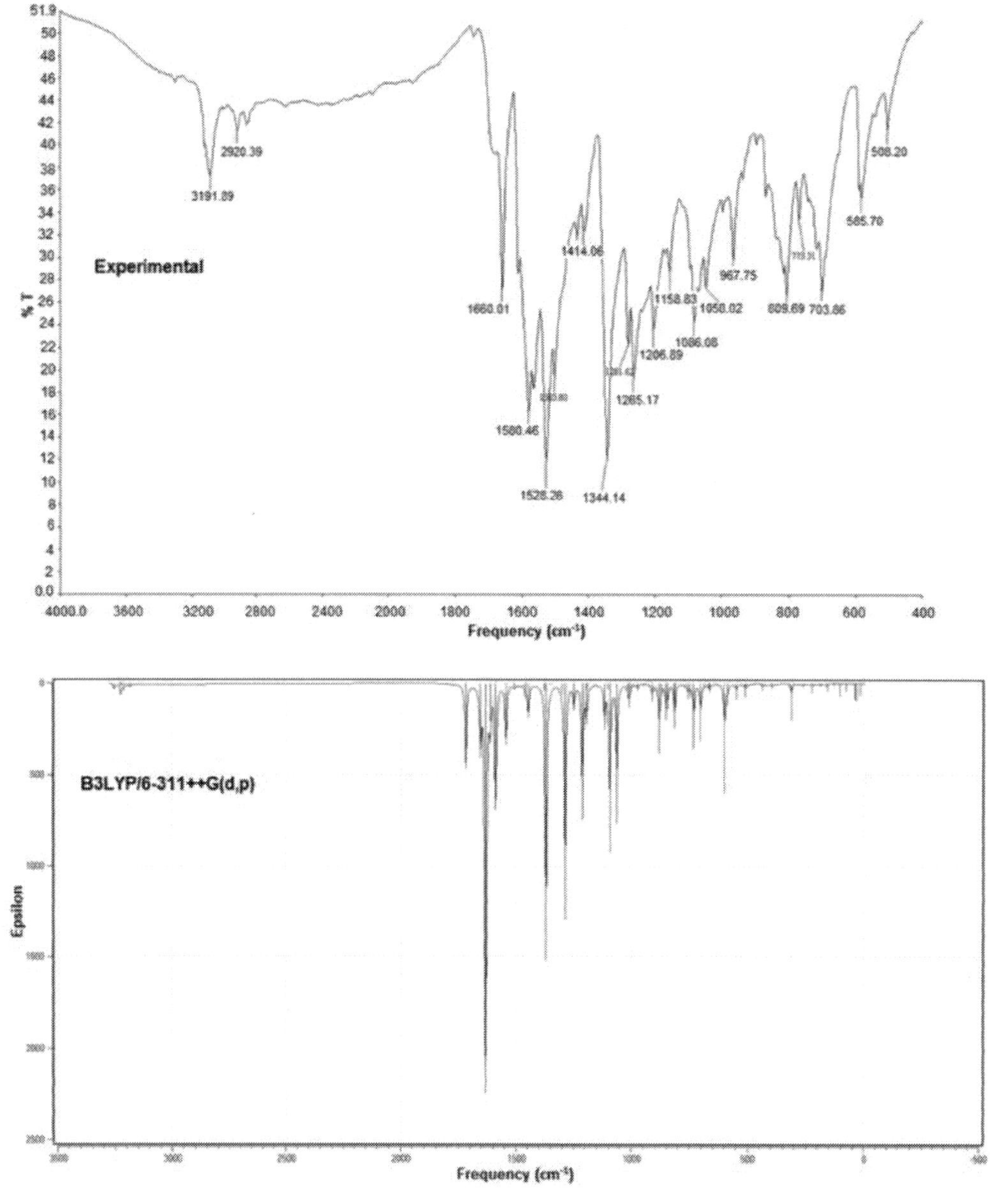

Figure 2. The experimental and simulated FTIR spectrum using DFT methods with B3LYP/6-311++G (d, p) functions.

Table 1. Summary of vibrational bands of FTIR

	Experimental (cm^{-1})	Theoretical (cm^{-1})	Literature (cm^{-1})
vC=O	1660.01 cm^{-1}	11724.1-1637.1 cm^{-1}	1750-1680 cm^{-1} (Panicker *et al.*,2015) 1659 cm^{-1} (Thanigaimani *et al.*,2015)
vC=C	1603.0 cm^{-1}	1637.1-1617.7 cm^{-1}	1667-1640 cm^{-1} (Roeges *et al.*,1994) 1600 cm^{-1} (Socrates *et al.*,1981)
vC-Br	585.70-508.20 cm^{-1}	558.3-507.6 cm^{-1}	1300-550 cm^{-1} (Krishnakumar *et al.* , 2005)
vC-S	809.69-703.86 cm^{-1}	818.36 cm^{-1}	722-710 cm^{-1} (Exp.) and 818.36 cm^{-1} (Calc.) (Kumar *et al.* , 2014)

3.3. UV spectral and FMO analysis

The calculated maximum absorption wavelengths (λcalc), oscillator strengths (*f*) and excitation energies were calculated employing Time Dependent method using B3LYP for which the optimized geometry obtained from DFT calculation were used. Then, the calculated values were compared with the experimental values. The calculations of the molecular orbital geometry show that the absorption maxima of this molecule correspond to the electron transition between frontier orbitals such as transition from HOMO to LUMO. The absorption maxima value has been found to be 344 nm (experimental) and 366 nm (theoretical). According to the investigation on the FMO energy levels of the title compound, the corresponding electronic transfer energy was found to be 3.3886 eV calculated at TD-DFT/B3LYP, which is related to the switch from ground state to second state. Based on the spectral data recorded showed the strong cut off at 388 nm. Through an extrapolation of the linear trend observed in the optical spectra, the estimated energy band gap show 3.26 eV. Furthermore, the energy gap between the HOMO and LUMO orbital is a critical parameter in determining molecular electrical transport properties because it is a measure of electron conductivity [34] and in the present case the energy gap of the molecule is 3.68 eV. The calculated energy transition is shifted from the experiment. This may be probably due to the fact that the calculations are confined to gaseous equivalent whereas the observations are from solution state.

3.4. Nuclear Magnetic Resonance Analysis

In this compound, the H-α (H-6) and H-β (H-5) protons of chalcone are resonated as two doublets in the ^{1}HNMR spectra at 7.878 ppm, 7.862 ppm and 7.961 ppm, 7.944 ppm, respectively. The coupling constants (*J*) of H_α and H_β are 8.2 and 8.3 Hz, respectively which is correspond to the *trans* double bond configurations. The theoretical chemical shift values are comparable to the experimental values of 7.34 (H6) and 8.394 (H5). The strong coupling constant mainly represent the *trans* coupling of the adjacent protons and further confirms the trans configuration of the molecules with respect to C5=C6 double bond [35]. In addition, the protons of aromatic ring resonate experimentally and theoretically in the range of 8.803-7.898 ppm and 7.5338-9.6433 ppm, respectively. While, thiophene ring resonate in the range of 7.809 -7.777 ppm and 7.2242-7.2744 ppm. It is found that aromatic ring appear at a much lower field. This is because of the strong electron with-drawing ability of the nitro group.

In ^{13}C NMR spectra, the experimental and theoretical carbonyl carbon of the chalcones appears at the most deshielded δ values area of 186.93 ppm and 187.107 ppm, respectively. The carbon of this functional group is found at the most shifted area compare to the environment factor and increasing of electronegativity from oxygen atom. Meanwhile α- and β- carbon atoms (C6 and C5) show characteristic signal at δC6/ δC5= 127.32 ppm/ 133.85 ppm (exp.) and 118.46 ppm/ 144.88 (DFT). Additionally, all other chemical shift of the carbon atoms for aromatic ring are found to be normal ranges.

4. Conclusion

The chalcone derivative, (*E*)-3-(4-Bromothiophen-2-yl-1-(3-nitrophenyl) prop-2-en-1-one has been succesfully synthesized and the single crystal has been grown by solution growth technique at room temperature. The functional groups present in this compound are identified using spectroscopic FTIR and NMR studies. The crystal structure, space group and the crystal parameter are found out using single crystal XRD. The UV-visible characteristic shows that the crystal is transparent in the entire visible range. The absorption at 344 nm (exp.) and 366 nm (DFT) were predicted as n-π* transition. The calculated HOMO, LUMO energy and frontier orbital energy gap values along with study for better understanding of charge transfer occur within the molecule.

Acknowledgements

The authors thank the Malaysian Government and Universiti Sains Malaysia (USM) for Short Term Grant Scheme (304/PFIZIK/6313336) to conduct this work. DAZ thanks to Malaysian Government for the My Brain15 scholarship.

Supplementary material

This data CCDC: 1564380 can be obtained free of charge at www.ccdc.cam.ac.uk.conts/retrieving.html/ or from the Cambridge Crystallographic Data Centre (CCDC), 12 Union Road, Cambridge CB2 IEZ, UK; fax: +44(0) 1223-336033; e-mail: deposit@ccdc.cam.ac.uk.

References

[1] Espinoza-Hicks J C, Rodriguez-Valdez L M, Nevarez-Moorillon G V and Camacho-Davila A 2012 *J. Mol. Struct.* 1030 88
[2] Lin Y M, Zhou Y, Flavin M T, Zhou L M, Nie W and Chen F C 2002 *Biorganic & Medicinal Chemistry* 10 2795
[3] Tavares L D C, Johann S, De Almeida Alves T M, Guerra J C, De Souza Fagundes E M, Cisalpino P S, Bortoluzzi A J, Caramori G F, De MattosPiccoli R, Braibante H T S, Braibante M E F and Pizzolatti M G 2011 *European Journal of Medicinal Chemistry* 46 4448
[4] Goswami P, Talukdar M, Bora T C, Phukan P and Sarma J C 2013 *CompresRendusChimie* 16 442
[5] Thanigaimani K, Arshad S, Khalib N C, Razak I A, Arunagiri C, Subashini A, Sulaiman S F, Hashim N S and Ooi K L 2015 *Spectrochim. Acta A* 149 90
[6] Bangal P R, Lahiri S, Kar S, Chakravorti S and Lumin J 1996 69 49
[7] Tarek F A 2006 *chem. Phys.* 324 631
[8] Demir S, Tinmaz F, Dege N and Ilhan I O 2016 *J. Mol. Struct.* 1108 637
[9] Alen S, Sajan D, Joseph L, Chaltanya K, Shettigar V and Jothy V B 2015 *Chemical Physics Letters* 636 208
[10] Kumar A, Deval V, Tandon P, Gupta A and D'Silva E D 2014 *Spectrochim. Acta Part A Mol. Biomol. Spectrosc.* 130 41
[11] Joseph L, Arunsasi B S, Sarjan D and Shettigar V 2014 *J. Mol. Struct.* 1076 687
[12] Bruker SADABS, APEX2 and SAINT 2009 *Bruker AXS Inc Madison WI,* USA
[13] Sheldrick G M 2008 *ActaCrystallogr* A64 112
[14] Spek A L 2009 *ActaCrystallogr* D65 148
[15] Frisch M J, Trucks G W, Schlegel H B, Scuseria G E, Robb M A, Cheeseman J R, Scalmani G, Barone V, Mennucci B, Petersson G A, Nakatsuji H, Caricato M, Li X, Hratchian H P, Izmaylov A F, Bloino J, Zheng G, Sonnenberg J L, Hada M, Ehara M, Toyota K, Fukuda R, Hasegawa J, Ishida M, Nakajima T, Honda Y, Kitao O, Nakai H, Vreven T, Montgomery G A, Peralta J J

E, Ogliaro F, Bearpark M, Heyd J J, Brothers E, Kudin K N, Staroverov V N, Kobayashi R, Normand J, Raghavachari K, Rendell A, Burant J C, Iyengar S S, Tomasi J, Cossi M, Rega N, Millam J M, Klene M, Knox J E, Cross J B, Bakken V, Adamo C, Jaramillo J, Gomperts R, Stratmann R E, Yazyev O, Austin A J, Cammi R, Pomelli C, Ochterski J W, Martin R L, Morokuma K, Zakrzewski V G, Voth G A, Salvador P, Dannenberg J J, Dapprich S, Daniels A D, Farkas O, Foresman J B, Ortiz J V, Ciosiowski J, Fox D J 2009 *Gaussian 09*, Revision A. 1 (Wallingford CT: Gaussian, Inc.)

[16] Arunagiri C, Subashini A, Saranya M, Muthiah P T, Thanigaimani K, Razak A I 2015 *Spectrochim. Acta Part A Mol. Biomol. Spectrosc.* 135 307

[17] Becke A D 1993 *J. Chem. Phys.* 98 5648

[18] Bernstein J, Davis R E, Shimoni L, Chang N L 1995 *Angew.Chem. Int. Ed. Engl.* 34 1555

[19] Karabacak M, Sinha L, Prasad O, Cinar Z, Cinar M 2012 *Spect. Acta Part A: Mol. And Bio. Spect.* 93 33

[20] Roeges N P G 1994 *A Guide to the Complete Intrepretation of IR Spectra of Organic Compounds*, (New York: Wiley)

[21] Romani D, Brandân S A 2015 *Predict. Their React. Comput. Theor. Chem.* 1061 89

[22] Xiao-Hong L, Xian-Zhou Z 2011*Comput. Theor. Chem.* 963 34

[23] Karnan M, Balachandran V, Murugan M 2013 *J. Mol. Struct.* 1039 197

[24] Kovacs A, Keresztury G, Izvekov V 2000 *Chem. Phys.* 253 193

[25] Sundaraganesan N, Ayyapan S, Umamaheswari H, DomnicJoshua B 2007 *Spectrochim. Acta A* 66 17

[26] Coates J 2000 *Interpretation of Infrared Spect. Practical approach* in: R. A. Meyers (Ed.) (Chichester: John Wiley and Sons Inc)

[27] Bayari S, Saglam S, Ustundag H F 2005 *J. Mol. Struct. (Theochem).* 726 225

[28] Mohan J 2001 *Organic Spectroscopy – Principle and Applications* second ed. (New Delhi: Narosa Publishing House)

[29] Varsanyi G 1969 *Vibrational Spectra of Benzene Derivatives* Academic Press New York

[30] Sundaragnesan N, Saleem H, Mohan S 2003 *Spectrochim.Acta Part A* 59 2511

[31] Panicker C Y, Varghese H T, Nayak P S, Narayana B, Sarojini B K, Fun H K, War J A, Srivastava S K, Alsenoy C V 2015 *Spectrochim. Acta Part A Mol. Biomol. Spectrosc.* 148 18

[32] Socrates G 1981 *Infrared Characteristic Group Freuencies* (New York: John Wiley and Sons)

[33] Krishnakumar V, Balachandran V 2005 *Spectrochim. Acta Part A Mol. Biomol. Spectrosc.* 61 1001

[34] Mary Y S, Raju K, Bolelli T E, Yildiz I, Nogueria H I S, Granadein C M, VanAlsenoy C 2012 *J.Mol. Struct.*1012 22

[35] Kumar P C R, Ravindrachary V, Janardhana K, Manjunath H R, Karegouda P, Crasta V, Sridhar M A 2011 *J. Mol. Struct.* 1005 1

IOP Conf. Series: Journal of Physics: Conf. Series **1083** (2018) 012047 doi:10.1088/1742-6596/1083/1/012047

Structural Properties and Quantum Chemical Analysis on a New Chalcone Derivative of (*E*)-3-(4-bromophenyl)-1-(4-fluorophenyl)prop-2-en-1-one

M F Zaini[1,a *], **S Arshad**[1,b], **A R Ibrahim**[1,c], **N Che Khalib**[1,d] **and D A Zainuri**[1,e]

[1]X-ray Crystallography Unit, School of Physics, Universiti Sains Malaysia, 11800 USM, Malaysia

[a]epy.pasoh1@gmail.com, [b]suhanaarshad@usm.my, [c]arazaki@usm.my, [d]yanti_198@yahoo.com, [e]dianalwani14@gmail.com

Abstract. The molecular structure of (*E*)-3-(4-bromophenyl)-1-(4-fluorophenyl)prop-2-en-1-one ($C_{15}H_{10}BrFO$) was successfully synthesized and crystallized in the monoclinic system of $P2_1/c$ space group. The unit cell dimensions are: a =27.795 (3) Å, b =3.9690 (4) Å, c =11.3703 (11) Å, α =90°, β =93.290 (2)°, γ =90° and Z =4. The physical properties of this compound were characterized by ^{1}H and ^{13}C Nuclear Magnetic Resonance (NMR) and UV-Vis spectroscopic methods. The molecular structure and crystal packing of the compound was structurally determined *via* single crystal X-ray diffraction method. Quantum chemical investigations have been employed to investigate the structural and spectral properties where the molecular geometry parameters obtained were further optimized by density functional theory (DFT) using B3LYP/6-311G++(d, p) basis set in the ground state. The optimized structure was used to calculate the geometrical parameters and chemical shifts of the title compound. TD-DFT/B3LYP method has been carried out to obtain electronic properties and HOMO-LUMO energy gap. There was a slight discrepancy between the experimental and theoretical results obtained due to the different environmental state in which the compounds were analyzed.

1. Introduction

Organic chalcone is the most attractive research studies and yet promising a significant role in multiple fields such as electrochemical sensing [1], non-linear optical [2] and luminescent properties [3]. Chemically, chalcones have a structure formed by two aromatic rings with a various array of functional groups linked by α, β-unsaturated carbonyl group. On both of benzene rings, chalcone possess a system of conjugated double bonds and completely delocalized electrons. Modification of C − H or C − OH to C − F were widely acknowledged since such substitutions showed alterations in physicochemical properties and biological activities of organic compounds, without introducing much major steric changes [4]. Due to this, fluorine substitution has become centre of attraction among researchers to develop wide range of various active materials. In addition, K. Thanigaimani et al. [5] reported bromo substitution at the forth-position on the acetophenone ring assists in enchancing the antimicrobial and antioxidant activities. This supports bioactive property of the molecule by lowering the HOMO-LUMO band gap.

In this study we designed and synthesized the bromo and fluoro substitutions at the 4-position of both benzene rings. The structure of the compound was characterized by NMR, UV-Vis and confirmed

Content from this work may be used under the terms of the Creative Commons Attribution 3.0 licence. Any further distribution of this work must maintain attribution to the author(s) and the title of the work, journal citation and DOI.
Published under licence by IOP Publishing Ltd

by single-crystal X-ray diffraction studies. The theoretical part also was employed to predict the structural and spectroscopic parameters of the compound in the gas phase with the aid of DFT/B3LYP methods in 6-311++G(d,p) basis set.

2. Experimental

2.1 Synthesis

The title compound of (*E*)-3-(4-bromophenyl)-1-(4-fluorophenyl)prop-2-en-1-one was successfully synthesized using standard Claisen-Schmidt condensation method at room temperature. The commercially available starting materials such as 4-fluoroacetophenone (0.1 mol) and 4-bromobenzaldehyde (0.1 mol) were used to prepare the title compound. The precursors were dissolved in 20 ml of methanol solution and stirred continuously. Catalytic amount of sodium hydroxide (NaOH) was added dropwise to the solution. The reaction was then stirred for about 5 – 6 hours at room temperature. The resultant crude products were filtered, rinsed successfully with distilled water and recrystallized from acetone to get the corresponding chalcone (Scheme 1). The obtained synthesized crystal was then characterized via single crystal XRD, ^{1}H NMR, ^{13}C NMR and UV-Vis spectroscopy. The colourless plate crystals (CCDC No.: 1567481) is formed and obtained.

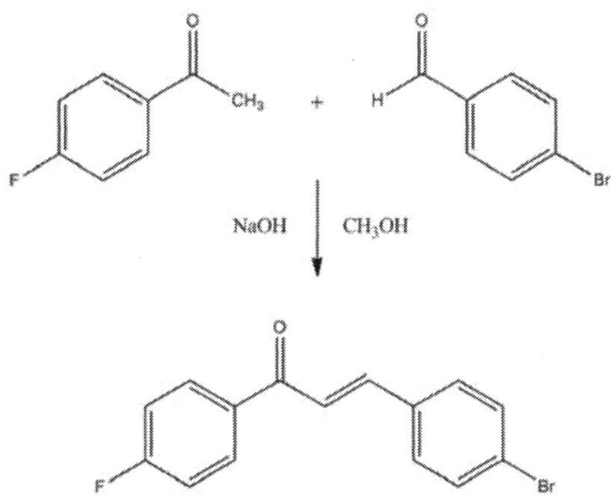

Scheme 1: Synthesis of the title compound.

2.2 X-ray crystallographic analysis

The single crystal obtained was proceeded for X-ray analysis which performed on APEXII Duo CCD area-detector using MoKα radiation (λ = 0.71073 Å). Data collection was performed using the APEX2 software [6], whereas the cell refinement and data reduction were performed using the SAINT software [6]. The crystal structure was solved by direct method using the program SHELXTL [7] and refined by full-matrix least squares technique on F^2. Absorption correction was applied to the final crystal data using the SADABS software [6]. All geometrical calculations were generated using the program PLATON [8]. The molecular graphics were drawn using SHELXTL [7] and Mercury [9] programs. The non-hydrogen atoms were refined anisotropically. In this compound, all the hydrogen atom were positioned geometrically (C – H = 0.93 Å) and refined using a riding model with isotropic displacement parameters. In the final refinement, the most disagreeable reflections was omitted (3 1 1). Details of crystal data and the structure refinement parameters of the title compound are presented in Table 1.

2.3 NMR and UV-Vis analysis

The ^{1}H and ^{13}C NMR spectra were recorded at 500MHz, in DMSO-d_6, on Bruker 500MHz Avanze III spectrometer. The chemical shifts (δ) are reported in parts per million (ppm) in the presence of tetramethylsilane (TMS) as internal reference. In addition, for electronic transition analysis, the compound was characterized via UV-Visible analysis using UV-Visible Spectrophotometer Model Cary 5000 in 1 cm^3 cuvette and the spectrum was recorded in the 800-200 nm region.

Table 1. Crystal data and structure refinement parameters.

Crystal data	Structure refinement parameters
CCDC deposition numbers	1567481
Molecular formula	$C_{15}H_{10}BrFO$
Molecular weight	305.14
Crystal system	Monoclinic
Space group	$P2_1/c$
a (Å)	27.795 (3)
b (Å)	3.9690 (4)
c (Å)	11.3703 (11)
α (°)	90
β (°)	93.290 (2)
γ (°)	90
V (Å)	1252.3 (2)
Z	4
D_{calc} (g cm^{-3})	1.618
Crystal dimensions (mm)	0.41 × 0.29 × 0.11
μ (mm^{-1})	3.28
Radiation, λ (Å)	0.71073
F(0 0 0)	608
T_{min}/T_{max}	0.174/0.282
Reflection measured	5690
Ranges/indices (h, k, l)	$h = -33 \rightarrow 33$ $k = -4 \rightarrow 4$ $l = -13 \rightarrow 13$
θ limit (°)	1.5–25.0
Unique reflections	2209
Observed reflections ($I > 2\sigma(I)$)	1820
Parameters	163
Goodness of fit (GOF) on F^2	1.05
R_1, wR_2 [$I > 2\sigma(I)$]	0.055, 0.157
R_{int}	0.043
Largest diff. Peak and hole (e/Å^{-3})	1.26 and −0.30

2.4 Computational details

In computational procedures, the molecular geometries of the compound was obtained from X-ray refinement data. The optimization of the molecule geometries was carried out with GAUSSIAN 09 program package [10], using the three-parameter hybrid functional of Becke based on the correlation functional of Lee, Yang and Parr (B3LYP) with the 6-311++G(d,p) as a basis set. The optimized structural parameters were used to calculate the geometrical parameters and isotropic chemical shifts.

3. Results and discussion

3.1 Molecular and optimized structure analysis

The molecular and optimized structure of the title compound [molecular formula: $C_{15}H_{10}BrFO$] is shown in Figure 1 and Figure 2 respectively. This compound crystallizes in the monoclinic crystal system of $P2_1/c$ space group [a = 27.795 (3) Å, b = 3.9690 (4) Å, c = 11.3703 (11) Å, α = 90°, β =

Table 2. Experimental and geometrical parameters of the title compound.

Bond lengths	Value (Å)		Bond angles	Value (°)		Torsion angles	Value (°)	
	Expt.	B3LYP		Expt.	B3LYP		Expt.	B3LYP
Br1–C13	1.905 (4)	1.91309	C2–C1–C6	120.5 (5)	120.97748	C6–C1–C2–C3	−1.1 (9)	−0.36348
F1–C3	1.36 (6)	1.35179	C2–C1–H1A	119.8	118.18209	C1–C2–C3–F1	−178.5 (5)	−179.68584
O1–C7	1.216 (6)	1.22418	C6–C1–H1A	119.8	120.83127	C1–C2–C3–C4	1.7 (10)	0.41562
C1–C2	1.372 (7)	1.39222	C3–C2–C1	118.3 (5)	118.4303	F1–C3–C4–C5	178.7 (5)	−179.88944
C1–C6	1.388 (6)	1.40187	C3–C2–H2A	120.8	119.83852	C2–C3–C4–C5	−1.5 (10)	0.00909
C1–H1A	0.93	1.08202	C1–C2–H2A	120.8	121.73107	C3–C4–C5–C6	0.6 (9)	−0.4958
C2–C3	1.366 (8)	1.38573	F1–C3–C4	118.4 (5)	118.79404	C4–C5–C6–C1	−0.1 (9)	0.54113
C2–H2A	0.93	1.08266	F1–C3–C2	118.5 (5)	118.78189	C4–C5–C6–C7	179 (6)	179.5736
C3–C4	1.36 (8)	1.38872	C4–C3–C2	123.2 (5)	122.42399	C2–C1–C6–C5	0.3 (8)	−0.10119
C4–C5	1.368 (8)	1.38751	C3–C4–C5	117.9 (5)	118.37526	C2–C1–C6–C7	−178.7 (5)	−179.0688
C4–H4A	0.93	1.08286	C3–C4–H4A	121	119.7761	C5–C6–C7–O1	−8.3 (9)	−10.25148
C5–C6	1.382 (7)	1.40366	C5–C4–H4A	121	121.84857	C6–C7–C8–C9	−170.8 (5)	177.27433
C5–H5A	0.93	1.08267	C4–C5–C6	121.4 (5)	121.14469	C7–C8–C9–C10	179.7 (5)	179.17221
C6–C7	1.485 (6)	1.50118	C4–C5–H5A	119.3	120.63696	C8–C9–C10–C15	12.9 (8)	−0.81565
C7–C8	1.473 (7)	1.48461	C6–C5–H5A	119.3	118.2183	C8–C9–C10–C11	−164.4 (5)	179.17148
C8–C9	1.318 (7)	1.34539	C5–C6–C1	118.7 (4)	118.64507	C15–C10–C11–C12	−0.1 (8)	0.12451
C8–H8A	0.93	1.08203	C5–C6–C7	118 (4)	117.54404	C9–C10–C11–C12	177.3 (5)	−179.86336
C9–C10	1.465 (7)	1.46045	C1–C6–C7	123.3 (4)	123.80324	C10–C11–C12–C13	0.4 (8)	−0.03392
C9–H9A	0.93	1.0874	O1–C7–C8	120.7 (5)	121.16729	C11–C12–C13–C14	0.3 (8)	−0.08073
C10–C15	1.385 (7)	1.40654	O1–C7–C6	119.9 (4)	119.77688	C11–C12–C13–Br1	178.3 (4)	179.95656
C10–C11	1.403 (6)	1.40467	C8–C7–C6	119.5 (4)	119.0513	C12–C13–C14–C15	−1.3 (8)	0.09971
C11–C12	1.367 (7)	1.39105	C9–C8–C7	121.5 (5)	120.18841	Br1–C13–C14–C15	−179.3 (4)	−179.93751
C11–H11A	0.93	1.08457	C9–C8–H8A	119.3	120.99652	C1–C6–C7–O1	170.7 (6)	168.72667
C12–C13	1.36 (6)	1.3908	C7–C8–H8A	119.3	118.80605	C5–C6–C7–C8	172.8 (5)	168.98369
C12–H12A	0.93	1.08224	C8–C9–C10	127 (5)	127.98168	C1–C6–C7–C8	−8.1 (8)	−12.03817
C13–C14	1.374 (6)	1.39532	C8–C9–H9A	116.5	116.12501	O1–C7–C8–C9	10.4 (9)	−3.50151
C14–C15	1.367 (7)	1.38759	C10–C9–H9A	116.5	115.89325	C13–C14–C15–C10	1.6 (8)	−0.00434
C14–H14A	0.93	1.08238	C15–C10–C11	117.5 (4)	117.83076	C11–C10–C15–C14	−0.9 (7)	−0.10485
C15–H15A	0.93	1.08339	C15–C10–C9	123.7 (4)	123.55678	C9–C10–C15–C14	−178.2 (5)	179.88238
			C11–C10–C9	118.7 (4)	118.61245			
			C12–C11–C10	121.3 (4)	121.61475			
			C12–C11–H11A	119.4	119.15194			
			C10–C11–H11A	119.4	119.2333			
			C13–C12–C11	118.8 (4)	118.96178			
			C13–C12–H12A	120.6	120.4837			
			C11–C12–H12A	120.6	120.55452			
			C12–C13–C14	122.1 (4)	121.00579			
			C12–C13–Br1	119.6 (3)	119.58431			
			C14–C13–Br1	118.3 (3)	119.40989			
			C15–C14–C13	118.7 (4)	119.31353			
			C15–C14–H14A	120.7	120.46136			
			C13–C14–H14A	120.7	120.2251			
			C14–C15–C10	121.6 (4)	121.27323			
			C14–C15–H15A	119.2	118.56328			
			C10–C15–H15A	119.2	120.16341			

93.290 (2)°, $\gamma = 90°$ *and* $Z = 4$]. The molecule exhibits an *E* configuration with respect to the C8=C9 double bond with the C7–C8–C9–C10 torsion angle being 179.7 (5)° and 179.17° for the experimental and calculated (B3LYP) respectively. The presence of the α-β-unsaturated ketone is indicated by the shorter O1–C7 and C8–C9 bond lengths of 1.216 (6) and 1.318 (7) Å, respectively. The calculated O1–C7 and C8–C9 bond lengths are found to be 1.2242 and 1.3454 Å for B3LYP. The bond angles of O1–C7–C8 and C9–C8–C7 (Table 2) within the enone moiety (O1/C7–C9) are 120.7 (5) and 121.5 (5)°, respectively. The corresponding bond angles are 121.17 and 120.19° for B3LYP.

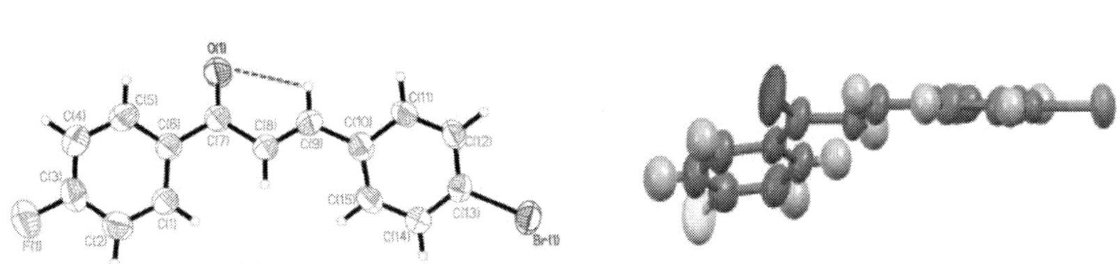

(a) (b)

Figure 1. (a) The molecular structure of the title compound showing the atomic numbering scheme. Displacement ellipsoids are drawn at the 50% probability level. Dashed line represent the intra-molecular hydrogen bond. (b) The view of the twisted fluoro-substituted phenyl ring.

Table 3. Hydrogen bonding geometry for the title compound.

Bond	Bond length (Å)			Angle (°)
D–H···A	D–H	H···A	D···A	D–H···A
C9–H9A···O1	0.93	2.46	2.789(6)	101

Table 4. Halogen ··· Halogen [I ··· J = Br ··· Br] interaction.

X–I···J	Bond length, *X–I*, (Å)	*X–I···J*, (°)
C13–Br(1)···Br(1)[(i)]	3.699	164.53(12)

[(i)]Symmetry code: 1-x, -1/2+y, 3/2-z
Van der Waals radii (Å): Br = 1.85

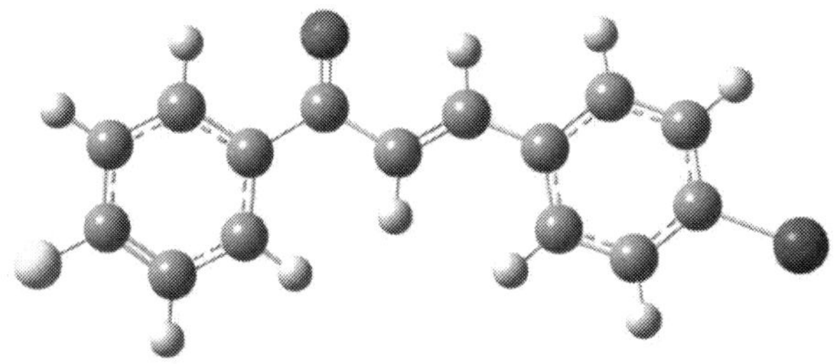

Figure 2. The optimized structure of the title compound.

The molecular conformation is stabilized by a weak intra-molecular C9–H9A···O1 hydrogen-bond interaction (Table 3), shown in Figure 1(a). This interaction forms an S(5) graph set notation [11] which stabilizes the molecular structure. In the structure, the enone moiety (O1/C7–C9) is slightly twisted at the C7–C8 bond where the experimental and calculated torsion angles of O1–C7–C8–C9 are 10.4 (9) and −3.50 (Table 2), respectively. Furthermore, the molecule structure is twisted about C6–C7 bond evidenced by the torsion angle value of −8.1 (8)° for C1–C6–C7–C8. In the crystal packing, the obvious existence of the Van der Waals inter-molecular contact by the short halogen···halogen interaction (Table 4) between bromine atoms further stabilized the crystal structure. The molecular structure is arranged like sandwich along the *c*-axis as shown in Figure 3(b).

In the crystal structure, the molecules are centrosymmetrically arranged, interconnected via short inter-molecular contact Van der Waals interactions between the bromine atom of the phenyl ring and the other bromine atom C13–Br(1)···Br(1). The observed Van der Waals interactions connect the molecules together into infinite supra-molecular one-dimensional chain arrangement along the *b*-axis as shown in Figure 3.

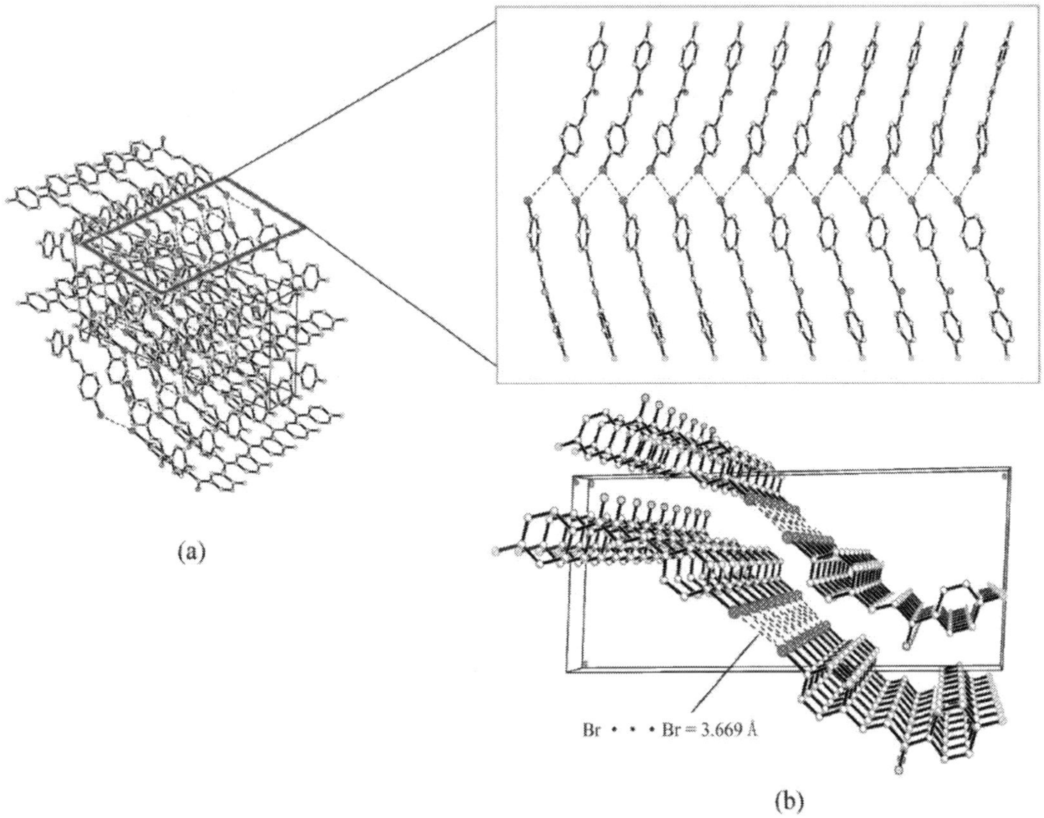

Figure 3. (a) Halogen-halogen interactions (Br ···Br contact) observed in the crystal structure, forming one-dimensional chain along *b*-axis. The hydrogen atoms in the crystal structure have been omitted for clarity. (b) The sandwich arrangement of crystal structure along *c*-axis.

3.2 NMR spectra analysis

GIAO ^{1}H and ^{13}C chemical shift values with respect to tetramethylsilane (TMS) were calculated at DFT/B3LYP/6-311++g(d,p) level of theory and compared to the corresponding experimental values. The ^{1}H NMR and ^{13}C NMR chemical shifts values are presented in Table 5 and Table 6 respectively. For the title compound, ^{1}H NMR chemical shift values were observed in the range 8.27−7.38 ppm, while the DFT values were calculated at 8.65−7.16 ppm. Two doublets observed at regions 7.86 and 7.65 ppm in the ^{1}H NMR spectrum was assigned to vinylene group hydrogen and corresponding values were calculated at 7.91 and 7.52 ppm. Signal observed at 187.56 ppm in the ^{13}C NMR spectrum was assigned to C7 atom of the carbonyl group and the calculated at 187.88 ppm. Signals recorded at 164.08 ppm and 166.09 ppm attribute to carbon atom in the bromine and fluorine atom respectively which are slightly conflict with the calculated at 150.80 ppm and 174.96 ppm respectively. ^{13}C NMR spectrum of the title compound show signals in the range 166.09−115.78 ppm corresponds to benzene rings (C1−C2−C3−C4−C5−C6) and 164.08−130.79 ppm for (C10−C11−C12−C13−C14−C15). The calculated signals for ^{13}C NMR spectrum corresponds to (C1−C2−C3−C4−C5−C6) and (C10−C11−C12−C13−C14−C15) in the range 174.96−118.50 ppm and 150.80−129.86 ppm respectively. Comparing the experimental data with the theoretical, it shows that DFT values are closer to the experimental data.

Table 5. Experimental and theoretical ^{1}H NMR isotropic shifts (ppm) for the title compound.

Atom	Chemical shifts, δ (ppm)	
	Experimental (DMSO)	DFT/B3LYP
H1A	8.00	8.10
H2A	7.39	7.16
H4A	7.41	7.30
H5A	8.27	8.65
H8A	7.73	7.75
H9A	8.25	8.17
H11A	7.67	7.52
H12A	7.65	7.52
H14A	7.70	7.60
H15A	7.97	8.07

Table 6. Experimental and theoretical ^{13}C NMR isotropic shifts (ppm) for the title compound.

Atom	Chemical shifts, δ (ppm)	
	Experimental (DMSO)	DFT/B3LYP
C1	131.52	133.94
C2	115.78	118.50
C3	166.09	174.96
C4	124.02	120.75
C5	131.84	138.23
C6	134.08	139.80
C7	187.56	187.88
C8	122.54	119.64
C9	142.74	148.84
C10	133.89	139.26
C11	134.08	139.74
C12	131.59	137.45
C13	164.08	150.80
C14	131.59	137.16
C15	130.79	129.86

3.3 UV-Vis analysis

The absorption spectrum of the title compound was carried out in acetonitrile by diluting the sample to obtain a solution with concentration of 10^{-5} M. Two major bands were observed as shown in Figure 4. The strongest band observed in the region of 313 nm which attributed to $\pi - \pi^*$ transition from HOMO to LUMO and this was expected to arise from phenyl ring connected to bromine atom [12]. The second band appeared in the region of 226 nm was assigned also as $\pi - \pi^*$ transition, which indicates the presence of phenyl rings. Consequently, the energy band gap (E_g) of the compound was determined from the UV-Vis absorption maximum (λ_{max}) within the range 200 – 500 nm. The optical band gap was further calculated using an equation (1) as shown [13].

$$E_g = \frac{1240}{\lambda_{edge}} \tag{1}$$

where λ_{edge} is the onset value of the absorption spectrum in the direction of longer wavelengths. From the UV-Vis data, the E_g for the title compound is 3.569 eV in which this value is responsible for the charge and electron transition within the molecule.

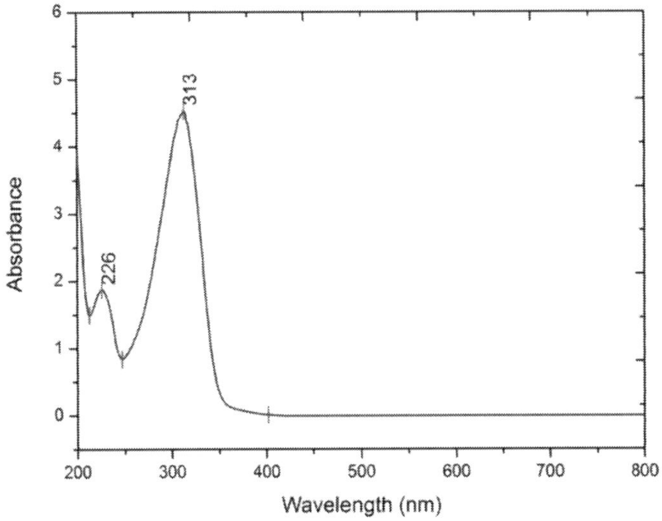

Figure 4. The absorbance spectrum for the title compound.

3.4 Frontier molecular orbitals

The frontier molecular orbitals analysis is significant in the development of the molecular electronic properties and yet it is crucial to determine the excitation energy of an organic compound. This is useful to study the movement of electrons and may has a substantial influence on the ICT. The TD-DFT/B3LYP/6-311++G(d,p) is employed to study the energy levels of the frontier molecular orbitals. The electron density plots of the HOMO and LUMO for the title molecule is illustrated in Figure 5 At the LUMO state, we seen that the molecular orbital covers over the entire molecular skeleton. LUMO is centred on the enone system and small contribution at the substituted phenyl rings from fluorine and bromine atoms. HOMO is more localized at enone system and phenyl ring attached to bromine atom. However, small molecular orbital contribution observed in phenyl ring that attached to fluorine atom. The energy values of the highest occupied molecular orbital (E_{HOMO}) and lowest unoccupied molecular orbital (E_{LUMO}) of the title compound are about -6.7638 eV and -2.7733 eV, respectively. The energy gap is calculated using $E_{LUMO} - E_{HOMO}$ and it is found to be -3.9905 eV. Indeed, the energy gap obtained from TD-DFT is 4.0 eV and thus slightly deviates with the maximum absorption wavelength obtained from experimental UV-Vis spectrum.

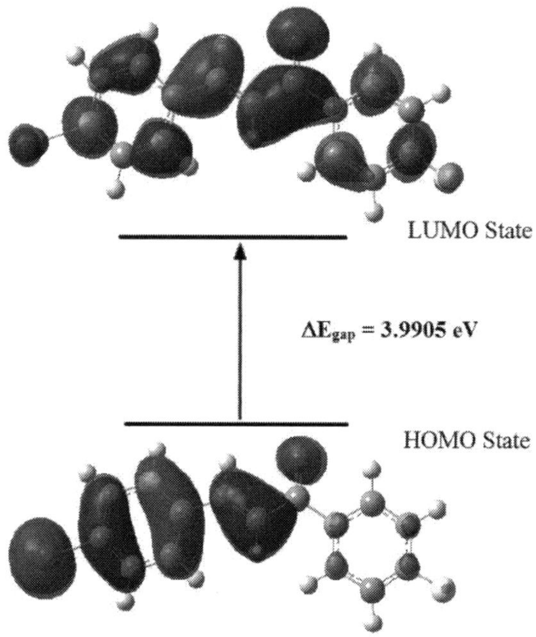

Figure 5. The molecular orbitals and energies for the HOMO and LUMO of the title compound.

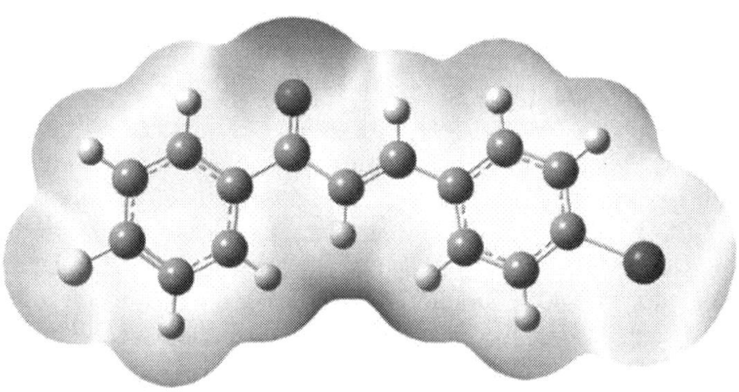

Figure 6. The molecular electrostatic potential surface of the title compound generated using Gaussian software.

3.5 Molecular electrostatic potential

MEP surfaces are very useful quantum molecular descriptor to visualize the charge distributions of the molecules particularly in locating their electron density. The spherical red region observed on the oxygen atom in the carbonyl group as shown in Figure 6 indicates the lowest electrostatic potential energy where the electron density is higher than the other atoms like fluorine and bromine. In addition, highest electrostatic potential energy is clearly can be seen by the blue region that appeared surrounds the hydrogen atoms where electron density is highest.

4. Conclusion

A novel chalcone derivative, *(E)-3-(4-bromophenyl)-1-(4-fluorophenyl)prop-2-en-1-one*, was synthesized and characterized by X-ray diffraction, ^{1}H and ^{13}C NMR spectra and UV-Vis spectrum. The compound crystallizes in monoclinic crystal system of $P2_1/c$ space group, unit cell parameters a = 27.795 (3) Å, b = 3.9690 (4) Å, c = 11.3703 (11) Å, α = 90°, β = 93.290 (2)°, γ = 90° and Z = 4. The

Density Functional Theory (DFT) calculations were performed to predict the structural geometry, chemical shifts and molecular orbitals HOMO-LUMO of the title compound. The experimental geometric parameters of the compound are comparable to the calculated values. The molecular structure was stabilized by the presence of intra-molecular and short contact interactions. The ^{1}H and ^{13}C NMR chemical shifts calculations showed an excellent agreement with the experimental data. Overall, the agreement between theoretical and experimental values is agreeable for every structural and spectroscopic calculations.

References

[1] Sarojini B, Narayana B, Ashalatha B, Indira J and Lobo K 2006 *J. Crsyt. Growth* 295 54-59
[2] Poornesh P, Shettigar S, Umesh G, Manjunatha K, Kamath K P, Sarojini B and Narayana B 2009 *Opt. Mater.* 31 854–59
[3] Komarova K G, Sakipov S N, Plotnikov V G and Alfimov M V 2015 *J. Luminescence* 164 57–63
[4] O'Hagan D 2008 *Chem. Soc. Rev.* 37 308–19
[5] Thanigaimani K, Arshad S, Che Khalib N, Abdul Razak I, Arunagiri C, Subashini A, Sulaiman S F, Hashim N S and Leong Ooi K 2015 *Spectrochimica Acta Part A: Molecular and Biomolecular Spectroscopy* 149 90–102
[6] Bruker SADABS, APEX2 and SAINT, Bruker AXS Inc Madison WI USA 2009
[7] Sheldrick G M 2008 *Acta Cryst.* 64A 112–22
[8] Spek A L 2009 *Acta Cryst.* 65D 148–55
[9] Macrae C F, Edgington P R, McCabe P, Pidcock E, Shields G P, Taylor R, Towler M and Van de Streek J 2006 *J. Appl. Cryst.* 39 453–57
[10] Frisch M J, Trucks G W, Schlegel H B, Scuseria G E, Robb M A, Cheeseman J R, Scalmani G, Barone V, Mennucci B, Petersson G A, Nakatsuji H, Caricato M, Li X, Hratchian H P, Izmaylov A F, Bloino J, Zheng G, Sonnenberg J L, Hada M, Ehara M, Toyota K, Fukuda R, Hasegawa J, Ishida M, Nakajima T, Honda Y, Kitao O, Nakai H, Vreven T, Montgomery Jr. J A, Peralta J E, Ogliaro F, Bearpark M, Heyd J J, Brothers E, Kudin K N, Staroverov V N, Kobayashi R, Normand J, Raghavachari K, Rendell A, Burant J C, Iyengar S S, Tomasi J, Cossi M, Rega N, Millam J M, Klene M, Knox J E, Cross J B, Bakken V, Adamo C, Jaramillo J, Gomperts R, Stratmann R E, Yazyev O, Austin A J, Cammi R, Pomelli C, Ochterski J W, Martin R L, Morokuma K, Zakrzewski V G, Voth G A, Salvador P, Dannenberg J J, Dapprich S, Daniels A D, Farkas O, Foresman J B, Ortiz J V, Cioslowski J and Fox D J 2009 Gaussian 09, Revision A.1, Gaussian Inc, Wallingford CT
[11] Bernstein J, Davis R E, Shimoni L, Chang N L and Angew 1995 *Chem. Int. Ed. Engl.* 34 pp 1555–73
[12] Xue Y, Mou J, Liu Y, Gong X, Yang Y and An L 2010 *Cent. Eur. J. Chem.* 8 928–36
[13] Yi Z, Qingguo H, Yi Y, Haizheng Z, Chang H, Guangyi S, Wei L, Chunhe Y, Fenglian B and Yongfang L 2008 *Adv. Funct. Mater.* 18 3299–3306

IOP Conf. Series: Journal of Physics: Conf. Series **1083** (2018) 012048 doi:10.1088/1742-6596/1083/1/012048

Hydrogen-bonded Supramolecular Association in Organic Acid-base Pyrimidine Salt and Hirshfeld Surface Analysis

Nuridayanti Che Khalib[1], Kaliyaperumal Thanigaimani[2], Suhana Arshad[1], Dian Alwani Zainuri[1] and Ibrahim Abdul Razak[1]

[1]X-ray Crystallography Unit, School of Physics, Universiti Sains Malaysia, 11800 USM, Penang, Malaysia.
[2]Department of Chemistry, Government Arts College, Trichy-620022, Tamil Nadu, India.

yantie_198@yahoo.com, Thanigaimani81@yahoo.com, suhanaashad@usm.my, dianalwani14@gmail.com, arazaki@usm.my

Abstract. The present study deals with the crystal structure of organic salt, namely, 2-amino-4-methoxy-6-methylpyrimidinium 2-chlorobenzoate, synthesized by slow evaporation technique. The structure of the grown crystal was elucidated by using single crystal X-ray diffraction technique. The colourless crystal structure of salt belongs to the monoclinic crystallographic system with space group $P2_1/c$, $Z = 4$, and $a = 7.3269$ (3) Å, $b = 25.2003$ (10) Å, $c = 7.9743$ (3) Å, $\beta = 112.795$ (2) °. The pyrimidine moieties are protonated at one of the nitrogen atoms of the pyrimidine rings. The carboxylate group of the anions (hydrogen 2-chlorobenzoate) interacts with the protonated pyrimidine moiety through a pair of N−H···O hydrogen bonds resulting the complementary DDAA (D = donor and A = acceptor in hydrogen bonds) arrays of quadruple hydrogen-bonding patterns. The -COO- acceptor consistently seeks out the -N-H+ donor generating the most important intermolecular interaction by comparison with similar structures in the Cambridge Structural Database (CSD). The combination of the strong hydrogen bonding (N–H···O and C–H···O) and weak hydrogen bonding (π–π stacking) associations in the crystal packing led to the formation of the 3-Dimensional network structures. A detailed analysis of Hirshfeld surfaces and 2-D fingerprint plots facilitate a comparison of intermolecular interactions in the supramolecular architecture. An evaluation of the Hirshfeld surfaces confirm the importance of intermolecular interactions involving oxygen atoms as well as the π—π stacking interactions.

1. Introduction

Supramolecular interactions have gained a lot of attentions for several past years since it involved the usage of intermolecular non-covalent interactions include classical/ non-classical hydrogen bond, stacking, electrostatic, hydrophobic and charge-transfer interactions [1–4]. Hydrogen bonding interaction plays a vital role in molecular recognition for design a new materials with highly specific features. With hydrogen bond interactions, cocrystals and organic salts can be generated [4]. In pharmaceuticals field, salt formation generally used to modify the properties of the compound such as solubility, stability, and hydgroscopicitybof the drug product [4, 5]. In this regard, the most frequently used moities with hydrogen bonding capability is carboxylic acid with N-containing compound, such as pyridine and pyrimidine [6, 7]. Pyrimidine and aminopyrimidine derivatives are biologically very

Content from this work may be used under the terms of the Creative Commons Attribution 3.0 licence. Any further distribution of this work must maintain attribution to the author(s) and the title of the work, journal citation and DOI.
Published under licence by IOP Publishing Ltd

important compounds since they occur in nature as components of nucleic acids, such as cytosine, uracil and thymine. Pyrimidine derivatives are very important molecules in biology and have many applications in the areas of pesticide and pharmaceutical agents [8] and also developed as antiviral agent (AZT), which is the most widely used as anti-AIDS drug [9]. The synthesis, crystal structure with hydrogen bonding and Hirshfeld surface analysis of the aminopyrimidine-carboxylate salt have been undertaken and are presented herein.

2. Synthesis

Both of the starting materials 2-amino-4-methoxy-6-methylpyrimidine and 2-chlorobenzoic acid were obtained from Sigma–Aldrich chemical suppliers. Analytical grade solvent was used for preparation of the salt. The 1:1 mixture of 2-amino-4-methoxy-6-methylpyrimidine (70 mg, 1 mmol) and 2-chlorobenzoic acid (78 mg, 1 mmol) was dissolved in 20 mL of hot methanol by heating magnetic stirrer hotplate for a few minutes. The resulting solution was allowed to cool slowly at room temperature and crystals of the title salt (**I**) appeared after a few days (Scheme 1).

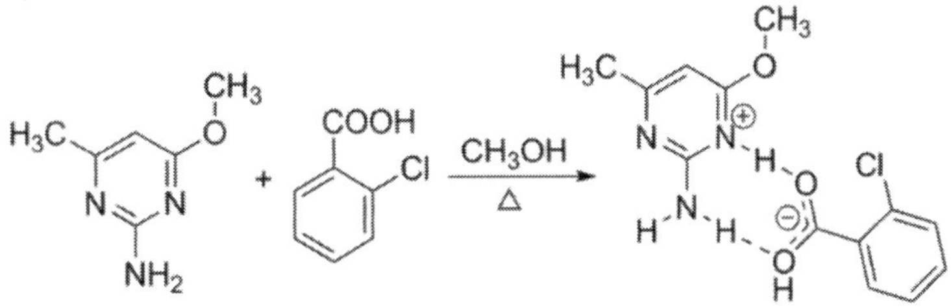

Scheme 1: Reaction scheme of **I**.

3. Experimental detailed

3.1. X-ray Crystallography and Data Collection

Single crystal suitable for X-ray analysis was performed on Bruker SMART APEX II CCD diffractometer using MoKα radiation (λ = 0.71073 Å) with φ and ω scans. The SAINT [10] software was used to intergrate the raw files to obtain useful data crystal, which was further solved and refined by the SHELXTL [11] software. Lastly, the data was enhanced by applying the absorption correction process through SADABS [10].The molecular graphics were drawn using SHELXTL [11] program. Anisotropic thermal factors were assigned to all non-hydrogen atoms. The N-bound hydrogen atoms were located in a difference Fourier map and refined freely [refined N—H distances 0.95 (2), 0.92 (2) and 0.86 (2) Å]. The remaining hydrogen atoms were positioned geometrically (C—H = 0.95—0.98 Å) and were refined using a riding model, with U_{iso} (H) = 1.2U_{eq}(C) or 1.5 U_{eq} (methyl C). A rotating group model was applied to the methyl groups. Crystallographic data for **I** is summarized in Table 1 and has been deposited with the Cambridge Crystallographic Data Centre as supplementary publications, namely CCDC 1024221. Copies of available material can be obtained free of charge, on application to CCDC, 12 Union Road, Cambridge CB2 1EZ, UK, (Fax: +44-(0) 1223-336033 or e-mail: deposit@ccdc.cam.ac.uk). Selected bond lengths and bond angles for salt I is listed in Table 2, the relevant hydrogen bond parameters are provided in Table 3.

3.2. Hirshfeld Surface Analysis

The Hirshfeld surface analysis [12, 13] of salt **I** was generated by CrystalExplorer 3.1 [14], which can be summarized with fingerprint plots mapped over d_{norm}. The contact distances to the closest atom inside (d_i) and outside (d_e) of the Hirshfeld surface analyzed the intermolecular interaction through the mapping of d_{norm}.

Table 1: Crystal Structure Parameters of salt **I.**

Compound	Salt **I**
CCDC	1024221
Empirical Formula	$(C_6H_{10}N_3O)^+ \cdot (C_7H_4ClO_2)^-$
Formula Mass	295.72
Crystal System and Space group	Monoclinic, $P2_1/c$
a (Å) , b (Å), c (Å)	7.3269 (3), 25.2003 (10), 7.9743 (3)
α (°), β (°), γ (°)	90, 112.795 (2), 90
V (Å^3)	1357.38 (9)
Z, Dc (g cm^{-3}), μ (mm^{-1})	4, 1.447, 0.29
F(000), θ *range*(°)	616, 2.90–30.0
Measured reflections	21861
Unique reflections	2650
Observed reflections (I>2σ(I))	2279
No. of parameters	195
$R^{(a)}$, $wR^{(b)}$, $GOF^{(c)}$	0.041, 0.085, 1.05
Residual peaks (e·Å^{-3})	0.24, −0.28
Colour, shape and Crystal size (mm)	Colourless, needle, 0.72 x 0.18 x 0.009

For **I**, $w = 1/[\sigma^2(F_o^2) + (0.0262P)^2 + 1.219P]$ where $P = (F_o^2 + 2F_c^2)/3$. [a] $R = \Sigma||F_o| - |F_c||/\Sigma|F_o|$, [b] $R_w = \{w\Sigma(|F_o| - |F_c|)^2/\Sigma w|F_o|^2\}\}^{1/2}$, [c] $GOF = \{\Sigma w(|F_o| - |F_c|)^2/(n-p)\}^{1/2}$, where n is the number of reflections and p the total number of parameters refined.

4. Results and Discussion

4.1. Molecular Features

Salt (**I**) of the composition 2-amino-4-methoxy-6-methylpyrimidine and 2-chlorobenzoic acid was prepared by reacting equal mole in 1:1 ratio, which crystallizes as monoclinic colourless needle crystals in the centrosymmetric space group $P2_1/c$. The structure of **I** with the atom numbering scheme is shown in Figure 1.

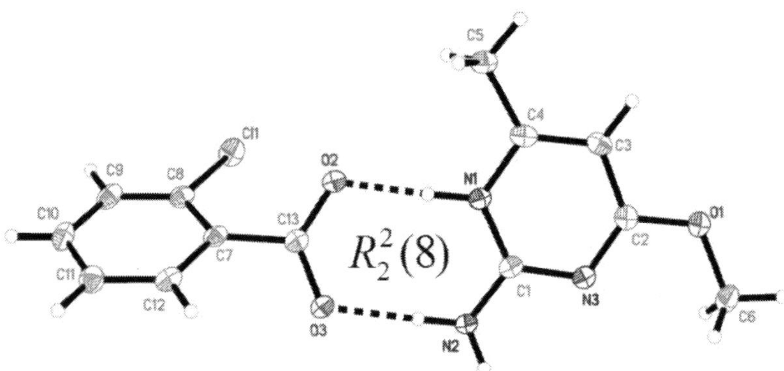

Figure 1: Molecular structure of **I** showing the atomic numbering scheme. Displacement ellipsoids are drown at the 50 % probability level. Dashed lines indicate intermolecular hydrogen bonds.

A search in the Cambridge Structural Database (CSD V5.38), updated October 2017; [15] revealed 9 crystal structures containing both 2-amino-4-methoxy-6-methyl pyrimidine and aromatic/aliphatic carboxylic acid, with such a supramolecular heterosynthon observed in all of them. In the structures with code of EMUMAC [16], EQAVOJ [17], NUQTOJ [18], VAQRUB [19], VAQSAI [19], VAQSEM [19], VAQSIQ [19], VAQSOW [19] and VACSUC [19], the 2-aminopyrimidine is a cation forming salts.

Table 2: Selected parameters (Å, °) of **I** with the similar 2-aminopyrimidine crystal structures.

Compounds	Selected parameters	Bond Length (Å)	Selected parameters	Angles (°)
I	N1—C1	1.363 (2)	C4—N1—C1	120.63 (16)
	C1—N2	1.326 (2)	N1—C1—N2	117.90 (17)
	O2—C13	1.262 (2)	N2—C1—N3	119.61 (16)
	O3—C13	1.254 (2)	O3—C13—O2	125.04 (17)
VAQSAI [19]	N1—C1	1.356	C4—N1—C1	120.06
	C1—N2	1.320	N1—C1—N2	117.53
	C13—O2	1.270	N2—C1—N3	119.68
	C13—O3	1.247	O3—C13—O2	124.59
VAQSIQ [19]	N1—C1	1.347	C4—N1—C1	120.49
	C1—N2	1.315	N1—C1—N2	118.04
	C13—O2	1.258	N2—C1—N3	119.10
	C13—O3	1.245	O3—C13—O2	124.55

Table 3: Hydrogen bonding geometry (Å, °) for **I**.

D—H···A	D—H	H···A	D···A	D—H···A
N1—H1N1···O1	0.95 (2)	1.71 (2)	2.662 (2)	176 (3)
N2—H1N2···O3	0.92 (2)	1.82 (2)	2.741 (2)	175 (2)
N2—H2N2···O3^{i}	0.86(2)	2.04 (2)	2.775 (2)	144 (2)
C9—H9A···O2^{ii}	0.95	2.49	3.423 (2)	168

Symmetry Code: (i) −*x*+1, −*y*+1, −*z*+1; (ii) *x*, −*y*+3/2, *z*+1/2. *Cg*1: C7−C12; *Cg*2: N1/C1/N3/C2−C4

The asymmetric unit organic salt **I** (Figure 1) comprises one 2-chlorobenzoate anion and one 2-amino-4-methoxy-6-methylpyrimidinium cation. The hydroxyl group of the 2-chlorobenzoic acid is deprotonated and proton-transferred to the nitrogen atoms of 2-aminopyrimidine moieties. This is confirmed by the bond length of O2—C13 being 1.262 (2) Å suggesting shortened bond length as compared to the neutral 2-chlorobenzoic acid [20]. In the cation, the protonated N1 atom leads to a slight widening in the C1—N1—C4 angle of the pyrimidine ring [Table 2: 120.63 (16) °], compared to the corresponding angle of 116.01 (18) ° in neutral 2-amino-4-methoxy-6-methylpyrimidine [21].

Table 2 also shows a comparison between the selected experimental structural parameters with the similar 2-aminopyrimidine crystal structures (CSD reference codes: VAQSAI [19] and VAQSIQ [19]). In general, the experimental bond length and bond angle values of the title salt **I** are comparable with the corresponding values obtained by similar crystal structures [19].

In the crystal packing of **I**, the protonated N1 atom and the 2-amino group (N2) are hydrogen-bonded to carboxylate oxygen atoms (O2 and O3) *via* a pair of intermolecular N1—H1N1···O2 and N2—H1N2···O3 hydrogen bonds forming a cyclic hydrogen-bonded motif (supramolecular heterosynthon) designated by the graph-set notation $R_2^2(8)$ [22]. The $R_2^2(8)$ motifs are centrosymmetrically paired *via* N2—H2N2···O3 (−*x*+1, −*y*+1, −*z*+1) (Table 3) hydrogen bonds resulting DDAA array (where D is a hydrogen-bond donor and A is a hydrogen-bond acceptor) of quadruple hydrogen bonds represented the graph-set notations of $R_2^2(8)$, $R_4^2(8)$ and $R_2^2(8)$ (Figure 2a). This type of motif has been reported in the crystal structures of and 2-amino-4-methoxy-6-methylpyrimidin-1-ium trifluoroacetate [17], 2-amino-4-methoxy-6-methylpyrimidinium 2-fluorobenzoate, 2-amino-4-methoxy-6-methylpyrimidinium 3-chlorobenzoate, 2-amino-4-methoxy-6-methylpyrimidinium 3-nitrobenzoate and 2-amino-4-methoxy-6-methylpyrimidinium benzoate [19]. The quadruple hydrogen-bonding motifs are further extended through a C9—H9A···O2 (*x*, −*y*+3/2, *z*+1/2) hydrogen bonds, leading to the formation of hydrogen-bonded supramolecular chain structure along *bc* plane. The crystal structure of **I** is further stabilized by π—π interactions between the pyrimidine (*Cg*2; N1/N3/C1–C4) rings (*Cg*2···*Cg*2 = 3.5392 (12) Å; 1–*x*, 1–*y*, –*z*) (Figure 2b) and resulting 3-Dimensional network.

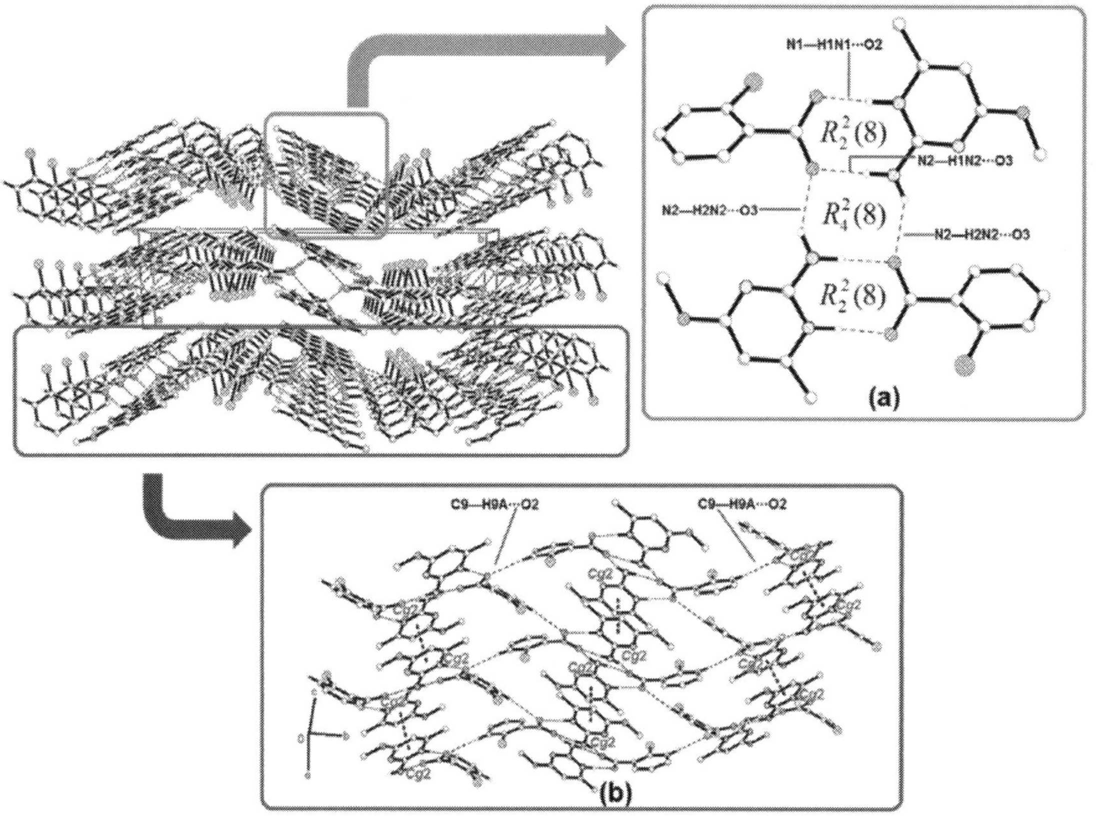

Figure 2: The packing diagram for salt **I** (a) DDAA array of quadruple hydrogen bonds (b) supramolecular chain and π—π interactions.

4.2. Hirshfeld Surface Analysis

The intermolecular interactions of the title salt **I** is using the Hirshfeld surfaces analysis. This analysis shows surfaces mapped over d_{norm}, shape index with the 2-D fingerprint plot showing the contributions of different kind of intermolecular contacts. In **I**, the N2—H2N2···O3 and C9—H9A···O2 interactions are shown on the Hirshfeld surfaces marked with bright red spot for short contacts (Figure 3a). The largest region of H···H interactions appear with high concentration in the middle region of the fingerprint plot, shown in light blue minimum at d_e=d_i ~1.5 Å (Figure 3ii) with the overall Hirshfeld surfaces of 38.3%. The O···H/H···O contacts (Figure 3iii) which comprise 16.3% of the total Hirshfeld surfaces are made up of N2—H2N2···O3 and C9—H9A···O2 interactions. The N2—H2N2···O3 interactions are represented by two symmetrical narrow pointed spikes with d_e+d_i ~ 1.95 Å, while the C9—H9A···O2 interactions are shown by the light blue spikes with d_e+d_i ~ 2.4Å. The C···C contacts assigned to π—π stacking interactions appear as a distinct triangle in the fingerprint plot, seen in figure 3(iv) at around d_e=d_i ~1.8 Å having 1.9% contribution. The presence of the π—π stacking interactions is also indicated by the appearance of red and blue triangles on the shape-indexed surfaces, identified with red circle in Figure 3b.

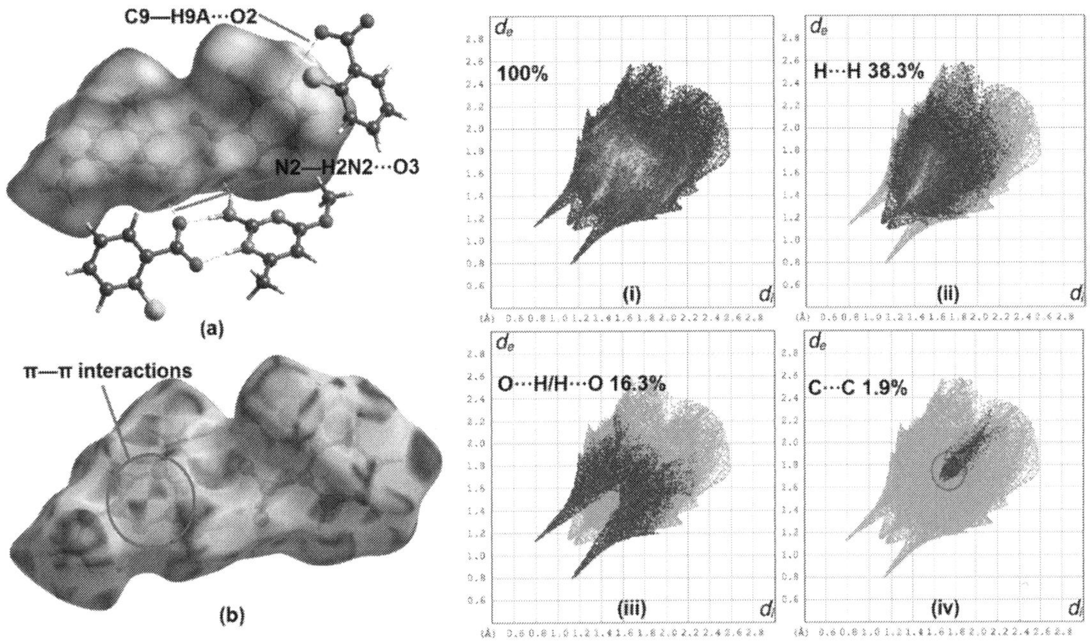

Figure 3: The Hirshfeld surface mapped on (a) dnorm, (b) shape index and (i-iv) fingerprint plot of the title salt **I**.

5. Conclusion

Supramolecular networks of salt **I** was investigated by single-crystal X-ray diffractions analysis. The bond lengths and angles of moieties which involved in proton transfer are differ from their corresponding neutral co-former or related molecule. In **I**, the 2-chlorobenzoate anion was hydrogen-bonded to the protonated pyrimidine ring to form supramolecular heterosynthons via N^{+}—H···O^{-} interactions. Bright red spots observed on the Hirshfeld surfaces visualized the intermolecular interactions of salt **I**. From the fingerprint plots, the close contacts contribution in salt **I** are dominated by H···H and O···H/H···O contacts and also π—π stacking interactions have clear signatures in the fingerprint plots. These interactions play a key role towards the stabilization of salt **I** structure in the solid state.

References

[1] L. Wang, R.A. Friesner, B.J. Berne, J. Phys Chem B. 114 (2010) 7294.
[2] D.Ž. Veljković, G.V. Janjić, S.D. Zarić, CrystEngComm. 13 (2011) 5005.
[3] G.R. Desiraju, Cryst Growth Des. 11 (2011) 896.
[4] J. Shouwen, W. Dang, Z. Hanfei, Z. Yixiang, L. Dong, R. Jiaqi, J. Chem. Crystallogr 44 (2014) 6–14.
[5] P. J. Gould, Int. J. Pharm. 33 (1986) 201.
[6] G.S. Nichol, W. Clegg, Cryst. Growth. Des. 9 (2009) 1844.
[7] Y.B. Men, J.L. Sun, Z.T. Huang, Q.Y. Zheng, CrystEngComm. 11 (2009) 978.
[8] M.E. Condon, T. E. Brady, D. Feist, T. Malefyt, P. Marc, L. S. Quakenbush, S.J. Rodaway, D.L. Shaner, B. Tecle, Brighton Crop Protection Conference on Weeds Alton, Hampshire, England: BCPC Publications (1993) 41–46.
[9] T.L. Gilchrist, Heterocyclic Chemistry, 3rd ed.Singapore: Addison Wesley Longman. (1997) 261–276.
[10] Bruker. APEX2, SAINT and SADABS. Bruker AXS Inc, Madison, (2009)
[11] G.M. Sheldrick, Acta Crystallogr. A.64 (2008) 112.

[12] J.J. McKinnon, D. Jayatilaka, M.A. Spackman, Chem. Commun. (2007) 3814–3816.
[13] M.A. Spackman, D. Jayatilaka, CrystEngComm, 11, (2009) 19–32.
[14] S.K. Wolff, D.J. Grimwood, J.J. Mckinnon, M.J. Turner, D. Jayatilaka, M.A. Spackman, CrystalExplorer. Universiti of Western Australia, Perth (2012).
[15] M. Jeevaraj, S. Palaniyappan, B. Edison, K. Thanigaimani, K. Balasubramani, I. A. Razak, Acta Cryst. E73 (2017) 1305–1307.
[16] R. Sanggetha, B. Edison, K. Thanikasalam, S.J. Kavitha, K. Balasubramani, IUCrData. 1 x160794 (2016).
[17] M. Jeevaraj, B. Edison, S.J. Kavitha, K. Thanikasalam, S. Britto, K. Balasubramani, IUCrData. 1 x161010 (2016).
[18] J.P. Jasinski, R.J. Butcher, H.S. Yathirajan, B. Narayana, K.K. Prakash, Acta Cryst. E66 (2010) o1189–o1190.
[19] C.B. Aakeröy, K. Beffert, J. Desper, E. Elisabeth, Cryst. Growth. Des. 3 (2003) 837.
[20] I.A. Guzei, L.C. Spencer, D. Appavoo, J. Darkwa, CSD Communication (Private Communication) (2010)
[21] C. Glidewell, J.N. Low, M. Melguizo, A. Quesada, Acta Cryst. C59 (2003) o9–o13.
[22] J. Bernstein, R.E. Davis, L. Shimoni, N.L. Chang, Angewandte Chemi International Edition, 34 (1995) 1555–1573.

IOP Conf. Series: Journal of Physics: Conf. Series **1083** (2018) 012049 doi:10.1088/1742-6596/1083/1/012049

Characterization of Acid Treated Activated Carbon From Oil Palm Empty Fruit Bunches (EFB)

Nurul Afifah Mahmud[1], Nafisah Osman[2] and Abdul Mutalib Md Jani[3]

[1] Faculty of Applied Sciences, Universiti Teknologi MARA, 40450 Shah Alam, Selangor, Malaysia
[2] Physics Department, Faculty of Applied Sciences, Universiti Teknologi MARA, 02600 Arau, Perlis, Malaysia
[3] Chemistry Department, Faculty of Applied Sciences, Universiti Teknologi MARA, 35400 Tapah, Perak, Malaysia

abdmutalib@uitm.edu.my

Abstract. Activated carbon (AC) has a high potential to act as dispersing agent as the surface of the activated carbon can be modified by chemical and/or physical treatments. In this work, several acid treatment methods have been conducted by using nitric acid (HNO_3), hydrogen peroxide (H_2O_2) and a mixture of nitric acid and sulfuric acid To functionalize the activated carbon from oil palm empty fruit bunches (AC-EFB), those treatments were followed by washing with distilled water until neutral pH and dried overnight in an oven at 80 °C. A dispersion test was conducted by dispersing a small amount of the treated AC-EFB in distilled water using a sonicator for 5 minutes. The modified AC-EFB has been characterized by using, Fourier transform infrared (FTIR) spectroscopy, scanning electron microscopy (SEM) and energy dispersive X-ray (EDX) analysis. Based on observation, the treated AC-EFB with a mixture of acids has a stable suspension compared to other methods which indicate the presence of a hydroxyl group attached to the surface of AC-EFB. FTIR results further confirmed the presence of the functional group of hydroxyl. The SEM micrograph shows the formation of unique microstructure on the AC-EFB structure after treated with method M3 whereas the number of pores developed was increased. The atomic percentage of oxygen of treated AC-EFB was higher than untreated AC-EFB, which indicates the hydroxyl group was attached to the surface of activated carbon as proven by EDX analysis. Thus, acid treated method using HNO_3/H_2SO_4 mixture shows a promising approach on synthesizing the activated carbon with unique properties and characteristics.

1. Introduction

Activated carbon is a porous solid material. It is made up to 85%-95% carbon and the remaining are hydrogen, nitrogen, sulfur, oxygen and other minerals [1]. The activated carbon usually produced from carbonaceous source materials at high temperature corncob, rice husk, coconut shell, bone, coal, coffee endocarp, and fibers of oil palm empty fruit bunches (OPEFB). Several benefits of using activated carbon were identified, such low cost, the high surface area up to 3000 m^2/g and a wide range of pore size. In general, the activated carbon can be produced by two stages which are carbonization and activation process [2].

Content from this work may be used under the terms of the Creative Commons Attribution 3.0 licence. Any further distribution of this work must maintain attribution to the author(s) and the title of the work, journal citation and DOI.
Published under licence by IOP Publishing Ltd

The carbonization process involved, enriching the carbon content to create initial porosity in the char by heating the precursor at high temperature in an inert atmosphere. While, the activation step can be either physical or chemical activation in which the physical activation is conducted by supplying the char with oxidizing gases such carbon dioxide. On the other hand, chemical activation involve impregnating the starting material with chemical reagent [3]. In this study, an activated carbon derived from oil palm empty fruit bunches (AC-EFB) is chosen due to the abundance of oil palm empty fruit bunches available from palm mill industry in Malaysia. The production of EFB in our country is about 8.5 million tons per year. Activated carbon is recently applied in many fields such act as an adsorbent in water treatment, food processing, pharmaceutical and chemical industries air pollution control and supercapacitor electrodes [4]. Due to high surface area and variety pore size properties, the activated carbon is commonly used as an adsorbent and catalytic support [5].

One of the promising applications of activated carbon is act as a dispersing agent. The activated carbon must be functionalized before it can be applied as a dispersing agent. This step is important as the surface of the activated carbon must have hydroxyl and carboxyl group which is an important characteristic of a dispersing agent. For this study, the AC-EFB was functionalized by chemical treatment. Several acid treatment methods have been conducted by using concentrated acid of HNO_3, treatment with H_2O_2 and treated with a mixture of concentrated acids of HNO_3 and H_2SO_4 to obtain better functionalized AC-EFB. The modified AC-EFB has been characterized by using Fourier transform infrared (FTIR) spectroscopy, scanning electron microscopy (SEM), energy dispersive X-ray (EDX) analysis and dispersion test.

2. Experimental procedure

Several acid oxidation methods were used to functionalize the activated carbon by using strong acids. For the first method (M1), about 1 g of untreated AC-EFB was functionalized using 15% H_2O_2 solution by stirring the mixture for 4 hours followed by washing with distilled water to obtain neutral pH. The treated AC-EFB was dried overnight in an oven at 60°C. While, for the second method (M2) 60% of the HNO_3 solution was used to treat AC-EFB by refluxing the mixture for 4 hours followed by washing and drying. Lastly, for the third method (M3) about 1 g of untreated activated carbon was functionalized by using a mixture of the strong acid (HNO_3, H_2SO_4) with the volume ratio of 1:3. Then, the mixture of the activated carbon and acids was sonicated in an ultrasonic bath (Branson 2800) at 40°C for 4 hours [6]. The functionalized activated carbon was stirred overnight after sonication step. After that, the functionalized activated carbon was washed with distilled water until neutral pH and was dried overnight in an oven at 60°C.

A dispersion test was conducted by dispersing a small amount of the treated AC-EFB and untreated AC-EFB in distilled water by sonicating the mixture for 5 minutes in an ultrasonic bath (Branson 2800). The modified AC-EFB was characterized by using dispersion test, Fourier transform infrared (FTIR) spectroscopy, scanning electron microscopy (SEM) and energy dispersive X-ray (EDX) analysis.

3. Results and discussion

3.1. Dispersion test.

Figure 1 shows a dispersion test conducted by dispersing a small amount of the AC-EFB in distilled water. The dispersion test was conducted to determine the dispersion capability of treated AC-EFB with various method of treatment. Based on observation, the treated AC-EFB following the method M3 has a stable suspension compared to the other approaches. It remains in a firm suspension for more than 30 days, while the untreated AC was settling down after 3 hours. Moreover, the treated AC-EFB from M1 and M2 also settle down after .several hours. These results indicate that a stable colloidal of functionalized AC-EFB was obtained due to the interaction between the distilled water and the functional group present on the surface of the activated carbon. The distilled water, which is polar in nature form a hydrogen bonding with the treated AC-EFB [7].

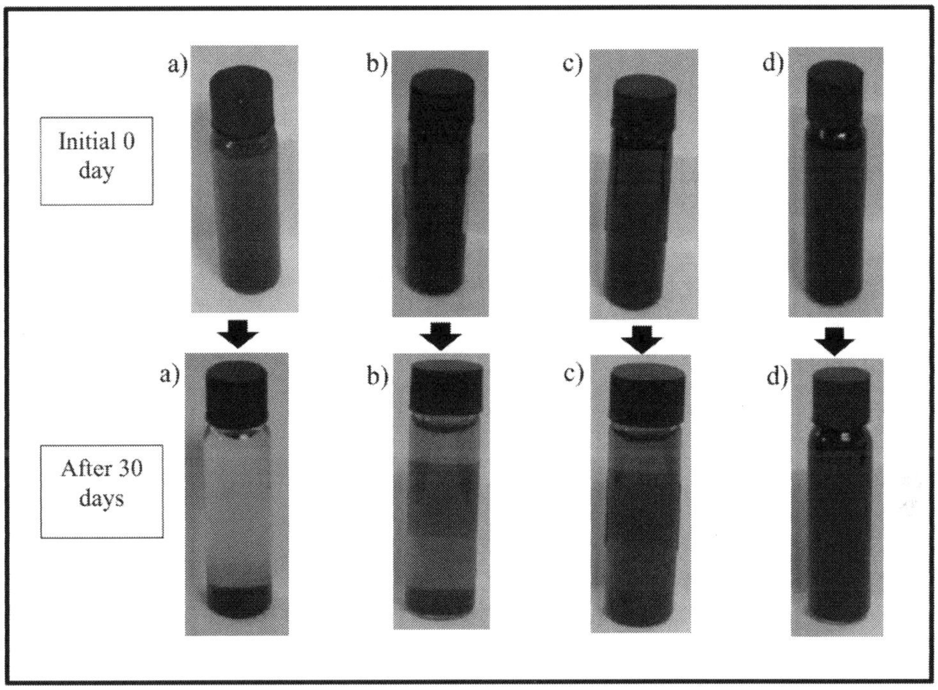

Figure 1. Photograph of AC-EFB: a) untreated AC-EFB; b) treated AC-EFB (M1); c) treated AC EFB (M2); d) treated AC-EFB (M3) initially and after 30 days disperse in distilled water.

3.2 Fourier Transform Infrared (FTIR) Spectroscopy.

To determine the functional group that presents in the sample of treated AC, FTIR analysis was conducted. In general, the absorption peaks in infrared spectroscopy determined by the energy of molecular vibration. Figure 2 depicted the FTIR data of the untreated AC-EFB and treated AC-EFB synthesized by different treatment. Based on the results, there are no transmittance peaks present for untreated AC-EFB as shown in Figure 2(a) while several peaks with various functional groups were observed on the treated AC-EFB as shown in Figure 2 (b-d). Meanwhile, Figure 2(d) shows the highest intensity of the functional peaks present compared to Figure 2(b) and (c). This may due to introducing a mixture of highly concentrated acids of HNO_3 and H_2SO_4, which will enhance the oxidation of the carbon and increase the intensity of the peaks.

Based on Figure 2(d), a strong and broad adsorption peak was observed between 2400 to 3600 cm^{-1}, which attributed to O-H stretching functional group and this indicated the hydroxyl group that attaches to the carbon after functionalization process [8]. The two sharp and weak peaks were observed at 1936 cm^{-1} and 2120 cm^{-1} belongs to the C-H stretching vibration of the surface of activated carbon. On the other hand, the bands at 1705 cm^{-1} corresponded to C=O functional group, which represents the carboxyl group due to oxidation of some carbon atom on the surface of AC-EFB by nitric acid [9]. A peak of C=C stretching was observed at 1586 cm^{-1} from the aromatic stretching of the activated carbon. While, a broad, small, and weak peak was observed at 1175 cm^{-1} from the C-O stretching in alcohols and at 978 cm^{-1} results from the out-plane C-H bending in benzene derivatives. The results presented here agrees with the previous finding using the carbon nanotube functionalization [10].

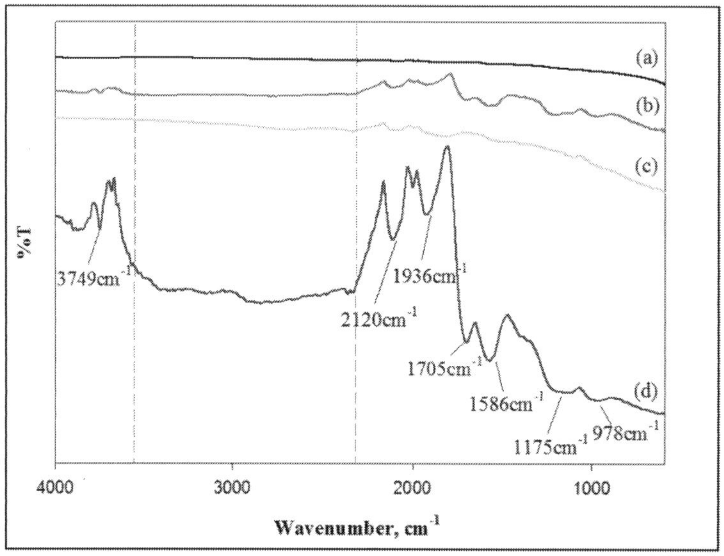

Figure 2. The FTIR plot for AC-EFB: a) untreated AC-EFB; b) treated AC-EFB (M1); c) treated AC-EFB (M2); (d) treated AC-EFB (M3).

3.3 Scanning Electron Microscopy (SEM).
To observe the morphological structure of the sample, the SEM analysis is carried out. For this research, SEM was used to characterize the morphological structures of the untreated and treated samples. The treated AC-EFB (M3) was proceed for SEM and EDX characterization as it shows better intensity and form a stable suspension compared to M1 and M2. Based on Figure 3(a), the microstructure of the untreated activated carbon has a large and less pore formed. While, Figure 3(b) showed smaller and significant number of pores was observed due to introducing of the strong acids, which will enhance the formation of the functional groups of hydroxyl and carboxyl at the surface of the activated carbon. Thus, more surface area was developed in the treated AC-EFB [11].

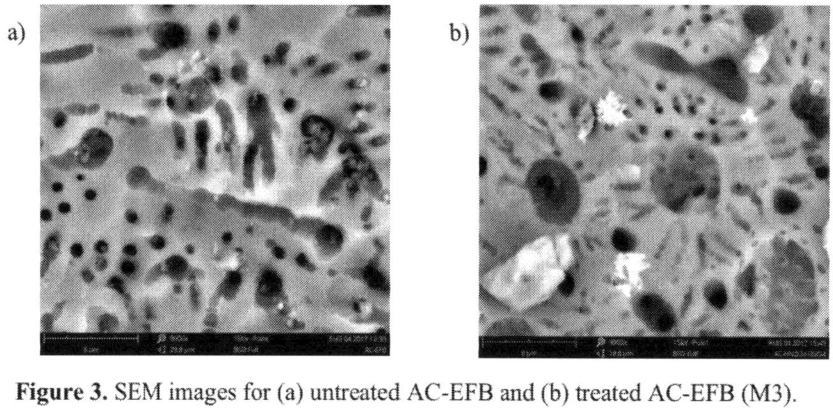

Figure 3. SEM images for (a) untreated AC-EFB and (b) treated AC-EFB (M3).

3.4 Energy Dispersive X-ray (EDX).

To determine the elemental composition of the AC-EFB, the EDX analysis was taken from a selected spot on the SEM which shown on Figure 4. It was proven that the carbon (C) and oxygen (O) element was present in the sample with a different atomic percentage. The treated AC-EFB (M3) shows a higher content of oxygen compared to the untreated AC-EFB due to more oxygen was bonded to the carbon after the treatment process [12]. Mineral such magnesium (Mg) was observed in untreated AC-EFB and not in treated AC-EFB. This may due to the introducing of the concentrated acids during the treatment which will eliminate the mineral and other impurities.

a)

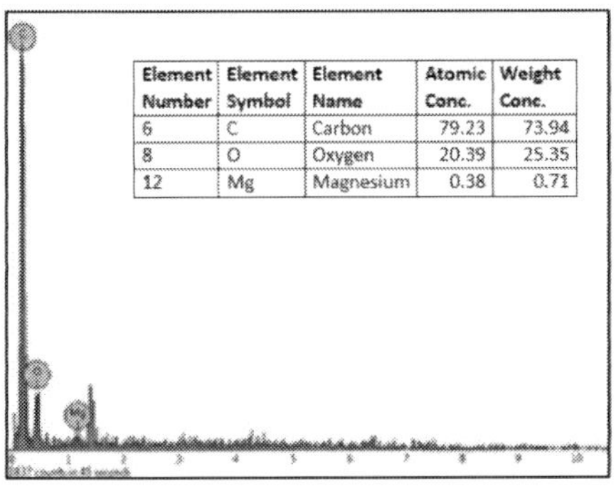

Element Number	Element Symbol	Element Name	Atomic Conc.	Weight Conc.
6	C	Carbon	79.23	73.94
8	O	Oxygen	20.39	25.35
12	Mg	Magnesium	0.38	0.71

b)

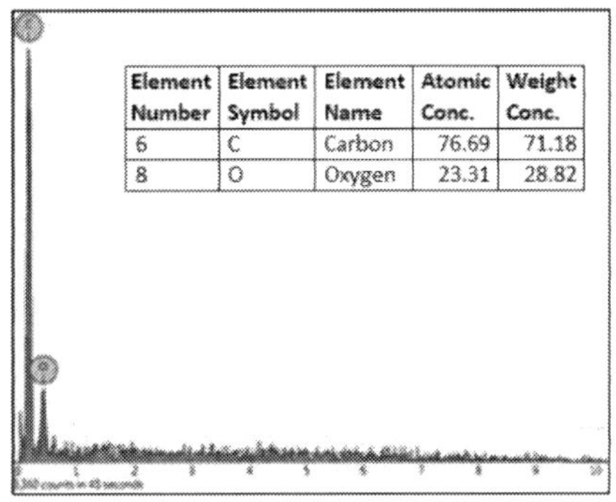

Element Number	Element Symbol	Element Name	Atomic Conc.	Weight Conc.
6	C	Carbon	76.69	71.18
8	O	Oxygen	23.31	28.82

Figure 4. EDX data for a) untreated AC-EFB and b) treated AC-EFB (M3).

4. Conclusion

The treatment of activated carbon from oil palm empty fruit bunches (AC-EFB) was done using HNO_3, H_2O_2 and HNO_3/H_2SO_4 mixture. The treated AC-EFB was characterized by FTIR, SEM, EDX and the dispersion test. Based on the dispersion test and FTIR data, the treated AC-EFB was successfully functionalized and the treatment with HNO_3/H_2SO_4 mixture has a better suspension performance compared to the others. While, the SEM micrograph and EDX analysis had shown that the treated AC-EFB by using HNO_3/H_2SO_4 mixture has better properties compared to untreated AC-EFB. As a conclusion, the treated AC-EFB using HNO_3/H_2SO_4 mixture shows suitable properties to act as dispersing agent for various applications such synthesizing perovskite electrolyte of $Ba(Ce,Zr)O_3$.

Acknowledgements

This work was financially supported by the Ministry of Higher Education (MOHE) through the (600-IRMI/TRGS 5/3(1/2016)-1). The authors thank to UiTM Cawangan Perlis for the facilities and also Supercapacitor Lab, School of Applied Physics, UKM for providing raw AC-EFB.

References

[1] Farma, R., M. Deraman, A. Awitdrus, I. Talib, E. Taer, N. Basri, J. Manjunatha, M. Ishak, B. Dollah and S. Hashmi 2013 *Bioresource technology* 132: 254-261.

[2] Rani, A., Hidayu, N., Mohamad, N. F., Matali, S., & Kadir, S. A. S. A. 2014 *Tech Publications.* 594:44-48.

[3] Hidayu, A., N. Mohamad, S. Matali and A. Sharifah 2013 *Procedia Engineering* 68: 379-384.

[4] Farma, R., Delika F., Awitdrus, and Mohamad D 2017 *AIP Conference Proceedings* 1801(1): 040001.

[5] Matos, J., Carol N., Laura R., and Maibelín R. 2011 *Journal of hazardous materials* 196: 360-369.

[6] Buang, N. A., F. Fadil, Z. A. Majid and S. Shahir 2012 *Digest Journal of Nanomaterials and Biostructures* 7(1): 33-39.
[7] Avilés, F., J. Cauich-Rodríguez, L. Moo-Tah, A. May-Pat and R. Vargas-Coronado (2009). *Carbon* 47(13): 2970-2975.
[8] Wirasnita, R., T. Hadibarata, A. R. M. Yusoff and Z. M. Lazim 2015 *Jurnal Teknologi* 74(11): 77-81.
[9] Ahmed, D. S., A. J. Haider and M. Mohammad 2013 *Energy Procedia* 36: 1111-1118.
[10] Ngo, C. L., Q. T. Le, T. T. Ngo, D. N. Nguyen and M. T. Vu 2013 *Advances in Natural Sciences: Nanoscience and Nanotechnology* 4(3): 035017.
[11] Hameed, B., I. Tan and A. Ahmad 2009 *Journal of hazardous materials* 164(2): 1316-1324.
[12] Abdullah, N., F. Sulaiman and H. Gerhauser 2011 *J. Phys. Sci* 22(1): 1-24.

IOP Conf. Series: Journal of Physics: Conf. Series **1083** (2018) 012050 doi:10.1088/1742-6596/1083/1/012050

Mechanical Behaviour of the Polypropylene-Devulcanized Natural Rubber Compound upon Natural Weathering

Kheng-Hooi Loo[1], Lee Tin Sin[1, a*], Soo-Tueen Bee[1, b*], C. T. Ratnam[2], Tiam-Ting Tee[1], A. R. Rahmat[3]

[1] Department of Chemical Engineering, Lee Kong Chian Faculty of Engineering and Science, Universiti Tunku Abdul Rahman, Jalan Sungai Long, Bandar Sungai Long, Cheras, 43000 Kajang, Selangor Malaysia
[2] Radiation Processing Technology Division, Malaysian Nuclear Agency, Bangi, 43000 Kajang, Selangor, Malaysia.
[3] Department of Polymer Engineering, Faculty of Chemical Engineering, Universiti Teknologi Malaysia, 81310 UTM Skudai, Johor, Malaysia.

[a]direct.tinsin@gmail.com, [b]direct.beest@gmail.com

Abstract. This study aims to investigate the mechanical behaviour of the polypropylene/devulcanized natural rubber (PP-DeVulcNR) compound upon natural weathering. The DeVulcNR was used to improve the mechanical properties of the PP-DeVulcNR compound at the range of 0 – 20 phr. These samples were subjected to natural weathering test at a duration range of 0 – 6 months to further study the weatherability of the compound. It was found that with the addition of 5 phr DeVulcNR in PP has the highest weatherability as it has obtained the optimum value of the tensile strength after 6 months outdoor exposure. Besides that, there is more fibrils formation on the fractured surface due to better dispersion of DeVulcNR particles in PP matrix. The Young's Modulus and elongation at break of the PP-DeVulcNR compounds reduced after 6 months outdoor exposure due to the embrittlement of the materials. The presence of the hydroxyl (O-H) and carbonyl (C=O) stretching vibration indicated that the photo-oxidation occurs in the compound during outdoor exposure.

1. Introduction

Most of the polymers are prone to degradation caused by the natural environment in which involved with ultraviolet and atmospheric oxygen and lead to chain scission in polymer chains. It may be accelerated by the elevated temperatures from the sun [1]. Photo-oxidation degradation occurs when the oxygen present by the activation of the polymer macromolecule with the absorption of photon from light in the polymer [2]. Polypropylene (PP) is well used in many applications due to its unique properties, such as high melting temperature, high chemical resistance and heat resistance. The superior properties of PP are mainly contributed by its high crystallinity structure and stability of the methyl group. However, the major drawback of the PP is poor in impact strength which gives limitation to several other applications [3]. Furthermore, PP has low resistibility towards degradation in the atmospheric environment. For instance, it is enough to initiate the degradation within the range of the sunlight wavelength at the earth's surface which is over 290 nm and caused discoloration, chalking and embrittlement of PP [4]. In order to decrease the effect of environmental degradation and

Content from this work may be used under the terms of the Creative Commons Attribution 3.0 licence. Any further distribution of this work must maintain attribution to the author(s) and the title of the work, journal citation and DOI.
Published under licence by IOP Publishing Ltd

expand the lifespan of PP polymer, devulcanized natural rubber (DeVulcNR) has been introduced to combine with the PP polymer to obtain better material. Hence, this paper focused on the mechanical behaviour of the polypropylene/devulcanized natural rubber (PP-DeVulcNR) compound upon natural weathering.

2. Experimental Procedure

2.1 Materials

Polypropylene (PP) with the grade of TITANPRO 6331 produced by Titan Chemicals Corporations Sdn Bhd, Malaysia was used in this study. The density and MFI of the PP were 0.9 g/cm^3 and 14 g/10 min respectively. Devulcanized natural rubber (DeVulcNR) was purchased from SRI Elastomers Sdn Bhd, Malaysia.

2.2 Sample Preparation

The various content of devulcanized natural rubber mix with 100 phr of PP was extruded by using twin screw extruder at 190 °C. The extruded compound cooled by chilled water. Then, the extruded compound was pelletized using pelletizer before compression molded at 170 °C into sheet form using 25 tons hot molding press. The compound was cooled to room air temperature for 20 min.

2.3 Natural Weathering Test

The natural weathering test was conducted according to ASTM D1435. The samples were placed on the rack by adjusted to face the equator at an angle of 45°. The rack was located in an open area where it was free from overshadowed by other objects and exposed to all environmental effects such as rain, sunlight, wind and etc. The samples were collected after 2, 4 and 6 months to determine the degree of degradation by UV source. at the surface of the samples was removed with a clean towel and left in air for 24 hours at room temperature before proceeding for further testing. The sample surface was cleaned by using a clean towel to remove water droplets due to condensation and leaved in air at room temperature for 24 hours before further testing.

2.4 Tensile Test

The tensile test was conducted by using Instron Universal Testing Machine 5582 series IX tensile tester. The 3 mm thick sheets were cut into dumbbell shapes accordance to the ASTM D638 (Type IV) standard. The tensile strength, elongation at break and Young's Modulus of the samples were obtained as the average of five specimens.

2.5 Scanning Electron Microscope (SEM) Analysis

The surface morphologies of the fractured surface for all the samples were scanned using Hitachi Scanning Electron Microscope model BS 340 TESLA. The fractured surfaces of the samples were cut into 3 mm height and placed on the copper stub with a fractured surface facing up. The samples were coated with a thin layer of gold before proceeding to SEM scanning. The SEM micrographs were recorded at a magnification of 1000x.

2.6 Fourier Transform Infrared Spectroscopy (FTIR)

The interaction of chemical functional group of the specimens was analyzed by using Perkin Elmer Spectrum EX1. The infrared spectra were recorded with 32 scans for each measurement over the wavelength range of 500 – 4000 cm^{-1}.

3. Results and Discussion

With reference to Figure 1(a), the tensile strength of the pristine PP was decreased drastically with around 45% of reduction after 6 months outdoor exposure. The photo-degradation process occurs which the high-energy of UV radiation from sunlight attacks the macromolecule and generates free radicals at the tertiary C-atom of the PP chain [5]. This effect leads to chain scissoring process which

IOP Conf. Series: Journal of Physics: Conf. Series **1083** (2018) 012050 doi:10.1088/1742-6596/1083/1/012050

molecular weight of the PP reduced. Thus, it reduced the tensile strength of the pristine PP. It is noteworthy that the addition of 5 phr DeVulcNR into PP matrix could give higher tensile strength than pristine PP after outdoor exposure. The existing of the carbon black particles in DeVulcNR acts as UV absorber and antioxidant by absorbing all the incident lights and inhibits photo-oxidation via its chemical properties [6]. Therefore, it increased the ability of PP matrix to resist the straining stress. However, further addition of DeVulcNR up to 20 phr into PP matrix could gradually reduce the tensile strength of the compounds after outdoor exposure. This is due to the DeVulcNR particles tend to agglomerate together at higher concentration and weaken the interaction between DeVulcNR and PP. Thus, this effect has deteriorated by the photo-degradation process and lower resistance against the straining stress.

With reference to Figure 1(b), the Young's Modulus of the pristine PP reduced from 422.38 MPa to 337.53 MPa after 6 months outdoor exposure. The photo-degradation process causes the molecular chain scission causing the breakdown of tie chain molecules and entanglements in the amorphous region [7]. Thus, it has weakened the rigidity of pristine PP and induced it to change from ductile to the brittle material. Moreover, it can be noticed that the addition of DeVulcNR into PP could slightly increase the Young's Modulus of the compounds when exposed to outdoor for 2 months. This might be caused by the photo-oxidation which increasing in crosslinking due to the high rate of radical termination in the bulk of the polymer where O_2 from photo-oxidation cannot gain access [8]. Thus, it resulted in increasing the rigidity of the PP-DeVulcNR compound. In contrast, the Young's Modulus of the PP-DeVulcNR compounds gradually reduced with longer exposure period up to 6 months. This might be due to the predominance of the chain scission occurs during photo-oxidation process has decreased the rate of radical termination in the bulk of the polymer. Therefore, the rigidity of the PP-DeVulcNR compounds reduced as the chain scission leads to embrittlement of the materials.

With reference to Figure 1(c), the elongation at break of the pristine PP and PP-DeVulcNR compounds reduced after exposed to outdoor for 6 months. The photo-degradation process caused the surface crack and defect as a result of disruption of crystalline order which in turn the crack become breakage concentrators [5]. Therefore, it reduced the elongation ability of the PP matrix and accelerated the breakage of the PP matrix by straining stress applied. On the other hand, it is noteworthy that the addition of DeVulcNR up to 15 phr into PP exhibits a higher value of elongation at break compared to pristine PP. The incorporation of DeVulcNR into PP matrix reduces resilience and toughness of the compounds due to its rubber-like properties [9]. Thus, it increased the elongation ability of the PP- DeVulcNR compound under tensile stress. As per contra, 20 phr DeVulcNR added in PP after 6 months outdoor exposure exhibits the lowest elongation at break (2.39%). The excessive of the DeVulcNR particles in PP matrix caused the uneven distribution of DeVulcNR particles in the compound. Hence, it limits the flow and mobility of the PP matrix when straining stress applied.

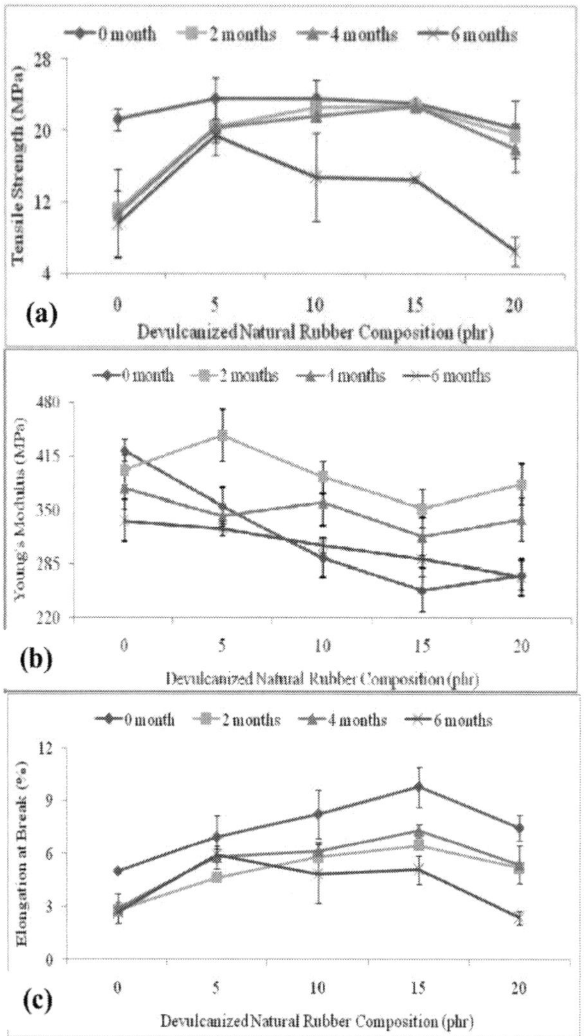

Figure 1: (a) Tensile strength (b) Young's Modulus (c) elongation at break for different DeVulcNR composition of PP-DeVulcNR compound subject to natural weathering.

Figure 2 represents SEM micrographs of the fractured surface for different DeVulcNR composition of PP-DeVulcNR compound after 6 months outdoor exposure. According to the Figure 2(a), it can be seen that the fractured surface of pristine PP exposed under sunlight for 6 months has severely deteriorated with large and long surface cracks formed. This might be due to the chemi-crystallization which release of entangled segments and tie chain molecules in the amorphous region that unable to crystallize during the original solidification process caused by molecular chain scission. Thus, these freed segments then rearrange into a crystalline phase to provide them with enough mobility resulted in a contraction of the surface layer by crack formation [7]. With reference to Figure 2(b), it can be seen that with the addition of 5 phr DeVulcNR in the PP could improve the fractured surface with more fibrils formation as a better dispersion of DeVulcNR particles in PP. Moreover, the presence of carbon black particles in DeVulcNR caused the compound more resistant toward weathering as it acts as UV absorber and antioxidant by absorbing all the incident lights and inhibits photo-oxidation with its phenolic and quinoid groups [10]. However, the fractured surface of the PP-DeVulcNR compound become more flake-like appearance with further increasing of DeVulcNR up to 15 phr as depicted in Figure 2(c) and Figure 2(d). The photo-oxidation process has deteriorated the interaction between DeVulcNR and PP as the predominance of chain scission occurs. According to Figure 2(e), it can be noticed that 20 phr DeVulcNR added in PP has a long crack surface formed after 6 months exposure. At the high concentration of DeVulcNR, it tends to agglomerate together and weakens the interaction between DeVulcNR and PP. Furthermore, photo-degradation process deteriorated the PP matrix and allowed a higher rate of oxygen penetration into the interior of the compound. As a result, long crack surface formed and weaken the structure of the compound.

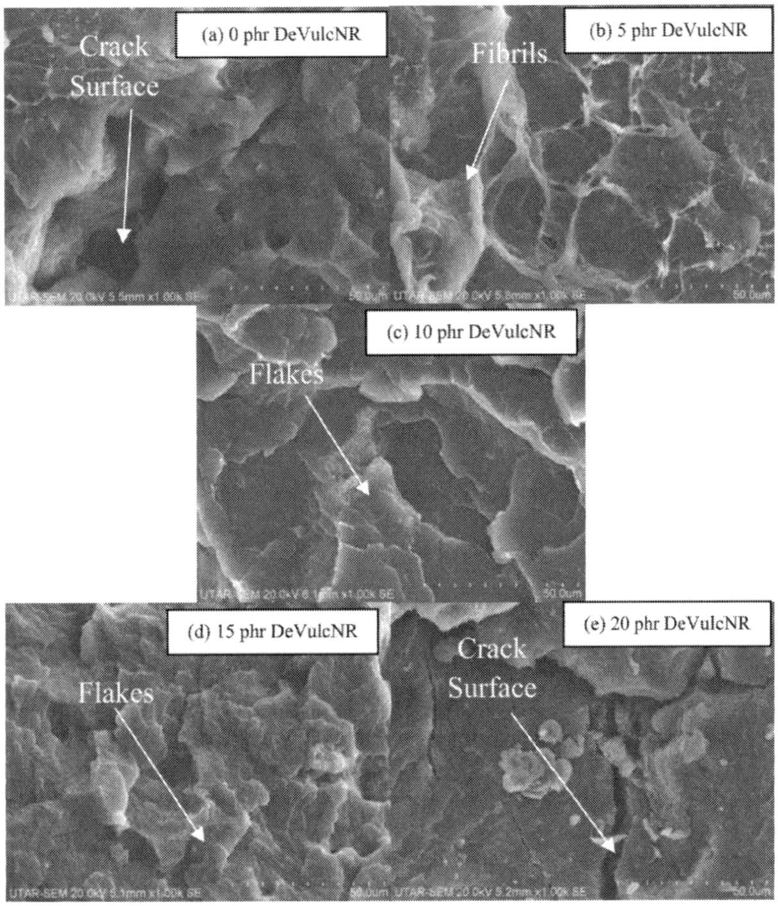

Figure 2: SEM micrographs of fractured surface for different DeVulcNR composition of PP-DeVulcNR compound after 6 months outdoor exposure under magnification of 1000x.

Figure 3 shows the FTIR absorption spectra of the PP-DeVulcNR compound with different composition of DeVulcNR upon natural weathering made in the range of 4000 - 650 cm^{-1}. The appearance of the absorption peak at 3300 cm^{-1} and 1710 cm^{-1}corresponding to hydroxyl (O-H) and carbonyl (C=O) stretching vibration respectively represented that the photo-degradation occurs after 6 months outdoor exposure. It can be seen that the intensity of absorption peak around 2915 cm^{-1} assigned to asymmetric C-H stretching vibration reduced significantly after 6 months outdoor exposure due to the oxidation occurs by photo-degradation lead to the formation of hydroxyl and carbonyl group. Furthermore, higher DeVulcNR composition could increase the intensity of the peak at 1640 cm^{-1} representing C=C stretching due to more C-C bonding formed by DeVulcNR particles with PP matrix. Other than that, the peak around 1454 cm^{-1} and 1372 cm^{-1} corresponding to CH_2 bending vibration and CH_3 bending vibration, respectively [11]. Moreover, 3 small peaks visible at 1160, 980 and 836 cm^{-1} were due to –CH3 symmetric deformation vibration, –CH3 rocking vibration and –CH2 rocking vibration of PP [12]. The wavenumber of asymmetric C-H stretching vibration for the addition of DeVulcNR composition up to 15 phr in PP gradually increased with longer exposure time. This result indicated that good distribution and dispersion of DeVulcNR in the PP matrix has improved the stiffness of the matrix. Furthermore, carbon black particles in DeVulvNR absorb all the incident lights and inhibit photo-oxidation via its chemical properties. Therefore, photo-oxidation required higher energy to enable the vibrational effect of the asymmetric C-H stretching. However, the wavenumber for 20 phr DeVulcNR added in PP reduced upon natural weathering. This might be due to the DeVulcNR particles tend to agglomerate together at high concentration and weaken the interaction between DeVulcNR and PP. Thus, O_2 from photo-oxidation can easily gain access to PP matrix and weaken the C-H group bonding.

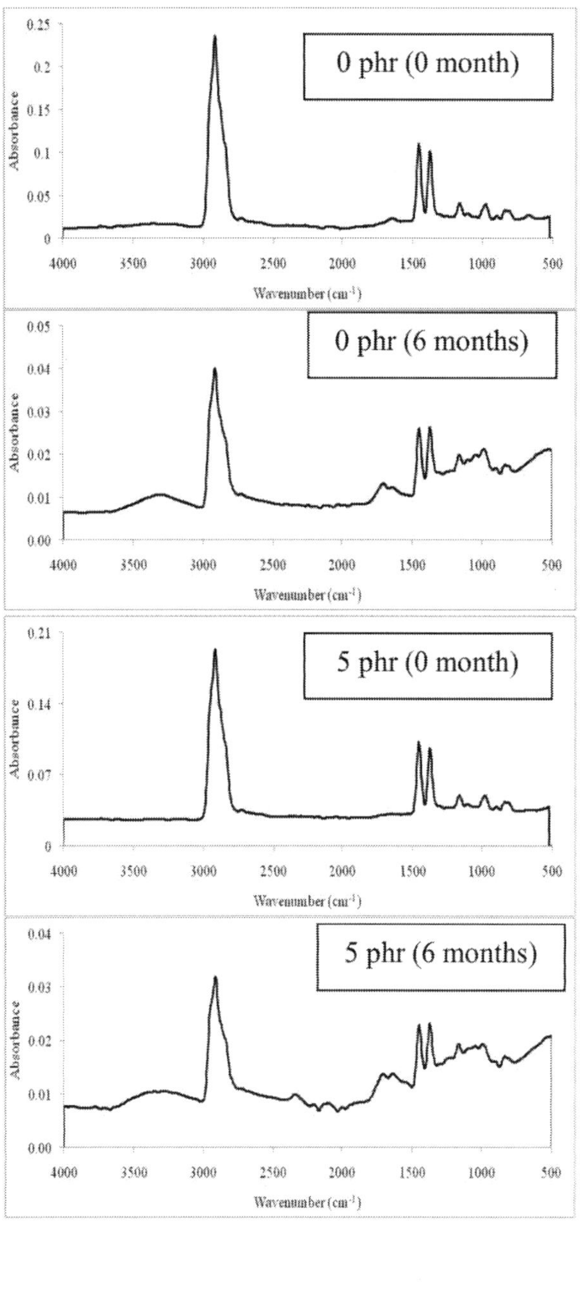

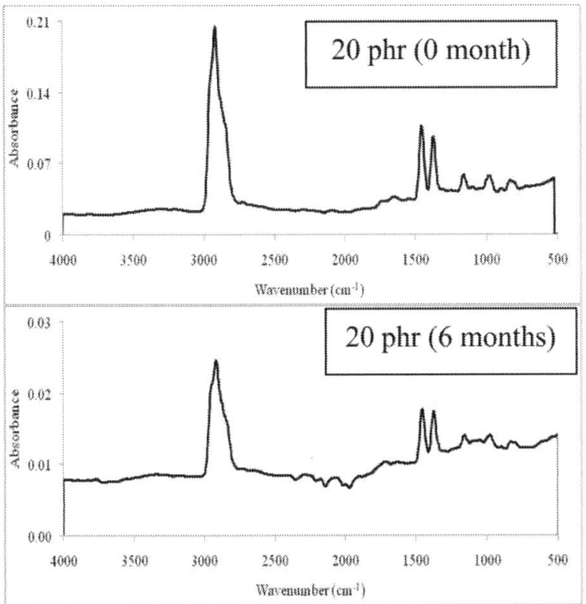

Figure 3: FTIR absorption spectra of PP-DeVulcNR compound with different composition of DeVulcNR before and after natural weathering.

4. Conclusions

The mechanical behaviour of the different DeVulcNR compositions added in PP upon natural weathering was investigated. It was found that 5 phr of DeVulcNR added in PP compound obtained the highest tensile strength after 6 months outdoor exposure. There is more fibrils formation on the fractured surface as better dispersion between DeVulcNR and PP at low DeVulcNR composition. The Young's Modulus and elongation at break of the PP-DeVulcNR compounds reduced after 6 months outdoor exposure due to the predominance of the chain scission occurs during photo-oxidation process has decreased the rate of radical termination in the bulk of the polymer. The presence of the hydroxyl (O-H) and carbonyl (C=O) stretching vibration indicated that the photo-oxidation occurs in the compound during outdoor exposure.

References

[1] White, J. R., Turnbull, A., 1994. Review Weathering of Polymers: Mechanisms of Degradation and Stabilization, Testing Strategies and Modelling. *Journal of Materials Science*, 29, pp. 584-613.
[2] Rabek, J. F., 1995. *Polymer Photodegradation: Mechanisms and Experimental Methods*. London: Chapman and Hill. Inconsistency in referencing
[3] Mohamad, N., Zainol, N. S., Rahim, R. R., Maulod, H. E., Rahim, T. A., Shamsuri, A. R., Azam, M. A., Yaakub, M. Y., Abdollah, M. F., Manaf, M. E., 2013. Mechanical and Morphological Properties of Polypropylene/Epoxidized Natural Rubber Blends at Various Mixing Ratio. *Procedia Engineering*, 68, pp. 439-445.
[4] Song, D., Gao, J., Li, X., Lu, L., 2014. Evaluation of Aging Behavior of Polypropylene in Natural Environment by Principal Component Analysis. *Polymer Testing*, 33, pp. 131-137.
[5] Jose, J., Nag, A., Nando, G. B., 2014. Environmental Ageing Studies of Impact Modified Waste Polypropylene. *Iranian Polymer Journal*, 23(8), pp. 619-636.
[6] Amin, A., 2009. Synergistic Effect of TNPP and Carbon Black in Weathered XLPE Materials. *Journal of Polymers and the Environment*, 17, pp. 267-272.
[7] Rabello, M. S., White, J. R., 1997. Crystallization and Melting Behaviour of Photodegraded Polypropylene - 1. Chemi-crystallization. *Polymer*, 38(26), pp. 6379-6387.

[8] Majid, R. A., Ismail, H., Taib, R. M., 2010. The Effects of Natural Weathering on the Properties of Linear Density Polyethylene (LDPE)/Thermoplastic Sago Starch (TPSS) Blends. *Polymer-Plastics Technology and Engineering*, 49, pp. 1142-1149.

[9] Mohamad, N., Zainol, N. S., Rahim, R. R., Maulod, H. E., Rahim, T. A., Shamsuri, A. R., Azam, M. A., Yaakub, M. Y., Abdollah, M. F., Manaf, M. E., 2013. Mechanical and Morphological Properties of Polypropylene/Epoxidized Natural Rubber Blends at Various Mixing Ratio. *Procedia Engineering*, 68, pp. 439-445.

[10] Ammala, A., Hill, A. J., Pas, S, J., Turney, T. W., 2002. Degradation Studies of Polyolefins Incorporating Transparent Nanoparticulate Zinc Oxide UV Stabilizers. Journal of Nanoparticle Research, 4, pp. 167-174.

[11] Medhat, M. H., Nagwa, A. B., Mona, Y. E., El-Sayed, A. H., 2013. Thermo-mechanical Properties of Devulcanized Rubber/High Crystalline Polypropylene Blends Modified by Ionizing Radiation. *Journal of Industrial and Engineering Chemistry*, 19, pp. 1241-1250.

[12] Wieslawa, U. D., 2012. *The Use of the Spectrometric Technique FTIR-ATR to Examine the Polymers Surface*. Rijeka: InTech, pp. 90.

IOP Conf. Series: Journal of Physics: Conf. Series **1083** (2018) 012051 doi:10.1088/1742-6596/1083/1/012051

Effects of Grain Sizes on Rice Husk Ash Composite as Microwave Absorber

N. Z. Yahaya[1*], S. R. Salim[2], I. S. Mustafa[2], N. N. A. Razak[2] and Liyana Zahid[3]

[1]School of Distance Education, University Sains Malaysia, 11800 Minden, Pulau Pinang, Malaysia
[2]School of Physics, University Sains Malaysia, 11800 Minden, Pulau Pinang, Malaysia
[3]School of Computer and Communication Engineering, Universiti Malaysia Perlis (UniMAP), Pauh Putra Campus, 02600 Arau, Perlis, Malaysia

*norzakiah@usm.my

Abstract. Electromagnetic interference affects human's health and interrupt the electrical equipment's performance. Metallic materials commonly used as microwave absorber is not cost-effective and too heavy. The agriculture waste from farming activities can be an alternative for inexpensive microwave absorbing material. Agriculture waste contain organic compound which has carbon element in their composition. Rice husk is one of the agriculture waste known to have high carbon content and also lightweight which are desired to be a good microwave absorber. Therefore, rice husk ash is used and the effect of the rice husk grains sizes as microwave absorber is investigated. The reflection loss for smallest grain size 2.5 × 10-3 cm was achieved -22.7 dB which is the highest which also possess smallest pore size which is 0.432 μm indicates high surface area.

1. Introduction

Electronic devices are widely use which lead to the undesired electromagnetic interference (EMI) affects the human health and disturb the performance of electrical devices or equipment. The exposure of EMI on human cell can lead to development of cancer or stimuli on electrically excitable tissue which can cause brain malfunction, headache and memory loss (1). Metallic composite materials are mostly used to manufactured as electromagnetic shielding material due to its effectiveness in shielding and high electrical conductivity. However, the huge penalty for this metallic material are heavy weight and not cost-effective (2).

The use of agricultural wastes produced through farming activities are highly demand in research as microwave absorber material. These are some of agricultural wastes which had been used as

Content from this work may be used under the terms of the Creative Commons Attribution 3.0 licence. Any further distribution of this work must maintain attribution to the author(s) and the title of the work, journal citation and DOI.
Published under licence by IOP Publishing Ltd

alternative for microwave absorber including rice husk, rice straw, sugarcane bagasse and banana leave (etc) that would add economic value to the existing microwave absorbing material which create inexpensive alternative while reducing the cost of waste disposal (3-5). About 97% of rice husk was produced from the processing of rice mills from developed country and usually was burned in order to disposed it. This will cause air pollution and lead to health problems (4, 5). Rice husk ash is highly porous, very high in surface area and lightweight (6). Porous carbon usually prepared from rice husk due to the cost - effective and cheap energy source instead of been discarded. It also low cost compared to others polymer composite and eco-friendly to the environment (7, 8). Not only that, the rice husk ash contains about 35-37 % of carbon which make the rice husk ash the potential useful as a microwave absorber and other applications. When the microwave passes through the rice husk ash, the carbon will reduce the effect of the wave (9). Carbon has high surface area and pore structure which contribute to the reducing the attenuation of microwave radiation (10). Not only that, carbon has good dielectric properties with high value of loss tangent (tan δ) which allow for dielectric heating (11, 12). The importance of carbon in the role of microwave absorption has been stated by several researchers where carbon can enhance the transformation of microwave energy to thermal energy contributed from their permeability and permittivity properties (12, 13). In this work, we are interested in studying the effect of the grain size of rice husk ash towards the microwave absorber performance based on the S-parameter using Agilent E8362B performance network analyser.

2. Methodology

2.1 Microwave Absorber Fabrication

The works begin with the collection of raw rice husk ash from the paddy and rice industry, BERNAS, Sungai Limau, Yan, Kedah. The rice husk ash is produced in an open air factory and uncontrolled burning. The temperature used in the burning process is 300 °C and the treated rice husk is black in colour which indicates high carbon content. Then, the rice husk ashes were dried in the oven in 70-80 °C for five hours to ensure they were completely dried. In order to get different grain sizes characteristic, the RHA was grinded for three hours. The grains obtained were sieved by using sieving machine in order to separate accordingly to the diameter of the sieve container. Subsequently, rice husk ash (RHA) was mixed with polyester and methyl ethyl ketone peroxide (MEKP). The function of polyester resin is to bind the RHA without creating any new bonding. Meanwhile, the MEKP act as hardening agent in order to harden the samples during the microwave absorber fabrication process. The RHA mixtures were then added to the mould until it fills the space with a dimension of 22.86 mm (length) × 10.16 mm (width) × 5 mm (thickness). After being added into the mould, the RHA mixture was compressed using a clamp in 30 minutes time before taking it out. Then the weight was measured to calculate the density of each samples. The RH samples were left in room temperature for three hours to ensure the sample was totally dried and hardened. The mixture of MEKP and Polyester was which act as control sample is also fabricated. The RHA and control samples were tabulated accordingly based on the grain size, weight and density. The characterizations were performed using Scanning Electron microscopy (SEM) and Energy Dispersive X-ray Spectroscopy (EDX) in order to determine the surface morphology and element composition.

2.2 Reflection and Absorption Measurement

The reflection loss, S_{11} of rice husk ash can be obtained using the S-parameter data measured by using rectangular waveguide method. The RHA sample will be fitted in the inner dimension of the waveguide at the frequency of 8.2 – 12.0 GHz. The thickness of the sample usually less than a quarter of the rectangular waveguide measurement wavelength. The samples should be fitted properly in order to get accurate result. The standard thru-reflect-line (TRL) calibration technique is required which to avoid unwanted losses/reflection in the inner wall of the waveguide as well as to achieve zero reference plane for the measurement. (13). An Agilent network analyser was used to measure the microwave absorption and reflection properties RHA and control samples, respectively.

3. Result and Discussion

3.1 Density

After the Control (C0) and rice hush ash samples with different grain sizes 2.5×10^{-3} cm, 1.0×10^{-1} cm, 3.0×10^{-1} cm and 5.0×10^{-1} cm of RHA have been compressed, the weights were measured in order to manually compute the density (Table 1) by using equation (1).

$$\text{Density } (\rho) = \frac{\text{Mass (g)}}{\text{Volume (cm}^{-3})} \quad (1)$$

The density is dependent on grain sizes and the weight of each samples. The weight and density for each sample decreases with grain sizes, respectively (Table 1). Smaller grain size will be more compact during the compression process and the grains will be closed together where the size of samples has been fixed with similar dimension. Therefore, it will reduce the buoyancy of the samples. The highest density of R1 which has smallest grain size (2.5×10^{-3} cm) has been obtained in this work. Reflection loss of the samples might be affected by the different of grain sizes of rice husk ash. As the grain size decrease, the surface area increase, therefore the interaction between the microwave with material increase. This will increase the absorption of microwave instead of been reflected or transmitted through the rice husk samples.

Table 1. The density value for different RHA samples and Polyester Resin.

Sample	Grain sizes (cm)	Weight (g)	Density (g/cm^{-3})
R1	2.5×10^{-3}	1.72	1.48
R2	1.0×10^{-1}	1.56	1.34
R3	3.0×10^{-1}	1.47	1.27
R4	5.0×10^{-1}	1.33	1.15
C0 (Control)	Polyester resin + MEKP	1.32	1.14

3.2 *SEM and EDX analysis*

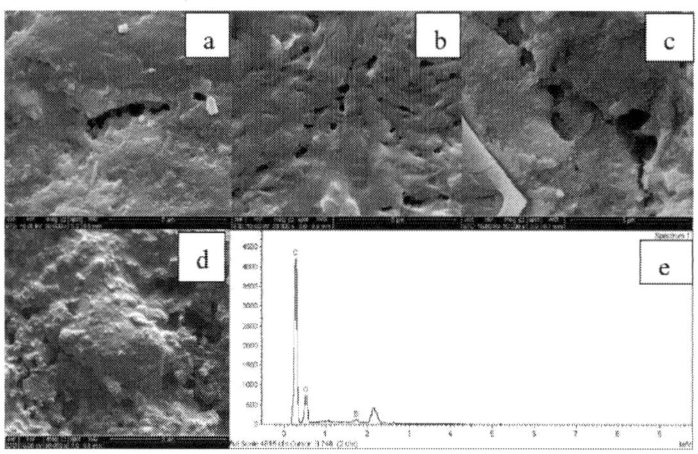

Figure 1. The surface morphology for (a) R1, (b) R2, (c) R3, (d) R4 at 30k magnifications and (e) EDX of the sample.

The surface morphology of R1, R2, R3 and R4 at 30k magnification were illustrated in Figure 1. The average pores sizes for R1, R2, R3 and R4 are 0.432 μm, 0.590 μm, 1.179 μm and 13.27 μm, respectively. R1 with smallest grains size of 2.5×10^{-3} cm is more compact and possess smaller pores size and smooth structure shown in SEM. Meanwhile, the R4 has the biggest grain size of 5.0×10^{-1} cm has rough surface and larger pore size structure. Smaller pores size indicates high surface area of grain size in total which can increase the interaction of microwave with the sample. Moreover, the R4 shows agglomeration in the sample and all rice husk ash samples structure are not uniform. From

EDX result (figure 1(e)), the carbon composition is higher compared to oxygen and silica which is essential in microwave absorber characteristic. The atomic percentage composition of carbon, oxygen and silica are 78.41 %, 21.25% and 0.35%, respectively. High carbon content of the rice husk ash samples can be related to high surface area and porous structure which is very useful in enhancing the microwave absorbing properties (10-12).

3.3 *Reflection Loss*

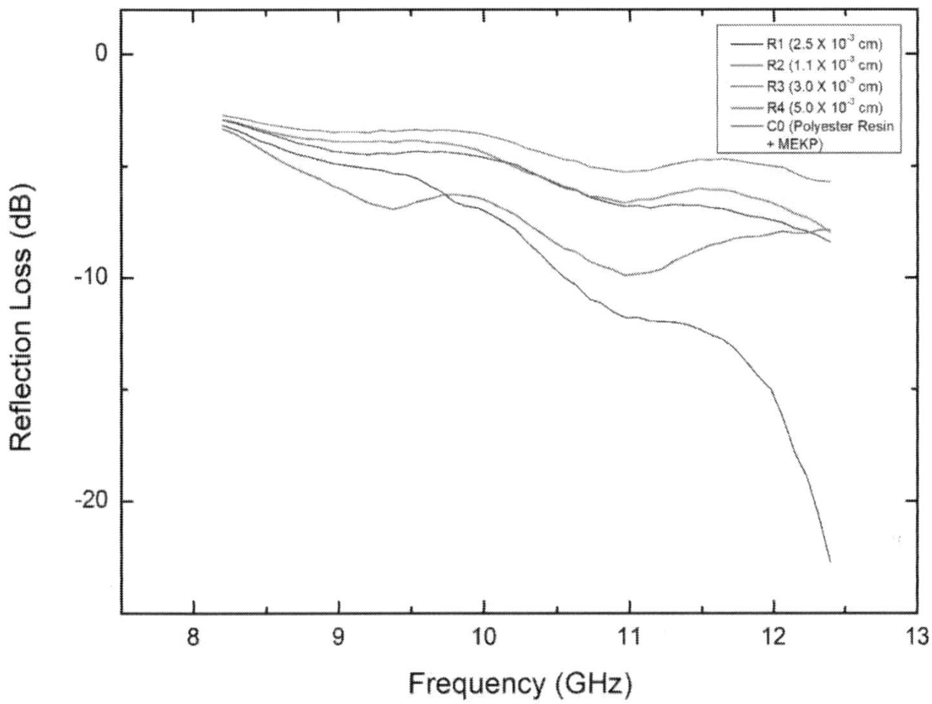

Figure 2. The reflection loss of different grains sizes of rice husk ash and polyester resin for microwave absorber

For reflection loss of samples, it is compared using different grain sizes of rice husk ash in the range of 8.2 – 12.0 GHz. The reflection loss achieved by R1, R2, R3 and R4 samples at -22.7 dB, -9.91 dB, -8.42 dB and -7.99 dB at frequency range from 8.2 GHz to 12.0 GHz, respectively (figure 2). In normal phenomenon occurs in microwave absorber, when the rice husk ash absorber is irradiated by the incident wave, the attenuation of incident wave occurred. The incident wave reflects and re-reflects by the rice husk structure as it passed through the absorber. The conversion of heat energy from electromagnetic wave occurs in this process therefore cause the decreasing or reducing of microwave intensity [2]. A material requires to possess reflection loss better than -10 dB, which denotes that the efficiency of the absorbing performance is about 90 percent, therefore indicate a good microwave absorber performance [10]. R1 shows largest value of reflection coefficient which indicates good microwave absorber performance. The best performance of R1 is at frequency 12.0 GHz and higher frequency ranges. This indicate that, the small grain size of rice husk ash have significant effect on the microwave absorber performance as R1 possess higher surface area for interaction with microwave radiation. Nevertheless, the process of internal reflection of microwave absorber of R1 also increased which reduce the microwave intensity and increase the absorption of the microwave in the sample.

4. Conclusion

Rice husk ashes with different grain sizes are investigated to determine their reflection loss and microwave absorption capabilities. The rectangular waveguide technique is used to measure the reflected, transmitted and absorbed of microwave radiation by the rice husk ash microwave absorber. The SEM and EDX results show R1 has smaller pore size and the composition of carbon in the all samples also high about 78.41% compared to silica composition. The smaller grain size of R1 give significant effect on the microwave absorbing properties. The smaller the grain sizes of rice husk ash, the smaller the pore sizes possess by the microwave absorber. This will increase the total surface area of R1 which lead to the increase of microwave radiation absorption.

5. Acknowledgement

This work was funded under Fundamental Research Grant Scheme, FRGS, Ministry of Higher Education (203/PJJAUH/6711489).

6. References

[1] Jha BK. 2012. Effects of Electromagnetic Fields on Human Beings and Electronic Devices. *Himalayan Physics*. p 3:38-9.

[2] Liu Q, Fan T, Zhang D. 2004. Electromagnetic shielding capacity of carbon matrix composites made from nickel-loaded black rice husk. *Journal of materials science*. p 39(20):6209-14.

[3] Se SM, Shaaban A, Ibrahim IM, editors. 2011. Microwave absorbing material using rubber wood sawdust. *Wireless Technology and Applications (ISWTA)*, 2011 IEEE Symposium on; 2011: IEEE.

[4] Kaur H, Aul GD, Chawla V. 2015. Enhanced Reflection Loss Performance of Square Based Pyramidal Microwave Absorber Using Rice Husk-Coal. *Progress In Electromagnetics Research M*. p 43:165-73.

[5] Zhang S, Dong Q, Zhang L, Xiong Y. 2015. High quality syngas production from microwave pyrolysis of rice husk with char-supported metallic catalysts. *Bioresource technology*. p 191:17-23.

[6] Nornikman H, Malek F, Soh P, Azremi A, Ghani S, editors. 2010. Potential of rice husk for pyramidal microwave absorber design. *The 2nd International Conference of Institution of Engineering and Technology (IET) Brunei Darussalam Network (ICIET 2010)*.

[7] Malek MFBA, Cheng EM, Nadiah O, Nornikman H, Ahmed M, Abdul Aziz MZA, et al. 2011. Rubber tire dust-rice husk pyramidal microwave absorber. *Progress In Electromagnetics Research*. p 117:449-77.

[8] Gao Y, Li L, Jin Y, Wang Y, Yuan C, Wei Y, et al. 2015. Porous carbon made from rice husk as electrode material for electrochemical double layer capacitor. *Applied Energy*. p 53:41-7.

[9] Iqbal MN, Malek MFBA, Ronald SH, Bin Mezan MS, Juni KM, Chat R. 2012. A study of the EMC performance of a graded-impedance, microwave, rice-husk absorber. *Progress In Electromagnetics Research*. p 131:19-44.

[10] Zahid L, Malek F, Nornikman H, Affendi NAM, Ali A, Hussin N, et al. 2013. Development of Pyramidal Microwave Absorber Using Sugar Cane Bagasse (Scb). *Prog Electromagn Res*. p 137:687-702.

[11] Lam SS, Chase HA. 2012. A review on waste to energy processes using microwave pyrolysis. Energies. p 5(10):4209-32.

[12] Shaaban A, Se S-M, Ibrahim IM, Ahsan Q. 2015. Preparation of rubber wood sawdust-based activated carbon and its use as a filler of polyurethane matrix composites for microwave absorption. *New Carbon Materials*. p 30(2):167-75.

[13] Lee YS, Malek F, Cheng EM, Liu WW, You KY, Iqbal MN, et al. 2013. Experimental determination of the performance of rice husk-carbon nanotube composites for absorbing microwave signals in the frequency range of 12.4-18 GHz. *Prog Electromagn Res*. p 140:795-812.

IOP Conf. Series: Journal of Physics: Conf. Series **1083** (2018) 012052 doi:10.1088/1742-6596/1083/1/012052

Effect of Metal Salts on Ionic Conductivity of Waste Cooking Oil Based Polyurethane Solid Polymer Electrolyte

Marina Abu Husin[1] and Syuhada Mohd Tahir[1]*

[1]Universiti Teknologi MARA Pahang, 26400 Bandar Tun Abdul Razak Jengka, Pahang, Malaysia

syuhadamohdtahir@uitm.edu.my

Abstract. The constant quest for green material to replace petroleum leads to utilization of waste cooking oil (WCO) as sustainable raw material to prepare polyurethane solid polymer electrolyte (SPE). In this study, pre-treated WCO was converted into polyol via epoxidation and hydroxylation reaction which aimed to introduce hydroxyl group on double bonds of free fatty acids in WCO. Then, WCO-based polyol was combined with 4,4-diphenylmethane diisocyanate to produce polyurethane (PU) as polymer host and 30 wt% metal salt was added to obtain SPE. Three types of metal salts were chosen to be incorporated in the SPE; lithium chloride (LiCl), lithium perchlorate ($LiClO_3$) and sodium chloride (NaCl). The SPE films were prepared using solvent casting method and characterized using Fourier transform infrared (FTIR) spectroscopy, electrochemical impedance spectroscopy (EIS), differential scanning calorimetry (DSC) and thermogravimetric analysis (TGA). FTIR spectra showed the presence of urethane linkage to verify that polyurethane has been successfully synthesized. EIS revealed the highest ionic conductivity value achieved at 3.02 x 10^{-5} S/cm for SPE with 30 wt% LiCl. The results showed the significance of metal salts cation and anion size on performance of SPE.

1. Introduction

In 1970s, Peter Wright and Michael Armand introduced the first new class of solid ionic conductors that led to the development of solid polymer electrolyte (SPE). SPE defined as solid ion conductors formed by dissolving salt in high molecular weight polymer [1]. SPE had gained substantial attention due to their advantages over liquid electrolyte especially in terms of thermal and mechanical stability as well as safety. These advantages make SPE widely applied in electrochemical devices such as batteries, fuel cells, solar cells supercapacitors and chemical sensors [2].

Polyurethane (PU) is one of the promising polymer host used in SPE. PU is a block copolymer that contains soft and hard segments made up of isocyanate and polyol. The soft segment will act as solvent to solvate cations while the hard segment give the mechanical stability to the PU structure [1,3]. Polyol is a polyether or polyester compound containing many hydroxyl groups. Polyol is conventionally prepared from petroleum. Due to constant increase in price and depletion in supply, preparation of polyols from vegetable oils has been the subject of many studies as a greener and more sustainable option. Castor oil, palm oil, soybean oil and rapeseed oil are the examples of vegetable oil that had been used [4-6].

Recently, the vegetable oil has been replaced with waste cooking oil (WCO) as raw material for polyurethane because it is cheaper and has organic structure almost similar to vegetable oil and contains

Content from this work may be used under the terms of the Creative Commons Attribution 3.0 licence. Any further distribution of this work must maintain attribution to the author(s) and the title of the work, journal citation and DOI.
Published under licence by IOP Publishing Ltd

high free fatty acid content [7]. However, studies on WCO-based PU as ion conductive solid are still very limited [8-10].

In this study, epoxidation and hydroxylation processes was used to synthesize WCO based polyol and reacted with 4,4-diphenylmethane diisocyanate (MDI) to form polyurethane and 30 wt% metal salts was added. Three types of metal salts; lithium chloride (LiCl), lithium perchlorate ($LiClO_3$) and sodium chloride (NaCl), chosen as charge carriers and their effect on ionic conductivity of WCO-based polyurethane SPE was observed.

2. Experimental

2.1. Collection and pre-treatment of waste cooking oil

Waste cooking oil (WCO) sample was collected from a restaurant located at Bandar Pusat Jengka, Pahang. Pre-treatment of WCO was carried out to remove food particles and water. The pre-treatment process involved filtration followed by heating for two hours at 95-125°C [7]. The sample was characterized using FTIR spectroscopy.

2.2. Synthesis of waste cooking oil based polyol

WCO based polyol was synthesized using one-pot epoxidation and hydroxylation reaction. 40 g of pretreated WCO and 5 g of formic acid were stirred in a three-necked round bottom flask equipped with a thermometer, magnetic stirrer bar and water condenser. Then, the flask was immersed in ice bath to reduce temperature to 10 °C - 15 °C and 80 g of 35% hydrogen peroxide was added drop wise with continuous stirring. Next, temperature was raised to 50 °C and reaction was continued for 5 hours and the mixture was then allowed to cool down to room temperature. Afterwards, 50 ml of distilled water was added followed by 5 ml of 11.6 M concentrated HCl. Temperature of the mixture increased gradually to 80 °C and reaction was continued for another 5 hr before allowing it to cool down at room temperature. Then, the product obtained was extracted in petroleum ether and the ether layer was washed away with aqueous sodium carbonate, followed by distilled water and sodium chloride solution. Finally, the ether was removed over rotary evaporator at 80 °C to obtain the WCO-based polyol. The polyol obtained was characterized using FTIR to confirm the formation of hydroxyl groups [4].

2.3. Production of polyurethane solid polymer electrolyte

In this study, the solution casting method was used to prepare SPE. Initially, waste cooking oil based polyol, 2,4-MDI, ethylene carbonate (EC) as plasticizer and metal salts were separately dissolved in acetone at room temperature. Then, MDI was added into polyol drop-wise followed by appropriate amount of metal salt and EC. WCO-based polyol to MDI ratio of 20:80 (NCO: OH) was used in this study. The mixture was stirred for more than an hour until homogenization and then it was casted onto a petri dish. The solvent was slowly evaporated in a fume hood at room temperature for 24 h [3]. The compositions of the SPE films prepared are shown in Table 1. The film produced were observed using FTIR, TGA, EIS and DSC.

Table 1 The compositions of PU SPE films

Sample	MDI	Polyol	$LiClO_4$	LiCl	NaCl	EC
PU-SPE 1	20	80	30%	0	0	10%
PU-SPE 2	20	80	0	30%	0	10%
PU-SPE 3	20	80	0	0	30%	10%

2.4. Characterizations

FTIR spectra were recorded using Spectrum 100, Perkin Elmer, in the range of 450-4000 cm^{-1}. Thermal stability was analyzed using Netzsch DSC model 214 Polyma with heating temperature from 20 °C to 250 °C at scanning rate of 10 °C /min under nitrogen atmosphere. Approximately 3 mg of the film samples were used for each measurement. The T_g of the sample was taken at mid-point of the endothermic peak using STARe software. The crystallinity of the sample was determined by using XRD model D-5000 Siemen. The data was collected for diffraction angle 2θ ranging from 5° to 60° at the rate

IOP Conf. Series: Journal of Physics: Conf. Series **1083** (2018) 012052 doi:10.1088/1742-6596/1083/1/012052

of 0.04 s^{-1}. The ionic conductivity was analyzed by EIS equipped with high frequency resonance analyzer (HFRA; Solartron 1260, Schlumberger), with applied frequencies ranging from 1 MHz to 0.1 Hz at the voltage of 1000 mV. The measurement was taken at room temperature using 1 cm^2 of SPE film sample that was sandwiched between two stainless steel block electrodes.

3. Results and Discussion

3.1. Waste cooking oil based polyol

FTIR spectra in Figure 1 shows the presence of hydroxyl (O-H) stretching peak at 3482.99 cm^{-1}. The presence of hydroxyl group indicated the successful formation of the polyester polyol in the WCO-based polyol using epoxidation and hydroxylation reaction. In addition, the carbonyl group (C=O) was observed at 1742.71 cm^{-1} confirmed the presence of ester group in the WCO-based polyol.

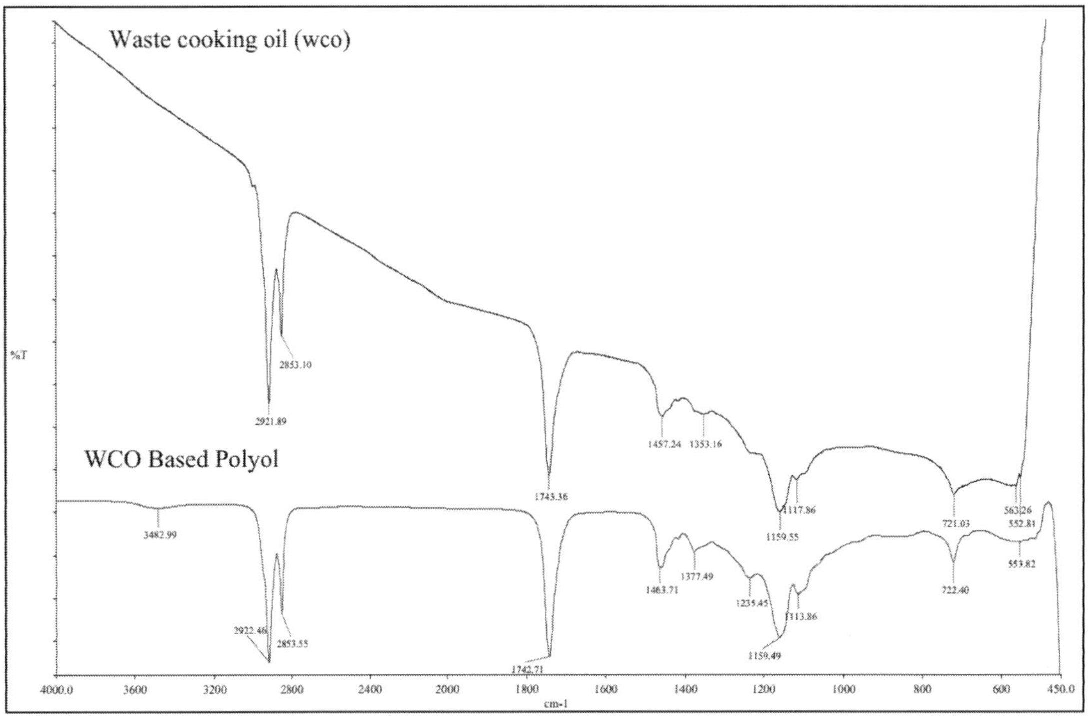

Figure 1 FTIR spectra of waste cooking oil and waste cooking oil based polyol

3.2. Waste cooking oil based polyurethane solid polymer electrolyte

Using WCO-based polyol, polyurethane solid polymer electrolyte (PU SPE) film was synthesized by incorporating three different types of metal salts. Ethylene carbonate (EC) as plasticizer was added to enhance the ionic conductivity of the polymer electrolytes [11].

Figure 2 presents the FTIR spectra of PU SPE with different type of metal salts. The FTIR analysis was conducted to observe the coordination and the interaction of the metal salts with polyurethane structure [1]. The peaks at ~1511 and ~3350 cm^{-1} refer to N-H vibration (-NH-C=O-) and the peaks at area ~1740 cm^{-1}confirm the presence of carbonyl group ester (C=O). The presence of the amine and carbonyl peaks indicates the conversion of monomers to urethane [12]. The band of C=O is shifted to lower wavenumber from 1741.53 to 1740.19 cm^{-1} at the time of addition of lithium and sodium chloride salts. It is due to the coordination of lithium and sodium ions with the carbonyl oxygen atoms and indicates the strong intermolecular interaction between the ions and electron donor atoms in the polymer matrix [12]. The effect observed for lithium salts is related with their anion structure. The anion size for perchlorate (ClO_4^-) ion is bigger than chloride (Cl^-) ion. Previous study shows that the $LiClO_4$ have much stronger interaction compared to LiCl and the salts with large size anions had low lattice energies and require less solvation [13].

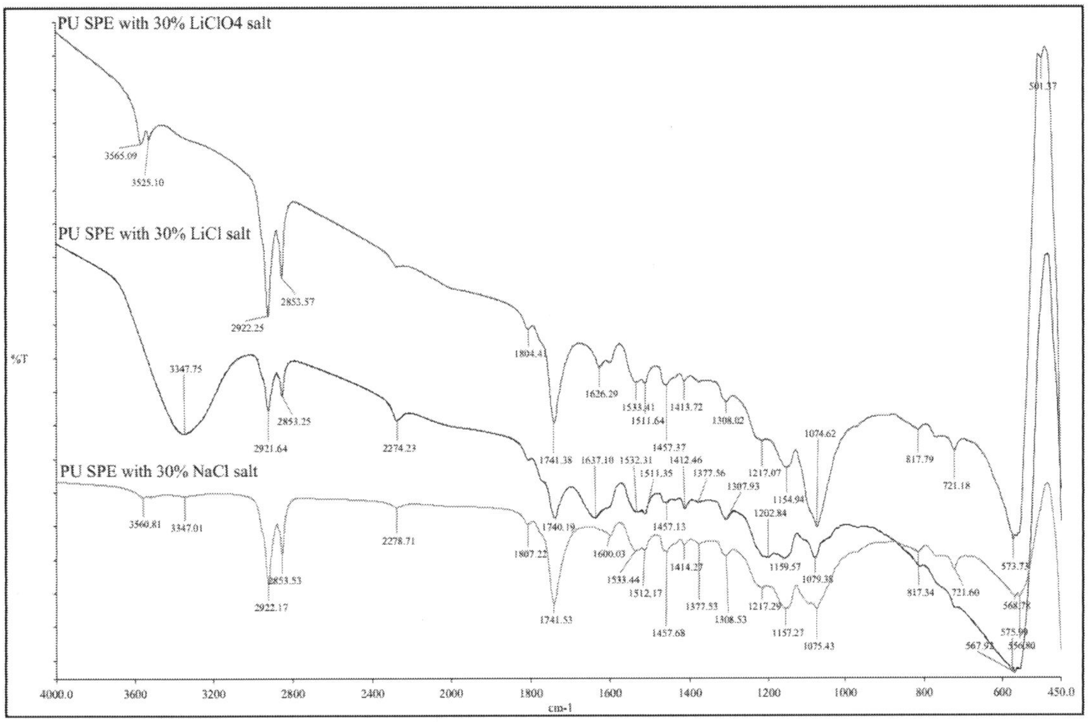

Figure 2 FTIR spectra of waste cooking oil-based polyurethane solid polymer electrolyte

In this study, the thermal stability is affected by the addition of cation and anion metal salts. Figure 3 shows the thermogravimetric curves for PU SPE with different type of metal salts. The thermal degradation stages (Td) of PU SPE containing ethylene carbonate (EC) and different type of metal salts were observed at four temperature regions; 25-110 °C, 150-230 °C, 240-370 °C and 370-520 °C. Table 2 lists the degradation temperatures and the percentage of weight loss. From the diagrams, it was observed that PU SPE with lithium perchlorate ($LiClO_4$) salt started the thermal degradation first at temperature of 25.74°C followed by PU-SPE with lithium chloride (LiCl) and sodium chloride (NaCl). The analysis shows the degradation of soft segments and hard segments. From the table, Td_1 indicate the degradation of soft segment and water at temperature range 25-110°C. At Td_2, the degradation of the hard segment of the urethane linkages occurred. The third step of the thermal degradation, Td_3, refer to the release of free isocyanate and last step of thermal degradation, Td_4, referred to breakage of the ester linkages [13].

Table 2 The degradation temperatures (Td) and the percentage of weight loss after degradation

Sample	Percentage of Weight Loss (wt. %)			
	Td_1 25-110°C	Td_2 150-230°C	Td_3 240-370°C	Td_4 370-520°C
PU-SPE $LiClO_4$	11.6	14.5	20.0	54.5
PU-SPE LiCl	9.9	21.2	26.1	70.3
PU-SPE NaCl	6.4	13.6	25.9	50.7

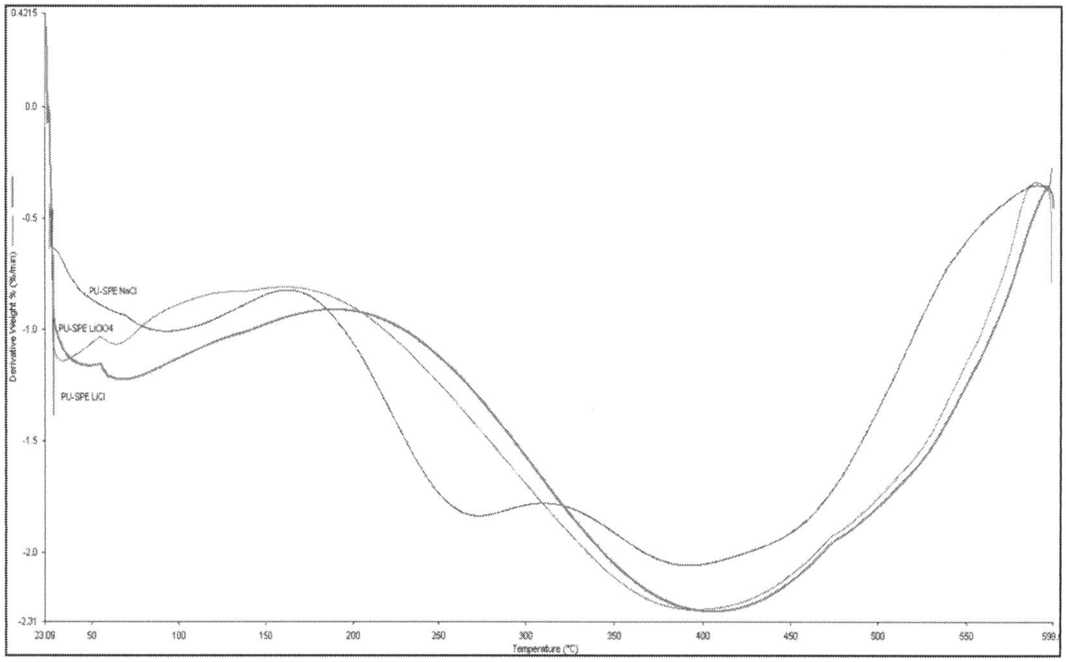

Figure 3 Thermogravimetric curves of waste cooking oil-based polyurethane solid polymer electrolyte

The ionic conductivity was calculated by using the formula of $\sigma = l / (R_b A)$, where l is the thickness of the film, A is the surface area and R_b is bulk resistance value that can be observed from the intercept of the impedance plot. Table 3 shows the ionic conductivity value based on different types of metal salts. The higher the concentration of the salt, the higher the ionic conductivity [3]. In this study, 30 wt. % of salt was used in order to achieve higher ionic conductivity value. It can be seen in Table 3 that PU-LiCl has the highest ionic conductivity followed by PU-$LiClO_4$ and PU-NaCl. An increase in the ionic conductivity is due to the interaction between the ions with the chemical group in the PU chains. PU-LiCl has the highest conductivity because of the dissociation of the salt in the polymer matrix. The Li^+ and Cl^- moved through the polymer matrix that created a flow of charges, with the help of the PU segments. The free mobile ions present in the polymer may increase the amorphous structure of the polymer host; making migration of ion easier [3]. Compared to PU-NaCl, the conductivity for PU-LiCl is greater because of the size of cations. Lithium ions have smaller size of cation as compared to sodium ion. Smaller cations have higher conductivity because of the greater mobility of the cation [12].

Table 3 Ionic conductivity value of PU-SPE with different type of metal salts

Sample	Ionic conductivity value (S/cm)
PU-LiCl	3.02×10^{-5}
PU-$LiClO_4$	9.00×10^{-7}
PU-NaCl	2.34×10^{-8}

Differential scanning calorimetry (DSC) analysis was carried out to determine the glass transition temperature of the synthesized polyurethane and polymer electrolytes. The glass transition temperature (T_g) gave information about the polymer chain dynamics and it plays a vital role in ionic conduction [14]. The DSC analysis can be related with the EIS analysis. Table 4 shows the value of the T_g for PU SPE with each type of metal salts. In the amorphous phase of SPE, the cation conduction is supported by the polymer segmental chain motion, while the anion moves in the created free space [14]. Based on the results, PU-$LiClO_4$ gave higher T_g value as compared to PU-LiCl followed by PU-NaCl. An electrolyte with low T_g indicate fast ionic conduction [12]. Therefore, between PU-LiCl and PU-NaCl electrolytes, the ionic conductivity of PU-LiCl is higher than PU-NaCl. In PU-LiCl and PU-$LiClO_4$ electrolyte, the T_g recorded explained the isolation of hydrogen bonds in the urethane soft segment which may contribute to the segmental motion of the polymer. The lithium cations were present as free ions and may act as transient crosslinkers between polyurethane chains segment by the coordinated interaction between ester oxygen and lithium cations, resulting in increase in T_g [3].

Table 4 Glass transition temperature (T_g) for PU-LiCl, PU-$LiClO_4$ and PU-NaCl

Sample	Glass transition temperature (°C)
PU-LiCl	157.4
PU-$LiClO_4$	164.2
PU-NaCl	148.3

4. Conclusions

In this study, the waste cooking oil-based polyol was successfully synthesized by one-pot epoxidation and hydroxylation process. By using the polyol, polyurethane (PU) film for solid polymer electrolyte (SPE) was successfully synthesized by solution casting method. Three kinds of metal salts with different cation and anion size were used to prepare the polyurethane-solid polymer electrolyte (PU-SPE). The metal salts are lithium perchlorate ($LiClO_4$), lithium chloride (LiCl) and sodium chloride (NaCl). From the analysis and chemical reaction, it can be concluded that the structure of the cation and anion of the metal salts affect the chemical interaction, thermal stability and ionic conductivity of the PU electrolytes. It was related with the ability of the cation and anion dissociation of metal salts in the PU as well as the interaction capability between the ion and the PU segments. The FTIR analysis proved the occurrence of chemical interaction between segmented PU and lithium ion as well as sodium ion to form a coordinate bond and subsequently forming polymer-salt complexes. EIS reveals the highest ionic conductivity value achieved at 3.02 x 10^{-5} S/cm for SPE with LiCl. These results indicated that the waste cooking oil- based PU posed potential for the application as host in polymer electrolytes.

Acknowledgements

The authors gratefully acknowledge the support from Universiti Teknologi Mara Pahang in terms of financial and facilities provided. A highly gratitude goes to the Ministry of Education, Malaysia for the financial support through Fundamental Research Grant Scheme, (FRGS/1/2016/STG07/UITM/03/1).

References

[1] Mustapa, S R, Aung M M, Ahmad A, Mansor A and Khoon L T 2016 *Electrochimica Acta* 222 293
[2] Wong C S, Badri K H, Ataollahi N, Law K P, Su M S and Hassan N I 2014 *International Journal of Chemical and Molecular Engineering* 8 1168–75
[3] Daud, F N, Ahmad, A and Badri, K H 2014 *International Journal of Polymer Science* 1-5
[4] Akintayo, C O, Akintayo, E T, Thomas Z and Babalola, B M 2013 3 984–993
[5] Ali, A, Yusoh, K and Hasany, S F 2014 *Journal of Nanomaterials* 2014
[6] Sun, L J, Yao, C, Zheng, H F and Lin, J 2012 *Chinese Chemical Letters* 23 919-922
[7] Ferraro L G, Alvarenga F B S, Gelfuso M V and Thomazini D 2014 *Materials Science Forum* 775 351-356.
[8] W Salleh, SM Tahir, NS Mohamed Polymer Bulletin 75 (1), 109-120
[9] Enderus N F and Tahir S M 2017 *IOP Conference Series: Materials Science and Engineering* 271 012062
[10] Huzaizi R M, Tahir S M and Mahbor K M 2017 *AIP Conference Proceedings* 1901 090002
[11] Su'ait, M S, Ahmad, A, Badri, K H, Mohamed, N S, Rahman, M Y A, Ricardo, C L A and Scardi, P 2014 *International Journal of Hydrogen Energy* 39 3005-3017

[12] Ibrahim, S, Ahmad, A and Mohamed, N S 2015 *Polymers* 7 747-759
[13] Wang, S and Min, K 2010 *Polymer* 51 2621-2628
[14] Boschin, A and Johansson, P 2015 *Electrochimica Acta* 175 124-133

Measuring The Acoustical Properties of Fluids and Solid Materials Via Dealing With A-SCAN (GAMPT) Ultrasonic

Ammar A. Oglat [1, a], MZ Matjafril [b, *], Nursakinah Suardi [1, c], Mohammad A. Oqlat [d, 2], Mostafa A Abdelrahman [3, e], Ahmad A. Oqlat [3, f], Raed Abdalrheem [1, g]

[1] Department of Medical Physics and Radiation Science, School of Physics, Univirsti Sains Malaysia, 11800 Penang Malaysia
[2] Department of Biological Sciences, School of Science, Yarmouk University, Irbid Jordan
[3] Faculty of Medicine, Jordan University of Science and Technology, Irbid Jordan

mjafri@usm.my

Abstract. This article displays a general overview about the measurements of acoustical properties (speed of sound and attenuation) for both the fluids and solid materials via dealing with A-scan or A-mode GAMPT mbH ultrasound technique (ultrasound transmission and reception). In addition, measuring the thickness of protective layer of the probe and its influence on the speed of sound and attenuation of fluids and solids materials by applying the Pulse Echo ultrasound speed method and its formula. However, this method used to measure the distilled water and Doppler dummy fluid as a liquid items and acrylic plate with different sizes as solid materials. The outcomes resulted from the study were the speed of sound and attenuation measurements of both the fluids and solid materials via dealing with A-scan GAMPT mbH ultrasound technique were suitable. Furthermore, the protective layer thickness of the probe was influenced on the speed of sound measurements. So, we can remove the influence of inaccuracy by knowing the protective layer thickness and this way can increase the precise in ultrasonic speed.

1. Introduction

The fixing of the acoustical properties of fluids and solid materials is fundamental for medical process innovation. For instance, measuring the acoustical properties of blood or blood mimicking fluids (BMF) and its components. Also, measuring the acoustical properties of tissue mimicking material (TMM) as a solid material [1]. Theoretically, fluids and solids content can be measured by applying the analysis test of the ultrasonic features of the materials. The interaction of fluids or solids with ultrasonic wave alters the speed of the ultrasonic wave because of the reflection and absorption which occurs during this interaction. This alteration was studied and searched to measure the acoustical properties of fluids and solids. Studies on the relationship between pulse echo ultrasonic speed and fluids like BMF and solids like TMM have been carried out [2-7].

Generally, measurement of liquids and solids applying ultrasound method is done via the Amplitude modification analysis (A-mode). In A-scan, determination of ultrasound speed in materials can be done by Pulse Echo (PE) mode of ultrasound traveling mode. The ultrasonic speed of item is known as the proportion of the distance propagated by the ultrasound signal and the time of traveling

Content from this work may be used under the terms of the Creative Commons Attribution 3.0 licence. Any further distribution of this work must maintain attribution to the author(s) and the title of the work, journal citation and DOI.
Published under licence by IOP Publishing Ltd

that is also called the time of flight (TOF). The equation that is used to measure the ultrasound speed for pulse echo (PE) represented is

$$SS=\frac{2xl}{t} \tag{1}$$

Where SS indicates the ultrasonic speed, L is the travel distance of ultrasonic wave of the sample, while t represents the time of flight. The digit 2 is two times of echo go and return during the sample.

Typically, the precision of acoustical measurement applying ultrasonic speed rely on two main factors; the method and equipment of ultrasonic features. Ultrasonic wave propagates or travels two times in distance for pulse echo (PE) mode. Because of the much longer distance, the influence of both the refraction and reflection of the ultrasonic signal pulse is more important for PE [8].

Ultrasound probe made of piezoelectric crystal layer and then protective layer above the piezoelectric crystal layer as a coating layer to protect it. The presence of the protective layer increases attention on its influence during the determination of ultrasonic speed. The influence of the protective layer on ultrasonic speed measurements has been debated by Konrad W. Nowak [9] in his research on ultrasonic speed measurement way. The ultrasonic signal wave will be generated then travel within the protective layer of the probe before going to the medium or region of interest (experiment specimen). The protective layer of the probe with a constant thickness will produce the overvaluation of the present time of flight (TOF) measurement with fixed time delay [10, 11]. For precise ultrasonic speeds measurement, the influence of the protective layer must be completely removed through the measurement.

2. Materials and Methods

The initial researches were done by applying two acrylics plate structure with thickness or depth of 40 and 80 mm respectively and speed of sound 2700 m/s. The acrylic plate which utilized in this experiment was a homogeneous material. In addition, both the distilled water and Doppler dummy fluid were tested and examined their acoustical properties.

2.1 Handled and Processed Ultrasonic Pulse Echo (Preliminary Experimental Set-Up)

The echoscope and its parts were prepared. Then, the echoscope has been connected to the PC and switched on. The front panel of the device was controlled and examined. Echoscope supplied the electrical signal to send and receive ultrasound echo pulses. To begin with, reflection transducer plug connected with 1-5 MHz (red transducer) with 2.7 cm diameter. The centre frequency of this probe is 2 MHz. The knob which is responsible for reflection pulse in the device switched on to "Reflection" not "Transmission" mode, which authorizes the sonar ranging mode (Pulse echo). In the pulse echo (A-Scan) mode, the same probe sends brief or concise ultrasound pulses and then works as a receiver to discover the reflecting or returning echo signal. The amplification of transmission and receiving can be set. A low value of transmitter power was chosen, and the receiver amplifier power was increased until the highest signal peak amplitude, which is about 80% of maximum scale.

2.2 Ultrasonic Transducer (probes) used in Ultrasonic GAMPT Scan

The ultrasonic transducers used in this research was red colour transducers with frequency 1-5 MHz Figure 1. The ultrasonic transducers were coupled with powerful snap-in-connectors. The frequency of the transducer is automatically known from the appliance. With the assistance of the adjustable sending and receiving power, it can appropriate the ultrasound signal to almost each arbitrary realization object [12].

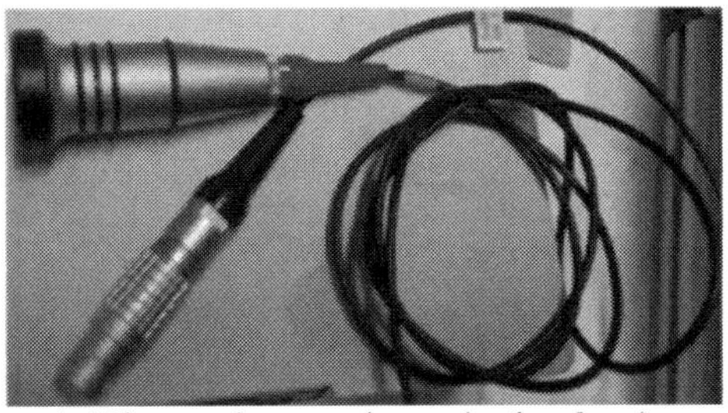

Figure 1. Ultrasonic transducer (red colour).

2.3 Measure the Velocity (Speed) of Sound and Protective Layer Thickness of the Probe

The typical formula for determining the speed of ultrasound using pulse echo (PE) method was displayed in equation 1. Thus, the speed of ultrasound can be measured by using PE method, by measuring the time of flight (TOF) between the highest two following peaks (peak to peak) of transmitted and received wave and then use the depth or distance of the sample. In addition, the speed of ultrasound can be measured by measuring the time of flight (TOF) between two identical peak signal pulses. Furthermore, the average measurements between the highest two following peaks and two identical peak signal pulses of transmitted and received wave were taken in account. However, the protective layer of the transducer can affect the TOF, because the measured value of TOF does not just point to the travelled time of ultrasound signal wave in the specimen, but it also represents the travelled time in the protective layer of transducer [13]. Figure 2. For good precision of ultrasonic speed measurement, protective layer thickness should be measured first by following the Equation 2. To measure the thickness of transducer protective layer, we have used two similar homogeneous acrylic transparent plates with a thickness of 40mm and 80mm. To correct the A-scan system before measuring the acoustic properties of fluid samples and solids, the speed of sound and attenuation in distilled water and in Doppler dummy fluid in our experimental setup system was measured. To measure the velocity of sound of distilled water without the thickness of the protective layer by applying Equation 1. Moreover, for measuring the velocity of sound of distilled water with thickness of protective layer by applying Equation 3.

$$SS = \frac{2xl}{t} \qquad (1)$$

$$d_{pl} = \frac{(\frac{t1}{t2}\,da2) - da1}{2(1 - \frac{t1}{t2})} \qquad (2)$$

$$SS = \frac{2\,(l + dpl)}{t} \qquad (3)$$

Where l is sample thickness or distance, dpl is the thickness of transducer protective layer, da1 is the thickness of acrylic plate (40mm), da2 is the thickness of acrylic plate (80mm), t1 is the time of flight of acrylic plate (40mm) and t2 is the time of flight of acrylic plate (80mm).

For measuring the protective layer thickness of the probe, two acrylic plates have been used. Because the ultrasonic probe was used as an emitting and receiving at the same time, so, it was placed above the 40 and 80 mm of the acrylic plate. TOF for both two acrylic plates were registered. By applying Equation 2., the thickness of probe protective layer (dpl) was measured and the calculated value of the thickness of the protective layer was 0.515 mm.

For attenuation measurements, the frequency-based attenuation of the individual signal was measured by proceeding a Fast Fourier Transform (FFT) on the radio frequency (RF) signal from the reflector. Attenuation unit in dB, it was calculated by the natural log (In) difference between the signal waves by using Equation 4. After knowing the speed of sound of the sample, inserting the value in pulse-echo techniques (A-scan system) to calculate the attenuation coefficient of the sample (distilled water) by the following equation:

$$\alpha = \frac{1}{x1 - x2} \ln \frac{A2}{A1} \quad (4)$$

Where_ α is the attenuation coefficient of sample, x1 and x2 is the difference in distance or depth of sample in mm, A1 is the power signal amplitude at frequency f and x, y position of (reference signal) with no presence sample (amplitude of transmitted signal wave), A2 is the power signal amplitude at frequency f and x, y position through the sample (amplitude of received signal wave).

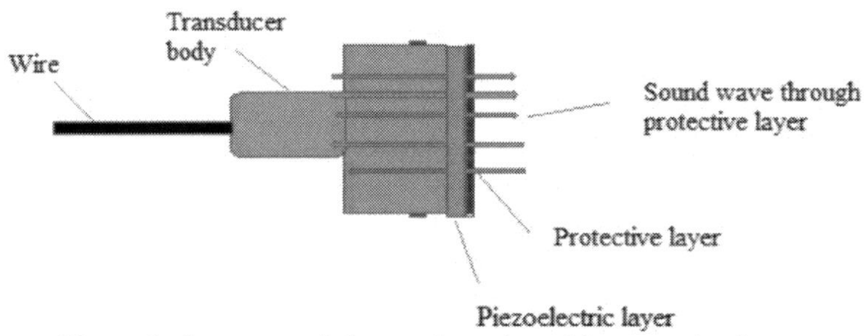

Figure 2. Structure of ultrasonic probe with protective layer.

3. Results and Discussion

From the initial study, the overall calculated value of the protective layer thickness is 1.03 mm, because the ultrasonic signal wave travelled two times into the protective layer (go and return), thus the half distance was wrapped by the ultrasonic wave signal in the protective layer was 0.515 mm. In addition, the speed of sound measurements that resulted of TOF by using PE method between the average highest two following peaks of transmitted and received wave were the same speed of sound measurements that resulted of TOF between the average identical or similar peaks. See Figures 1. and 2. Furthermore, the protective layer thickness of this probe in our research study (0.512 mm) different from protective layer thickness of previous literature review which was around (0.57 mm) [13]. Thus, the protective layer thickness of each probe should be measured. Moreover, in this research study, we found that the effectiveness of the protective layer thickness of the probe can be removed by measuring the two-similar material but difference in the thickness.

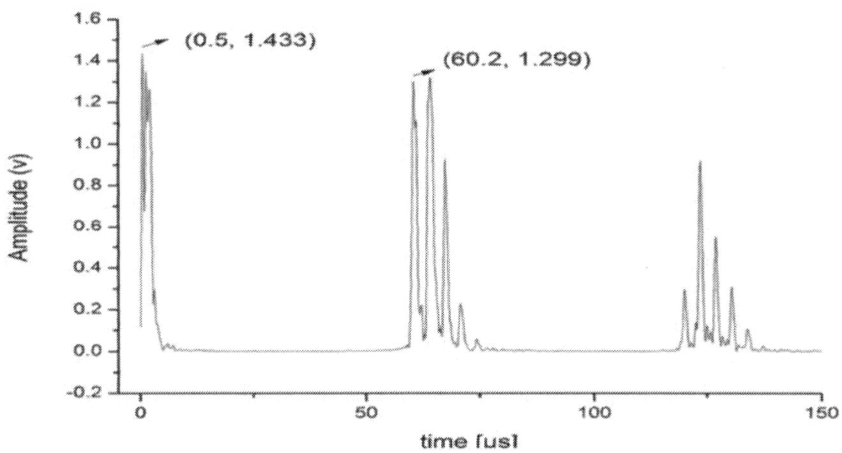

Figure 3(a). Transmission and reflection wavelength of distilled water at temperature 37±0.8°C for measurement of TOF by using PE method between the highest two following peaks of transmitted and received wave.

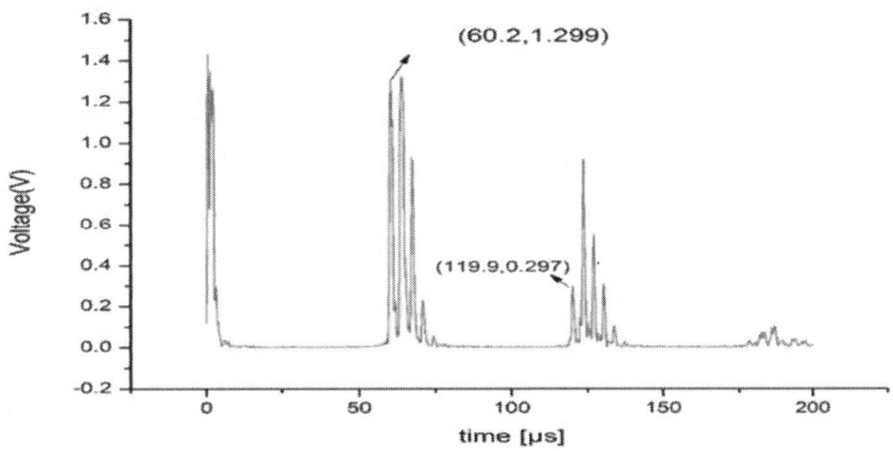

Figure 3(b). Transmission and reflection wavelength of distilled water at temperature 37±0.8°C. Measurement of distilled water ultrasonic speed using PE between two identical signal waves.

The speed sound in distilled water was measured in water tank container with using of ultrasonic echoscope. The ultrasonic transducer was fixed on the middle region of the water tank (Plexiglas) with using ultrasonic gel or directly above the sample without using ultrasonic gel. The transmitted signal wave is reflected due to the inhomogeneities of sound impedance and compressibility in water or fluids in general. Then, the reflected signal wave was discovered by the same transducer. Since the distance (d) (depth) of sample (water tank container) and the time of flight also measured, the longitudinal speed of sound can be measured by following Equation 1. using PE methods between the highest two following peaks of transmitted and received wave or by using the distance of the sample between two identical signal pulses. The water had been measured initially before the velocity of sound of mixture samples was measured. Firstly, we have measured speed of sound in water without taking into consider the protective layer thickness of the probe. The result was 1474 m/s without the protective layer thickness and 1508m/s with the thickness of the protective layer, and the difference is 34 m/s (see Figs. 1 and 2). From this result we can notice that the protective layer of the probe can affect the speed of sound result for the samples. However, the calculated value of the thickness of the protective layer is 0.515 mm. Because the ultrasonic signal wave propagates two times in the protective layer of the probe, the overall distance wrapped by the ultrasonic signal wave in the protective layer is 1.03 mm. Furthermore, the velocity of sound in Doppler dummy fluid measured in Plexiglas tank container with using of ultrasonic echoscope and using the same method of speed of sound measurement in distilled water. Thus, the velocity of sound was 1687 m/s without measuring the protective layer thickness of probe while it was 1810 m/s with measuring the protective layer thickness of probe, and the difference was 123 m/s. However, speed of sound measurement of distilled water and Doppler dummy fluid was repeated many times with adjusting temperature of mixture fluid to be like the human body temperature as much as possible, because the velocity of sound changes with temperature changes. In addition, the attenuation coefficient measurements changes with temperature changes. Thus, the acoustical properties in blood mimicking fluid were measured at temperature like the human body temperature.

From previous measurements of both distilled water and Doppler dummy we can note that the speed of sound measurements with measuring the thickness of the protective layer of probe are better than without the thickness of protective layer since the speed of sound in distilled water is1480 ±30 m/s [6]and in Doppler dummy fluid is 1800 ±30 m/s [14]. Moreover, in this research study, we found that the acoustical properties can be measured not only through the two identical signal waves, but between the two highest signal waves.

4. Conclusion

Results from the initial search study pointed out that we can measure the acoustical properties (speed of sound and attenuation) of both solid materials and liquids by dealing with A-scan GAMPT technique. As a conclusion, the acoustical features of the liquid and solid materials were corresponding to the standard values. In addition, we can measure the longitudinal speed of sound by using PE method, not only between the highest two following peaks of transmitted and received wave but also between two identical signal pulses. Furthermore, we can remove the influence of inaccuracy by knowing the protective layer thickness. Thus, we can increase the precise in ultrasonic speed.

References

[1] Browne, J.E., et al., Validation of a sensitivity performance index test protocol and evaluation of colour Doppler sensitivity for a range of ultrasound scanners. Ultrasound in medicine & biology, 2004. 30(11): p. 1475-1483.
[2] Ramnarine, K.V., et al., Validation of a new blood-mimicking fluid for use in Doppler flow test objects. Ultrasound in medicine & biology, 1998. 24(3): p. 451-459.
[3] Zhou, X., et al., Fabrication of Two Flow Phantoms for Doppler Ultrasound Imaging. IEEE transactions on ultrasonics, ferroelectrics, and frequency control, 2017. 64(1): p. 53-65.
[4] Kenwright, D.A., et al., Wall-less Flow Phantom for High-Frequency Ultrasound Applications. Ultrasound in medicine & biology, 2015. 41(3): p. 890-897.
[5] Teirlinck, C.J., et al., Development of an example flow test object and comparison of five of these test objects, constructed in various laboratories. Ultrasonics, 1998. 36(1-5): p. 653-660.
[6] Samavat, H. and J. Evans, An ideal blood mimicking fluid for doppler ultrasound phantoms. Journal of Medical Physics, 2006. 31(4): p. 275.
[7] Boote, E. and J. Zagzebski, Performance tests of Doppler ultrasound equipment with a tissue and blood-mimicking phantom. Journal of ultrasound in medicine, 1988. 7(3): p. 137-147.
[8] Krautkrämer, J. and H. Krautkrämer, Ultrasonic testing of materials. 2013: Springer Science & Business Media.
[9] Nowak, K.W. and M. Markowski, A comparison of methods for the determination of sound velocity in biological materials: A case study. Ultrasonics, 2013. 53(5): p. 923-927.
[10] Sarvazyan, A., A. Tatarinov, and N. Sarvazyan, Ultrasonic assessment of tissue hydration status. Ultrasonics, 2005. 43(8): p. 661-671.
[11] Norli, P., Sound velocity cell for gas characterization. 2007, PhD thesis, Department of Physics and Technology, University of Bergen, Bergen, Norway.
[12] Michael, S., and Grit, O., GAMPT Ultrasound cataloggue experemintal manual. 2009.
[13] Halim, M.A., N. Buniyamin, and Z. Mohamad. Improving intramuscular fat measurement by considering the thickness of protective layer in ultrasonic transducer. in Electrical, Electronics and System Engineering (ICEESE), 2014 International Conference on. 2014. IEEE.
[14] Michael, S., and Grit, O., GAMPT A-Scan Cataloge 2009.

Determination of Source's Location of a Dipole Field using Euler Relation

Nuraddeen Usman[1, 2, a*], Khiruddin Abdullah[1], and Mohd Nawawi[1]

[1] School of Physics, Universiti Sains Malaysia, 11800 Pulau Pinang
[2] Umaru Musa Yar'adua University P.M.B 2218, Katsina, Nigeria

[a]nu14_phy055@student.usm.my.

Abstract: Magnetic materials such as rocks and other geological structures can be treated as an assortment of magnetic dipoles. This concept can be used to simulate magnetic response of a body if its parameters are known. Likewise, source location of a concealed body can be retrieved if its field response is acquired and adequate magnetization contrast with the surrounding material exists. In this paper, a magnetic response of dipole source is simulated from the assumed model parameters in order to test a technique for estimating source location. The technique is based on the assumption that the structural indices of some simple sources are related to the geology of the solid Earth. The effect of inclination on the magnetic field was studied. The magnetic field depends on magnetization contrast and inclination of the location, this corroborate that the simulation carried out is valid. The simulated field of the dipole was processed and inversed using a new technique which is based on Euler relation. The result indicated that the technique performed very well in the estimate of horizontal position and depth of the dipole source despite the effect of inclination of the field. The technique is fast means of source's location estimate of concealed magnetized body. Filtering is recommended for the real field data because of the presence of so many sources and the Earth's magnetic field.

1. Introduction

Geophysical techniques have been applied to investigate the subsurface of the earth in order to explore geological structures of economic interest (in most cases) in areas of environmental studies [1,2], hydrology, hydrocarbons, geochemical [3], engineering [4], geothermal studies [5], geo-hazard assessment, and solid minerals [6]. With the aid of techniques used for inversion, it is possible to determine the horizontal and vertical positions of concealed metallic objects in the near vicinity of the earth's surface in addition to the delineation of deep-seated structures [7, 8, 9, 10, 11].

The pattern of magnetic anomaly depends on its position on the earth surface. The same structure placed at different geographical locations would give different anomaly's shape because of the variation in magnetic latitude. The dipolar nature of the magnetic field causes distortion in the anomaly's shape and as a result of this effect, error will be introduced to the data and thereby affecting the estimate of the anomaly's location [12]. Reduction to the pole (RTP) and equator (RTE) are the traditional techniques to correct the effect of inclination. While the use of RTP is recommended to be applied on the data prior to the application of Euler deconvolution [13], other researchers [14] are of the opinion that it should not be applied. In the present work, the main objective is to estimate source's location of a dipole field using fully automated Euler relation and to determine the effect of inclination on the technique. The significance of this research is that the unknown parameters in the Euler deconvolution

Content from this work may be used under the terms of the Creative Commons Attribution 3.0 licence. Any further distribution of this work must maintain attribution to the author(s) and the title of the work, journal citation and DOI.
Published under licence by IOP Publishing Ltd

can be accurately estimated simultaneously. Moreover, the effect of inclination on the automated technique can be studied using simulated data of a dipole model. Modeling is useful in both testing of new technique, survey design and educational setting. In this case, the possibility of applying RTP or any other related technique prior to the deconvolution will be deduced.

2. Methodology

Dipole model has been frequently used in geophysical literatures to test filters, performance assessment of depth estimation methods and source parameter imaging. A dipole model can be interpreted geologically as compact object such as material buried in a ferro-metallic container/drum, boulder, void, cavity, buried metallic materials, iron pilling and ancient features. A magnetic response due to a point dipole can be modeled in 3D Cartesian coordinate using the elements given as [15]:

$$Bx = K\frac{(3Dr_x - r^2 l)}{r^5} \quad (1)$$

$$By = K\frac{(3Dr_y - r^2 m)}{r^5} \quad (2)$$

$$Bz = K\frac{(3Dr_z - r^2 n)}{r^5} \quad (3)$$

In Equation (1, 2 and 3), Bx, By and Bz are the elements of the magnetic field due to a point dipole source, K is the dipole moment. The vector in the direction of point dipole towards the observation point (r) is given by the relation (Equation 4)

$$r = \left(r_x^2 + r_y^2 + r_z^2\right)^{1/2} \quad (4)$$

r_x, r_y and r_z are the components of r in x, y and z directions respectively. The quantities that represent the magnetization directions (l, m, n and D) are defined as:

$$l = \cos(a)\cos(b)$$
$$m = \cos(a)\sin(b)$$
$$n = \sin(a)$$
$$D = lr_x + mr_y + nr_z$$

where a and b are the inclination and declination in the direction of magnetization respectively. The magnetic declination (d) is the angle between the horizontal component and geographic north. This implies that the magnetic north is along the horizontal component (from the definition of declination).

The total magnetic anomaly due to a point dipole (ΔT) is related to the quantities representing the direction of geomagnetic field as (Equation 5):

$$\Delta T = LB_x + MB_y + NB_z \quad (5)$$
$$L = \cos(A)\cos(B) \quad (6)$$
$$M = \cos(A)\sin(B) \quad (7)$$
$$N = \sin(b) \quad (8)$$

In the above equations, A and B are the inclination and declination of the geomagnetic field respectively. In this case, B is the angle between horizontal component and geomagnetic north. The geographic north is along x-axis. To simplify the derivation, B is set to be equal to zero (B = 0) and it is assumed that the magnetization is formed by induction. This implies that a = A and b = B.

Substitute Equation 6, 7 and 8 into Equation 5 to obtain Equation 9. Equation 9 is used to simulate the field used in this research.

$$\Delta T = \cos A \cos B \,.\, K\frac{(3Dr_x - r^2 l)}{r^5} + \cos(A)\sin(B)\,.\,K\frac{(3Dr_y - r^2 m)}{r^5} + \sin A\,.\,K\frac{(3Dr_z - r^2 n)}{r^5}$$

$$\Delta T = K\frac{(3Dr_x l - r^2 l^2)}{r^5} + K\frac{(3Dr_y m - r^2 m^2)}{r^5} + K\frac{(3Dr_z n - r^2 n^2)}{r^5}$$

$$\Delta T = \frac{K}{r^5}\left(3Dr_x l + 3Dr_y m + 3Dr_z n - r^2 l^2 - r^2 m^2 - r^2 n^2\right)$$

But $D = lr_x + mr_y + nr_z$

$$\Delta T = \frac{K}{r^5}\left(3D^2 - r^2(l^2 + m^2 + n^2)\right)$$

And $l^2 + m^2 + n^2 = 1$ (trigonometry)

$$\Delta T = \frac{K}{r^5}(3D^2 - r^2) \tag{9}$$

2.1 Theory of the method

Euler deconvolution method is based on Euler's homogeneity relationship [16] it was first initiated to solve 2D magnetic field by [13]. Euler deconvolution equation is normally used in order to find the source location (x_0, y_0, z_0). Equation (10) forms the basis for the methodology used in this research and it is given as

$$(x - x_0)\frac{\partial F}{\partial x} + (y - y_0)\frac{\partial F}{\partial y} + (z - z_0)\frac{\partial F}{\partial z} = N\,(B - F) \tag{10}$$

where x, y and z are the observation point coordinates; x_0, y_0, and z_0 are the source locations; $\frac{\partial F}{\partial x}, \frac{\partial F}{\partial y}, \frac{\partial F}{\partial z}$ are the potential derivatives; N is the structural index; B is the background of field F [13].

2.2Multiple Linear Regression

The Multiple Linear Regression (MLR) method has been extensively applied in statistics to establish relationships among multiple variables (more than two) by fitting a straight line to the observed data. MLR model [17] can be expressed in Equation 11:

$$Y_i = \beta_0 + \beta_1 X_{1i} + \beta_2 X_{2i} + \cdots + \beta_k X_{ki} + \varepsilon_i \tag{11}$$

where β_0 is the intercept of Y_i; β_1, β_2 and β_k denote the slope of regression lines for each variable respectively, ε_i is the regression error term and Y_i is the dependent variable. To apply the MLR on 3D Euler deconvolution technique, Equation (10) can be written in the following form (Equation 11) as MLR method (Equation 12):

$$x\frac{\partial F}{\partial x} + y\frac{\partial F}{\partial y} + z\frac{\partial F}{\partial z} = NB + x_0\frac{\partial F}{\partial x} + y_0\frac{\partial F}{\partial y} + z_0\frac{\partial F}{\partial z} - NF + \varepsilon_i \tag{12}$$

By comparing Equation 11 and Equation 12, it could be observed that:

$Y_i = x\frac{\partial F}{\partial x} + y\frac{\partial F}{\partial y} + z\frac{\partial F}{\partial z}$; $\beta_0 = NB$; $\beta_1 = x_0$; $\beta_2 = y_0$; $\beta_3 = z_0$; $\beta_4 = N$; $X_1 = \frac{\partial F}{\partial x}$; $X_2 = \frac{\partial F}{\partial y}$; $X_3 = \frac{\partial F}{\partial z}$ and $X_4 = -F$.

All $\beta$is (β_0, β_1, β_2, β_3, β_4) are the coefficients (unknown) that need to be solved and all Xi values are the independent variables, in which the values are known from the data (input). Assuming that

Equation 3 is linear, the equation could solve 5 unknowns which are horizontal positions (x_0, y_0), depth (z_0), SI (N) and background (B).

2.3 Dipole model

The analysis of the model was carried out using 2D and 3D forms. In 2D form, the analysis is qualitative whereas in 3D form the analysis is quantitative.

2.3.1. 2D analysis

This model was simulated with the assumption that the surrounding is sandstone and the target is granite. The susceptibility contrast was 30 and magnetization was 0.00126 A/m. It was located at 12.5km in both x and y directions and the depth was 1 km. The radius of the cylinder was 0.4 km. It is designed to study the effect of inclination by analyzing the data obtained in profile and map forms. The profile passes directly at the centre (12.5km along y-direction) of the dipole in x direction. Using the mentioned parameters, the model is simulated using 7 inclinations (I = 0°, 15°, 30°, 45°, 60°, 75°, 90°) and a fixed declination (D = 0). In each case, the data along the mentioned profile is taken and plotted.

2.3.2. 3D analysis

This model is designed to simulate magnetic response of a granite boulder buried in sandstone environment. The magnetization contrast was 1 A/m. The depth to the centre and radius of the boulder were 1 and 0.5 km respectively. The horizontal coordinates in positive x directions was 12.4 km while that of y was 12.5 km. The anomalous magnetic field response using different inclinations (I = 0°, 15°, 30°, 45°, 60°, 75° and 90°) have been computed. The data was inverted using the window size of 9×9 grid points. The maximum distance to accept from the centre of convolution window and SI deviation were 9 units and 0.1 respectively. The maximum accepted regression error was 2%. The derivatives of the field (in x, y and z directions) were computed. The computed derivatives were used as an input to inverse the data using Equation 14. The estimated parameters were subjected to filtering and the final solutions are obtained. The mean and standard deviation of each of the parameter of the dipole source were tabulated. The source location was plotted and superimposed on analytic signal (AS).

3. Result and Discussion

The results obtained from the inversion of the simulated data is presented in both profile and map analysis.

3.1 2D Analysis

The anomalous field simulated using different inclinations are given in Figure 1. It could be observed that at 90°, the field is symmetrical directly above the position of the target (x = 12.4km). At this inclination, the field is positive (similar to the sign of magnetization). When the inclination is 0°, the field is also symmetrical, but the field takes negative value at peak. Other angles of inclination show different curve pattern along the profile. The pattern obtained in Figure 1, coincided with the ones obtained in the literature and this validates the modeling carried out in this study. Different pattern obtained from different inclination affect geophysical interpretation negatively and this effect needs to be corrected. In the case of techniques that make use of Euler relation, the field and its derivatives are used.

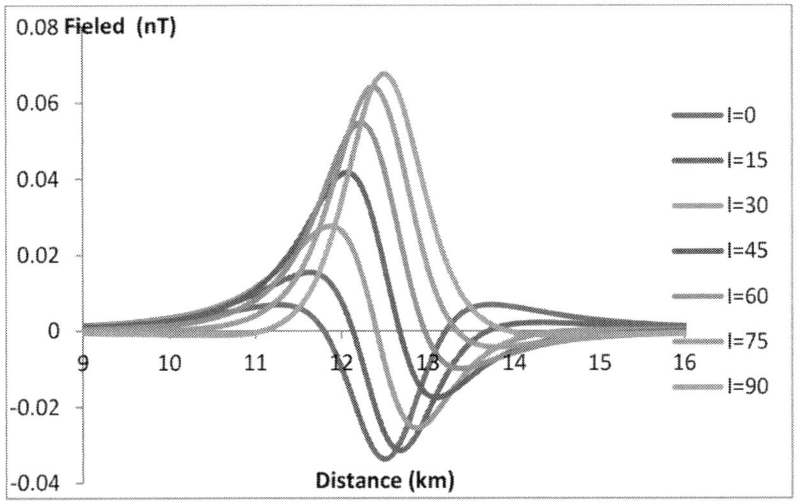

Figure 1: Anomalous magnetic field along a Profile (12.5 km along y-axis) for the 7 different inclinations

3.2. 3D Analysis

The anomalous magnetic field of isolated sphere model was simulated using different values of inclination (I = 0°, 15°, 30°,45°,60°,75° and 90°). The anomaly is positioned appropriately when the inclination is 0° and 90°, while for the rest of the values of I (15°, 30°, 45°, 60° and 75°) the positions have been slightly deviated from the true position. The most prominent effect of inclination could be observed at 45°, the centre of the anomaly is deviated from the true centre of the sphere. This effect of inclination is caused by the dipolar nature of the geomagnetic field [12] and it depends on the inclination as it has been observed in the plots (Figure 1). The mean estimate of each parameter together with its corresponding deviation are given in Table 1. It could observe from Table 1 that the parameters are estimated with good accuracy. The mean estimated values are about the same with the true parameters and the deviations of the estimates are minimum.

Table 1: Euler solutions of isolated sphere model

I	x km	$\hat{x}$ km	y km	$\hat{y}$ km	z km	$\hat{z}$ km	N	$\hat{N}$
0° (375)	12.5	12.5 ± 0.1	12.4	12.4	1	1.01 ± 0.03	3	3 ± 0.1
15° (375)	12.5	12.5 ± 0.1	12.4	12.4	1	1.005±0.035	3	3 ± 0.1
30° (367)	12.5	12.5	12.4	12.4	1	1 ± 0.03	3	3 ± 0.1
45° (334)	12.5	12.5	12.4	12.4	1	1.005 ± 0.035	3	3 ± 0.1
60° (303)	12.5	12.5	12.4	12.4	1	1.005 ± 0.035	3	3 ± 0.1
75° (219)	12.5	12.5	12.4	12.5	1	1 ± 0.03	3	3±0.1
90° (189)	12.5	12.5	12.4	12.4	1	1 ± 0.03	3	3 ± 0.1

4. Conclusion

A new approach of solving Euler relation that compute depth, horizontal position, structural index and background level has been tested using theoretical modeling. The simulated field of the dipole was processed and was inversed using a new technique that based on Euler relation. The result indicated that the technique performed very well in the estimate of horizontal position and depth of the dipole source despite the effect of inclination of the field. The mean values of the estimated parameters were about the same with the true parameters, the highest deviation obtained is ±0.1. The technique was fully automated, unlike the conventional technique that was semi-automated. The technique was fast means of source's location estimated of concealed magnetized body. Filtering is recommended for the real field data because of the presence of so many sources and the Earth's magnetic field. Many theoretical models shall be used to ascertain the effect of inclination.

References

[1] Loke M.H., Chambers, J.E., Rucker, D.F., Kuras, O., and Wilkinson, P.B., 2013. New developments in the direct-current geoelectrical imaging method. Journal of applied geophysics 95, 135-156.

[2] Nuraddeen Usman, Khiruddin Abdullah., Mohd Nawawi and Amin Esmail K, 2018. New approach of solving Euler deconvolution relation for the automatic interpretation of magnetic data. Terrestrial Atmospheric and Oceanographic sciences Journal (In press). Available at http://tao.cgu.org.tw/index.php/articles/archive/geophysics/item/1570-2017100501t

[3] Yang J., Agterberg F.P and Cheng Q., 2015. A novel filtering technique for enhancing mineralization associated geochemical and geophysical anomalies. Computers and Geosciences 27 (2001), pp. 1019-1029.

[4] Nuraddeen Usman, Khiruddin Abdullah., Mohd Nawawi, and Amin Esmail K., 2017: "Magnetic Survey for the Delineation of Concrete Pillars for Site Characterization" Electronic Journal of Geotechnical Engineering (22.10), pp 4011-4030. Available at ejge.com.

[5] Khalil, A.E., Nawawi, M., Arifin, M.H., Abdullah, F.M., Kayode, J.S., Usman, N. and Arisona, 2017. Soil Investigation at Wet World Hot Spring Complex for Future Development using Active Multichannel Analysis of Surface Waves. Sains Malaysiana, 46(4), pp.537-543.

[6] Arisona, A., Nawawi, M., Nuraddeen, U.K. and Hamzah, M., 2016. A preliminary mineralogical evaluation study of natural asphalt rock characterization, southeast Sulawesi, Indonesia. Arabian Journal of Geosciences, 9(4), p.272.

[7] Gerovska, D. and Araúzo-Bravo, M.J., 2003. Automatic interpretation of magnetic data based on Euler deconvolution with unprescribed structural index. Computers & Geosciences, 29(8), pp.949-960.

[8] Gerovska, D., Araúzo-Bravo, M.J., Stavrev, P. and Whaler, K., 2010. MaGSoundDST—3D automatic inversion of magnetic and gravity data based on the differential similarity transform. Geophysics, 75(1), pp.L25-L38.

[9] Cooper, G.R., 2014. Using the analytic signal amplitude to determine the location and depth of thin dikes from magnetic data. Geophysics, 80(1), pp.J1-J6.

[10] Cooper, G.R. and Whitehead, R.C., 2016. Determining the distance to magnetic sources. Geophysics, 81(2), pp.J25-J34.

[11] Salem, A., Williams, S., Fairhead, D., Smith, R. and Ravat, D., 2007. Interpretation of magnetic data using tilt-angle derivatives. Geophysics, 73(1), pp.L1-L10.

[12] Araffa, S.A.S., Santos, F.A.M. and Arafa-Hamed, T., 2012. Delineating active faults by using integrated geophysical data at northeastern part of Cairo, Egypt. NRIAG Journal of Astronomy and Geophysics, 1(1), pp.33-44.

[13] Thompson, D.T., 1982. EULDPH: A new technique for making computer-assisted depth estimates from magnetic data. Geophysics, 47(1), pp.31-37.

[14] Reid, A. B., Allsop, J. M., Granser, H., Millett, A. T., & Somerton, I. W. ,1990. Magnetic interpretation in three dimensions using Euler deconvolution. Geophysics, 55(1), 80-91.

[15] Salem, A., Ravat, D., Gamey, T.J. and Ushijima, K., 2002. Analytic signal approach and its applicability in environmental magnetic investigations. Journal of Applied Geophysics, 49(4), pp.231-244.

[16] John, R.C.F: "Introduction to Calculus and Analysis" 1965 (XXIII), 661.

[17] Levine, D.M., Ramsey, P.P. and Smidt, R.K., 2001. Applied statistics for engineers and scientists: using Microsoft Excel and Minitab. Pearson.

pH Sensing Characteristics of CuS/ZnO Thin Film Implemented as EGFET

Fayroz A. Sabah[1,2*], Naser M. Ahmed[1], and Z. Hassan[3]

[1]School of Physics, Universiti Sains Malaysia, 11800 USM, Penang, Malaysia
[2]Department of Electrical Engineering, College of Engineering, Al-Mustansiriya University, Baghdad, IRAQ
[3]Institute of Nano Optoelectronics Research and Technology (INOR), Universiti Sains Malaysia, 11800 USM, Penang, Malaysia

*fayroz_arif@yahoo.com

Abstract. Copper sulphide (CuS) is one of the most important semiconductor materials used in many applications for its semiconducting characteristics. CuS nanoscale thin films were deposited on various substrates using various methods of deposition; recently CuS thin films were used as extended gate field effect transistor (EGFET) and implemented as a pH sensor. In this work, CuS thin film was deposited on ZnO layer using spray pyrolysis deposition (SPD). CuS solution (0.4 M) was prepared from copper(II) chloride and sodium thiosulfate dissolved in deionised water. the precursors used to prepare CuS solution were copper chloride and sodium thiosulfate with 0.4 M concentration, and these precursors were solved using deionized water. The structural characteristics of this thin film show two phases for CuS; covellite and chalcocite with grain size of 31.2 nm. Nanoplate structure with a lot of aggregations was achieved from this deposition and it is confirmed by morphological examination, which estimates the roughness of the film to be 0.145 μm. CuS/ZnO thin film was used as EGFET and applied as pH sensor; the sensitivity and hysteresis were measured for this sensor to be 23.3 mV/pH and 17.5 mV, respectively.

1. Introduction

Determination of pH value was very necessary for a lot of biological and biochemical processes [1]. The ion-sensitive field-effect transistor (ISFET) was first proposed by Bergveld in 1970 in neurophysiological measurement [2]. An ISFET was modified to metal oxide field-effect transistor (MOSFET) which includes metal electrode gate [3], while in case of ISFET the reference electrode plays the role of the gate [4]. Furthermore, the MOSFET was alternated to extended gate field-effect transistor (EGFET) which is investigated by Spiegel et al. in 1983 [5] to isolate the FET from the chemical environment, EGFET can be described as the structure that contains chemically sensitive thin film at the end of the signal line which extended from the FET gate electrode [3]. The EGFET was distinguished from ISFET by many advantages such as; low cost, less insensitivity induced by temperature and light fluctuations, high shape flexibility of the extended gate structure, simpler to passive and package, better long-term stability [6], and disposable gate [7]. Furthermore, the method of using a disposable thin film enables the exchange of the sensing thin film when a chemical damage occurs, it will be easier than fabricating a new device [3].

Content from this work may be used under the terms of the Creative Commons Attribution 3.0 licence. Any further distribution of this work must maintain attribution to the author(s) and the title of the work, journal citation and DOI.
Published under licence by IOP Publishing Ltd

The chemical reaction depends on the number of the surface sites per unit area (Ns) which is also called surface site density, according to the site-binding model [1].

The sensitivity parameter β(italic) was estimated using the equation [8] expressed below:

$$\beta = \frac{2q^2 N_S \left(\frac{K_a}{K_b}\right)^{1/2}}{kTC_{DL}} \tag{1}$$

where q is the electron charge, N_S represents the surface site density, K_a and K_b are the acidic and basic constants, respectively, k is the Boltzmann constant, T indicates the absolute temperature and C_{DL} is the capacitance of the electrical double layer.

A lot of thin films of different materials were used to implement as a pH sensor, some of these materials which exhibit high pH sensitivity were PdO [9] and ZnO [6] thin films. Recently CuS thin films also implemented as pH sensors under various conditions and parameters [3, 10-12]. Due to the high sensitivity of ZnO sensor and good semiconducting characteristics of CuS material, the CuS/ZnO thin film was fabricated in this work to implement as pH sensor to modify the sensitivity of CuS sensor.

To the best knowledge of authors, there is no previous application for the CuS/ZnO as a pH sensor. In this work, the structural, morphological and sensing characteristics were studied, and the sensing mechanism was analyzed to highlight the main parameters which affect the pH sensor quality.

2. Methodology

2.1 Thin Film Synthesis

$CuCl_2.2H_2O$ (0.4 M) and $Na_2S_2O_3.5H_2O$ (0.4 M) were dissolved separately in deionized water using ultrasonic instrument. Then these two solutions were mixed at room temperature to form CuS solution (0.4 M), this solution was sprayed onto 100 nm thickness of ZnO layer (deposited by RF sputtering onto glass substrate) by using spray system. This system was used to synthesis CuS thin film in nanoscale; it is very simple and low-cost system to assemble. The substrate temperature used for deposition was fixed at 200 °C and the distance between substrate and airbrush nozzle was 30 cm, this distance was suitable for solvent evaporation and helps the chemical reaction to form the CuS nanoplate structure on the substrate. The spray rate was (1:1) s (ON: OFF) for 10 times then switch off the spray for 1 minute to keep the temperature of the substrate at the specified range, this cycle was repeated 6 times. Number of cycle was used to determine the thickness of thin film, after completing the process. The thin film was kept on the heater for about 1 hour to complete the growth of the nanostructure particles. The CuS/ZnO thin film was analysed to check its structural and morphological characteristics before the sample was being used as EGFET for pH sensing application.

2.2 Characterization Techniques

Phase purity and crystal structure were analysed by using PANalytical X-ray diffractometer (XRD) equipped with CuKα source (λ=0.15418 nm). Morphological observations were obtained by NOVA NANOSEM 450 field emission scanning electron microscopy (FESEM). Surface roughness of the thin films was examined by BRUKER, Dimension edge atomic force microscopy (AFM). And finally, for sensing measurements; pH sensitivity and hysteresis were measured using Keithley, Semiconductor Characterization System (2400-SCS).

2.3 pH Sensing and Hysteresis Systems

The system used to measure pH sensitivity of the CuS/ZnO thin film was sketched in Figure 1(a). It consists of two Keithley instruments; one was used to set the gate-source voltage (V_{gs}) at 3 V and measure the drain-source current (I_{ds}) response versus drain-source voltage (V_{ds}) for pH buffer solution (2, 4, 6, 8, 10 and 12), this measurement gives saturation regime. And the second Keithley instrument was used to set (V_{ds}) at 0.3 V and measure (I_{ds}) response versus (V_{gs}), the output curves formed the linear regime. The hysteresis was measured using the setup shown in Figure 1(b), this setup is working without

IOP Conf. Series: Journal of Physics: Conf. Series **1083** (2018) 012055 doi:10.1088/1742-6596/1083/1/012055

MOSFET and have only one Keithley instrument to find the potential difference between the reference electrode and the sensing thin film. The output signal can be expressed as [13]:

$$V_{OUT} = V_{IN}^{+} - V_{IN}^{-} = V_{ERF} - V_{SENSING-FILM} \quad (2)$$

where V_{IN}^{+} and V_{IN}^{-} are the two-input terminal voltage of the Keithley instrument, V_{ERF} and $V_{SENSING-FILM}$ are the reference electrode voltage and the sensing film voltage, respectively. For this measurement, Keithley instrument was used as a voltmeter to measure the voltage difference between reference electrode and sensor surface at acidic, basic and neutral environment as the following sequence of pH (7-4-7-10-7). The hysteresis was measured by calculate the difference between the last value and first value of pH7.

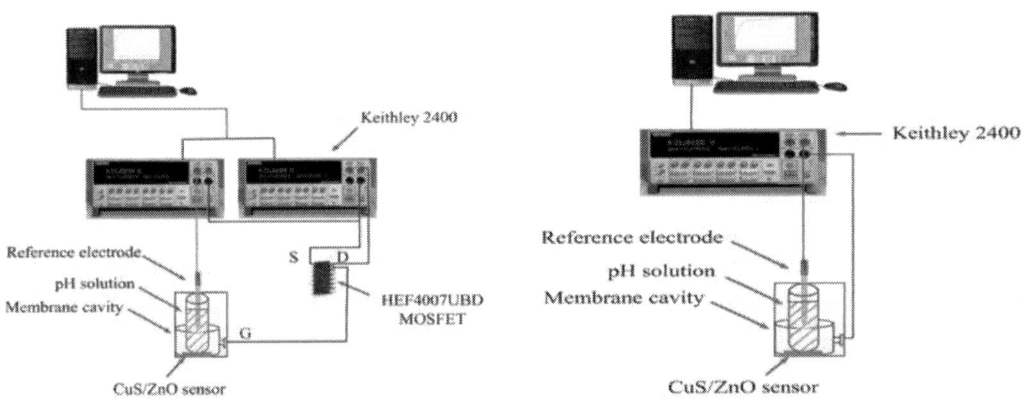

(a) (b)

Figure 1: pH sensing and hysteresis systems

3. Results and Discussion

3.1 Structural Analysis

Structural characteristics of CuS/ZnO thin film were characterisedusing X-ray diffractometer as shown in Figure. 2. The XRD image illustrates that this film includes two phases of CuS; covellite CuS phase with orientations indexed as (103) and (206) and chalcocite Cu_2S phase with single peak indexed as (002), with a hexagonal unit cell. Furthermore, there is a weak peak indexed as (002) belongs to ZnO material; among all these peaks and phases the covellite CuS phase is the dominant phase.

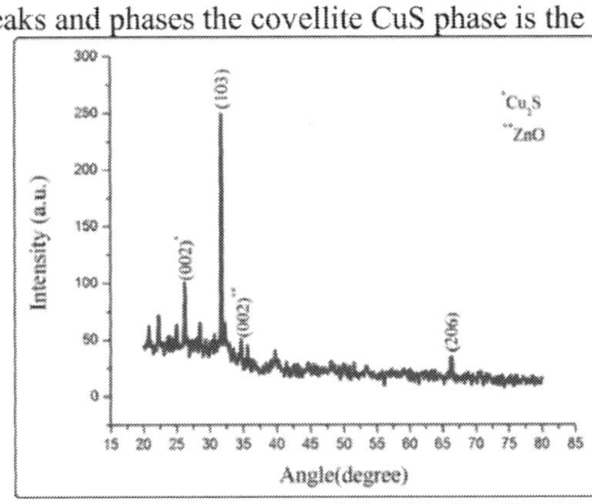

Figure 2: XRD image of CuS/ZnO EGFET.

The grain size was calculated from the Scherrer equation [14] as expressed below:

$$D = \frac{0.9\lambda}{\beta cos\theta} \qquad (3)$$

where λ is the wavelength of the CuKα radiation, β is the FWHM in radius, and θ is the Bragg angle. The average grain size of CuS nanoparticles was estimated to be 31.2 nm.

3.2 Morphological and Roughness Characteristics

FESEM image of CuS/ZnO thin film described the formation of nanoparticles on the surface of thin film, as illustrated in Figure 3. These nanoparticles include nanoplates and large amount of agglomerations, which is due to uneven heating on the substrate. The surface roughness of the CuS/ZnO thin film was measured using AFM to be 0.145 μm. The image of surface roughness was illustrated in Figure 4.

Figure 3: FESEM image of CuS/ZnO EGFET.

Figure. 4: AFM image of CuS/ZnO EGFET.

3.3 pH Sensitivity, Linearity and Hysteresis

Two regimes were obtained from the measurement using system shown in Figure. 1.a. These regimes represent the response of drain current according to drain and gate voltages. The essential concept of pH sensor is the reaction between hydrogen ions available within the electrolyte and metal ions available on the sensor surface. This reaction leads to ion-exchange between electrolyte and sensor surface and that affects the gate charge, which is the sensor itself. Metal oxide could be hydrolyzed in the presence of water, forming hydroxide layers (M-OH) throughout its surface. Water molecules that available in the electrolyte could be adsorbed onto the sensor surface of the scattered oxide particles; the sensor surface may attract and adsorb additional layers of polar water molecules due to the polar hydroxyl groups (-OH). The oxide or hydroxide surface may be charged by reacting with H^+ or OH^- ions according to the surface amphoteric reactions, which can be expressed in the following two equations [15]:

$$M - OH + H^+ \leftrightarrow M - OH_2^+ \qquad (4)$$

$$M - OH + OH^- \leftrightarrow M - O^- + H_2O \qquad (5)$$

In acidic environment (high amount of H^+ and less pH value), hydroxide sensor surface adsorbs protons [H^+] producing positive charge surface ($M - OH_2^+$). Whilst, in basic environment (high amount of OH^- and high pH value), the surface loses protons producing negative charge surface ($M - O^-$). The sensing process depends on these sites and the surface charge which were controlled by the pH of the solution [15]. Since the sensor is the extended gate of the MOSFET, then the positive charge accumulated on it causes increasing conduction channel (drain and source), which increase the current flow through this channel.

The relation between V_{gs} and pH value was plotted in Figure. 5 to calculate the pH voltage sensitivity which is 23.3 mV/pH and linearity 96.1%. The hysteresis of the pH sensor indicates the delay in pH response due to the surface defect of the sensor, this defect belongs to the nanocrystals in the sensor surface or underneath the surface, low hysteresis value denotes the quality of the sensor surface [3]. The result of hysteresis measurement was shown in Figure 6, its value was estimated using Equation 2, to be 17.5 mV, which consider reasonable value compared to a-Si:H and a-WO_3 sensors which have hysteresis of 17.9 and 26.0 mV, respectively [16, 17]. The sensitivity and hysteresis values of CuS/ZnO sensor and other pH sensors were listed in Table 1. As can be seen from the table, CuS/ZnO sensor provides good sensing characteristics compared to many other pH sensors. As obtained in the previous paper [18], the pH sensing characteristics for CuS thin film can be affected by the type of the substrate used, because the substrate can modify the structure and crystal orientation of the thin films. Glass, Si and tungsten substrates have been used in the previous paper [18], while in this paper, ZnO was used as a substrate and the pH sensing characteristics have been discussed for this type of substrate and its effect on the structure and crystal orientation.

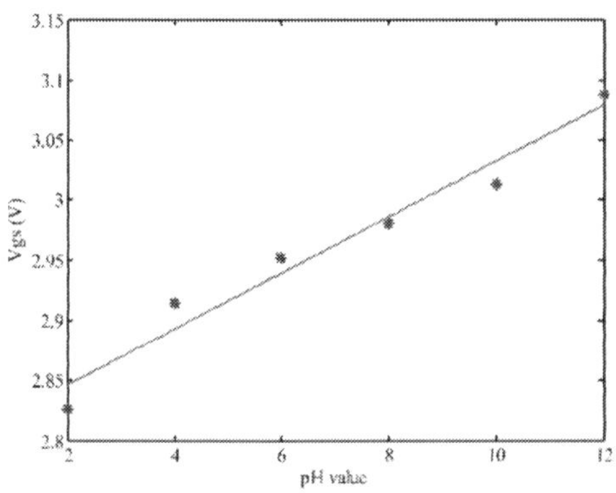

Figure 5: Sensitivity plot of CuS/ZnO EGFET.

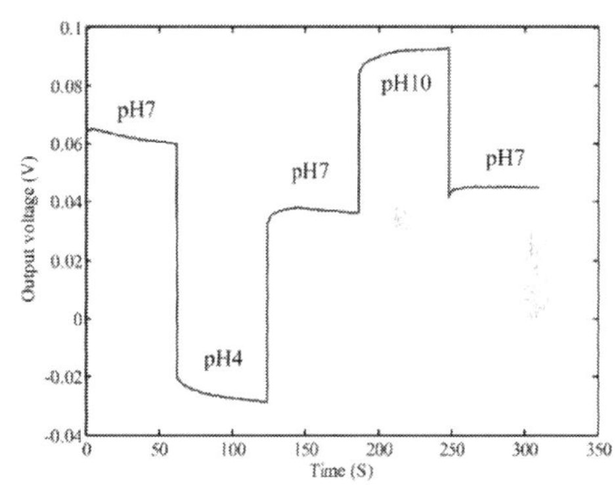

Figure 6: Hysteresis of CuS/ZnO EGFET.

Table 1: Comparison between sensing parameters of CuS/ZnO with other pH sensors.

pH sensor	Method of fabrication	pH solution range	pH voltage Sensitivity [mV/pH]	Linearity [%]	Hysteresis [mV]	Loop time [minute]	Ref.
CZTSe	Solution-based method	4-10	9	-	-	-	[14]
CuS using (DIW:Et as solvent)	Spray pyrolysis deposition	2-12	22.9	93.8	23.75	5	[12]
CuS in dark	Spray pyrolysis deposition	2-12	23	96.5	2.66	5	[11]

CuS/glass	Spray pyrolysis deposition	2-12	24	98.18	12.8	5	[3]
CuS/ZnO	Spray pyrolysis deposition	2-12	23.3	96.1	17.5	5	This work

4. Conclusion

CuS/ZnO thin film was deposited onto glass substrate using spray pyrolysis method at substrate temperature 200°C. XRD analysis exhibited two CuS phases; covellite CuS and chalcocite Cu_2S phases with weak peak of ZnO. The average grain size of these nanoparticles was 31.2 nm. FESEM and AFM results show difference in nanoplate's diffusion on the surface with many agglomerations, the surface roughness was found to be 0.145 µm. The pH voltage sensitivity and linearity were calculated in this study to be 23.3 mV/pH and 96.1%, respectively. The hysteresis of this sensor which described the delay in pH response was 17.5 mV. These results were seemed to be reasonable to use CuS/ZnO thin film as a pH sensor compared to many various pH sensors.

Acknowledgement

The authors gratefully acknowledge the financial support of the University Sains Malaysia and Grant number RU Top- Down grant 1001/CINOR/870019.

References

[1] H. H. Li, C. E. Yang, C. C. Kei, C. Y. Su, W. S. Dai, J. K. Tseng, P. Y. Yang, J. C. Chou, H. C. Cheng, Coaxial-structured ZnO/silicon nanowires extended-gate field-effect transistor as pH sensor, Thin Solid Films 529 (2013) 173-176.
[2] P. Bergveld, Development of an ion-sensitive solid-state device for neurophysiological measurements, IEEE Trans. Biomed. Eng. 1 (1970) 70-71.
[3] F. A. Sabah, N. M. Ahmed, Z. Hassan, M. A. Almessiere, N. H. Al-Hardan, Sensitivity of CuS membrane pH sensor with and without MOSFET, JOM 69 (2017) 1134-1142.
[4] K. D. Sattler, Fundamentals of Picoscience, Taylor & Francis Group, LLC, 2014.pages
[5] J. van der spiegel, I. Lauks, P. Chan, D. Babic, The extended gate chemically sensitive field effect transistor as multi-species microprobe, Sensors and Actuators 4 (1983) 291-298.
[6] P. Y. Yang, J. L. Wang, P. C. Chiu, J. C. Chou, C. W. Chen, H. H. Li, H. C. Cheng, pH sensing characteristics of extended-gate field-effect transistor based on Al-doped ZnO nanostructures hydrothermally synthesized at low temperatures, IEEE Electron Device Lett. 32 (2011) 1603-1605.
[7] Y. S. Chiu, C. Y. Tseng, C. T. Lee, Nanostructured EGFET pH sensors with surface-passivated ZnO thin-film and nanorod array, IEEE Sens. J. 12 (2012) 930-934.
[8] N. H. Al-Hardan, M. A. Abdul Hamid, N. M. Ahmed, A. Jalar, R. Shamsudin, N. K. Othman, L. Kar Keng, W. Chiu, H. N. Al-Rawi, High sensitivity pH sensor based on porous silicon (PSi) extended gate field-effect transistor, Sensors 16 (2016) 839.
[9] A. Das, D. H. Ko, C. H. Chen, L. B. Chang, C. S. Lai, F. C. Chu, L. Chow, R. M. Lin, Highly sensitive palladium oxide thin film extended gate FETs as pH sensor, Sensor Actuat B- Chem 205 (2014) 199-205.
[10] F. A. Sabah, N. M. Ahmed, Z. Hassan, N. H. Al-Hardan, Sensitivity of CuS and CuS/ITO EGFETs implemented as pH sensors, Appl. Phys. A. Mater. Sci.Process 122 (2016) 839.
[11] F. A. Sabah, N. M. Ahmed, Z. Hassan, M. A. Almessiere, Effect of light on the sensitivity of CuS thin film EGFET implemented as pH sensor, Int. J. Electrochem. Sci. 11 (2016) 4380-4388.
[12] F. A. Sabah, N. M. Ahmed, Z. Hassan, M. A. Almessiere, Study effects of thin film thickness on the behavior of CuS EGFET implemented as pH sensor, Dig. J. Nanomater Bios. 11 (2016) 787-

793.
[13] J. L. Wang, P. Y. Yang, T. Y. Hsieh, C. C. Hwang, M. H. Juang, pH-sensing characteristics of hydrothermal Al-doped ZnO nanostructures, Journal of Nanomaterials 2013 (2013) 13.
[14] P. Y. Lee, S. P. Chang, P. J. Kuo, E. H. Hsu, S. J. Chang, S. C. Shei, Sensing performance of EGFET pH sensors with CZTSe nanoparticles fabricated on glass substrates, Int. J. Electrochem. Sci. 8 (2013) 3866-3875.
[15] A. Degen, M. Kosec, Effect of pH and impurities on the surface charge of zinc oxide in aqueous solution, JECS 20 (2000) 667-673.
[16] J. C. Chou, C. N. Hsiao, The hysteresis and drift effect of hydrogenated amorphous silicon for ISFET, Sensor Actuat B- Chem 66 (2000) 181-183.
[17] J. L. Chiang, S. S. Jan, J. C. Chou, Y. C. Chen, Study on the temperature effect, hysteresis and drift of pH-ISFET devices based on amorphous tungsten oxide, Sensor Actuat B- Chem 76 (2001) 624-628.
[18] F. A. Sabah, N. M. Ahmed, Z. Hassan, M. A. Almessiere, Influences of substrate type on the pH sensitivity of CuS thin films EGFET prepared by spray pyrolysis deposition, Mater Sci Semicond Process, 63 (2017) 269-278.

IOP Conf. Series: Journal of Physics: Conf. Series **1083** (2018) 012056 doi:10.1088/1742-6596/1083/1/012056

Hierarchical Multiple Linear Regression for Fast Estimation of Subsurface Resistivity from Apparent Resistivity Measurements Based on Dipole-Dipole Array

S B Muhammad[1,2] and R Saad[1]

[1] Geophysics Section, School of Physics, Universiti Sains Malaysia, 11800 Pulau Pinang, Malaysia
[2] Department of Physics, Usmanu Danfodiyo University, Sokoto, PMB 2346 Sokoto, Nigeria

sabiubala@gmail.com

Abstract. Multiple linear regression (MLR) models for fast estimation of true subsurface resistivity from apparent resistivity field measurements are developed and assessed in this study, based on Dipole-dipole array. The parameters that have been investigated are apparent resistivity (ρ_a), horizontal location (X) and depth (Z) of measurement as the independent variables; and true resistivity (ρ_t) as dependent variable. To achieve linearity in both resistivities, datasets were first transformed into logarithmic domain following diagnostic checks of normality of the dependent variable and heteroscedasticity to ensure accurate models. Four MLR models namely; DD1, DD2, DD3 and DD4 were developed based on hierarchical combination of the independent variables. The generated MLR coefficients were applied to another dataset to estimate ρ_t values for validation. The accuracy of the models was assessed using coefficient of determination (R^2), standard error (SE) and weighted mean absolute percentage error (wMAPE). It is concluded that the MLR models can estimate ρ_t for the Dipole-dipole array with high level of accuracy, with the DD4 model being the best.

1. Introduction

The importance of resistivity surveying to solving many practical problems cannot be overemphasized. It plays vital role in water exploration [1,2], sites characterization [3,4] and mineral prospecting [5,6]. However, the measured apparent resistivity data from such surveys cannot be used directly to describe the subsurface [7]. Therefore, there is a need to develop resistivity models that can portray the true subsurface condition based on the apparent resistivity measurements, typically by inverse modeling exists [8]. Resistivity inversion has received enormous attention from various researchers over the years. Investigations are still ongoing to further mitigate the problems associated its speed and efficacy. Several researches have been conducted to develop and control resistivity inversion mechanisms.

The least – squares optimization based techniques are the most common and widely explored by various researchers [9,10] compared to other approaches. They have been applied to solve many practical inversion problems of both simple and complex structures with recorded remarkable success [11]. They are however not without problems. Some of their major drawbacks are: non-uniqueness of solutions, consume a lot of time and memory space for the iterative computations to converge at

Content from this work may be used under the terms of the Creative Commons Attribution 3.0 licence. Any further distribution of this work must maintain attribution to the author(s) and the title of the work, journal citation and DOI.
Published under licence by IOP Publishing Ltd

minimum data misfit [12,13], inability to produce accurate models when the starting initial model is far from the actual solution, emergence of false anomalies at intermediate and later iterations [14] and difficulties in dealing with nonlinear relationships among the modeling (subsurface) parameters [15].

Multiple regression coefficients obtained from linear least squares method can provide alternative means for rapidly estimating the true subsurface resistivity whenever apparent resistivity site measurements are conducted. The nonlinearity in the nature underground resistivity distributions can be taken care of by using certain mathematical transformations, rather than using the data directly in the regression process. The objective of this research is to develop and assess mathematical models for fast estimation of true subsurface resistivity from apparent resistivity measurements. This will aid in cutting down the data processing time and computer memory capacity needed to analyze the apparent resistivity data using conventional algorithms.

2. Methodology

2.1 Resistivity Data

A typical electrical resistivity method measures underground potential contrast (V) upon injecting specific amount of current (I) into the ground (Fig.1). Resistance (R) is then automatically computed from Ohm's law (Eq. 1) and subsequently the apparent resistivity when array geometrical factor is known [16]. The data utilized for the regression modelling of this research was acquired from ground resistivity survey conducted along two profiles, one for model calibration (Line 1) and the other for model validation (Line 2), using multi-electrode resistivity meter (ABEM Terrameter SAS4000 instrument). Each line consisted of 41 stainless steel electrodes at 1 m minimum spacing connected to a cable take – out using jumper with dipole-dipole array as the chosen protocol (Fig. 2). Dipole-dipole array configuration was particularly selected for its high sensitivity to lateral resistivity variations compared to the remaining arrays. The current was set between 1 mA and 100 mA. A computer – controlled system Terrameter SAS4000 was used to automatically select the four active electrodes for each voltage measurement and consequently compute the apparent resistivity (Eq. 2) using the array geometrical factor (k) defined in Figure 2. Apparent resistivity values are arranged systematically such that each datum point is specified with its horizontal (X) and vertical (Z) coordinates (Fig.3).

$$R = \frac{V}{I} \tag{1}$$

$$\rho_a = k\frac{\Delta V}{I} \tag{2}$$

For dipole-dipole array, $k = \pi n(n+1)(n+2)a$, where a is the dipole length and n is the dipole separation factor.

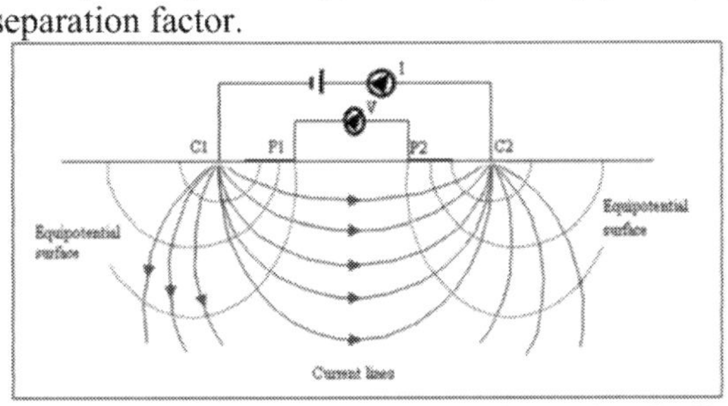

Fig. 1. Four-point electrodes configuration [16].

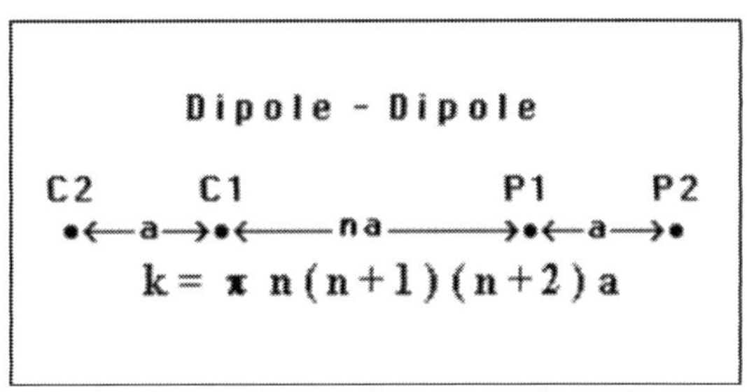

Fig. 2. Electrode configuration for Dipole – dipole array [16].

Standard constrained least-squares inversion was applied to the apparent resistivity data with help of Res2Dinv software package to obtain the true resistivity subsurface model (at 2.4% RMS error), which was later digitized to obtain ρ_t values at same coordinates (X and Z) as the ρ_a values. Datasets (X, Z, ρ_a and ρ_t) were then arranged in columns in preparation for the multiple linear regression modelling.

2.2 Multiple Linear Regression

Least squares based multiple linear regression model with n independent variables (x) in the observations and a dependent variable (y) can be written as

$$y = \beta_0 + \beta_1 x_1 + \cdots + \beta_n x_n + \varepsilon \tag{3}$$

where β_0 is the effective intercept of the n regression lines and ε is the error term. In this approach, the model converges when the squared sum of the differences between the estimated and observed values is minimum. From Eq. 3, the values of the dependent variable can be estimate for given values of independent variable(s) if the error terms are assumed to be normally distributed with zero mean and constant variance. The distribution of the y_i is thus symmetrical and bell-shaped, with a constant standard deviation at x levels [17].

In this research, true resistivity (ρ_t) is considered as dependent variable, while apparent resistivity (ρ_a), horizontal location (X) and depth (Z) are the independent variables, with ρ_a being the most important variable. X and Z are only added to verify whether they can add any statistical significance to the resulting model. The independent variables were not entered into the regression modelling at once. Thus, four models namely; DD1, DD2, DD3 and DD4 were developed based on hierarchical combination of the independent variables to ascertain their contributions. Logarithmic transforms of ρ_a and ρ_t, denoted as ρ_A and ρ_T respectively, were rather used instead of using the actual variables directly following pre-modelling diagnoses such as tests of normality, linearity, heteroscedasticity, axes balance and outliers. First datasets (Line 1) were imported into SPSS to run the regression modelling (model calibration). Regression coefficients obtained from model calibration were applied to the second datasets (Line 2) for validation. Estimated ρ_T values for the various models were transformed back to ρ_t and plotted using Golden Surfer 11 at same color scale and blanking as the observed true resistivity subsurface model ρ_t model obtained from the inversion (with Res2Dinv) for visual assessment.

The accuracy of each model was assessed using coefficient of determination (R^2), standard error (SE) and weighted mean absolute percentage error (wMAPE) for model calibration; and SE and wMAPE for validation. According to [18], an added variable is statistically significant if it brings about an increase in R^2 (or decrease in SE) of at least 5% and above. The R^2 SE and wMAPE were calculated from Eq. 4, 5 and 6 respectively.

$$R^2 = \frac{\Sigma(\hat{\rho}_{T_i} - \bar{\rho}_T)^2}{\Sigma(\rho_{T_i} - \bar{\rho}_T)^2} \tag{4}$$

$$SE = \left[\frac{1}{N}\Sigma(\rho_{T_i} - \hat{\rho}_{T_i})^2\right]^{1/2} \tag{5}$$

$$wMAPE = \frac{1}{N}\Sigma\left|\frac{\rho_{T_i} - \hat{\rho}_{T_i}}{\rho_{T_i}}\right| \times 100 \tag{6}$$

where ρ_{T_i}, $\hat{\rho}_{T_i}$ and $\bar{\rho}_T$ are observed, estimated and average Log ρ_t values respectively; and N is the number of data points.

3. Results and Discussion

Table 1 presents summary of the four regression models developed in this study. The R^2 is found to increase considerably from DD1 to DD4. Similarly, the calibration SE and wMAPE are found to decrease from DD1 to DD4, even though the decrease in wMAPE is not all that significant. It can be deduced from the table that there is only about 1.4 % increase in R^2 and 1 % decrease in SE for DD2 model as result of adding X variable to the original (DD1) model, rendering it insignificant. When Z is added to original model (to create DD3) however, an increase of about 11 % in R^2 and 13 % drop in

SE occurred, making DD3 a statically significant model going by the assertion made by [18] that adding a variable into a regression model is only significant if bring about at least 5 % rise in R^2 or 5 % drop in SE. When both X and Z variables were added to form the overall (DD4) model, there was about 19 % increase in R^2 and 26 % decrease in SE.

The validation SE and wMAPE showed comparable results (Table 2), but has slightly favoured DD3 (with 0.220 and 5.56 %) to DD4 (with 0.224 and 5.68 %) respectively. This might be for to the fact that outliers were not removed from the validation data (Line 2), as was the case for the model calibration data (Line 1). However, this should not be a problem, considering that these differences are just so small to warrant any significant preference between the two models. For this reason, DD4 is therefore regarded as most preferred among the four models (as pointed out previously) for the rapid estimation true subsurface resistivity whenever apparent measurements are conducted using Dipole – dipole array.

Table 1. Models summary table

Models	Estimated MLR Models	R^2	SE_c	$wMAPE_c$ [%]	SE_v	$wMAPE_v$ [%]
DD1	$\rho_T = 1.008\,\rho_A + 0.065$	0.47	0.210	5.35	0.242	6.07
DD2	$\rho_T = 1.003\rho_A - 0.002X + 0.124$	0.48	0.208	5.34	0.242	6.12
DD3	$\rho_T = 0.485\rho_A + 0.108Z + 1.898$	0.58	0.183	4.61	0.220	5.56
DD4	$\rho_T = 0.432\rho_A - 0.002X + 0.118Z + 2.125$	0.66	0.155	4.02	0.224	5.68

*subscripts c and v designate calibration and validation respectively

4. Conclusions

Four hierarchical multiple linear regression models for estimating true subsurface resistivity rapidly are developed and validated in this study. Numerical accuracy assessment carried out suggested DD4 as the best model having the highest R^2 of 0.659 and lowest SE and wMAPE of 0.155 and 4.02 % respectively. It is therefore concluded that the DD4 MLR model can estimate ρ_t for the Dipole-dipole array with within allowable accuracy limits. It is recommended that the procedure be applied to the remaining resistivity array configurations for the same objective as outlined in this study. Other subsurface parameters which were not investigated in this study are also recommended be examined to achieve better estimation models.

References

[1] Saad R, Adli I and Mohamad A S 2012 The study of iron ore prospect using 2-d resistivity and induced polarization (IP) method *Electron. J. Geotech. Eng.* 17 2981–2988
[2] Saad R, Syukri M, Nordiana M M and Ismail N A 2013 Water table delineation for leachate identification using 2-D electrical resistivity imaging and seismic refraction at Gampong Jawa, Banda Aceh *Electron. J. Geotech. Eng.* 18 1529–1535
[3] Dahlin T, Wisén R and Zhang D 2007 3D effects on 2D resistivity imaging – modelling and field surveying results. Near Surface 2007, 13th European Meeting of Environmental and Engineering Geophysics, Istanbul, Turkey.
[4] Bery A A and Saad R 2012 Clayey sand soil's behaviour analysis and imaging subsurface structure via engineering characterizations and integrated geophysicals tomography modeling methods *Int. J. Geosci.* 3 93–104
[5] Saad R, Nawawi M N M and Mohamad E T 2012 Groundwater detection in alluvium using 2-D electrical resistivity tomography (ERT) *Electron. J. Geotech. Eng.* 17 369–376

[6] Song W, Wang Y and Shao Z 2017 Categorical modeling on electrical anomaly of room-and-pillar coal mine fires and application for field electrical resistivity tomography, J. *Appl. Geophys.* 136 474–483
[7] De Donno G and Cardarelli E 2017 Tomographic inversion of time-domain resistivity and chargeability data for the investigation of landfills using a priori information. *Waste Manage.* 59 302–315
[8] Narayan S, Dusseault M B and Nobes D C Inversion techniques applied to resistivity inverse problems *Inverse Problems* 10 669
[9] Olayinka A I and Yaramanci U 2000 Use of block inversion in the 2-D interpretation of apparent resistivity data and its comparison with smooth inversion *J. Appl. Geophys.* 45 63–81
[10] Loke M H and Dahlin T 2002 A comparison of the Gauss-Newton and quasi-Newton methods in resistivity imaging inversion *J. Appl. Geophys.* 49 149–162
[11] Pain C C, Herwanger J V, Loke M H and Worthington C R E 2002 De Oliveira, Effective multidimensional resistivity inversion using finite-element techniques *Geophys. J. Int.* 151 710–728
[12] Gyulai Á, Ormos T and Dobroka M 2010 A quick 2-D geoelectric inversion method using series expansion *J. Appl.Geophys.* 72 232–241.
[13] Vachiratienchai C and Siripunvaraporn W 2013 An efficient inversion for two-dimensional direct current resistivity surveys based on the hybrid finite difference-finite element method *Phys. Earth Planet. Inter.* 215 1–11.
[14] Liu B, Nie L, Li S, Xu L, Liu Z, Song J and Lin C 2012 3D Electrical resistivity inversion tomography with spatial structural constraint, Yanshilixue Yu Gongcheng Xuebao/Chinese *J. Rock Mech. Eng.* 31 2258–2268.
[15] Singh U K, Tiwari R K and Singh S B 2010 Inversion of 2-D DC resistivity data using rapid optimization and minimal complexity neural network, Nonlinear Process. *Geophys.* 17 65–76
[16] Loke M H 2016 Tutorial: 2-D and 3-D Electrical Imaging Surveys, Revised Edition.
[17] Balarabe M, Abdullah K, Nawawi M N M and Khalil A E 2016 Monthly temporal-spatial variability and estimation of absorbing aerosol index using ground-based meteorological data in Nigeria *Atmos. Climate Sci.* 6 425–444
[18] Noori R, Hoshyaripour G and Ashrafi K 2010 Araabi, Uncertainty analysis of developed ANN and ANFIS models in prediction of carbon monoxide daily concentration *Atmos. Environ.* 44 476-482

Regional Structural Architecture of Part of Northern Nigeria Basement Complex inferred from Upward Continuation of Magnetic Field Intensity Data

S B Muhammad[1,2] and R Saad[1]

[1] Geophysics Section, School of Physics, Universiti Sains Malaysia, 11800 Pulau Pinang, Malaysia
[2] Department of Physics, Usmanu Danfodiyo University, Sokoto, PMB 2346 Sokoto, Nigeria

sabiubala@gmail.com

Abstract. 2–D upward continuation of total magnetic field intensity data at Katsina area Northern Nigeria Basement Complex, aimed at identifying major regional features, has been carried out and interpreted in this study. The significance is to update the existing geology map of Nigeria and to guide further exploration studies in the area. Data for six aeromagnetic field maps were merged to form the composite data. 2–D upward continuation at various planes of observation was applied to the gridded magnetic field intensity data after coordinates were first converted (from degrees to meters) to understate anomalies due to shallow features. The plane was 15 km above the ground level. The continuation results revealed a high magnetic region at center surrounded by low magnetic parts, especially at the southern part. Corroboration with geology suggested that the high magnetic region could be a central structural uplift of the bedrock, highly rich in magnetic minerals. The low magnetic area at southern part is interpreted as sedimentary deposit, extended from the prominent Bornu Basin. The south – eastern part of the study area is moderately magnetic in nature, and corresponds to the metamorphic rocks of the Kazaure schist belt.

1. Introduction

Magnetic method provides a means for probing the Earth's subsurface by taking advantage of the contrast in magnetic susceptibility of underlying materials [1, 2]. Techniques are now available to acquire magnetic data on land [3], at sea [4] and in air [5]. Cost effectiveness and operational speed obtained in aeromagnetic surveys, especially for large scale investigations, have made it a very attractive tool for mineral prospecting [6], ground water [7] and structural studies [8]. The difficulty associated with the effect of shallow ensembles and noise in grid in interpretation of potential field (magnetic and gravity) data to probe deeper subsurface structures of the earth is well known in geophysics. A couple of filters capable of isolating anomalies of interest are readily available to mitigate some these problems. Of these filters, upward continuation is proven to be suitable and straightforward, for it does not other step operation to be carried out. The technique reduces the effect of short wavelength features due shallow sources and noise in grids to a certain level, thereby enhancing large scale, primary features of the study area [9].

Content from this work may be used under the terms of the Creative Commons Attribution 3.0 licence. Any further distribution of this work must maintain attribution to the author(s) and the title of the work, journal citation and DOI.
Published under licence by IOP Publishing Ltd

It is convenient to implement upward continuation on magnetic field data for the fact that potential fields generally obey Laplace's equation. This permits us to determine the field over an arbitrary surface if the field is known completely over another surface. Following the method of Grant and West [10], we have for upward continuation, the following relation (Equation 1)

$$F(x', y', -h) = \frac{h}{2\pi} \iint \frac{F(x,y,0)dxdy}{\{(x-x')^2+(y-y')^2+h^2\}^{1/2}} \tag{1}$$

The left side of the equation is the total field at the point $F(x', y', -h)$ above the surface on which $F(x, y, 0)$ is known. The calculation procedure is to replace the integral with a weighted sum of values taken on a regular grid.

The area under investigation exactly match one such complex cases that can be resolved using upward continuation, for it encompasses granitic rocks, schist belt and sedimentary formation with no clear boundaries. Detailed information about the area is limited as previous studies solely concentrated on depth to basement estimation [2] and surface geological mapping [11]. More efforts to exactly delineate the different geological features of the area are required to update the existing geology map and to guide further exploration studies in the area. This forms the major objective this research.

2. The study area

The study area is part of the Northern Nigerian Basement Complex (Fig. 1) and covers an estimated area of about 4170 km^2 bounded by longitudes 7^{0}00' and 8^0 30'; and latitudes 12^{0}30' and 13^{0}30'. It includes Katsina, Jibiya, Dankama, Daura, Ruma, and Kazaure towns; represented covered respectively by sheets No. 14, 15, 16, 33, 34 and 35 of aeromagnetic data maps published by the Nigerian Geological Survey. It is characterized by two distinct climatic conditions; dry season which usually heralded by the dry Harmattan wind, occurs between November and March. Rainy season lasts between June and October. Vegetation is typically savannah with isolated trees among stunted shrubs [2].

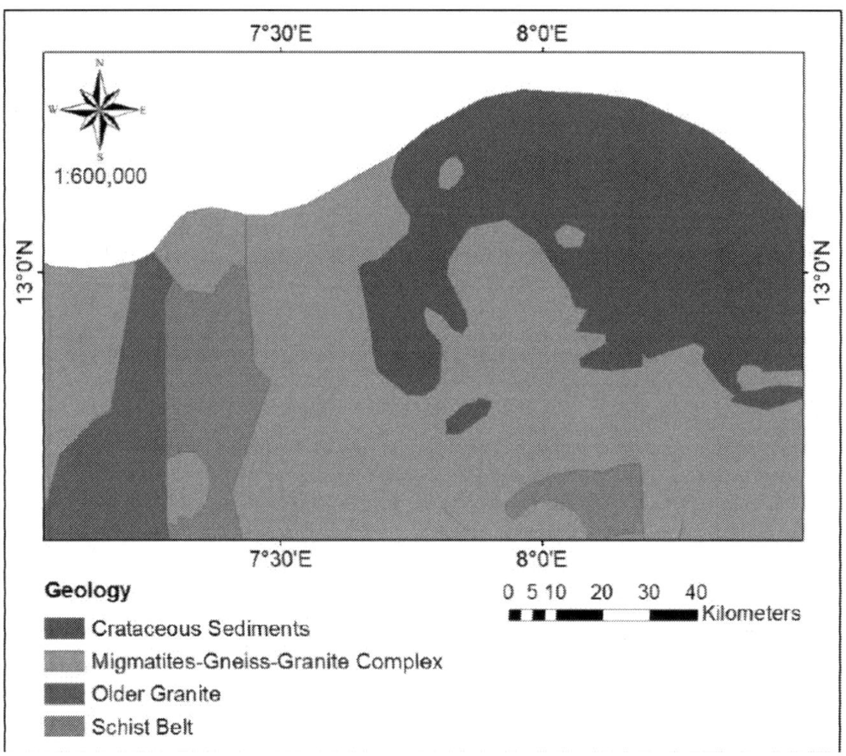

Figure 1. Geology map of the study area.

The area is dominated mainly by crystalline basement rocks as migmatite–gneiss–granite complex with older granite at the western part of the area. Cretaceous sediments of Gundumi formations are also noticeable at extreme northern part of the study area. Also present in the study area, are schist rocks of the Wonaka and Kazaure Shist Belts that appear at the western and south–eastern parts of the area respectively. The top soil is merely light brown to dark grey with topography which is generally undulating on a height of about 576 meters above sea level [12].

3. Methodology

Total magnetic field intensity of sheets 14, 15, 16, 33, 34 and 35 published by the Geological Survey Agency of Nigeria were obtained compiled and interpreted in this study. Each map was digitized on a 3 x 3 km^2 grid system on a tracing paper to form a 19 x 19 square grid layout. At grid points where contour lines do not cross, visual interpolation to the nearest contour lines was made to estimate the magnetic value of the grid point. This method allowed the coordinates of each grid point to be determined very easily considering the manner they were spaced at regular intervals. The data generated each of the six maps were collected and recorded in a 19 by 19 coding sheet. Each sheet was specified by its boundary coordinate, the map's number and name of the town it represented. These values were then recontoured and the emerging maps each was then put side by side with the corresponding fundamental map to trace and correct errors introduced in the process of digitization. After digitizing the data, a composite map of the study area was then generated by joining the six aeromagnetic maps forming the study area. The maps were arranged in such a way that the adjacent coordinates of any two neighboring maps were the same and therefore in merging these maps, all the magnetic values with same coordinates were averaged in order not allow more than one data point with same coordinates. These entries were carried out in *Microsoft Excel* environment and a total of two thousand and thirty-five (2035) data points were entered in 37 (rows) by 55 (columns).

The total magnetic field data was then gridded and plotted after coordinates conversion (degrees to meters) in preparation for the upward continuation, using the Oasis montaj software package. The technique reduces the effect of short wavelength features due shallow sources and noise in grids to a certain level, thereby enhancing large scale features of the study area. Upward continuation of the magnetic field intensity data was carried out at various planes of observation. The planes were at 2 km, 5 km, 10 km 15 km and 20 km above the Earth's surface.

4. Result and discussion

Figure 2 shows the composite map of the total magnetic field intensity of the study area. Portions on this map that are colored red represent areas of high magnetic intensity, whereas those in blue correspond to the areas of low magnetic field intensity. Superimposed with these portions are areas of intermediate field intensities represented by colors of yellow and pale green that appear everywhere in the study map. Low magnetic field intensity at northern (top) part of the map confirms the presence of a part of the Cretaceous sediments of Gundumi formations in the area. Traces of red colors in this region, which are an indication of high magnetic field intensity, may be as a result of shallower disturbances and noise in the data. The red colour at eastern region of the map indicate the presence of high magnetic intensity rocks. They correspond to the highly magnetized crystalline basement rocks (magmatite-gneiss-granite complex) in the area. Deep blue at the south-eastern part of the map may indicate the presence of part of the Kazaure Schist Belt. To the western parts of the map, a superposition of intermediate to high magnetic intensities is noticed, which is also due to the presence of crystalline basement rocks.

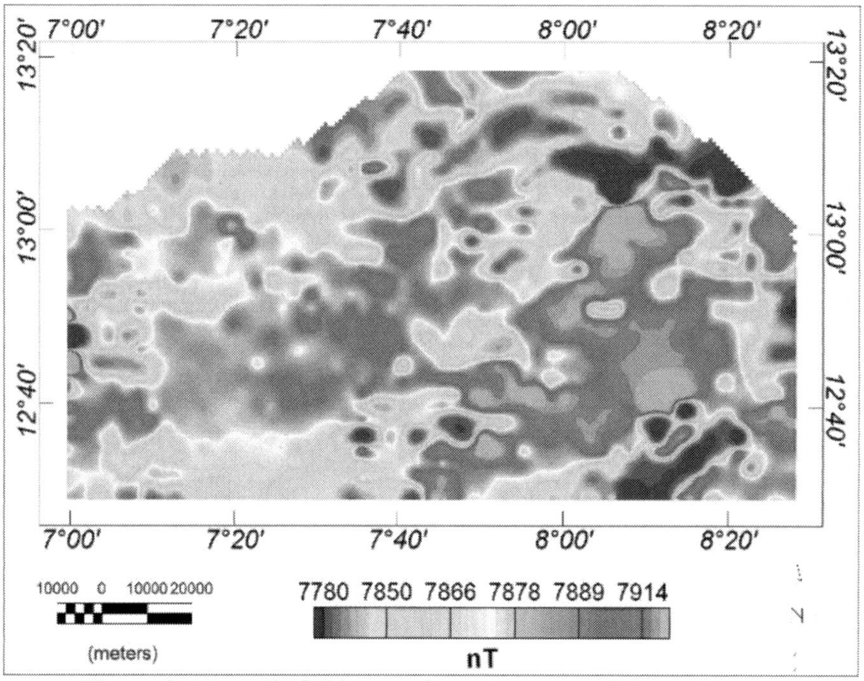

Figure 2. Magnetic field anomaly contour map of the study area.

To identify the regional structures, the field was continued at 15 km plane of observation. The method highlights the effect of deeper sources and suppresses those of shallower sources such as scattered local disturbances and noise. Figures 3 gives the result of the filtering process. The result indicates a structural upliftment of high magnetic field intensity at the central part of the map. The magnetic effect is more pronounced at the eastern part of the structure. This may be attributed to the presence of migmatite-gneiss complex in the crystalline basement rocks at the region [13]. This may be interpreted as thus the mineral deposits are either at considerably larger scale or relatively close to the surface. In the same manner, the blue colored region at extreme northern part of the map indicates the signature of low magnetic sediments of Gundumi formations [12]. Another low magnetic (blue colored) portion is also noticeable at the south-eastern part of the figures, this could be due to the presence of part of the metamorphic rocks of Kazaure Schist Belt in the region [11].

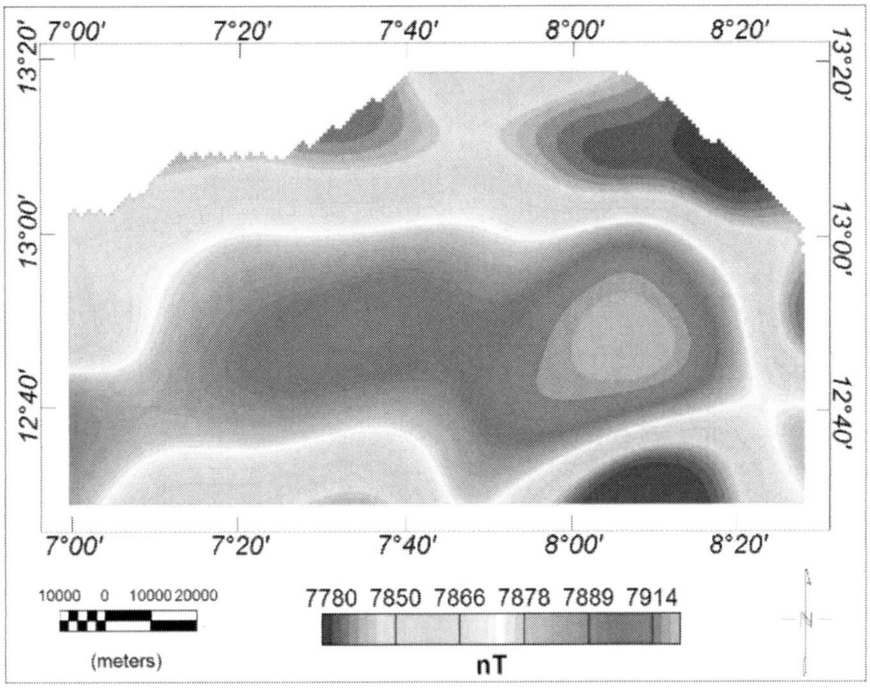

Figure 3. Upward continuation of the magnetic field data at 15 km.

5. Conclusion

Six aeromagnetic data maps covering part of Northern Nigeria basement complex were analyzed and interpreted in this study. Magnetic closures of both high and low were connected to the presence of ore bodies in the area and regional anomalies trend in the ENE-WSW direction. The total magnetic field was continued at 15 km above the surface of the Earth to remove noise and local disturbances, in which a critical was observed at 15 km. The continuation process showed well seated structural of high magnetic intensity at the central part of the study area. The persistence of this high intensity closure, especially around latitude 12.9^{0}N and longitude 8^{0}E may indicate an upliftment of basement or the presence of ferrigeneous minerals. This information will help Nigerian Geological Survey to update the existing geology map of country. It will also provide guide for future exploration studies in the area.

References

[1] Suleiman T, Udensi E E and Muhammad S B 2014 Analysis of aeromagnetic data across Kebbi State, Nigeria *Int. J. Mar. Atmos. Earth Sci.* 2 41-45.

[2] Muhammad S B, Udensi E E, Momoh M, Sanusi Y A, Suleiman T and Ajana O 2014 Spectral analysis and estimation of depth to magnetic rocks below the Katsina area, Northern Nigerian basement complex *JoPHY STM J.* 3 13-23.

[3] Anand S P, Erram V C, Patil J D, Pawar N J, Gupta G and Suryavanshi R A 2016 Structural mapping of chikotra river basin in the Deccan Volcanic Province of Maharashtra, India from ground magnetic data *J. Earth Syst. Sci.* 125 301-310.

[4] Yu X, Xue C, Shi H, Zhu W, Liu Y and Yin H 2017 Expansion of the South China Sea Basin: constraints from magnetic anomaly stripes, sea floor topography, satellite gravity and submarine geothermics *Geosci. Front.* 8 151-162.

[5] Ajana O, Udensi E E, Momoh M, Rai J K and Muhammad S B 2014 Spectral depths estimate of subsurface structures in parts of Borno basin, Northeastern Nigeria, using aeromagnetic data *IOSR J. Appl. Geol. Geophys.* 2 55-60.

[6] Rabeh T 2016 Tracing the manganese ore accumulations in Sinai Peninsula, Egypt, using magnetic method *Environ. Earth Sci.* 75 1-12.

[7] Joel E S, Olasehinde P I, De D K and Maxwell O 2016 Regional groundwater studies using aeromagnetic technique. Paper Presentation given at AAPG/SEG International Conference & Exhibition, Cancun, Mexico, September 6-9, 2016, Search and Discovery Article.

[8] Saad R, Nordiana M M and Saidin M 2014 The conclusion of searching Bukit Bunuh Crater using magnetic method *Electron. J. Geotech. Eng.* 19 4429-4437.

[9] Henderson R G 1970 On the validity of the use of the upward continuation integral for total magnetic intensity data *Geophys.* 35 916-919.

[10] Grant F S and West G F 1965 Interpretation Theory in Applied Geophysics, McGraw-Hill, New York.

[11] Danbatta U A 2003 The structural framework underlying the metamorphic evolution of the Kazaure schist belt, NW Nigeria *Global J. Geol. Sci.* 1 73-84.

[12] Gandu A H Ojo S B and Ajakaiye D E 1986 a gravity study of the Precambrian rocks in the Malumfashi area of Kaduna State, Nigeria *Tectonophys.* 22

[13] Nyikwagh D M, Ugodulunwa F X O and Ukaegbu U V 2008 Geology of the Pan-African basement complex in Ube-Wulko area of Akwanga, North Central Nigeria *Sci. Afr.* 7 100-110.

IOP Conf. Series: Journal of Physics: Conf. Series **1083** (2018) 012058 doi:10.1088/1742-6596/1083/1/012058

Effect of Particle Size on Maximum Energy Product of Waste-derived Ferrites

H S Woon[1], L S Ewe[1] , K P Lim[2], I Ismail[2], C Y Tan[3] and Y W Wong[1]

[1]College of Engineering, Universiti Tenaga Nasional,43000 Kajang, Selangor, Malaysia
[2]Materials Synthesis and Characterization Laboratory, Institute of Advanced Technology, Universiti Putra Malaysia, 43400 Serdang, Selangor, Malaysia.
[3]Advanced Manufacturing and Material Processing (AMMP) Centre, University of Malaya, 50603, Kuala Lumpur, Malaysia

hwoon@uniten.edu.my

Abstract. Ferrite is an important component for electric motors, loud-speakers, EMI suppressors and transformer core. Hematite (Fe_2O_3) and strontium carbonate are commonly used to prepare ferrite. In this work, waste-derived hematite powder was milled using ball mill, high-energy milling and high-energy blender and the powder size was observed. The samples was further filtered according to different particle size, namely mesh 20, mesh 100 and mesh 400, and used as starting materials in the synthesis of strontium hard ferrites. Maximum energy product $(BH)_{max}$ of the ferrites was analyzed by a B-H tracer. The results showed that the maximum energy product is significantly affected by particle size of the starting materials. As anticipated, the maximum energy product is inversely proportional to the particle size and the hard ferrite produced by high-energy blender obtained 0.99 MGOe. With that, it is confirmed that permanent magnets derived from waste materials is also affected by its particle size. Other magnetic properties such as remanence and coercivity are also reported in this paper.

1. Introduction

Hot rolling, i.e. processing a target metal above its recrystallization temperature, is a metal-working procedure that requires heating, rolling and forming of the target metal. Various steel products can be manufactured through hot forming, namely primary, section, flat, wire, rebar, profile, pipe and tube. These products can be generally categorized as long products. Billets or blooms are hot-rolled to produce such long products, of which are subjected to further rolling and drawing processes to form wire rods, tubes or coatings. Upon the completion of hot rolling, a layer of mill scale is formed on the surface of the steel due to reaction between the hot steel and atmospheric oxygen during the natural cooling process. Mill scale has a flaky appearance and consists of FeO, Fe_3O_4 and Fe_2O_3 [1]. Further processing of the steel (i.e. cold rolling) is preceded by the acid pickling process. Pickling liquor, a solution of strong acids such as HCL and H_2SO_4, is used as a cleaning agent to remove the mill scale layer from the surface. Alternative measures to clean the surface are electrolytic pickling and salt pickling. The scale-free steel is subsequently subjected to cold rolling below its recrystallization temperature to produce sheets and strips with superior surface finishing [2,3].

Content from this work may be used under the terms of the Creative Commons Attribution 3.0 licence. Any further distribution of this work must maintain attribution to the author(s) and the title of the work, journal citation and DOI.
Published under licence by IOP Publishing Ltd

The setup of continuous acid pickling requires several tanks that serve different purposes. The hot-rolled steel is first submerged into tanks contained pickling liquor for to remove the mill scale layer. Subsequently, the steel is cleaned in water rinsing tanks to expel the pickling liquor. In Malaysia, spent pickling acid (SPL) is collected and recovered as recycled pickling liquor for further pickling operation or neutralized with low quality alkali to produce solid waste. The spray-roaster acid regeneration process and fluidized-bed process are the two main regeneration techniques utilized by the Malaysia steel industry players.

Strontium hard ferrite ($SrFe_{12}O_{19}$) used in this work is synthesized by utilizing waste products collected from local cold rolling steel factory. The waste products are subjected to milling, magnetic separation and heat treatment to produce hematite (Fe_2O_3). Hematite is then further processed to prepare strontium hard ferrite ($SrFe_{12}O_{19}$).

Iron oxide and strontium oxide are used to produce hard ferrite magnet. Mixing and pre-sintering of the raw materials (powder) are conducted to create the magnetic phase in the powder. Subsequently, the powder is subjected to pressing in an isotropic environment (absence of magnetic field). Finally, it is sintered to obtain the final product. From the market supply of magnets, hard ferrite magnets are the most economically competitive magnets due to the cheap cost of raw materials. Based on past literature, ferrites are determined to possess high electrical isolation effect and strong capability of resisting demagnetization in strong external magnetic fields [4-7]. High maximum energy product $(BH)_{max}$, high residual induction (remanence) and high coercivity are crucial characteristics of a material to be capable of producing and maintaining its magnetic field [8, 9]. Materials in such category are termed oxide or ceramic magnets. The low cost of these materials contribute towards its wide usage in the market and industry. For the application of permanent hard magnet, the most commonly used material is hexagonal ferrite [10, 16].

In this work, the particle size of the oxide waste and the pre-formed powder was studied to confirm its effects on maximum energy product of the waste-derived hard ferrite.

2. Methodology

As aforementioned, waste products collected from local cold rolling steel factory were used as starting materials. Firstly, the waste products was divided into 3 samples and milled by using ball mill, high-energy milling (Model PM100) and high-energy blender (Model DM6), respectively. The milling period chosen was 24 h, 18 h and 10 minutes as suggested by the machine manufacturers. It was observed to be a useful process to reduce the size of large pieces into ultrafine powder. After drying and magnetic separation, heat treatment within the temperature range of 400°C to 600°C was executed on the powder. X-ray diffraction (XRD) analysis concluded that the Fe_2O_3 phase was present in all the samples. To segregate the powder according to different particle sizes, sieving was done with mesh 20, mesh 100 and mesh 400.

Strontium ferrite powder with magnetic phase was then produced by wet mixing and pre-sintering (at 1150°C) the Fe_2O_3 samples and strontium carbonate ($SrCO_3$) powder mixture. 6:1 was the selected mixing ratio of Fe_2O_3 to $SrCO_3$. The calcined samples were milled again via ball mill, high-energy milling and high-energy blending to produce single-domain particles. The 3 samples were sieved using mesh 20, mesh 100 and mesh 400 respectively before pelletized through pressing. The pellets were subjected to sintering for 7 hours at 1200°C. B-H tracer (Model Linkjoin MATS-2010H) was used to obtain the magnetic measurements of the pellets. Comparison of the maximum energy product, remanence and coercivity of the strontium hard ferrites was done. Figure 1 illustrates the procedure in synthesizing strontium hard ferrite sintered magnets.

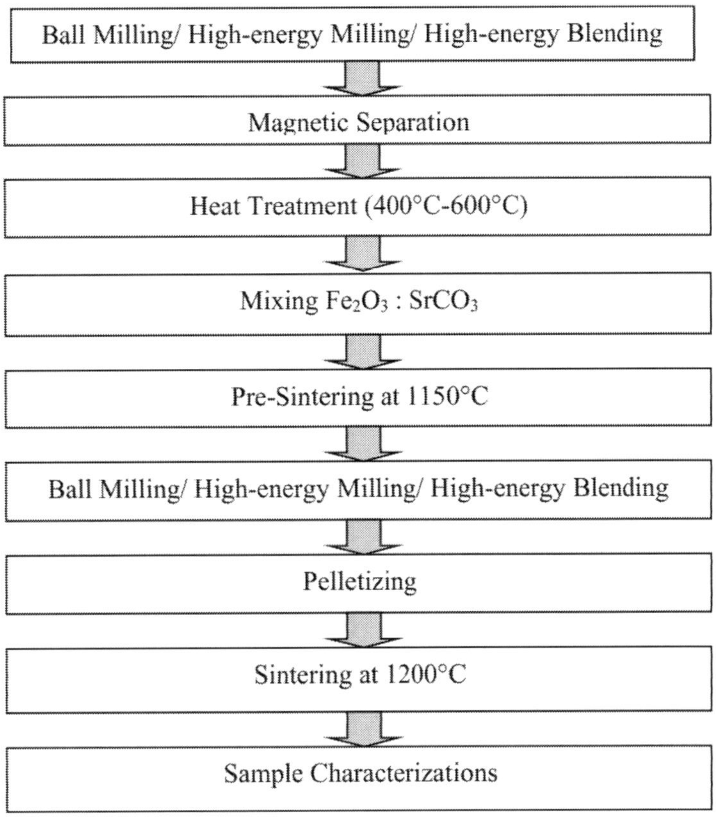

Figure 1: Procedure of synthesis of Strontium Hard Ferrite Sintered Magnets.

3. Results and Discussion

Three different iron oxide layers can be observed in the waste products from the steel industry, namely wuistite (FeO), magnetite (Fe_3O_4) and hematite (Fe_2O_3). The waste represents a rich source of iron due to its high iron (Fe) content of more than 70%. In this work, ultrafine single phase particles were obtained from the waste via milling. Defective FeO was collected via magnetic separation and oxidized into Fe_2O_3 that has more stability. The oxidation process is illustrated via Equation (1) [11]:

$$4FeO + O_2 \rightarrow 2\,Fe_2O_3 \qquad (1)$$

X-ray diffraction analysis was employed to determine the presence of Fe_2O_3.

From Figure 2, the phase exhibited by the treated waste correlate to standard Fe_2O_3 phase, providing evidence that the Fe_2O_3 phase exists in the waste sample. On another spectrum, due to the absence of quantitative data of the sample from the XRD analysis, the purity of the Fe_2O_3 phase could not be determined.

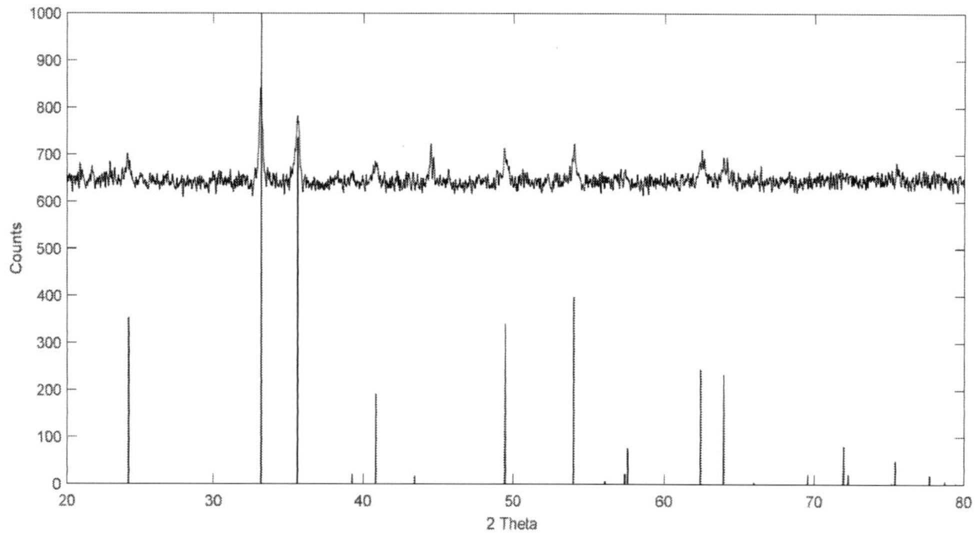

Figure 2: X-ray diffraction pattern of waste-derived Fe_2O_3.

To prepare strontium hard ferrite ($SrFe_{12}O_{19}$), the waste-derived Fe_2O_3 was subjected to the following the chemical reaction:

$$SrCO_3 + 6Fe_2O_3 \rightarrow SrFe_{12}O_{19} + CO_2 \tag{2}$$

The presence of $SrFe_{12}O_{19}$ phase was determined via X-ray diffraction analysis.

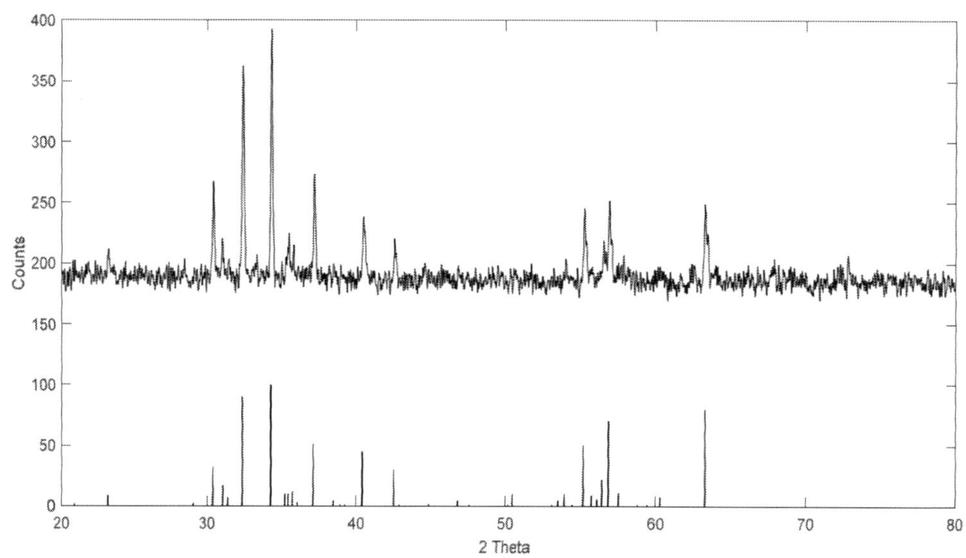

Figure 3: X-ray Diffraction Pattern of Waste-derived $SrFe_{12}O_{19}$

The good correlation between the XRD traces in Figure 3 indicates that the $SrFe_{12}O_{19}$ phase is observed in the sample. The selected mixing ratio of Fe_2O_3 to $SrCO_3$ and sintering temperature are found to be suitable to produce $SrFe_{12}O_{19}$.

IOP Conf. Series: Journal of Physics: Conf. Series **1083** (2018) 012058 doi:10.1088/1742-6596/1083/1/012058

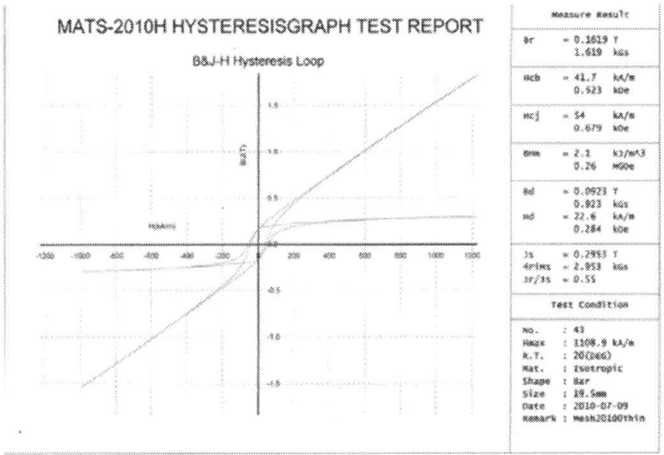

Figure 4: Hysteresis-graph of hard ferrite with Mesh 20 raw material.

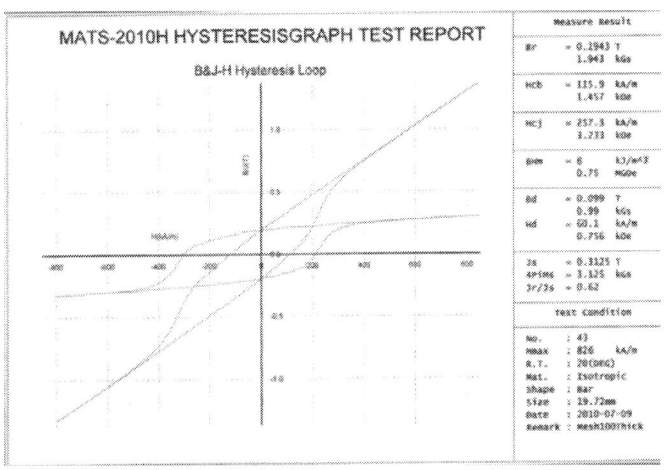

Figure 5: Hysteresis-graph of hard ferrite with Mesh 100 raw material.

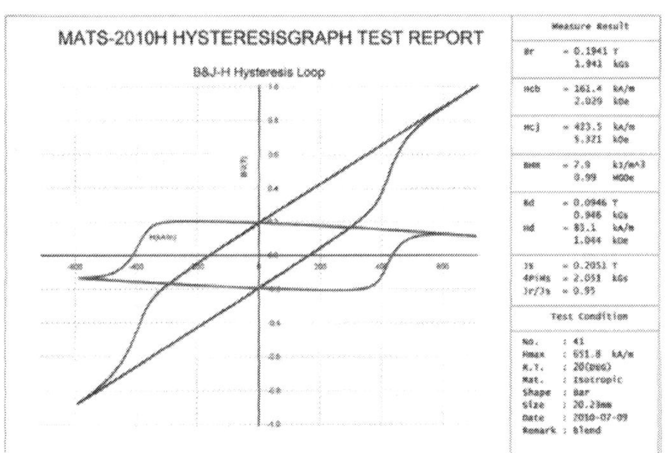

Figure 6: Hysteresis-graph of hard ferrite with Mesh 400 raw material

The magnetic properties of the hard ferrite samples produced are illustrated by Figures 4-6. The column on the right displays the maximum energy product $(BH)_{max}$, remanence (B_r) and coercivity (H_{cj}) of the samples, whereas J_r/J_s is the squareness of the graph.

The highest maximum energy product is exhibited by mesh 400 sample at 0.990 MGOe, while mesh 20 and mesh 100 exhibit 0.260 MGOe and 0.750 MGOe, respectively. $(BH)_{max}$ defines magnet's quality and amount of stored energy. On another spectrum, remanence represents the ability of a magnet to attract. Mesh 100 sample and mesh 400 sample have similar remanence at 1.943 kG and 1.941 kG respectively, indicating approximate equivalence in the samples' attraction and repulsion abilities. In terms of coercivity, mesh 400 sample is rated at 5.321 kOe while mesh 100 sample is rated at 3.233 kOe. Such difference implies that mesh 400 sample has better resistance against demagnetization in an external magnetic field. In contrast, mesh 20 sample is determined to be a weak magnet due to its low remanence (1.619 kG) and coercivity (0.679 kOe) values.

According to the Stoner-Wohlfarth model, hard ferrite with good magnetic characteristics should satisfy the single-domain-particle concept [12-14]. In a single-domain magnet, its magnetization is oriented uniformly in one direction. However, in a multi-domain magnet (commonly larger samples), domain walls segregate the magnetic domains, creating a different direction of magnetization in each domain. Transition from domain to domain requires the magnetization to rotate according to the direction in the next domain. Such characteristics allow the multi-domain magnet to minimize the magnetostatic energy or internal energy of the magnet. When it is subjected to demagnetization (i.e. process of minimizing magnetic moment), the multi-domain magnet that possesses minimal internal energy has lower coercivity or resistance towards randomized direction of magnetization. In contrast, the single-domain magnet has much higher magnetostatic energy. Its uniformly oriented magnetization has to be rotated entirely for it to be demagnetized, hence having larger coercivity or resistance towards demagnetization. As such, the aforementioned magnetic characteristics of mesh 20 sample imply that it is not a single-domain magnet; whereas the mesh 100 sample is determined to have moderate magnetic properties as certain particles are sufficiently small to act as single-domain magnet. The mesh 400 sample, with the smallest particle size, is determined to be a single-domain magnet (according to the Stoner-Wohlfarth model) due to its high maximum energy product, good attraction ability and high coercivity [15].

4. Conclusions

The X-ray diffraction analysis was essential in this work to determine the phase stability of the samples. The XRD traces of Fe_2O_3 obtained from cold-rolled factory waste products was found to correlate with the JCPDS standard Fe_2O_3 phases. Preparation procedure, i.e. magnetic separation and heat treatment, was proven effective in obtaining hematite, Fe_2O_3, from the waste products. In addition, the $SrFe_{12}O_{19}$ magnetic phase was observed in the samples, providing evidence that the selected ratio of 6:1 for Fe_2O_3 to $SrCO_3$ was suitable to produce strontium hard ferrite sintered magnets. The variable factor, namely different particle sizes (mesh 20, mesh 100 and mesh 400), is concluded to be influential in determining the magnetic properties of the samples. The mesh 20 sample was found not a single-domain, whereas the mesh 100 sample only partially satisfy the single-domain sample. Only the mesh 400 sample (smallest grain size in this work) satisfied the single-domain-particle concept, therefore having the highest maximum energy product. In a nut shell, high-energy blender is recommended in this work in order to produce single-domain-particle for the synthesis of ferrite materials.

Acknowledgments

This project is supported by UNITEN BOLD Grant Code 1289176/B/9/2017/62.

References

[1] Arumugam M 2002 *Materials Science* (Anuradha Ag Press) p 7.1-7.42.
[2] Vladimir B G 2002 *Steel-Rolling Technology* (Marcel Dekker Inc) p 483-491
[3] Brama D and Rob B 2017 *Fundamentals of Steelmaking Metallurgy* (Prentice Hall Int.) p 161-185
[4] Goldman A 2006 *Modern Ferrite Technology* (Springer) p 273-286.
[5] Wohlfarth E P 1980 *Ferromanegtic Materials Volume 2* (North-Holland Publishing Company) p 55-189
[6] Sugimoto M 1999 The Past, Present and Future of Ferrites *Journal of America Ceramic Soc* 82[2] 269-280.
[7] Giorgio B 1998 *Hysteresis In Magnetism* (Academic Press) p 255-294
[8] Ralph S 1999 *Permanent Magnetism* (Institute of Physics Publishing) p 303-337
[9] McCaig M and Clegg A G 1987 *Permanent Magnets in Theory and Practical* (Pentch Press)
[10] James P S and Ashok S 1999 *The Science and Design of Engineering Materials* (McGraw-Hill Int.) p 666-672.
[11] Trbble R S and Craik D J 1969 *Magnetic Materials* (John Wiley & Sons)
[12] Parker R J 1989 *Advances In Permanent Magnetism* (John Wiley and Sons) p 101-135
[13] Peter Campbell 1996 *Permanent Magnet Materials and Their Application* (Cambridge University Press) p 25-56.
[14] McCaig V 1967 *Attraction and Repulsion* (Oliver and Boyd) p 11-18
[15] David J 2015 *Introduction To Magnetism and Magnetic Materials* (Chapman & Hall) p 229-323.
[16] Woon H S, Lim K P, Puteri N, Rozaimah R and Tan C Y 2009 Synthesis and Characterization of Hot-roll and Cold-roll By-product-derived Strontium Hard Ferrites *American Journal of Engineering and Applied Sciences* 2(4):580-583
[17] Waje S B, Hashim M and Ismail I 2012 Effects of Sintering Temperature on Grain Growth and the Complex Permeability of $Co_{0.2}Ni_{0.3}Zn_{0.5}Fe_2O_4$ Material Prepared using Mechanically Alloyed Nanoparticles *Journal of Magnetism and Magnetic Materials* 323(11) 1433-1439

IOP Conf. Series: Journal of Physics: Conf. Series **1083** (2018) 012059 doi:10.1088/1742-6596/1083/1/012059

Grain Size Effect on Electrical and Magnetotransport Properties of $La_{0.67}Ca_{0.33}MnO_3$ synthesized via Sol-Gel Method

K. P. Lim*, S. A. Halim, S.K.Chen, M. M. Awang Kechik, S. W. Ng and W. N. W. Wan Jusoh

Superconductor Laboratory, Department of Physics, Faculty of Science, Universiti Putra Malaysia, 43400 Serdang, Selangor.

limkp@upm.edu.my

Abstract. The effect of grain size on electrical resistivity and magnetoresistance (MR) effect on $La_{0.67}Ca_{0.33}MnO_3$ prepared via sol-gel method was reported with different sintering temperature starting from 600°C to 1200°C. X-ray diffraction (XRD) patterns show that $La_{0.67}Ca_{0.33}MnO_3$ was in single phase and had orthorhombic structure with space group Pbnm (62). The crystallite size as well as particle size show strong dependence on sintering temperature (Ts). The metal-insulator transition (T_{MI}) and the magnitude of magnetoresistance are significantly influenced by the effective grain boundaries in LCMO polycrystalline. The T_{MI} and its resistivity increases as the grain size and crystallite size increases (Ts increase). The highest low-field magnetoresistance (LFMR), which extrinsic percentage of MR is more dominant, observed in LC-SG8 (-12.89%) with nano-sized distribution that showing double crystallite structure. However, the highest CMR is found in LC-SG10 (-23.48%) at 80 K with 1 Tesla. These percentage variations of MR for all samples were influenced by the grain size and crystallite size variations which consequently affect the spin polarized-tunneling scattering at the grain boundary layer.

1. Introduction

Perovskite manganites with general formula $La_{1-x}A_xMnO_3$ where A is a divalent alkaline earth cation (e.g Ca, Ba, Sr) has attracted extensive attention in the last few decades due to its special physical properties and potential applications in magnetoresistive devices [1]. It is proved that Jahn-Teller effect and double exchange (DE) mechanism that attributed to paramagnetic insulator state or antiferromagnetic insulating state and ferromagnetic metallic state is mostly conceived to be the relation of the rich physical properties in CMR materials. The effect of grain size on magnetic behaviors of different types of manganites had been studied by many researchers [2,3]. Goutam et al., (2006) [4] studied the effect of Ts from 600°C to 1000°C in nanophase $La_{0.7}Ca_{0.3}MnO_3$ manganites synthesized through polymeric precursor route and they observed that as Ts increased, both crystallite size and the particle size were increased due to the congregation effect. The decrease in magnetoresistance in these oxides becomes extremely large near the transition temperature and this phenomenon gives huge interest to information storage applications. The magnitude of MR which depends on the magnitude of external magnetic field applied becomes colossal only in the presence of great fields at the transition temperature, which in depends on the composition. Therefore, in order to

Content from this work may be used under the terms of the Creative Commons Attribution 3.0 licence. Any further distribution of this work must maintain attribution to the author(s) and the title of the work, journal citation and DOI.
Published under licence by IOP Publishing Ltd

enhance the magnitude of MR in the presence of relatively low fields and at high temperatures, grain size reduction [5], substitution doping [6], distribution of the manganite grains in a non-magnetic insulating matrix and magnetic insulating matrix are used as alternative approaches [7]. Thus, sol-gel technique has more advantages than solid state method to obtained ultrafine nano sized powders with precise stoichiometry and better uniform size. Besides, sol-gel method has the advantage of lower processing temperatures, short annealing times, high purity of materials, good control of size and shape of the particles and particle size well below 100 nm at the lower processing temperature [8]. The effect of sintering temperature on structure, microstructure, electrical and magneto-transport properties of single phase $La_{0.67}Ca_{0.33}MnO_3$ were synthesized using sol-gel reaction method was reported in this work.

2. Materials and Method

$La_{0.67}Ca_{0.33}MnO_3$ compound were prepared via sol-gel method by dissolving a stoichiometric mixture of nitrates cation ($La(NO_3)_3.6H_2O$, $Ca(NO_3)_2.4H_2O$ and $Mn(NO_3)_2.4H_2O$) in distilled water. Stoichiometric amount of the nitrate solution was stirred and heated at 70°C for 2 hours. Then, the aqueous solution of citric acid was added to the cationic mixture and after a while the ethylene glycol was added to the mixed solution. Finally, the whole solution was heated at 80°C with 250 rpm using magnetic stirrer until the gel is formed. On further heating, the gel was dried in the oven at 120°C until the gel was decomposed and produced a fluffy dried mass. The fluffy dried mass was ground and calcined at 500°C in ambient atmosphere for 5 hours with the heating and cooling rate of 2°C/min. Then, this powders were sintering with various temperature starting from 600°C until 1200°C [1,2]. The structural and microstructure of the samples were characterized by X-ray diffraction (XRD, Phillips PW 3040/60 Xpert Pro) using a CuKα radiation at room temperature and field emission scanning electron microscopy (FESEM, LEO1455 VSPEM, with an OXFORD INCA ENERGY 300EDXattachment). The magnetic behaviors were measured by vibrating sample magnetometer (VSM, LakeShore 7440) at room temperature. The Lake Shore Hall Effect measurement system 7604 was used to characterized and analyzed the metal insulator transition temperature (Tp) and magneto transport properties with temperature control (80 – 300 K) and magnetic field (0 and 1 T).

3. Results and Discussions

Figure 1(a) shows the XRD spectrums of LCMO with different sintering temperature from 600°C to 1200°C via sol-gel method labeled as LC-SG6, LC-SG7, LC-SG8, LC-SG9, LC-SG10, LC-SG11 and LC-SG12, respectively. Overall, all samples are completely formed into polycrystalline single phase without any detectable impurities with orthorhombic structure having Pnma space group (62) (ICSD collection code: 82820). The crystallite size was determined from the most intense of (121) peak using Scherrer's formula. The intensity of the X-ray peaks for the LCMO perovskite phase increases as sintering temperature increases from 600°C to 1200°C reflecting the growth of crystalline phase in all samples. Rietveld Refinement analysis conformed all samples were in orthorhombic system, where not changes of lattice constant (a, b, c) and cell volume, V were observed. However, crystallite size is increased with sintering temperature.

The micrographs revealed an increment of sintering temperatures (Ts) starting from 600°C to 1200°C significantly promotes grain growth and the porosity decreases. The variation of the crystallite sizes (CS) obtained from the width of the X-ray diffractions peaks and the grain size (GS) for the all LCMO samples were plotted in Figure 1(b). The growth rate of the crystallite size is slower as compared to grain size and reach saturated after 1000°C. The difference between CS and GS was more pronounced at higher sintering temperature. Both crystallite sizes (CS) and the grain size (GS) increases with sintering temperature (Ts) due to congregation of the grains because a grain may subsists of several crystallite domains and these domains arises due to twinning and other structure defects, such as vacancies, dislocations, and stacking faults in the grain [9]. Overall, at low sintering temperature (600°C and 700°C), it is expected that only single crystallite on each grain. Meanwhile, at higher temperature (1000°C-1200°C), the grain might consist of multi crystallite.

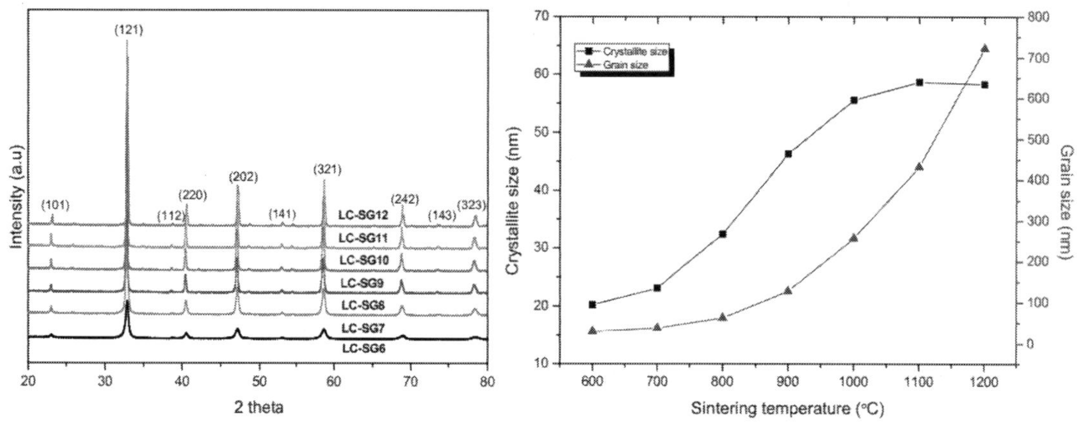

Figure 1: (a). XRD patterns for LCMO with different sintering temperature and (b) crystalline size and grain size dependence on sintering temperature

In this section, metal-insulator transition temperature (T_{MI}) was measured in the temperature range of 80 – 300 K by four probe technique with/without an external applied magnetic field. All samples of LCMO at 0 and 1 T were plotted in Figure 2. The T_{MI} of polycrystalline samples is an extrinsic property that strongly depends on the synthesis conditions and microstructure (e.g. grain boundaries or magnetic disorder layers). As lowering sintering temperature, a large dropped of the T_{MI} and a substantial increase in resistivity at $T < T_{MI}$. Siwach et al. (2008) [10] emphasized the importance of a spin-polarized tunnelling below phase temperature and the domain wall contribution. They suggested that in small particles and noticed electrical resistivity increase at low temperatures is due to the enhanced scattering of the charge carriers through the higher density of magnetic disorder in grain boundaries. Thus, the T_{MI} at lowest sintering labelled as LC-SG6 goes down to 155 K (0 T). The strong suppression of the T_{MI} is caused by induced disorders and also by the increase in the non-magnetic phase fractions, due to enhanced grain boundary densities as a consequence of lower sintering temperature. This is also causes the increase in the carrier scattering leading to a corresponding enhancement in the resistivity. Therefore, lowering of sintering temperature reduces the metallic transition temperature and hence, increased in resistivity. However, by apply an external magnetic field 1 T, the transition temperature shifts to higher values with increasing the particle size leading to decrease in the GB's and the associated disorder. Thus, resulting decrease in scattering of the carriers expressed by decreasing in the resistivity.

The simultaneous electronic transition can be clarify by double exchange (DE) mechanism model, the shifted T_{MI} is most probably due to strong interaction between the Mn^{3+} and Mn^{4+} spin. Goutam et al., (2006) [11] reported that in the presence of applied magnetic field, the localized electron spins (t_{2g}) are favorably oriented along the field direction and therefore, the electron can transferred more easily from Mn^{3+} to Mn^{4+}, hence resulting in the improved conductivity and consequently decrease in the resistivity and T_{MI} shifted to the higher temperature has been observed.

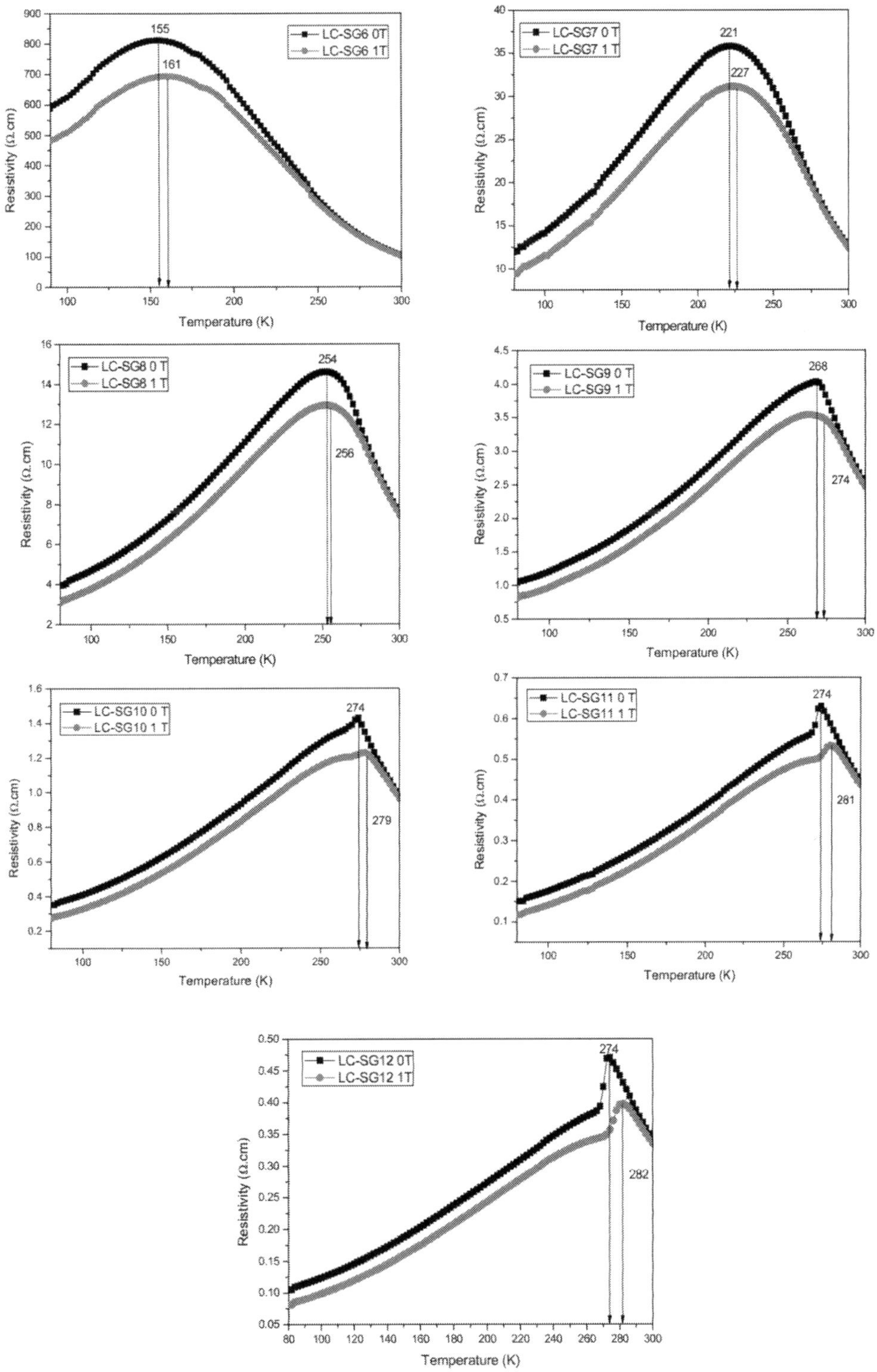

Figure 2: Temperature dependence of the resistivity for LCMO at sintering temperature from 600°C to 1200°C.

The percentage of MR curve for all LCMO samples were measured from 80 – 300 K with an external magnetic field up to 1 Tesla as shown in figure 3. At 80 K, the MR values were measured to be (-7.62%), (-11.61%), (-12.89%), (-12.08%), (-12.31%), (-11.21%), (-11.27%) for sample LC-SG6, LC-SG7, LC-SG8, LC-SG9, LC-SG10, LC-SG11, LC-SG12, respectively at the field of 0.1 T. The highest LFMR value, which extrinsic MR is more dominant, is observed for LC-SG8 with nano-sized distribution that showing pseudo-domain structure as suggested in. However, the highest CMR is found in LC-SG10 (-23.48%) at 80 K with 1 Tesla applied field.

The MR curves gave two significant different regions or gradient, which is faster drop of MR value from 0 – 0.1 T and follow by a linear or slow decrease of resistance at 0.1 T – 1.0 T. Generally, low field magnetoresistance (LFMR) is a significant MR, which observed at low magnetic field with always related to the extrinsic MR and Radaelli et al. (1997) [12] reported that commonly this kind of behavior is show in polycrystalline form that can be interpreted in terms of spin dependent tunneling and spin dependent scattering through electron barriers across the grain boundaries. The large scattering and tunneling effects occurred when the spin alignment disordered at the grain boundaries. When a low field is applied, the disorder spin among neighboring will align parallel with the magnetic field. Thus, this will enhanced the electron hopping and leads to the steep decrease in resistance at low fields. All LCMO systems with different sintering process have not achieved saturation with a magnetic field up to 1 Tesla, meaning that higher magnetic field is required to fully align all localized spin. This might be due to obtain of some disordered region at the grain boundary.

The appearance of peak in the MR-T curve around T_{MI} depicts in which all samples as shows in figure 4, there is a presence of the intrinsic component of MR, which arises due to the double exchange mechanism around T_{MI}. However, around T_{MI} the peak in the MR-T curve of the sample LC-SG12 (bigger in particle size) is significantly higher at all applied magnetic fields as compared to other samples. The magnitude of MR peak around T_{MI} decreases as we lower the sintering temperatures (smaller in particle size) as well as the magnitude of the applied magnetic field is reduced. The peak MR values are (-23.94%), (-19.47%), (-14.76%), (-12.69%), (-11.53%) at 1 Tesla applied magnetic field for LC-SG12, LC-SG11, LC-SG10, LC-SG9 and LC-SG8, respectively whereas for sample LC-SG6, and LC-SG7 there is a hump in the MR variation around T_{MI}. In the case of nanostructure manganites, magnetotransport studies reveal that the volume fraction of highly disordered grain boundaries play an important role in these materials. As the surface to volume ratio become larger, that is, grain size reduced, the effective of the grain boundaries will increase and as a result the intrinsic properties reduced, thus, humped the MR peaks. These studies clearly depict that sol-gel technique is an effective method to enhance ferromagnetic ordering temperature along with improving of low field magnetoresistance. Thus, the intrinsic of magnetoresistance follows the expected DE behavior with steady increase in temperature, meanwhile, the extrinsic of magnetoresistance improved that the nature of the surface region of nanosize grains plays a very crucial role in enhancement of magnetotransport behavior of nanodimensional system [13].

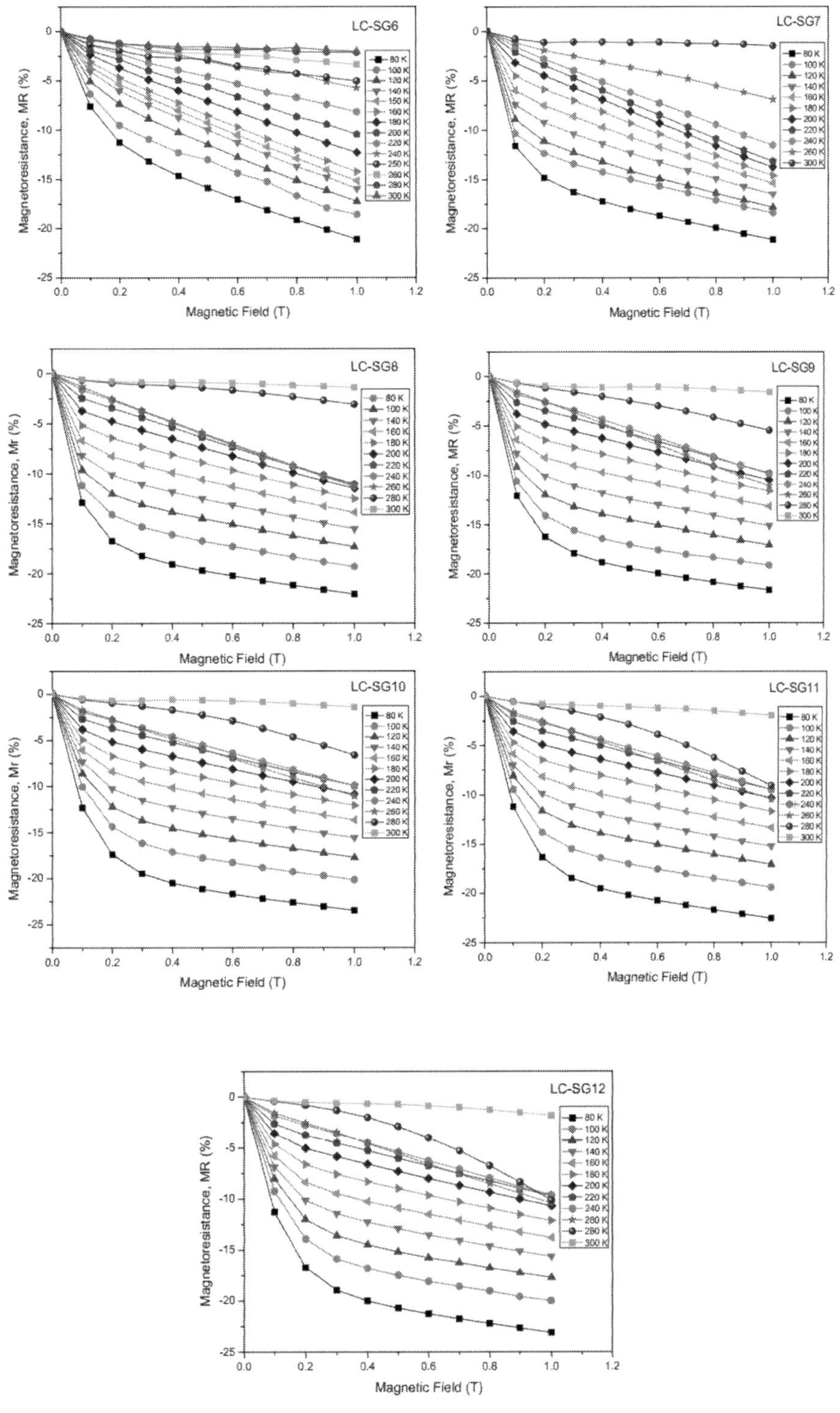

Figure 3: The percentage of MR curve for LCMO with different sintering temperature from 600°C to 1200°C as a function of magnetic field at various temperatures.

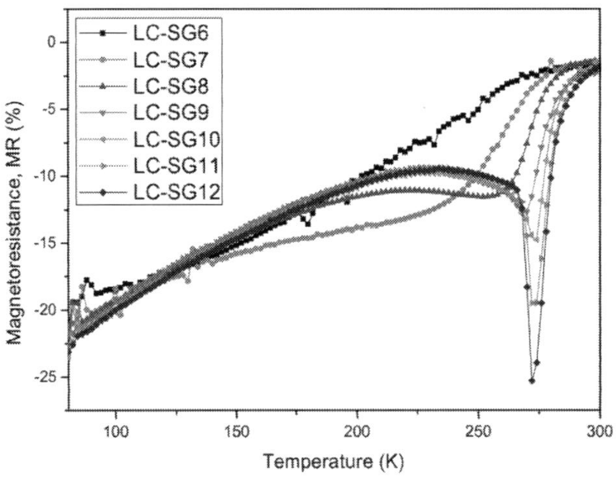

Figure 4: The percentage of MR versus temperature curve for LCMO with different sintering temperature from 600°C to 1200°C.

4. Conclusion

The effects of sintering temperature (grain size) on the electrical and magneto-transport properties of nanoparticles of polycrystalline $La_{0.67}Ca_{0.33}MnO_3$ ceramics synthesized by sol-gel method was demonstrated. The T_{MI} at lowest sintering labelled as LC-SG6 goes down to 155 K (0 T). The strong suppression of the T_{MI} is caused by induced disorders and also by the increase in the non-magnetic phase fractions, due to enhanced grain boundary densities as a consequence of lower sintering temperature. The highest LFMR value, which extrinsic MR is more dominant, is observed for LC-SG8 with nano-sized distribution that showing double crystallite. However, the highest CMR is found in LC-SG10 (-23.48%) at 80 K with 1 Tesla applied field. All LCMO systems with different sintering process have not achieved saturation with a magnetic field up to 1 Tesla, meaning that higher magnetic field is required to fully align all localized spin. This might be due to obtain of some disordered region at the grain boundary.

Acknowledgements

The authors gratefully acknowledge the support from the Ministry of Education (MOE) for the grant under the Fundamental Research Grant Scheme (FRGS) vote: 02-01-14-1469FR: Enhancement of Magnetoresistance Effect in La-(Sr, Ca)-Mn-O/Sb_2O_3, CeO_2, Al_2O_3 Composites for Low-Field Magnetic Sensor.

References

[1] J. Ma, M. Theingi, Q. Chen, W. Wang, X. Liu, and H. Zhang, "Influence of synthesis methods and calcination temperature on electrical," Ceram. Int., vol. 39, pp. 7839–7843, 2013.
[2] L. E. Hueso, F. Rivadulla, R. D. Sa, D. Caeiro, and C. Jardo, "Influence of the grain-size and oxygen stoichiometry on magnetic and transport properties of polycrystalline LaCaMnO perovskites," J. Magn. Magn. Mater., vol. 189, pp. 321–328, 1998.
[3] G. Venkataiah, D. C. Krishna, M. Vithal, S. S. Rao, S. V. Bhat, V. Prasad, S. Subramanyam, and P. Reddy, "Effect of sintering temperature on electrical transport properties of LaCaMnO," Phys. B Condens. Matter, vol. 357, pp. 370–379, 2005.
[4] U. K. Goutam, P. K. Siwach, H. K. Singh, R. S. Tiwari, and O. N. Srivastava, "Colossal Magnetoresistance Study in Nanophasic $La_{0.7}Ca_{0.3}MnO_3$ Manganite," J. Phys. D Appl. Physics,

vol. Volume 39, no. 1, pp. 1–21, 2006.

[5] Liu X, Jiao Z, Nakamura K, Hatano T and Zeng Y 2007 The grain size dependence of the resistance behaviors in doped lanthanum manganite polycrystalline films The grain size dependence of the resistance behaviors in doped lanthanum manganite polycrystalline films 2431.

[6] Lim, K. P., Ng, S. W., Halim, S. A., Chen, S. K., and Wong, J. K., 2009 Effect of divalent ions (A = Ca, Ba and Sr) Substitution in La-A-Mn-O manganite on structural, magnetic and electrical transport properties Am. J. Appl. Sci. 6 1153–7.

[7] Nayak B B, Vitta S and Bahadur D 2007 Synthesis and properties of nanograined La-Ca-manganite–Ni-ferrite composites Mater. Sci. Eng. B 139 171–6.

[8] J. Ma, M. Theingi, Q. Chen, W. Wang, X. Liu, and H. Zhang, "Influence of synthesis methods and calcination temperature on electrical," Ceram. Int., vol. 39, pp. 7839–7843, 2013.

[9] P. K. Siwach, R. Prasad, A. Gaur, H. K. Singh, G. D. Varma, and O. N. Srivastava, "Microstructure-magnetotransport correlation in La 0.7 Ca 0.3 MnO 3," J. Alloys Compd., vol. 443, pp. 26–31, 2007.

[10] Siwach, P. K., Prasad, R., Gaur, A., Singh, H. K., Varma, G. D., and Srivastava, O. N., 2007. Microstructure-magnetotransport correlation in La0.7Ca0.3MnO3. J. Alloys Compd. 443: 26–31.

[11] Goutam, U. K., Siwach, P. K., Singh, H. K., Tiwari, R. S., and O. N. Srivastava., 2006. Colossal Magnetoresistance Study in Nanophasic La0.7Ca0.3MnO3 Manganite. J. Phys. D Appl. Physics, Volume 39: 1–21.

[12] Radaelli, P., Iannone, G., Cox, D., Marezio, M., Hwang, H., and Cheong, S-W., 1997. Competition between charge ordering and ferromagnetism in manganese perovskites. Phys. B Condens. Matter 241-243: 295–302.

[13] Phong, P. T., 2008. Influence of sintering temperature on low-field spin-polarized tunneling magnetoresistance of $La_{0.7}Ca_{0.3}MnO_3$. Commun. Phys. Vol. 18: pp. 48–57.

IOP Conf. Series: Journal of Physics: Conf. Series **1083** (2018) 012060 doi:10.1088/1742-6596/1083/1/012060

Influence of Oxalic Acid Concentrations on The Growth of Molybdenum Disulfide via Spin Coating Technique

A L Tan[1,a], S S Ng[2,b] and H Abu Hassan[1,c]

[1] School of Physics, Universiti Sains Malaysia, 11800 USM, Penang, Malaysia.
[2] Institute of Nano Optoelectronics Research and Technology (INOR), Universiti Sains Malaysia, 11800 USM, Penang, Malaysia.

[a]elwinharrisontan92@gmail.com, [b]shashiong@usm.my, [c]haslan@usm.my

Abstract. In this context, the molybdenum disulfide (MoS_2) films were grown by spin coating technique and thermal vapour sulfurization (TVS) under the condition of different oxalic acid concentrations. Atomic force microscopy (AFM) surface topography shows the improved homogeneity and higher compactness of films with the increasing acid concentration. However, high concentration favoured the formation of larger grains and poor continuity of films. These observations were further proven by the particle size distribution profile. Apart from that, two distinct Raman phonon modes of the MoS_2 can be detected in all the deposited films. In addition, molybdenum dioxide (MoO_2) was detected at higher acid concentration due to the slower rate of diffusion of sulfur atoms during the surface sulfurization. From the direct analysis of frequency difference between in-plane (E_{2g}^1) and out-of-plane (A_{1g}) phonon modes, it proves that multilayers of MoS_2 films were synthesized. Ultraviolet-visible (UV-Vis) specular reflectance results also reveal that crystallite MoS_2 films with the absorption peaks of 612 nm and 660 nm were detected.

1. Introduction

Research advances in two dimensional (2-D) graphene have triggered the research interest in other 2-D materials, such as boron nitride [1], black phosphorus [2], silicene [3], and semiconducting 2-D transition metal dichalcogenides (TMDs) [4]. There is a wide range of semiconducting 2-D TMDs where molybdenum disulfide (MoS_2) certainly becomes the rising star among the family. The layered structure of MoS_2 consists of an atomic plane of molybdenum (Mo) sandwiched between two planes of sulfur (S) atoms. With the analogous layered structure to graphene, the weakly bonded interlayer enable to be isolated into single or multi-layers of MoS_2. In such the way, the energy bandgap experiences transition from indirect (bulk, ~1.2 eV) to direct (monolayer, ~1.8 eV) [4]. Hence, the semiconducting behaviour of MoS_2 has significant advantages over the zero bandgap graphene which is potentially to be applied in optoelectronic and electronic switching devices.

Major reports in the early research on the production of mono or multilayers of MoS_2 were based mainly on exfoliations, either by mechanical or chemical technique [5, 6]. This approach was also extended with the implementation of electrochemical and surfactant-based solution [7, 8]. However, the resulting exfoliated MoS_2 was structurally and electronically different from the bulk MoS_2. Thus, the MoS_2 experienced the conversion of the Mo atom molecular geometry from trigonal prismatic to octahedral, and change of electronic structure from semiconducting to metallic [9]. Besides, this method

Content from this work may be used under the terms of the Creative Commons Attribution 3.0 licence. Any further distribution of this work must maintain attribution to the author(s) and the title of the work, journal citation and DOI.
Published under licence by IOP Publishing Ltd

also resulted in poor uniformity and lower yield of MoS_2 flakes amount. In contrast, the growth of MoS_2 by chemical vapour deposition (CVD) of MoS_2 was rather a better approach in wafer-scale fabrication [10-13]. In this work, thermal vapour sulfurization (TVS) method and performed spin coating technique was applied as the pre-deposition of MoS_2-based compound films, followed by sulfurization of the deposited films as surface treatment process. Moreover, we suggest different acid concentrations have significant effect on the growth of molybdenum disulfide. Up to now, there have no reported works for the growth of MoS_2 with different acid concentrations.

Herein, we demonstrate the growth and characterizations of MoS_2 thin films with the variation of oxalic acid concentrations. Through this study, a better understanding of the growth MoS_2 under different acid concentration conditions. The research findings in this work, therefore, it significantly contributes to the research in the growth of 2-D semiconducting TMDs.

2. Methodology and characterizations

In this work, 4 inches of silicon, Si (111) substrate was cleaved into squared shaped with the dimension of 1 cm x 1 cm. Prior to the spin coating process, the cleaved Si (111) substrates were cleaned with acetone and diluted hydrofluoric acid (HF) solution. Next, the cleaned substrates were subjected to the oxygen plasma treatment as to improve the wettability of prepared precursors. The precursor was first prepared by mixing the powders, which contained ammonium molybdate tetrahydrate [$(NH_4)_6Mo_7O_{24}$•$4H_2O$, QReC^{TM} 99.5%], thiourea [$(NH_2)_2S$, UNILAB 99.0%] and oxalic acid (OA) [$C_2H_2O_4$, Brendosen 99.5%] with a drop of distilled water (DW) in ultrasonic agitation. After that, ethylenediamine [$C_2H_8N_2$, Merck 99.0%] solvent was then added into the precursor after the dark-blue colloid was observed.

To investigate the effects of acid concentrations on the growth of MoS_2 films, the precursor solutions with different oxalic acid concentrations (M, molar) of 2 M, 6 M and 10 M were prepared. A series of precursor solution with colour contrast are shown in Figure 1. The prepared precursors were dropped and spin for 25 s at the speed of 4000 rpm. To transform the deposited films into crystalline MoS_2 thin films, they were first annealed at 500°C, next, the system was slowly ramped up to 750°C for the sulfurization process in sulfur-rich environment. The experimental parameters were chosen based on the references [12] and [13]. This process was conducted by continuous bubbling carbon disulfide, CS_2 vapour with nitrogen gas into the quartz tube. At the end of the process, the furnace system was purged to remove excessive CS_2 and sulfur, S vapour and cooled passively before removed the samples.

To access the material properties, tapping mode atomic force microscopy (AFM, Bruker Dimension Edge) was used to study the changes of surface topography of the grown MoS_2 thin films. The dimension of 10 μm x 10 μm was obtained with 256 lines/scan. Next, the optical phonon modes of MoS_2 were determined by Raman spectroscopy (Horiba Jobin Yvon HR800UV), argon (Ar) ion laser with excitation wavelength of 514.5 nm and excitation power of 20 mW was used. Finally, the absorption measurements were performed by using ultraviolet-visible-near infrared (UV-Vis-NIR) spectrometer (Agilent Cary 5000 UV-Vis-NIR spectrometer) to determine the optical absorbance of MoS_2 films.

Figure 1: A series of precursor solutions with oxalic acid concentrations of 2 M, 6 M and 10 M.

3. Results and discussion

Through the investigation of AFM imaging, different aggregation processes and growth of MoS_2 grains were observed with the variation of OA concentrations. Particle-like MoS_2 grains are formed and well dispersed on the surface of substrate for all acid concentrations. At 2 M acid concentration, fine and homogenous MoS_2 grains were observed as shown in Figure 2(a). By referring to Figure 2(b), the smaller grains start to aggregate and form highly compact MoS_2 films. This observation leads to the lower formation of pinholes density when the OA concentration increases from 2 M to 6 M. However, at 10 M acid concentration, the aggregation of MoS_2 grains, which in turns causes the decomposition of MoS_2 films. As can be seen in Figure 2(c), films decomposition results in the formation of isolated larger grains with poor continuity of MoS_2 films. With increasing acid concentration, the root-mean square surface roughness of the films increases from 11.6 nm [Figure 2(g)] to 20.6 nm [Figure 2(i)].

Apart from that, the particle size also increases with increasing OA concentration, as evidenced from the particle size distribution profiles. For the 2 M OA concentration, the statistical analysis reveals that the distribution of particle sizes was narrow and centred at about 80 nm [Figure 2(d)]. As the OA concentration increases to 6 M, wide distribution of particle sizes with the highest maximum at around 75 nm and a shoulder position at 100 nm was observed [Figure 2(e)]. At 10 M concentration, the particle size increases to around 140 nm and a sharp maximum corresponds to about 170 nm [Figure 2(f)].

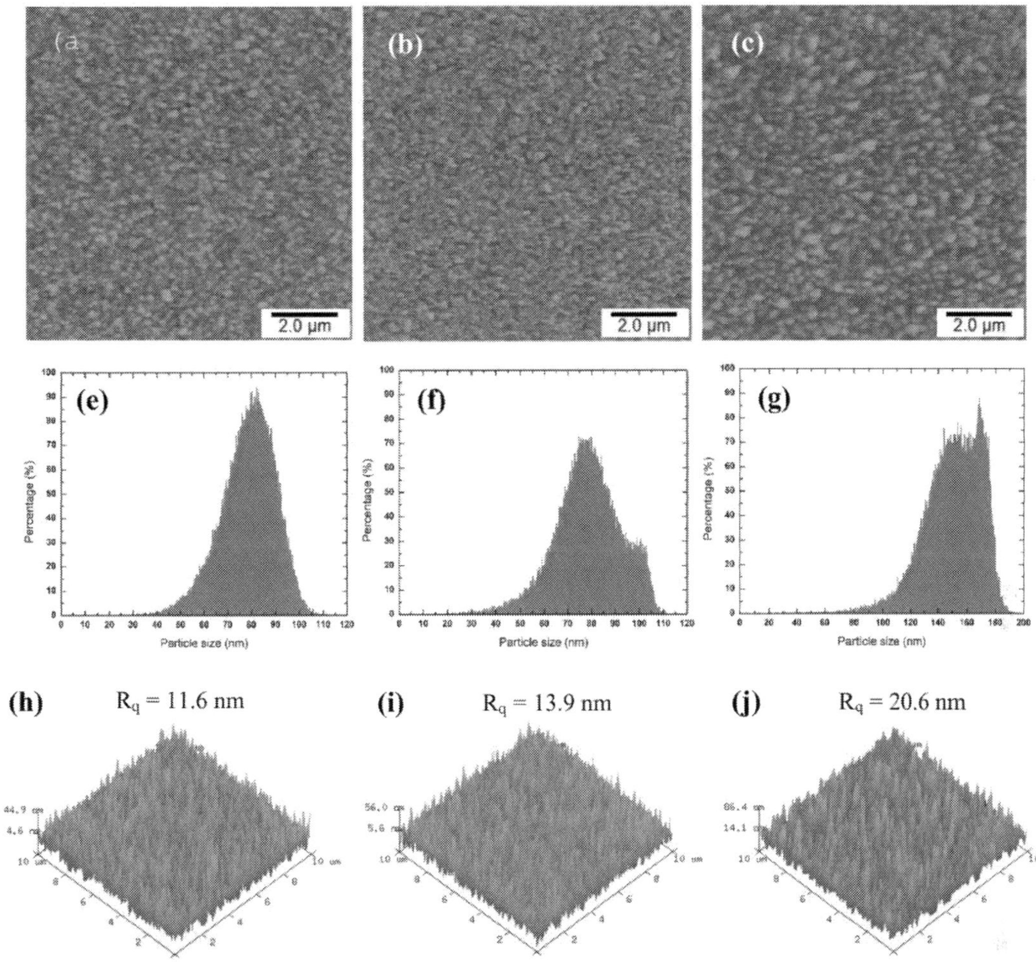

Figure 2: Surface topographies [(a) 2 M, (b) 6 M, and (c) 10 M], particle size distribution profiles [(e) 2 M, (f) 6 M and (g) 10 M], and 3D surface roughness image [(h) 2 M, (i) 6 M, and (j) 10 M] of the deposited MoS_2 with the varying acid concentrations. R_q indicates the root-mean-square surface roughness of the deposited films.

The changes in the growth of MoS_2 grains and surface topography can be realized by hydrolysis and polycondensation process in the intermediate steps of the solution preparation process. Through this study, it was found that the more acidic condition enables the growth of larger MoS_2 grains. The dissociation of OA in water which results in the partially formation of oxalate, $C_2O_4^{2-}$ ions and hydrogen, H^+ ions [14] as proposed by the equations (1) and (2) below.

$$C_2O_4H_2 \rightleftharpoons HC_2O_4^- + H^+ \quad (1)$$
$$HC_2O_4^- \rightleftharpoons C_2O_4^{2-} + H^+ \quad (2)$$

From these equations, higher acid concentration produces higher effective density of H^+ ions in the colloid. Therefore, in these conditions, it facilitates the higher rate of hydrolysis and condensation process [15], which leads to larger aggregates of MoS_2 grains.

Figures 3(a) to (c) show the typical Raman spectra for the MoS_2 thin films prepared using different oxalic acid concentrations. Two significant peaks correspond to the in-plane (E_{2g}^1) and out-of-plane phonon modes (A_{1g}) of MoS_2 located at ~384 cm^{-1} and ~409 cm^{-1}, respectively were detected. These values were in good agreement with that reported by Chakraborty *et al* [16]. Further increase of OA

IOP Conf. Series: Journal of Physics: Conf. Series **1083** (2018) 012060 doi:10.1088/1742-6596/1083/1/012060

concentration, vibrational phonon mode originated from the molybdenum dioxide, MoO_2 was detected at around ~365 cm^{-1} [17]. This probably due to the higher acid concentration will favour the polycondensation process of Mo-O-S compounds at the primary stage, and hence, the larger Mo-O-S based compounds were formed. This observation was in line with the AFM imaging results as discussed previously.

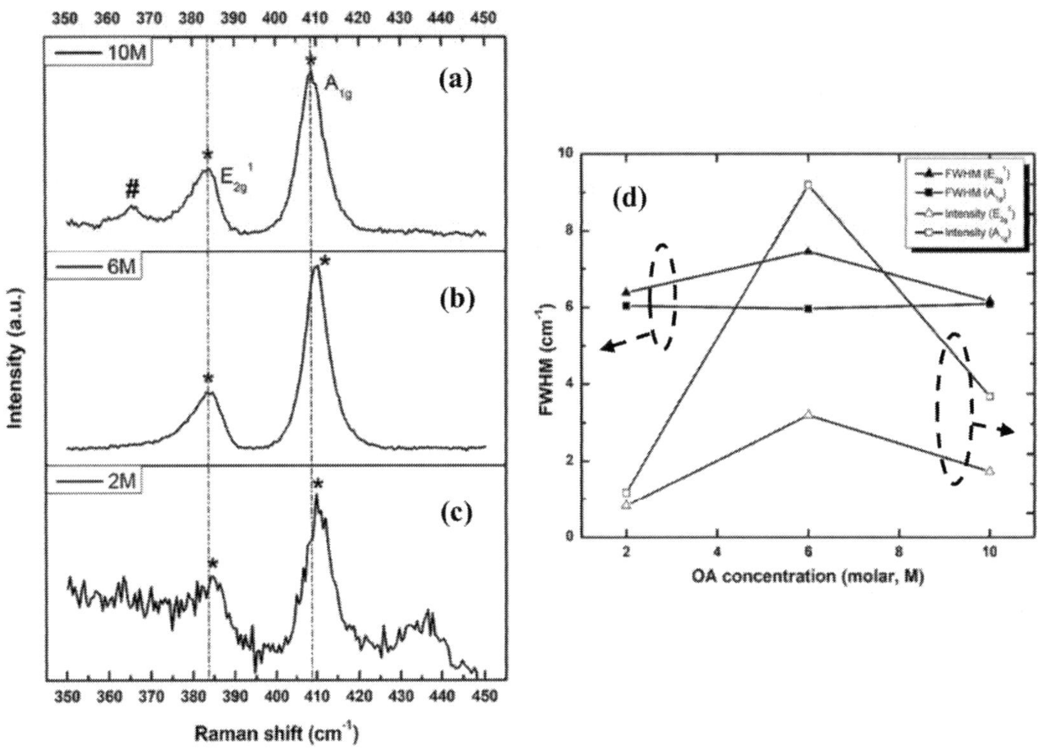

Figure 3: Raman spectra for MoS_2 thin films prepared using different oxalic acid concentrations, (a) 2 M, (b) 6 M, and (c) 10 M. The variation of the FWHM and intensity of Raman peaks as a function of the OA concentration was summarized in (d). The * represents MoS_2, while # indicates MoO_2.

The formation mechanism of the MoS_2 films can be briefly explained as below. Firstly, the decomposition of CS_2 vapour occurred at elevated temperature and the suggested chemical reaction is as shown in equation (3) [18]. The decomposed radicals species and sulfur gases cause the partial reduction of Mo-O-S based compounds and convert into MoS_2 crystalline.

$$CS_2 \rightleftharpoons CS + S \quad (3)$$

During the sulfurization process, the oxygen, O atoms in the Mo-O-S compounds are slowly replaced by the S atoms by diffusion. The layers on the surface of Mo-O-S compounds were sulfurized quickly in the beginning of the sulfurization process. Increasing formation of MoS_2 outer surface layers limited further diffusion of S atoms into the Mo-O-S compounds thus MoO_2 exists below the MoS_2 layers. As a result, the significantly larger Mo-O-S based compounds slowing down the rate of diffusion which leads to incomplete conversion into MoS_2 crystalline state [17]. The above explanations are summarized in the schematic drawing in Figure 4.

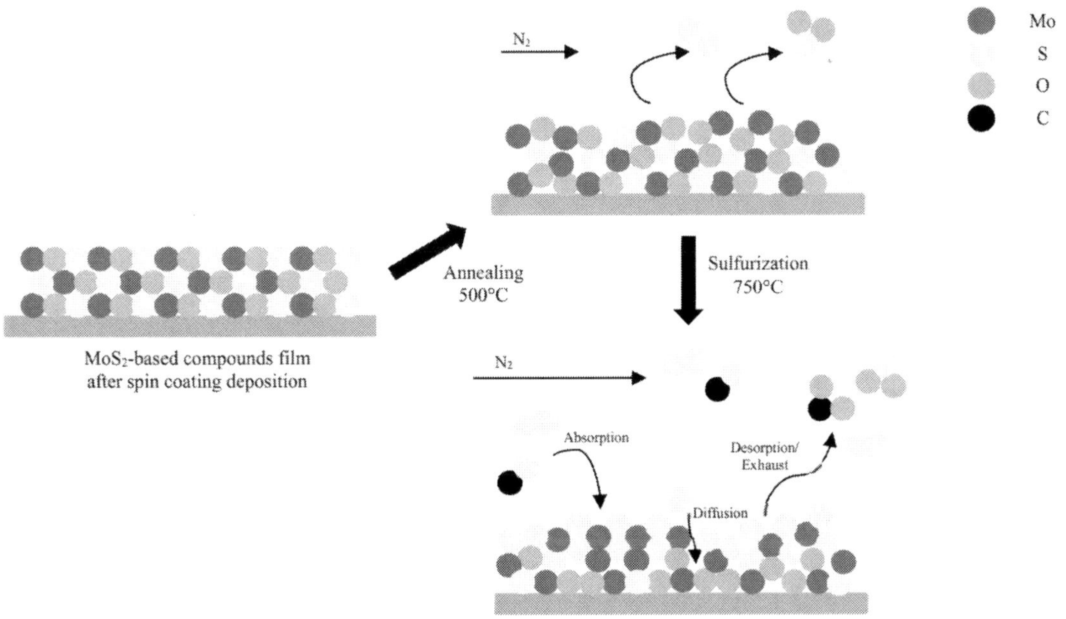

Fig. 4 The schematic drawing of the formation of the MoO_2 below the MoS_2 layers.

It is known that the direct analysis of the frequency different (Δk) between the two phonon modes ($\Delta k = A_{1g} - E_{2g}^{1}$) can estimate the thickness of the MoS_2 films. According to Ye *et al* [19], $\Delta k \approx 20$ cm^{-1} corresponded to the monolayer of MoS_2, while $\Delta k \approx 25$ cm^{-1} indicated the multi-layers or bulk MoS_2 was obtained. From the calculation, the value of $\Delta k = \sim 25$cm^{-1} are obtained which represents the grown samples for all acid concentrations are multi-layers of MoS_2 films.

Further analyses on the full width at half maximum (FWHM) and the intensity of the Raman peaks of the MoS_2 were carried out in which the Raman spectrum was fitted using Lorentz function. The obtained results were summarized in Fig. 3(d). It was found that the FWHMs for both phonon modes are insensitive to the changes of acid concentrations. Yet, the intensity of the Raman signal increases for the 6 M of OA concentrations. One can explain that higher compactness of MoS_2 films which results in higher volume to surface ratio. Thus, more MoS_2 phonon will be excited by the incident laser beam. Under the higher acidic conditions (i.e., >6 M), the intensity of MoS_2 Raman signal shows significant reduction. This is mainly attributed to the incomplete conversion of MoS_2 crystalline as discussed previously.

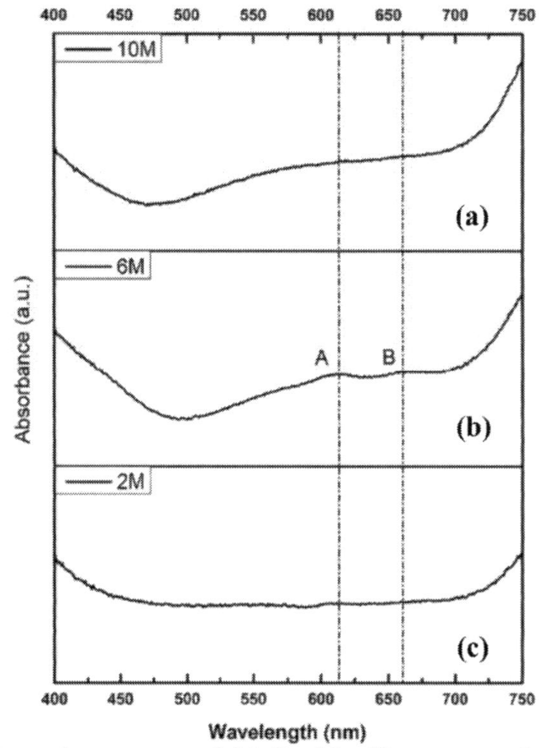

Figure 5: The UV-Vis absorption spectra of MoS_2 thin films prepared at different oxalic acid concentrations (a) 2 M, (b) 6 M, and (c) 10 M. A and B represent the absorption peaks of MoS_2.

Figure 5 shows the absorption spectra of MoS_2 thin films. Two weak features located at around 612 nm and 660 nm which originated from the exciton absorption peaks of the MoS_2 can be observed for samples prepared with 2 M and 6 M acid concentrations. While no absorption feature was observed for sample prepared with 10 M acid concentration. This is likely due to the lower formation MoS_2 at 10 M OA concentration, which leads to weak and untraceable absorption.

4. Conclusion

In summary, the MoS_2 films prepared using different OA concentrations were successfully grown by TVS and spin coating as pre-deposition technique. Due to changes in acid concentration, the surface topography and the grain size of the MoS_2 films were greatly influenced where the grain size and the root-mean-square surface roughness of MoS_2 films increased with the OA concentration. The optical properties were accessed by Raman analysis, which revealed the in-plane (E_{2g}^1) and out-of-plane phonon modes (A_{1g}) of MoS_2 crystalline. Two significant MoS_2 Raman peaks were detected for all the samples. At higher acid concentration, MoO_2 was detected due to the larger aggregate of MoS_2 grains, which resulted in slower rate of diffusion of sulfur atoms during the sulfurization process. Apart from that, multi-layers of MoS_2 films were obtained via the direct analysis of the frequency different ($\Delta k \approx 25$ cm^{-1}). Finally, the two exciton energies of MoS_2 was detected, however it is undetectable for higher acid concentration due to the lower formation of MoS_2.

Acknowledgement

The authors gratefully acknowledge the financial support from Ministry of Higher Education (MOHE) Malaysia for granting the Fundamental Research Grant Scheme (FRGS) (account no. : 203.PFIZIK.6711561). The corresponding author would like to offer gratitude to the MOHE for offering the MyMaster scholarship to pursue his Master's degree.

References

[1] Kim K K, Hsu A, Jia X, Kim S M, Shi Y, Hofmann M, Nezich D, Rodriguez-Nieva J F, Dresselhaus M and Palacios T 2011 *Nano Lett.* 12(1) pp 161-166.
[2] Smith J B, Hagaman D and Ji H F 2016 *Nanotech.* 27(21) 215602.
[3] Tao L, Cinquanta E, Chiappe D, Grazianetti C, Fanciulli M, Dubey M, Molle A and Akinwande D 2015 Nat. *Nanotech.* 10(3) pp 227-231.
[4] Duan X, Wang C, Pan A, Yu R and Duan X 2015 *Chem. Soc. Rev.* 44(24) pp 8859-8876.
[5] Eda G, Yamaguchi H, Voiry D, Fujita T, Chen M and Chhowalla M 2011 *Nano Lett* 11(12) pp 5111-5116.
[6] Li H, Wu J, Yin Z and Zhang H 2014 *Acc. Chem. Res.* 47(4) 1067-1075.
[7] Zeng Z, Yin Z, Huang X, Li H, He Q, Lu G, Boey F and Zhang H 2011 *Angew. Chem. Int. Ed.* 50(47) pp 11093-11097.
[8] Smith R J, King P J, Lotya M, Wirtz C, Khan U, De S, O'Neill A, Duesberg G S, Grunlan J C and Moriarty G 2011 *Adv. Mater.* 23(34) pp 3944-3948.
[9] Wang Q H, Kalantar-Zadeh K, Kis A, Coleman J N and Strano M S 2012 *Nat. Nanotechnol.* 7(11) pp 699-712.
[10] Lee Y H, Zhang X Q, Zhang W, Chang M T, Lin C T, Chang K D, Yu Y C, Wang J T W, Chang C S and Li L J 2012 *Adv. Mater,* 24(17), pp 2320-2325.
[11] Van Der Zande A M, Huang P Y, Chenet D A, Berkelbach T C, You Y, Lee G H, Heinz T F, Reichman D R, Muller D A and Hone J C 2013 *Nat. Mater.*, 12(6), pp 554-561.
[12] Liu K K, Zhang W, Lee Y H, Lin Y C, Chang M T, Su C Y, Chang C S, Li H, Shi Y, Zhang H, Lai C and Li L J 2012 *Nano lett.* 12 1538.
[13] Yang J, Gu Y, Lee E, Lee H, Park S H, Cho M H, Kim Y H, Kim Y H and Kim H 2015 *Nanoscale* 7 9311
[14] Kettler R M, Palmer D A and Wesolowski D J 1991 *J. Solution. Chem.* 20(9) pp 905-927.
[15] Houng B, Huang C L and Tsai S Y 2007 *J. Cryst. Growth.* 307(2) pp 328-333.
[16] Chakraborty B, Matte H, Sood A and Rao C 2013 *J. Raman Spectrosc.* 44(1) pp 92-96.
[17] Wang X, Feng H, Wu Y and Jiao L 2013 *J. Am. Chem. Soc.* 135(14) pp 5304-5307.
[18] Bamford C H 1972 *Decomposition of Inorganic and Organometallic Compounds* vol 4, ed Bamford C H, and Tipper C F H (Elsevier Publishing Company) chapter 2 pp 58
[19] Ye M, Winslow D, Zhang D, Pandey R and Yap Y K 2015 *Photonics.* 2 pp 288-307.

IOP Conf. Series: Journal of Physics: Conf. Series **1083** (2018) 012061 doi:10.1088/1742-6596/1083/1/012061

Effect of Nitrogen Gas on the Growth of Magnesium Doped Indium Nitride Thin Films via Sol-gel Spin Coating Method

H.S. Lee[1, a], S.S. Ng[2,b *] and F.K. Yam[1]

[1]School of Physics, Universiti Sains Malaysia, 11800 Penang, Malaysia
[2]Institute of Nano Optoelectronics Research and Technology (INOR), Universiti Sains Malaysia, 11800 Penang, Malaysia

[a]leehuisan530@yahoo.com, [b]shashiong@usm.my

Abstract. The growth of magnesium doped indium nitride (InN:Mg) thin films via sol-gel spin coating method followed by nitridation process was reported in this paper. In this work, the nitridation processes were carried under ammonia with and without nitrogen ambiences. X-ray diffraction patterns reveal that InN:Mg thin film grown under nitrogen ambiences show formation of InN (101) preferential orientation wurtzite structure. Field emission scanning electron microscopy results show that both deposited films exhibit coalesced island morphology with hexagonal like structure. Energy dispersive X-ray spectroscopy revealed that sample grown under ambient with N_2 gas has lower oxygen atomic percentage and higher ratio of indium to nitrogen. Two allowed Raman modes of wurtzite InN, namely, E_2(High) and A_1(LO) modes, were clearly detected for the deposited films under N_2 gas ambient. Nevertheless, the film grown under the present of N_2 gas shows an additional feature corresponding to v_4 vibration of the MgN_4 tetrahedron at around 564 cm^{-1}. The presence of this feature indicates that the magnesium acceptors were activated and the compensation of Mg_{In}-N (LVM) was occured. Finally, all the results suggest that present of N_2 gas during nitridation process will induce better grow of InN:Mg thin films.

1. Introduction

In 1986, Tansley and Foley achieved very high electron mobility and high crystalline quality n-type indium nitride (InN) thin film grow by reactive sputtering at low temperature [1]. Since then, the InN-based electronic and light emitting devices have made a great development. However, the poor hole mobility and electrical conductivity of p-type InN remained unsolved and continue to hinder the development of InN-based electronic devices.

Up to date, magnesium (Mg) is a common acceptor impurity for p-type InN and most of the studies used molecular beam epitaxy (MBE) [2] and metal organic vapor phase epitaxy (MOVPE) techniques [3] to grow Mg doped InN. However, these techniques require the ultrahigh vacuum system, high maintenance cost, as well as complicated in setup and procedures. Therefore, the cost spent in the equipment and maintenance can be noticed. In present work, the sol-gel spin coating could be an alternative method to grow Mg-doped InN. In general, this technique is relatively simpler, cheaper, safer, and scalable as compared with the conventional techniques as mentioned previously. Recently, Fong et al. (2015) and Lee et al. (2017) have successfully demonstrated the growth of GaN and InN thin films by means of this technique [4][5].

Content from this work may be used under the terms of the Creative Commons Attribution 3.0 licence. Any further distribution of this work must maintain attribution to the author(s) and the title of the work, journal citation and DOI.
Published under licence by IOP Publishing Ltd

Moreover, the growth of high quality p-type InN thin films is challenging due to the difficulties compensation of Mg in InN [6] and passivation effect of hydrogen toward magnesium [7]. There are two mechanisms which can cause the passivation of Mg acceptors: one is self-compensation caused by the formation of a deep donor, Mg_{In}-V_N, namely a nearest-neighbor association between Mg acceptor and nitrogen vacancy. The nitrogen vacancy (V_N) which functions as a donor for InN, thus the V_N and Mg acceptors, which are oppositely charged tend to attract each other to form Mg_{In}-V_N. The other mechanism is the hydrogen passivation effect which results in electrically inactive Mg–H complex defects in InN during deposition. The present of too much of Mg-H complex can not only degrade the crystalline quality but also affect the electrical properties of the deposited thin film [8]. Nitrogen (N_2) is a common gas for Mg doped GaN thermal annealing activation, and it has been considered to be capable of compensating nitrogen vacancies during annealing. Therefore, similar approach can be applied to InN:Mg thin films. Nevertheless, sol–gel spin coating growth of Mg doped InN thin films with N_2 thermal treatment remains unexplored and depth investigation is absolutely necessary.

2. Experimental Detail

Commercially available aluminum nitride on silicon [AlN/Si(111)] templates was used. The substrates were cleaned through immersion into the chemical solution of hydrofluoric acid and distilled water (with a ratio of 1:50) for 20 s to remove the native oxide layer. The substrates were then rinsed with distilled water, dried with nitrogen gas, and subjected to plasma treatment to enhance the wetting process. Indium nitrate hydrate powder (99.99% purity) and magnesium chloride 6-hydratre (98% purity) powder were disolved in ethanol (99.7% purity) and subjected to the ultrasonically agitation. The prepared precursor was dropped onto the substrates and rotated for 25 s at 2000 rpm by using spin coater. The spin coating process was repeated for several times until the desired thickness was achieved. Annealing was performed at 300°C for 2 hours to refine and dry the deposited films. Finally, nitridation was performed in a furnace system under ammonia and nitrogen or without nitrogen atmosphere for 45 minutes with a constant flow rate of 400 and 60 sccm, respectively, to transform the deposited film into InN polycrystalline. The furnace was set at 600°C to break ammonia bonds and to generate active nitrogen radicals for chemical reactions. The samples were then cooled to room temperature.

The crystallite quality, crystallographic structure, and growth orientation of the deposited films were investigated through high-resolution X-ray diffraction spectroscopy (HR-XRD, PANalytical X'Pert PRO MRD) analysis with $Cuk_{\alpha 1}$ radiation source of 0.154 nm. Field-emission scanning electron microscopy (FESEM, NOVA NANOSEM 450) was used to study the surface morphologies of the deposited films. Furthermore, energy dispersive X-ray spectroscopy (EDX) was used to analyze the chemical composition of the deposited films. The lattice vibrational properties of the films were evaluated through Raman spectroscopy (Horiba Jobin Yvon HR800UV) with a 514.5 nm Ar ion laser as excitation light source.

3. Results and discussion

The XRD patterns are shown in Figure 1. The polycrystalline structure can be clearly observed for InN: Mg film grown on both samples. Three dominant diffraction peaks can be observed at $2\theta = 29.1°$, $2\theta= 31.1°$ and $2\theta = 33.1°$ in both samples. All the diffraction peaks were ascribed to the formation of wurtzite structure InN (JCPDS file ID 00-050-1239). All the samples somewhat preferred orientation along the axis has been observed similar to the reported work by Lee et al. (2017) [5], Chen and Kuo (2012) [9] and Mann et al. (2008) [10]. Nevertheless, the results of θ-2θ angular scans for InN: Mg film that grows under N_2 gas ambient show highest InN (101) peak intensity with smallest FWHM which indicating a strong orientation and better crystallinity along the axis (refer Table 1). This result suggests that nitrogen addition during the nitridation process favors the preferential growth orientation of InN crystallites along (101) direction. To calculate the average crystallite size of the samples, the Debye-Scherrer equation was used,

$$D = \frac{K\lambda}{B\cos\theta} \tag{1}$$

where D = crystalline size, K = shape factor (0.94), and λ= wavelength of Cukα radiation. The values of crystallite sizes for with and without N_2 gas samples for InN (101) diffraction peak were found to be around 51.72 and 39.14 nm, respectively (refer Table 1). It is observed that the InN:Mg grown in N_2 gas ambient has the largest crystallite growth compare with absent N_2 gas condition. This indicates that N_2 gas has successfully activating the Mg atoms in InN and increase the crystallite growth of InN [11]. Furthermore, lattice parameter a and c values for the films were calculated using following equation,

$$\frac{1}{d^2} = \frac{4}{3}\frac{h^2+hk+k^2}{a^2} + \frac{l^2}{c^2} \tag{2}$$

here d is the inter-planar distance and h, k, l are the Miller Indices. It was also found that lattice parameters (a and c) increase as the InN:Mg thin film grown under N_2 gas condition, as shown in Table 1. From table 1, it can be clearly see that the lattice parameters a and c values for Mg doped InN were smaller than the undoped InN (a = 3.55 Å, c = 5.71 Å) as reported by Amirhoseiny et al. [12]. This phenomenon can be correlated to the compensation of Mg into InN. Since the ionic radius of Mg^{2+} (0.065 nm) is smaller than In (0.080 nm) and N (0.146 nm) the decreasing of the lattice parameter implies the substitution of Mg into In and N lattice site [13]. Hence, the calculated lattice constant smaller than that of undoped InN film, confirming that Mg was introduced into the film and compensated In or N lattice site. On the other hand, the a and c values for InN:Mg grown without N_2 gas condition was found smaller than with N_2 gas ambient. Mg^{2+} and H^+ has ionic radius of 0.065 nm and 0.012 nm, respectively. When there is no assistance from N_2 gas, the substitution of H atom into In and N lattice site occurred and passivation Mg-H or Mg-O complexes formed which cause the decreasing of lattice parameter.

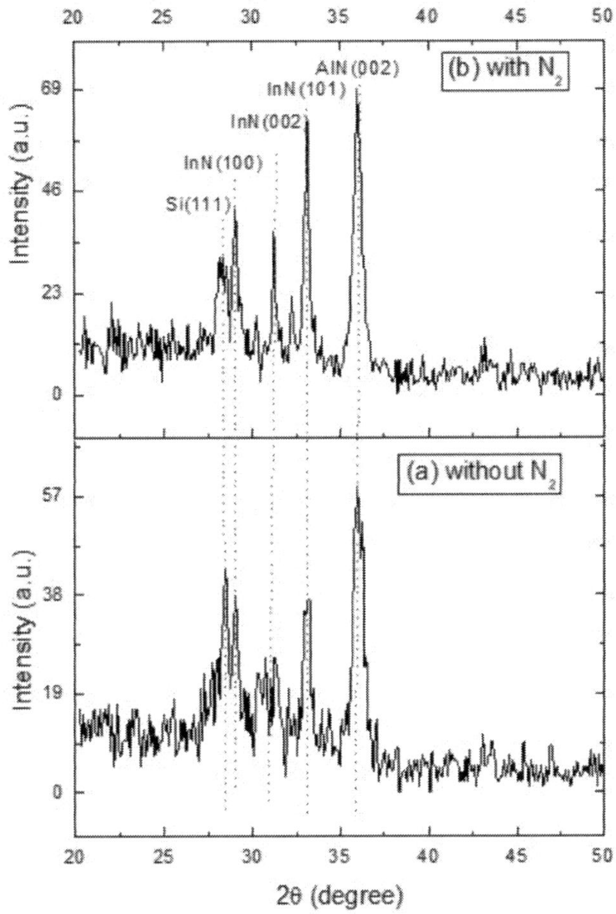

Figure 1: XRD patterns for InN:Mg thin films grow (a) without N_2 gas and (b) with N_2 gas ambient.

Table 1. Full–width at half maximum (FWHM) of InN (101) diffraction peak, crystallite size, lattice constant for InN:Mg thin films grown without N_2 gas and with N_2 gas ambient.

Sample	FWHM (°)	Crystallite size, (nm)	Lattice Parameter (Å)	Reported Lattice Parameter (Å)
With N_2 gas	0.28	51.72	*a*= 3.5475 *c*= 5.6680	a= 3.55 c= 5.71
Without N_2 gas	0.37	39.14	*a*= 3.5475 *c*= 5.6533	

Figure 2 shows surface morphology FESEM images of the InN:Mg thin films grown on AlN/ Si(111) substrate captured with 100kx magnifications. It can be seen that the crystallite size increases after grown under N_2 gas ambient which is consistent with the results indicated in Table 1. Moreover, the images show that both samples were shaped in nearly hexagonal structure; while for sample grown without N_2 gas some flat and plane surface still observed. The absent of N_2 gas might lead to inactivated the Mg atoms in InN, which impede the crystallite growth rate. On the other hand, most of the crystals were form hexagonal structure when grow under of N_2 gas ambient.

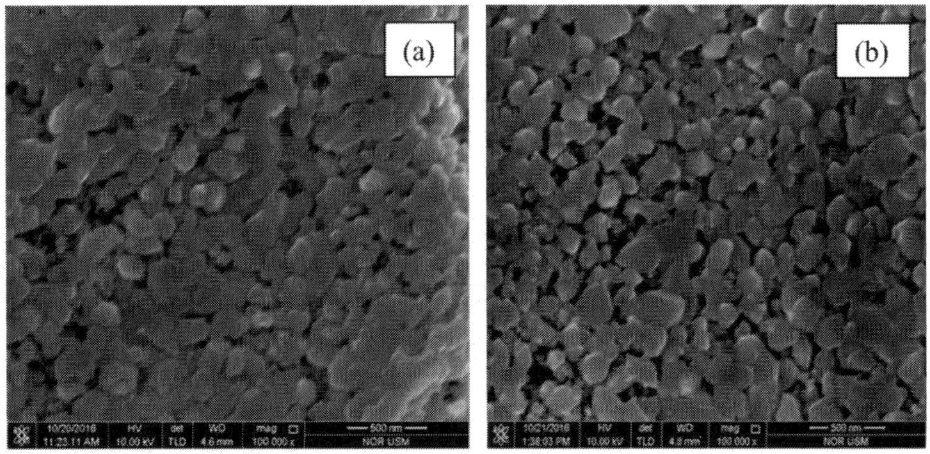

Figure 2. FESEM images for InN: Mg thin films grow (a) without N_2 gas and (b) with N_2 gas ambient.

EDX analysis was conducted on the Mg doped InN thin films. The quantitative amounts of each element as measured with the EDX are shown in Figure 3. From the EDX measurements, it was noticed that the atomic percentage for In, N and Mg atoms were detected. Similar results were obtained from different areas for several times. It can be clearly see that the atomic percentage ratio of indium to nitrogen was almost 1:1 for both samples. This observation confirmed the formation of InN. Other than that, the sample with N_2 gas showed the highest percentage ratio of indium to nitrogen. This indicated that N_2 gas able to promote formation of InN. Moreover, there were existed of 6% of oxygen (O) in the sample grown under N_2 gas ambient, this might be due to oxygen contaminated during cooling sample at atmosphere ambient. On the other hand, for the sample grown without N_2 gas showed higher O atomic percentage. The chemical reaction during the nitridation process are as below [5][14]:

$$2NH_3 \rightarrow N_2 + 3H_{2(g)} \qquad (3)$$

$$NH_{3(g)} \rightarrow H + NH_2 \qquad (4)$$

$$In_2O_{3(s)} + 2NH_{3(g)} \rightarrow 2InN_{(g,s)} + 3H_2O_{(g)} \qquad (5)$$

In-O broken and the N atom tend to react with In to form InN during the nitridation process. Consequently, the released O atom tends to combine with H atom to form H_2O as a side product. Without the N_2 gas during nitridation process, the excessive O atoms tend to stay in furnace and might diffuse into the thin films again which cause the increase of the O atomic percentage in EDX measurement. Therefore, addition N_2 gas during the growth of InN:Mg thin films act as a role to purge out the side product, excessive O atoms and impurities in furnace.

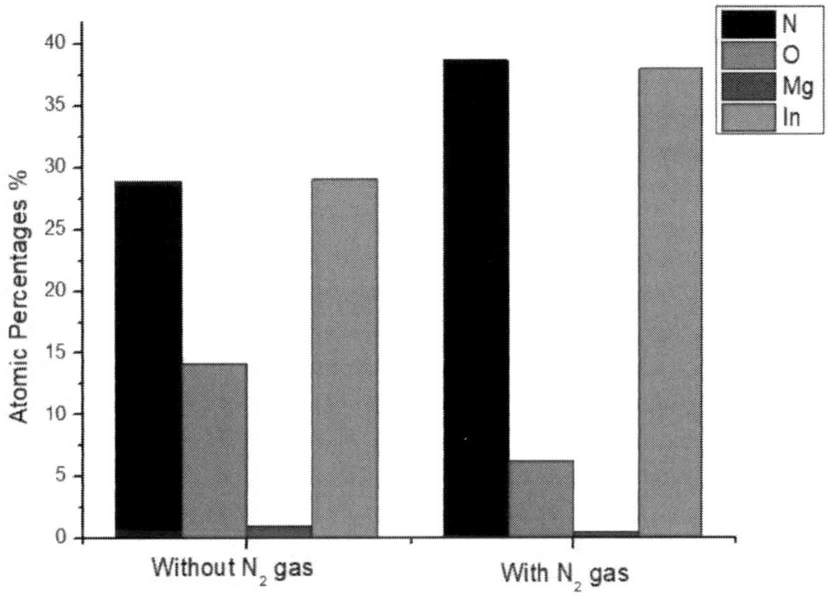

Figure 3. EDX spectra for InN:Mg thin films grown at without N_2 gas and with N_2 gas ambient.

Figure 5 shows the Raman spectra of Mg doped InN thin films with In:Mg thin films grown on AlN/Si(111) template. The Raman measurements were conducted under backscattering geometry $z(x,x)\bar{z}$ configuration, where the z-axis is parallel to the *c*-axis of InN. From the Raman results, the E_2 (high) and A_1 (LO) modes of wurtzite structure InN were observed for the sample grown with N_2 gas which located at 495 cm^{-1} and 576 cm^{-1}, respectively. For the sample grown without N_2 gas, only A_1 (LO) modes was observed. This is mainly due to low crystalline quality of the deposited InN films. For Mg-doped InN thin film grown under N_2 gas ambient, a peak around 564 cm^{-1} was detected. According to Davydov et al.(2008) [15], this mode can be assigned to the local vibration mode (LVM) of Mg-N. To further verify the origin of this peak, the theoretical LVM frequency was calculated by equation as shown in below [13],

$$\frac{\omega_{InN}}{\omega_{LvM}} = \sqrt{\frac{\mu_{LvM}}{\mu_{InN}}} \tag{6}$$

The reduced mass of LVM and InN (μ_{LVM} and μ_{InN}) were calculated using following equation,

$$\mu_{LVM} = \frac{m_{Mg}\, m_N}{m_{Mg} + m_N} \tag{7}$$

$$\mu_{InN} = \frac{m_{In}\, m_N}{m_{In} + m_N} \tag{8}$$

where the m_{Mg} is the atomic mass of Mg (24.305 u), m_N is the atomic mass of N (14.0067 u) and m_{In} is the atomic mass of In (114.818 u). Here, the Mg is assumed to be incorporated on In lattice site and the $\omega_{InN} = \omega[E_1(TO)] = 477$ cm^{-1}[16]. The calculated theoretical value of the LVM for Mg-N is 565 cm^{-1}, which is in good agreement with the experimentally found value of 564 cm^{-1}. These results imply that the Mg atoms were successfully incorporated into the InN. This might be due to the present of N_2 gas has activated the Mg dopant in the InN, which can be seen in figure below [11].

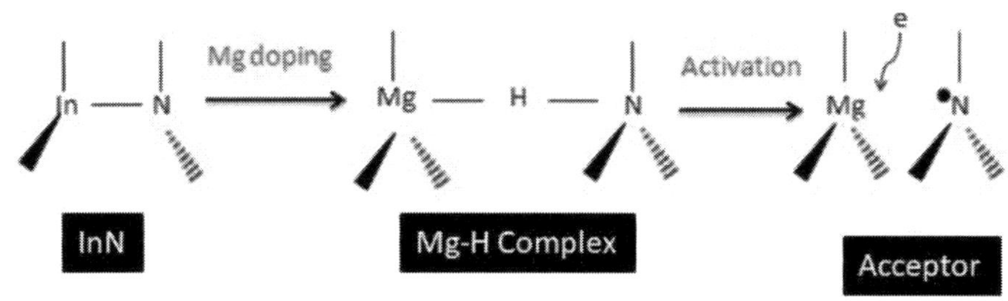

Figure 4. Compensation of Mg into InN.

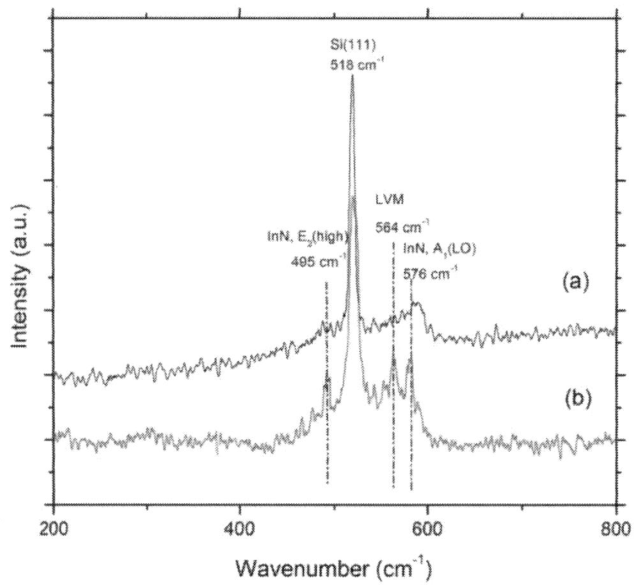

Figure 5. Raman spectra for InN:Mg thin films grown (a) without N_2 gas and (b) with N_2 gas ambient.

4. Conclusion

In this work, Mg-doped InN (In:Mg) thin films were deposited on AlN/Si(111) substrates by sol-gel spin coating method in combination with the nitridation process in different N_2 gas ambients. The XRD results revealed the formation of wurtzite InN:Mg thin films for both samples. The growth of the InN:Mg thin film with additional nitrogen carrier gas resulted smallest FWHM and highest intensity for the InN (101) diffraction peak. FESEM results revealed that the film grown under N_2 gas ambient exhibits largest coalesced island morphology with hexagonal symmetry. The EDX results showed that the InN:Mg film grown under N_2 gas ambient formed with almost 1:1 atomic percentage ratio of indium to nitrogen and possesses small amount of oxygen atomic percentages. Raman results revealed Mg–N mode was detected which indicting the existed of compensation between Mg–InN. In conclusion, further works are needed to improve the crystalline quality of the Mg-doped InN thin films as well as to investigate its electrical properties. Nevertheless, the results obtained here are vital for researcher to explore the sol-gel spin coating technique for the growth of InN:Mg.

Acknowledgement

This work was supported by the Ministry of Higher Education of Malaysia through Fundamental Research Grant Scheme (Account no: 203/CINOR/6711453), Ministry of Science, Technology and

Innovation of Malaysia (MOSTI) through Science Fund (Account no: 305/CINOR/613329), and the Nippon Sheet Glass Foundation Japan (Account no: 304/CINOR/660807/N100). The corresponding author would like to thank the Ministry of Higher Education Malaysia for offering MyMaster scholarship scheme to pursue her studies. Finally, the support from Universiti Sains Malaysia is gratefully acknowledged.

References

[1] Tansley T L and Foley C P 1984 *Electronics Letters* 20 1066-1068.
[2] Anderson P A, Swartz C H, Carder D, Reeves R J, Durbin S M, Chandril S and Myers T H 2006 *Applied Physics Letters* 89 184104-1-3.
[3] Wang X, Che S B, Ishitani Y and Yoshikawa A 2007 *Applied physics letters* 90 201913-1-3.
[4] Fong C Y, Ng S S, Yam F K, Hassan H A and Hassan Z 2015 *Journal of Crystal Growth* 413 1-4.
[5] Lee Z Y, Ng S S and Yam F K 2017 *Surface and Coatings Technology* 310 38-42.
[6] Miceli G and Pasquarello A 2016 *Physical Review B.* 93 165207-1-11.
[7] Alves H, Böhm M, Hofstaetter A, Amano H, Einfeldt S, Hommel D, Hofmann D M and Meyer B K 2001 *Physica B.* 308 38-44
[8] Lu W, Aplin D, Clawson A R and Yu P K 2013 *Journal of Vacuum Science & Technology A: Vacuum, Surfaces, and Films* 31 011502-1-5.
[9] Chen W C and Kuo S Y 2012 *Journal of Nanomaterials* 2012 1-5.
[10] Mann A K, Varandani D, Mehta B R, Malhotra L K, Mangamma G and Tyagi A K 2008 *Bulletin of Materials Science* 31 233-240.
[11] Azimah E, Zainal N, Shuhaimi A and Hassan Z 2015 *Superlattices and Microstructures* 82 592-598.
[12] Amirhoseiny M, Hassan Z, Ng S S and Alahyarizadeh G 2013 *Surface Review and Letters* 20 1350008-1-5.
[13] Long X, Li X, Lin P T, Cheng X W, Liu Y and Cao C B 2010 *Chinese Physics B.* 19 027202-1-4.
[14] Ghazali, N M, Yasui K and Hashim A M 2014 *Nanoscale research letters* 9 685-1-8.
[15] Davydov V Y, A. Klochikhin A, Smirnov M B, Kitaev Y E, Smirnov A N, Lundina E Y, Lu H, Schaff W J, Lee H M, Lin H W, Hong Y L and Gwo S 2008 *Physica Status Solidi (c)* 5 1648-1651.
[16] Kasic A, Schubert M, Saito Y, Nanishi Y and Wagner G 2002 *Physical Review B.* 65 115206-1-115206-7.

IOP Conf. Series: Journal of Physics: Conf. Series **1083** (2018) 012062 doi:10.1088/1742-6596/1083/1/012062

Fabrication and Characterization of Cu-doped ZnO Films Using RF Reactive Magnetron Sputtering

A. S. Yusof[1, a], **Z. Hassan**[2, b]

[1] School of Physics, Universiti Sains Malaysia, 11800 USM, Penang, Malaysia
[2] Institute of Nano Optoelectronics Research and Technology (INOR), Universiti Sains Malaysia, 11800 USM, Penang, Malaysia

[a]opi_yusof@yahoo.com, [b]zai@usm.my

Abstract. ZnO thin films have emerged as an interesting research area owing to its useful properties. Recently, lots of attention have been given to doped ZnO with Cu atom due to its favourable potential in semiconductor devices. Pure and Cu-doped ZnO (CZO) thin films were deposited on the glass, p-GaN/Al_2O_3 and n-GaN/Al_2O_3 substrates using radio frequency magnetron sputtering of Cu/ZnO alloy target with a ratio of 10/90 at 200 °C. The crystal structure, optical properties, surface morphology and electrical properties were investigated by using X-ray diffraction (XRD), ultraviolet-visible (UV-VIS) spectrophotometer, atomic force microscopy (AFM) and Hall measurement with four-point Van der Pauw configuration respectively. XRD analysis showed that single phase ZnO with hexagonal wurtzite structure and c-axis orientation was fabricated. The transmittance of all films deposited on glass in the visible region were more than 85%. The optical band gap of the films were calculated by using transmittance data obtained from UV-VIS spectrophotometer. Optical band gap reduction occurred when Cu is introduced into ZnO. Deposited CZO films showed smoother surface compare with ZnO films. Hall measurement results revealed that CZO film deposited on n-GaN/Al_2O_3 had higher mobility and conductivity than pure ZnO films.

1. Introduction

Gallium nitride (GaN) and zinc oxide (ZnO) have been recognized as semiconductor materials with a wide direct band gap of 3.39eV and 3.37eV respectively [1-2]. However, ZnO has larger exciton binding energy (60meV) compared to GaN (21-25 meV) which gives an advantage to ZnO to have more efficient exciton emission at room temperature. In recent years, ZnO has been deposited on many different types of substrates such as glass, silicon carbide (SiC) and mostly sapphire (Al_2O_3). Even though Al_2O_3 has been widely used as a substrate for ZnO films, the lattice mismatched between them was considerably large (18.4%) that can cause a strain misfit in ZnO films. Thus, to overcome the problem, a new substrate needs to be searched. Among the candidate, GaN was a potential substrate due to smaller lattice mismatched between ZnO and GaN (1.89%). Currently, doped and un-doped ZnO thin films are the alternative material for indium and tin oxide based films have been studied extensively due to their high potential as a transparent conductive oxide (TCO) in optoelectronic devices and flat panel display [3]. Despite all the promising features, the major problem in the development of ZnO was to produce a stable p-type conductivity. ZnO experiences a doping asymmetry limitation problem [4] causing a major step back in the development of ZnO films. N-type conductivity can be easily produced by doping group 13 element such as Aluminum (Al) into lattice size of ZnO [5] but it is challenging to dope a p-type. For

Content from this work may be used under the terms of the Creative Commons Attribution 3.0 licence. Any further distribution of this work must maintain attribution to the author(s) and the title of the work, journal citation and DOI.
Published under licence by IOP Publishing Ltd

application purpose, the electrical, optical and microstructure of ZnO can be changed by introducing an element of impurities into ZnO such as gallium (Ga), titanium (Ti), aluminium (Al), silver (Ag), Cu, tin (Sn) and gold (Au) [6-10]. Among the impurities elements, Cu is a low-cost, non-toxic and naturally abundant element which makes Cu a high potential dopant. Cu has the same radius and electronic shell with Zn, thus making doping easier. Cu also came from group 11 element which can perform as acceptor state in ZnO when doped into Zn interstitial site. Various different thin film deposition techniques have been reported before such as sputtering, pulse laser deposition, molecular beam epitaxy (MBE) and metal organic chemical vapour deposition (MOCVD) [11-14]. The highest quality of ZnO thin film that has been reported was fabricated on Al_2O_3 substrate using MBE and MOCVD techniques [13-14]. To reduce the practical cost and the compatibility for different types of substrate, a low-cost deposition technique such as RF magnetron sputtering is used. It offered high deposition rate, uniform coating and the deposition thickness was easier to be controlled. In recent study, a different sputtering setup have been utilized to deposit Cu-doped ZnO (CZO) thin film such as using RF magnetron sputtering of ZnO ceramic target with Cu foils or chips incorporated in it [15-16] or using RF and direct current (DC) co-sputtering system of ZnO ceramic and Cu metal target [17-18]. In this study, Cu/ZnO alloy target was used to fabricate CZO, which offers higher uniformity dispersion of Cu atom on the substrate surface. The effect of using different substrate in the sputtering process on the microstructure, optical properties and electrical properties were investigated.

2. Experimental details

ZnO and CZO thin films were deposited on glass, p-GaN/Al_2O_3 and n-GaN/Al_2O_3 substrates by using radio frequency (RF) magnetron sputtering (HHV Auto 500) of Cu/ZnO alloy target (99.9% purity) with a ratio of 10/90 respectively at room temperature. The glass substrates were ultrasonically cleaned with ethanol and acetone in a sequence of 5 minutes each then rinsed with deionized water and dried using nitrogen gas. For p-GaN/Al_2O_3 and n-GaN/Al_2O_3 substrates, the cleaning methods were done by immersing the substrates in a boiling aqua regia which is a mixture of nitric acid and hydrochloric acid (HCl: HNO3 = 3: 1) and dilute hydrofluoric acid (HF: H20 = 1: 50) in a sequence of 15 minutes and 10 minutes, then rinsed with deionized water and blown dry using nitrogen gas. Prior to deposition, the target was pre-sputtered for 10 minutes to clean and remove any impurities on the target surface. Deposition thickness of all films were fixed at 200 nm, which were controlled by the deposition time. The details of the sputtering parameters are summarized in Table 1. The crystallization and structural analysis were carried out by using X-ray diffraction (XRD) with generator voltage and tube current of 40 kV and 30 mA respectively, Cu-Kα (λ = 0.1541 nm) with the 2theta scanning angle ranging between 20θ to 60θ. The XRD data were recorded using PANalytical X'PERT PRO MRD PW 3040/60. The transmittance spectra were investigated by using an ultraviolet-visible (UV-VIS) spectrophotometer (Agilent Technology carry series UV-Vis-NIR spectrometer). The surface morphology and roughness of the films were observed by using field emission scanning electron microscopy (FE-SEM, FEI Nova NanoSEM 450) and atomic force microscopy (AFM, Dimension EDGE Bruker). The electrical properties were obtained by using Hall measurement with a four-point Van der Pauw configuration.

Table 1. Sputtering parameters of the experiment.

Base pressure (Pa)	~$2x10^{-5}$
Working pressure (Pa)	~$2x10^{-3}$
Argon flow (sccm)	12
RF power (W)	100
Thickness (nm)	200
Substrate temperature (°C)	200

3. Results and Discussion

Fig. 1 and Fig. 2 shows the XRD pattern of ZnO and CZO thin films deposited on different types of substrate (glass, p-GaN/Al_2O_3, n-GaN/Al_2O_3). For the sample deposited on glass, un-doped ZnO displayed single diffraction peak of ZnO (002) at $2\theta = 34.348°$. Thus, monocrystalline films with c-axis hexagonal wurtzite structure have been fabricated. When Cu was introduced into ZnO structure, a secondary diffraction peak was detected which corresponded to ZnO (100). In the case of films deposited on p-GaN/Al_2O_3 and n-GaN/Al_2O_3 substrates, all films exhibited a single diffraction peak of (0002) that corresponded to ZnO and GaN planes. Both ZnO and GaN (0002) peaks are hard to be distinguished from one another due to their close lattice matching. The deposited films were monocrystalline with strong preferential orientation along c-axis that belongs to the hexagonal wurtzite structure. No Cu related phase was detected within the sensitivity of the instrument. As observed in Fig. 1 and Fig. 2, the (0002) diffraction peak were shifted to a higher angle with a reduced intensity when Cu was introduced into ZnO structure. From the calculation using Bragg formula, $\lambda = 2d \sin \theta$, the plane spacing d_{002} and lattice constant c decreased when Cu was doped in ZnO. The ionic radii of Cu could be assumed to be Cu^0 (0.060nm), Cu^{+1} (0.054nm) or Cu^{+2} (0.060nm). Thus, the smaller size of c suggests that Cu^{2+} ions have become the major substitutional for Zn^{2+} ions that act as interstitials in the crystal lattice of CZO films. This could suppress the intensity of diffraction angle compared to un-doped ZnO films.

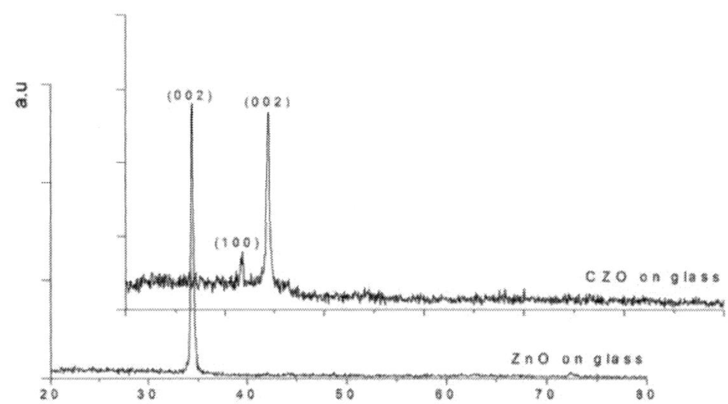

Figure 1. XRD pattern of ZnO and CZO films deposited on glass substrates

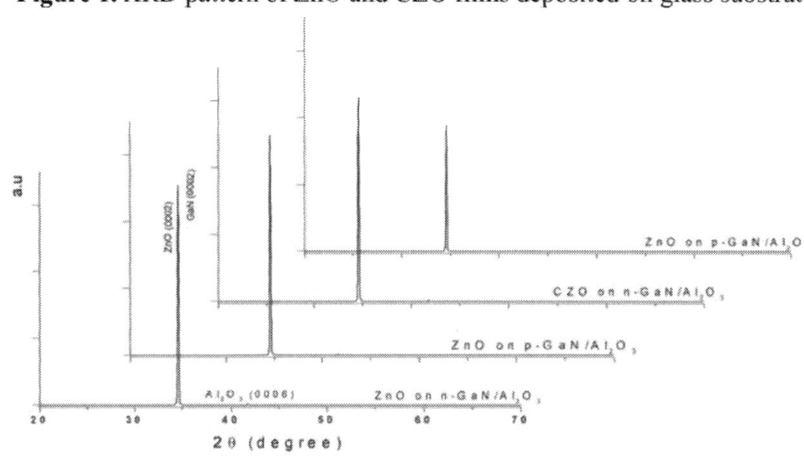

Figure 2. XRD pattern of ZnO and CZO films deposited on p-GaN/Al_2O_3 and n-GaN/Al_2O_3 substrates.

Table 2. Parameters of ZnO and CZO films diffraction peaks.

	(002) peaks					
	ZnO			CZO		
	Glass	p-GaN/Al_2O_3	n-GaN/Al_2O_3	Glass	p-GaN/Al_2O_3	n-GaN/Al_2O_3
2θ (°)	34.348	34.589	34.588	34.435	34.592	34.590
d (nm)	0.2609	0.2592	0.2592	0.2603	0.2591	0.2591
FHWM (°)	0.246	0.133	0.138	0.295	0.115	0.126
D (nm)	35.31	62.52	62.97	29.43	75.90	69.04
c	0.5218	0.5184	0.5184	0.5206	0.5182	0.5182
stress	-0.618	1.012	1.012	-0.050	0.985	0.967

UV-Vis characterization as shown in Fig. 3 were carried out with an aim to determine the optical band gap of the films. The relation of optical band gap, E_g and absorption coefficient are summarized by using the Tauc's relation for direct transition semiconductor [19],

$$(\alpha h\nu) = A(h\nu - Eg)^{1/2} \quad (1)$$

where $h\nu$ is the photon energy, A is an energy independent constant and E_g is an optical energy band gap. The optical band gap was obtained by extrapolating the straight line of $(\alpha h\nu)^2$ versus $h\nu$ graph as shown in Fig. 3(b). Band gap narrowing occurred when Cu was doped into ZnO. The substitution of Cu^{2+} with Zn^{2+} resulted in increasing number of oxygen vacancies due to ionic radii and electronegativity of Cu^{2+}. The band gap narrowing was caused by the defect created by Cu, which induced the Fermi level to move up towards the conduction band [20] as well as the interaction and hybridization exchange of the Cu-3d band and O-2p band [21].

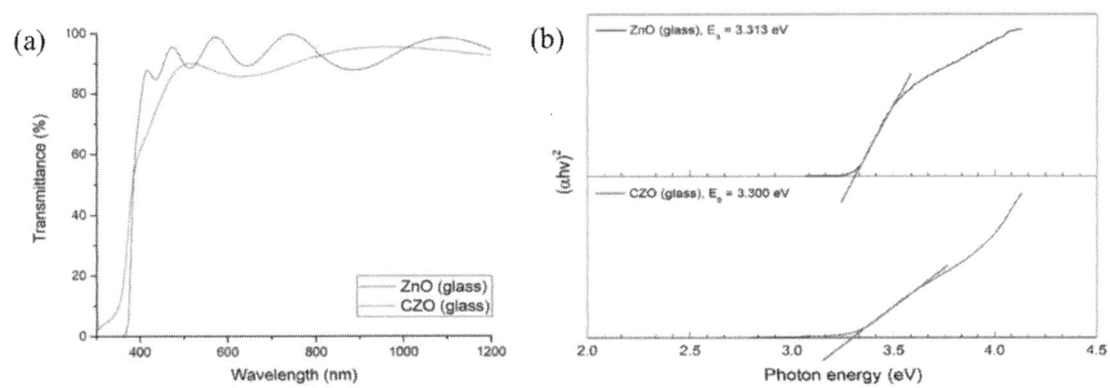

Figure 3. (a) Transmittance spectra at room temperature, (b) Tauc plot and the optical band gap of the films.

The surface morphology of un-doped and Cu-doped films analysis were carried out by using FE-SEM and AFM. Fig. 4 shows the typical FE-SEM image of ZnO deposited on glass and CZO deposited on glass and n-GaN/Al_2O_3. Both CZO films deposited on glass and n-GaN/Al_2O_3 were shown to be dense and had homogenous grain distribution without any precipitation, dislocation, void and pinholes. This suggests that high-quality films have been fabricated. ZnO films are composed of dense rounded grain while CZO films have elongated grain mainly parallel to the substrate surface. Thus it indicates that CZO grain grows c-axis preferentially parallel to substrate surface (c-axis) which is consistent with the XRD results in Fig. 1. Surface roughness analysis was carried out using AFM tapping mode over a

5 μm x 5 μm as shown in Fig. 5 and Fig. 6. As summarized in Table 3, CZO films show smoother surface compared to un-doped films that were calculated by using the software in the AFM system. This is due to nucleation of centres during incorporation of Cu into the ZnO systems. Electrical properties as summarized in Table 4 were determined by using Hall measurement with four-point Van der Pauw configuration at room temperature. The overall electrical properties of CZO films deposited on n-GaN/Al_2O_3 were enhanced compared to the un-doped films due to the improvement of carrier concentration by the Cu-doping related defect (zinc interstitials and oxygen vacancy) in the ZnO system. Many reported that conduction properties of ZnO were primarily caused by oxygen vacancy and Zn interstitial [22-23] contributed by Cu^{2+} ion on the substitutional site of the Zn^{2+} ion. The doped Cu atom act as an acceptor type impurities by supplying holes for each Cu atom [24]. The deteriorating electrical properties were observed for CZO deposited on p-GaN/Al_2O_3. This was due to the self-compensation in CZO films that creates a deep acceptor level, thus trapping electron from the conduction band.

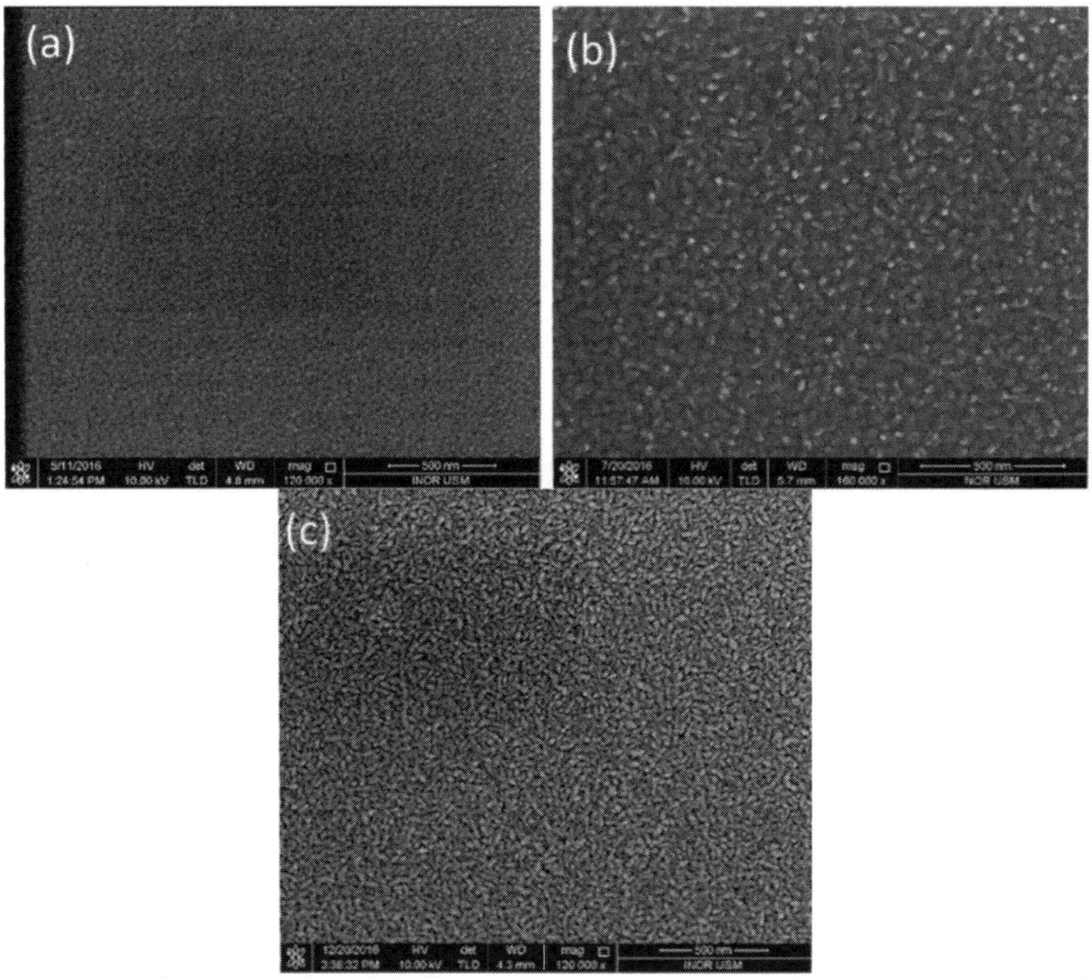

Figure 4. FE-SEM image of; (a) ZnO films deposited on glass, (b) CZO films deposited on glass and (c) CZO deposited on n-GaN/Al_2O_3 substrates.

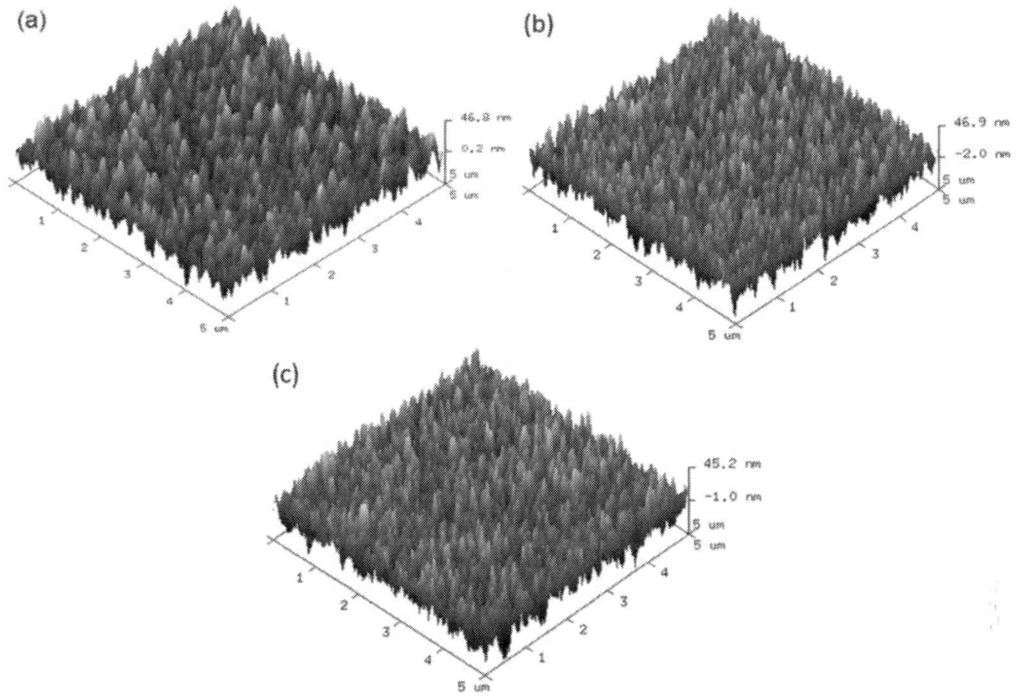

Figure 5. AFM images of ZnO films (5μm x 5μm) deposited on different substrates; (a) glass, (b) n-GaN/Al_2O_3, (c) p-GaN/Al_2O_3.

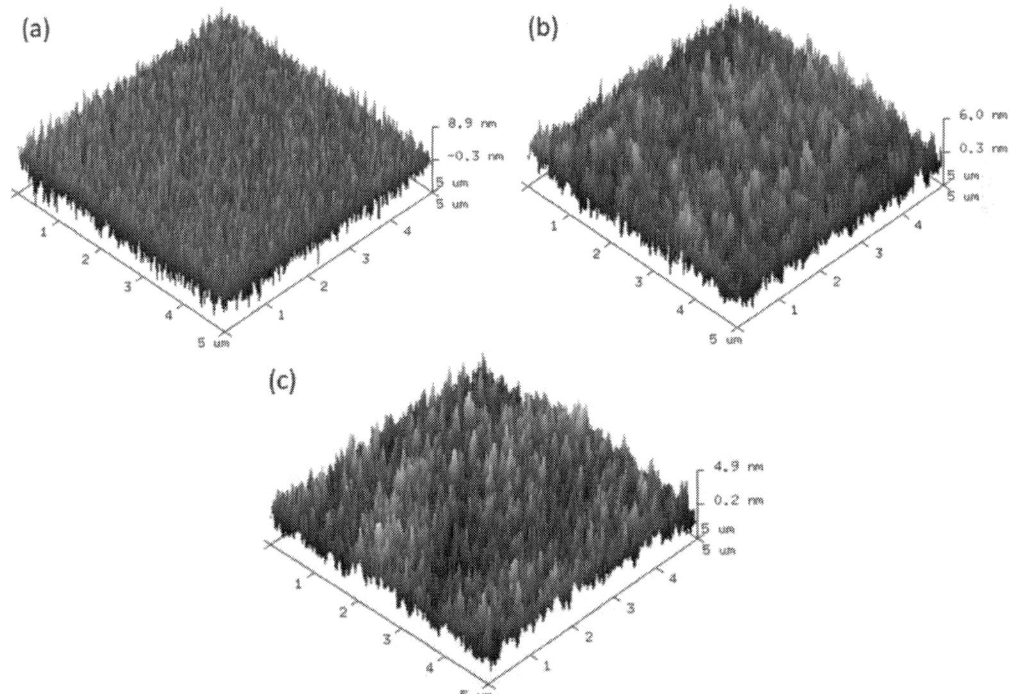

Figure 6. AFM images of CZO films (5μm x 5μm) deposited on different substrates; (a) glass, (b) n-GaN/Al_2O_3, (c) p-GaN/Al_2O_3.

Table 3. Summary of RMS roughness value estimated from AFM.

Substrate	Deposited films	RMS (nm)
Glass	ZnO	13.1
	CZO	2.7
n-GaN/Al_2O_3	ZnO	14.3
	CZO	1.4
p-GaN/Al_2O_3	ZnO	13.4
	CZO	1.3

Table 4. Electrical properties of ZnO and CZO thin films deposited on different substrates.

	ZnO		CZO	
	p-GaN/Al_2O_3	n-GaN/Al_2O_3	p-GaN/Al_2O_3	n-GaN/Al_2O_3
Bulk concentration, (cm^{-3})	-4.61 x 10^{19}	-4.87 x 10^{19}	3.38 x 10^{20}	-1.02 x 10^{20}
Sheet concentration, (cm^{-3})	-1.38 x 10^{15}	-1.51 x 10^{15}	2.72 x 10^{15}	-3.07 x 10^{15}
Sheet resistance, R_{sh} (Ω/sq.)	6.55 x 10^2	6.64 x 10^2	4.72 x 10^2	9.19
Resistivity, ρ (Ω cm)	1.82 x 10^{-2}	1.84 x 10^{-2}	3.50 x 10^6	2.76 x 10^{-4}
Conductivity, σ (1/Ω cm)	54.97	54.21	8.57 x 10^{-6}	3.63 x 10^3
Mobility, (cm^2/Vs)	7.43	6.95	2.22 x 10^{-7}	2.21 x 10^2
Hall coefficient	-0.18	-0.16	0.73	-0.61

4. Conclusion

Monocrystalline CZO thin films deposited on p-GaN/Al_2O_3 and n-GaN/Al_2O_3 substrates have been obtained using RF magnetron sputtering. The effect of Cu doping on the microstructure, optical and electrical properties were investigated. Un-doped and Cu doped ZnO films showed good transparency in the visible region. There was a band gap narrowing in CZO films due to hybridization of the Cu-3d band and O-2p band. RMS roughness value of CZO films was lower than ZnO films. Improved electrical properties were achieved when CZO was deposited on n-GaN/Al_2O_3. P-type conductivity was obtained for CZO on p-GaN/Al_2O_3 but with degrading electrical properties due to self-compensation in CZO films.

Acknowledgements

Support from Universiti Sains Malaysia and LRGS grant (203/CINOR/6720013) are gratefully acknowledged.

References

[1] S Tuzemen and E Gur *Opt. Mater.* 30 292-310
[2] S J Pearton, D P Norton, K Ip, Y W Heo and T Steiner 2004 *J. Vac. Sci. Technol.* B22 932
[3] R B H Tahar, T Ban, Y Ohya and Y Takahashi 1998 *J. Appl. Phys.* 83 2631
[4] V Avrutin, D J Silversmith, H Morkoc 2010 Proc. *IEEE* 98 1269-1280
[5] S H Jeong, B N Park, D G Yoo and J H Boo 2007 *J. Korean Phys. Soc.* 50 622-625
[6] W S Liu, S Y Wu, C H Tseng and C Y Hung 2014 *Thin Solid Films* 570 568-573
[7] C Guillen and J Herrero 2010 *Vacuum* 84 924-929
[8] Y Chen, X L Xu, G H Zhang, H Xue and S Y Ma 2009 *Physica B* 404 3645-3649
[9] Z Jiang, N I Jiamiao, Z Xiujian and X Yuli, J Wuhan 2011 *Univ. Technol.* 26 (3) 388-392
[10] V Perumal, U Hashim, S C B Gopinath, R Haarindraprasad, W W Liu, P Poopalan, S R Balakrishnan, V Thivina l and A R Ruslinda 2015 *PLOS ONE* 10 (12) e0144964
[11] Y M Tao, S Y Ma, H X Chen. J X Meng, L L Hou, Y F Jia and X R Shang 2011 *Vacuum* 85 744-748
[12] S Hayamizu, H Tabata, H Tanaka and T Kawai 1996 *J. Appl. Phys.* 80 787
[13] A E Shaer, A C Mofor, A Bakin, M Kreye and A Waag 2005 *Superlattices Microst.* 38 265

[14] J D Ye, S L Gu, S M Zhu, T Chen, L Q Hu, F Qin, R Zhang, Y Shi and Y D Zheng 2002 J. *Cryst. Growth* 243 151–156
[15] L Ma, S Ma, H Chen, X Ai and X Huang 2011 *Appl. Surf. Sci.* 257 10036-10041
[16] F M Li, C T Zhu, S Y Ma, A M Sun, H S Song, X B Li and X Wang 2013 *Mat. Sci. Semicon. Proc.* 16 1079-1085
[17] S M Chung, J H Shin, J M Lee, M K Ryu, W S Cheong, S H Park, C S Hwang and K I Cho 2011 *J. Nanosci. Nanotechnol.* 11 782-786.
[18] Y Liua, H Liua, Y Yua, Q Wanga, Y Lia and Z Wang 2015 *Materials Letters* 143 319-321
[19] S M Chung, J H Shin, J M Lee, M K Ryu, W S Cheong, S H K Park, C S Hwang, K I Cho 2011 *J. Nanosci. Nanotechnol.* 11 782–786
[20] A Ghosh, N Kumari and A Bhattacharjee 2014 *Pramana – J. Phys.* xx x
[21] P Zhou, H Liu, L Zhang, X Suo, Z Liang, Y Liu, Y Li, Z Jiang, Z Wang 2016 *J Mater Sci: Mater Electron* 27 7822-7828
[22] H Colak and O Turkoglu 2012 *Mater. High Temp.* 29 344-350
[23] K Y Cheong, N Muti, S R Raman 2002 *Thin Solid Films* 410 142-146
[24] M Bedir, M Oztas, A N Yazici, E V Kafadar 2006 *Chin. Phys. Lett.* 23 939

Gafchromic XRQA-2 film for Strontium-90/Yttrium-90 (Sr-90/Y-90) Detection

S Buang[1], E D Subri[1], S Kandaiya[1], N N A N A Razak[1], N Z Yahaya[2]

[1]School of Physics, Universiti Sains Malaysia, 11800 Penang, Malaysia
[2]School of Distance Study, Universiti Sains Malaysia, 11800 Penang, Malaysia

sakinahbuang@gmail.com

Abstract. Gafchromic XRQA-2 radiochromic film is an effective tool for quality assurance and dose assessment in kilovoltage radiotherapy and diagnostic applications. This work is done to study the response of Gafchromic XRQA-2 film for beta particle detection. Materials investigated are the beta particle source, Sr-90/Y-90 and Gafchromic XRQA-2 film. The film is exposed to Sr-90/Y-90 for different duration of time and scanned with flatbed scanner. The film is analyzed by using a ImageJ software in the red mode. The result shows that the net reflected optical density (Net ROD) increased linearly with increase in time. The Net ROD and time integrated specific surface activity increased linearly with time. The Net ROD for Sr-90/Y-90 is from 0.00 to 0.13. Thus, the Net ROD is highly related to the duration of time exposed to the beta source.

1. Introduction

Gafchromic XRQA film is a radiochromic film designed specifically to be used in radiology and dosimetry applications. It is a reflective type of films which change in color upon irradiation and does not require any chemical process for its development. The change in the optical density (OD) of the film is directly proportional to the absorbed dose of ionizing radiation. XR-QA2 Gafchromic film model is a new successor to the previous XR-QA film model. This film consists a radiation sensitive layer 25 μm thick and attached to the 97 μm thick white reflective polyester surface layer. On top of the sensitive layer, a 20 μm thick pressure sensitive adhesive which a 97 μm protective yellow polyester is over-coated [1]. The atomic composition of the active layer of the film is made up of H, N, O, C, Li, Br, Bi and Cs [2]. The XR-QA2 film contains Bi2O3 in its sensitive layer which aids to boost the photoelectric effect at low energies. The addition of bismuth lowered the otherwise noteworthy energy dependent response of XR film models [3]. The dose range for this XR-QA2 film is 0.1-20 cGy [5]. Figure 1 shows the layers of the Gafchromic XRQA-2 film. The numerous advantages of the radiochromic film, which have high spatial resolution, relatively small energy dependent response [4,5] near to tissue-equivalence and high dose distribution measurement have make it a powerful dosimetry tool for radiotherapy treatment verification and quality assurance [1].

Beta particle is emitted during a radioactive process called beta decay. Beta ray is a high speed of electron or positron emitted in the radioactive decay of an atomic nucleus. When a radioactive nucleus undergoes beta decay, the daughter nucleus contains the same number of nucleons, A as the parent nucleus, but changed by 1 in atomic number, Z which means the number of protons has changed [6]. Beta source, Strontium 90 (Sr-90) disintegrates to the ground state of Yttrium 90 (Y-90) by emitting

Content from this work may be used under the terms of the Creative Commons Attribution 3.0 licence. Any further distribution of this work must maintain attribution to the author(s) and the title of the work, journal citation and DOI.
Published under licence by IOP Publishing Ltd

beta minus (β-1) radiation. Y-90 is then decay to Zr-90 by β-1 with energy of 2.28 MeV. Thus, the maximum beta energy of this radioactive substances is 2.28 MeV. 100% of beta is emitted from this source. There is no gamma and alpha rays emitted from this source. The physical half-life is 28.2 years. Strontium 90 should be shielded with Plexiglas with a thickness of 9.2 mm. Figure 2 shows the decay scheme of the Strontium 90. Special precautions need to be taken during handling this source. The use of tools such as tongs, forceps and plastic books is crucial for indirectly handling unshielded sources [7]. The Strontium-90 source that is available in the Universiti Sains Malaysia (USM) labhasory have an initial activity of 5μCi. The source was bought in 1971.

A lot of studies have been conducted to measure the beta particles detection. However, the problem arises where a specific beta detector is not available to detect the beta particles. Therefore, this study is done to measure the beta particles from beta emitting nuclides using Gafchromic XRQA-2.

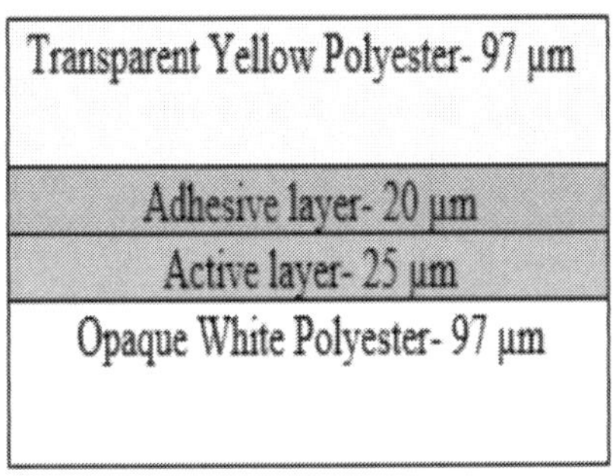

Figure 1. Gafchromic XRQA-2 film composition.

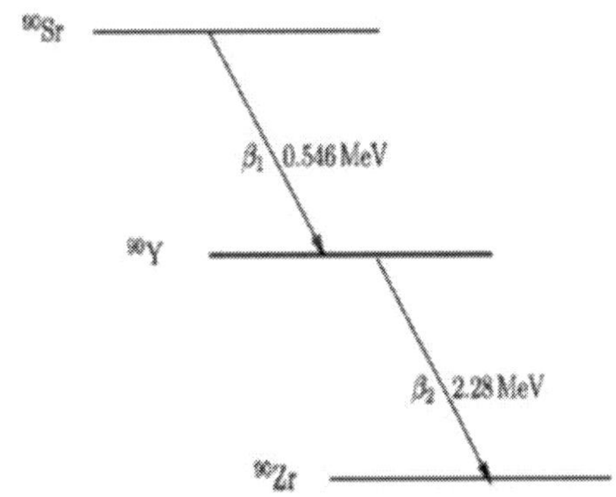

Figure 2. Decay scheme of Sr-90.

2. Materials and Methods

2.1. Film irradiation

Beta source Sr-90/Y-90 is placed on the cardboard washer of 2 mm thick to prevent direct contact of the source with the film as shown in Figure 3. The film size is 2 cm × 3 cm dimension. The purpose of the cardboard washer is to prevent the crosswire formation on the film when the Sr-90 is placed on the film. The formation of the crosswire on the film affect the mean pixel value reading. The film is exposed for 16 hours. After 16 hours, the film is observed to see whether there are any color changes on the film. The procedures are repeated by changing the time of the exposure to 24 hours, 36 hours, 48 hours, 72 hours and 96 hours using different films.

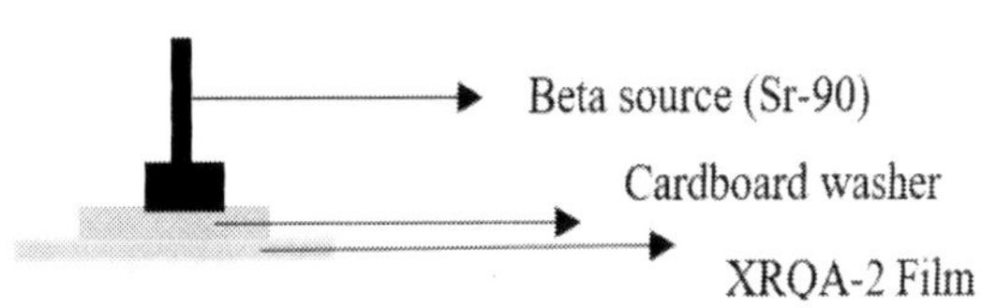

Figure 3. Schematic diagram of Sr-90/Y-90 irradiation on XRQA-2 film.

2.2. Film scanning and analysis

After the irradiation of beta source to the film, the film is left for 24 hours before scanning procedure can be done. The polymerization process in radiochromic film lasts for extensive time after irradiation [8,9,10]. The significance of the continued polymerization is that the optical density (OD) grows over time. Therefore, films should be left for at least 6 hours and preferably 24 hours before scanning. This is to minimize uncertainties related with post-irradiation coloration in both the calibration films and experimental films. The film should be scanned after approximately the same time interval following irradiation [11]. In this study, the film is scanned using Epson Perfection V700 professional flatbed scanner. The scanner is set with resolution of 72 dpi, image pixel of 48-bits, reflective mode is chosen since it contains 16 bits per colour mode in red, green and blue (RGB) mode which sufficient for the analysis [2]. The scanner is warmed up several times without any object placed in the scanner bed before the actual scanning of the film [12]. The film images analyzedlysed with Image-J using 16-bit images in TIFF format [2]. These images were analysed using the red mode in the ImageJ software as it gives a higher mean pixel value compared to the green and blue mode [13].

The main quantities that can describe the response of XRQA-2 Gafchromic film to radiation are the mean pixel value (MPV), net reflectance optical density in equation 1 and reflectance in equation 2. The base of XRQA-2 film is opaque white polyester. For opaque films, the quantity of choice would be reflectance [1]. The mean pixel value is a quantity which the value is obtained directly from the ImageJ software.

$$\text{Net ROD} = \log\left(\frac{MPV_u}{MPV_t}\right) \quad (1)$$

Where MPVu is the mean pixel value of the reflected intensity of unexposed film where the maximum pixel value is found and MPVt is the mean pixel value of the reflected intensity of irradiation [13].

$$\text{Reflectance} = \frac{MPV_t}{2^{16}} \quad (2)$$

Orientation and positioning for film scanning process is crucial. The artifacts associated with the radiochromic films with flatbed scanner are known as the orientation effect. The film is scanned in landscape mode because of the lateral response artifacts on Charge-Coupled Device (CCD) scanner is smaller compared to portrait orientation. It is recommended that the same orientation be used for Gafchromic EBT film analysis for accurate film dosimetry apply to Gafchromic XRQA film [13].

3. Results and discussion

3.1. 5μCi Strontium 90/Yttrium 90

Table 1 shows the main quantitate response of beta particle from Strontium 90/Yttrium 90 on Gafchromic XRQA-2 film in terms of mean pixel value, reflectance and net reflectance optical density. According to Figure 4, the MPVt is exponentially decreased as the time of irradiation increased. On contrary, Figure 5 shows that Net ROD is linearly increased with the irradiation time. The Net ROD is a quantity that shows the difference in pixel value before irradiation and after irradiation where the film changes its colour upon irradiation. The higher values of Net ROD show higher degree of film blackening due to the absorption of beta particle.

Table 1. Variation of Reflectance, Net ROD and MPV with irradiation time.

Time (hours)	Average MPV_t	Average MPV_t	Reflectance	Net ROD
0.00	0.00	65073	1.00	0.00
16.00	61879	65535	0.94	0.02
24.00	59626	65073	0.91	0.04
36.00	56701	65535	0.87	0.06
48.00	52361	65076	0.80	0.10
72.00	50703	64947	0.77	0.11
96.00	48240	65507	0.74	0.13

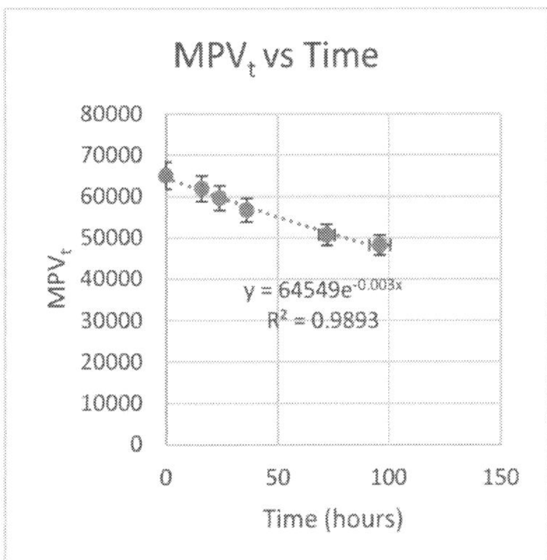

Figure 4. The MPV value with respect to irradiation time.

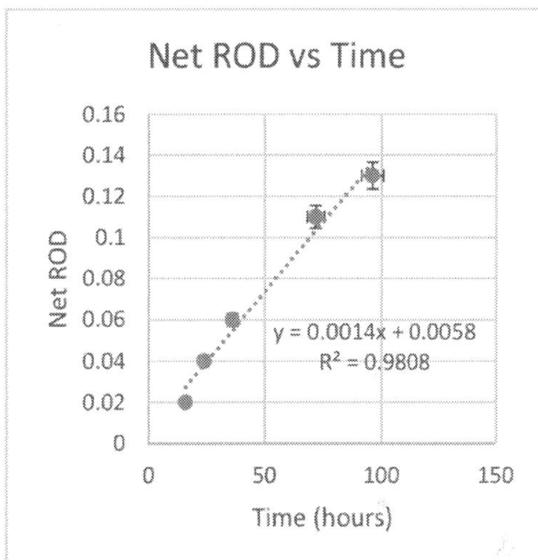

Figure 5. Response of MPVt and Net ROD against irradiation time.

Another quantity that can be analysed from the film which is the time-integrated specific surface activity (A) which is given by equation 3 [14]. This quantity describes the distribution of the source in isotropically dimension at a fixed area.

$$A\ (\text{kBq cm}^{-2}\ \text{h}) = Qt/\pi r^2 \quad (3)$$

Where Q is the source activity, r is the source radius, t (hour) is the exposure time and πr^2 is the area of the circle.

The current activity of Sr90 is 1.65 μCi or 61.05 kBq. The area of the exposed film is 0.19 cm^2 with a radius of 0.25 cm. Table 2 shows the variation of A with irradiation time and Net ROD. From the Figure 6, it can be concluded that the Net ROD is linearly increased to A with a regression coefficient of 0.9808.

Table 2. Variation of A and Net ROD with irradiation time.

Time (hours)	A (kBq cm^{-2} h)	Net ROD
0.00	0.00	0.00
16.00	5141.05	0.02
24.00	7711.58	0.04
36.00	11567.37	0.06
48.00	15423.16	0.10
72.00	23134.74	0.11
96.00	30846.32	0.13

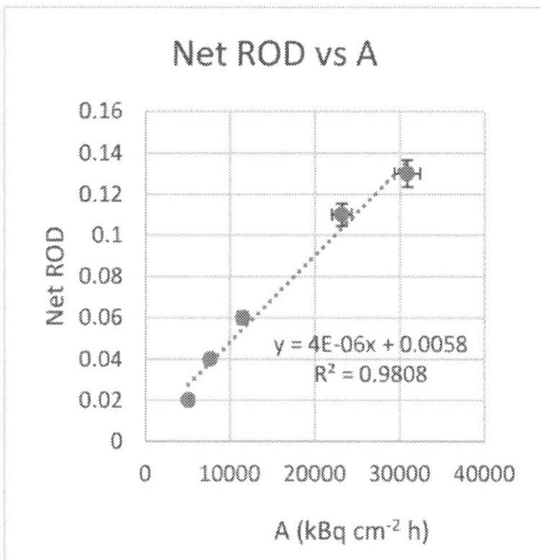

Figure 6. Net ROD against A.

4. Conclusion

This project shows that Gafchromic XRQA-2 radiochromic film is sensitive to beta particles. The MPVt and Net ROD shows the exponential relationship against time, whereas the relationship between Net ROD and time-integrates specific surface activity shows a linear relationship with respect to irradiation time.

Acknowledgements

This work was funded under Fundamental Research Grant Scheme, FRGS, Ministry of Higher Education (203/PJJAUH/6711489). This work was also conducted under the Grant No: 304/PFIZIK/6313323 support from Universiti Sains Malaysia.

References

[1] Devic S, Tomic N and Lewis D 2016 *Physica. Medica* 22 541-556
[2] Ahmed A A., Abbas S., Kandaiya S, Nik Noor Ashikin Nik Abd Razak N N A and Qaeed M A 2017 *Appl. Radiat. Isot* 129 130-134
[3] Tomic N, Quintero C, Whiting B R, Aidelaijan S, Bekerat H and Liang L 2014 Characterization of the calibration curves and energy dependence GafchromicTM XR-QA2 model based radiochromic film dosimetry system *Med. Phys* 41:062105
[4] Butson M J, Cheung T, and Yu P K N 2006 *Appl. Radiat. Isot* 64 602

[5] Rink A, Vitkin I A and Jaffray D A 2007 Energy dependence (75 kVp to 18 MV) of radiochromic films assessed using a real-time optical dosimeter *Med.Phys* 34:458-63
[6] Serway R A and Jewett J W 2014 *Physics for Scientists and Engineers with Modern Physics 9th Edition* (Singapore:Cengage Learning Asia Pte Ltd) p1400
[7] Delacroix D, Guerre J P, Leblanc P and Hickman C 2002 *Radionuclide and Radiation Protection Data Handbook 2nd Edition 2002*. (England: Nuclear Technology Publishing) p 80
[8] Niroomand R A, Blackwell C R, Coursey B M,Gall K P,Galvin J M, McLaughlin W L, Meigooni A S, Nath R, Rodgers J E and Soares C G 1998 *Med. Phys 25 (11)* 2093-2115
[9] Ali I, Costescu C, Vicic M, Dempsey J F and Williamson J F 2003 *Med.Phys 30 (8)* 1958-1967
[10] Cheung T, Butson M J and Yu P K 2005 *Phys. Med. Biol 50 (20)* N281-285
[11] Williams M. and Metcalfe P 2011 *Radiochromic film dosimetry and its applications in radiotherapy. 4th SSD Summer School: Concepts and Trends in Medical Radiation Dosimetry* (Wollongong: AIP) p 75-99
[12] Ferreira B C, Lopes M C and Capela M 2009 Evaluation of an Epson flatbed scanner to read Gafchromic EBT films for radiation dosimetry *Phys. Med. Biol* 2009:54 1073-85.
[13] Alnawaf H, Butson M J, Cheung T and Yu P K N 2010 *Physica. Medica* 26 216-219.
[14] Mukherjee B, Gholami Y H, Bhonsle U, Hentschel R and Khachan J 2015 A unique alpha dosimeter technique using Gafchromic EBT3 film and feasibility study for an activity calibrator for alpha-emitting radiopharmaceuticals *Br. J. Radiol* 2015;88(1056):20150035

Micromechanical Properties of Solder Joint Towards Air Blast Exposure

W.Y.W Yusoff[1,a*], N.S. Safee[1,b], A. Ismail[1,c], M.A. Bakar[2,d] and A. Jalar[2&3,e]

[1]Center for Defence Foundation Studies, Universiti Pertahanan Nasional Malaysia, Kem Sg. Besi, 57000 Kuala Lumpur, Malaysia.
[2]Institute of Microengineering and Nanoelectronics, Universiti Kebangsaan Malaysia, 43600 Bangi, Selangor, Malaysia.
[3]School of Applied Physics, Faculty of Science and Technology, Universiti Kebangsaan Malaysia, 43600 Bangi, Selangor, Malaysia.

[a*]yusmawati@upnm.edu.my,[b]nshafiqasafee@gmail.com,[c]ariffin@upnm.edu.my,
[d]maria@ukm.edu.my, [e]azmn@ukm.edu.my

Abstract. In modern war fighting and conflict prevention, the important of electronics has been and continues to be the fastest moving in technology areas. The key process in industrial manufacture of electronic device is soldering. Sn-Ag-Cu solder have been recommended to replace Sn-Pb solder in the chip-joint process because of low cost and large volume consumer product. In order to study the micromechanical properties of SAC 0307, the blast test was conducted using 500 g and 1000 g of trinitrotoluene (TNT) and the samples were fixed to two meter distance. The effect of micromechanical properties of SAC 0307 were investigated using nanoindentation technique. The hardness and reduced modulus of SAC 0307 were decreased while the plastic work increased when the source of blast increased. The characterization especially after air blast exposure in order to accurately predict the reliability of solder joints there were very large discrepancies in both the tensile and creep properties provided in Sn-Ag-Cu current databases.

1. Introduction

In modern electronics packaging, the system interacts with users, other systems, the platform on which it operates, the signal or data acquisition, dispensing, processing and actuation functions while performing that electronics makes possible. These functions support communications, weapons delivery, electronic warfare, surveillance, platform control, logistics, associated with modern military forces and their missions. In order to protect environment, the adoption of lead free solder has become unavoidable trend among researcher. However, the properties of lead-free solders are stiffer, brittle, feeble to dynamic loads rather than solders containing lead, which are frequently applied in portable electronic devices during normal or extreme conditions [1]. Therefore, any lead-free solder alternative for military application must conform to this broad range of properties and characteristics to be accepted as suitable and reliable candidate to replace the eutectic Sn-Pb solder. Specifically, solder joints of high-density, high-performance electronic packages are expected to withstand increasing mechanical, electrical, and thermal loads, which can cause serious reliability issues and cases of failures during manufacturing or while in operation [2]. Lee & Mohamad [3] reported that Sn-Ag-Cu

Content from this work may be used under the terms of the Creative Commons Attribution 3.0 licence. Any further distribution of this work must maintain attribution to the author(s) and the title of the work, journal citation and DOI.
Published under licence by IOP Publishing Ltd

alloys is the most promising lead-free solder candidates due to their relatively low melting temperature, superior mechanical properties, and good compatibility with other components. Most studies of SAC solder alloys cover reliability, solderability, intermetallic compound (IMC) evaluation by shear strength behaviour, tensile and wettability properties or electromigration behaviour as well as at high current density and high temperature [4].

Nanoindentation has received an attention for measuring the hardness and Young's modulus of bulk materials and thin films microstructural phases also as valuable technique for penetrating the mechanical behavior of materials at narrow areas. According to Deng et al. [5], analysis of nanoindentation includes internal friction and damping properties such as activation energy, stress exponent for creep, the storage and loss modulus, and fracture toughness of brittle materials.

However, the study on micromechanical properties of solder alloy towards blast wave was not fully discover yet. In this research, trinitrotoluene (TNT) was used as the source of exposure. The explosive charge was detonated in the air, the blast waves propagated spherically outwards impinged directly onto the sample without prior interaction with other obstacles or the ground. The objective of this work is to investigate the micromechanical properties of SAC solder towards blast exposure using nanoindentation technique.

2. Materials and Methods

2.1 Sample Preparation

Lead free solder paste, SAC 0307 consist of Sn 99%, Ag 0.3% and Cu 0.7% was supplied by Redring Solder (M) Sdn. Bhd. Firstly, the SAC 0307 solder paste was deposited on the printed circuit board (PCB) using a stencil and undergo reflow soldering process at 215°C for 8 seconds. Then, the samples were exposed to TNT equivalent with 500 g and 1000 g and were placed at two meter distance from the blast sources and non-exposure sample was used as control. Prior the blast test, the SAC 0307 sample were covered with satin. After the exposure, the test board were cut approximately 0.5 cm × 0.4 cm × 0.2 cm dimension undergo cross-sectioned for cross-sectioned for standard metallographic technique such as (1) mounting using mixed of hardening resin powder and epoxy resin liquid, (2) grinding with silicon carbide abrasive paper from 400 until 1200 grits and (3) polished with DP Nap polishing cloth, liquid lubricant and 1μm of diamond spray to prepare the cross-section for nanoindentation test [6].

2.2 Nanoindentation Test

Nanoindentation test was carried out at room temperature using nanoindenter machine (Micro Materials, NanoTest^{TM}) with Berkovich tip as shown in figure 1. The indentation test was performed at the middle of solder cross section. A constant of loading and unloading rate of 0.5 mN/s was applied to the sample surface until reached at 10 mN (maximum load). The dwell time was 30 seconds at maximum load followed by the unloading process [7]. The graph from this experiment shows the hysteresis of load versus depth. This hysteresis consists of three segments, which were loading, dwell time at target load, and the unloading processes. The loading curve is a function of loading rate or strain rate. The unloading curve is tightly related to elastic recovery which used power-law model to the examination of reduced modulus. The dwell time at target load was used to investigate the creep deformation at a certain temperature.

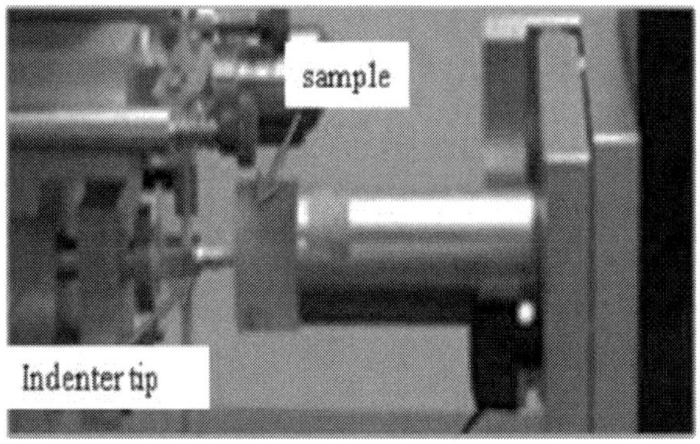

Figure 1. Nanoindenter machine

3. Results and Discussion

During indentation process, the load (*P*) and depth (*h*) was recorded concurrently. The *P-h* curves of SAC 0307 solder with different blast source are plotted as shown in figure 2. This *P-h* curve reflects the mechanical behaviour of solder [8]. From this figure, the SAC 0307 solder with 1000 g blast source displayed the presence more pop-in (indicates by arrow) as compared to control and 500 g blast source. The pop-in event indicates the local discontinuity during loading process. The shift from elastic to plastic deformation of the SAC 0307 solder can be observed at the very earliest stages (initial pop-in event) of the mechanical contact known as 'incipient plasticity'. The following pop-ins related to the microstructure of materials [9].

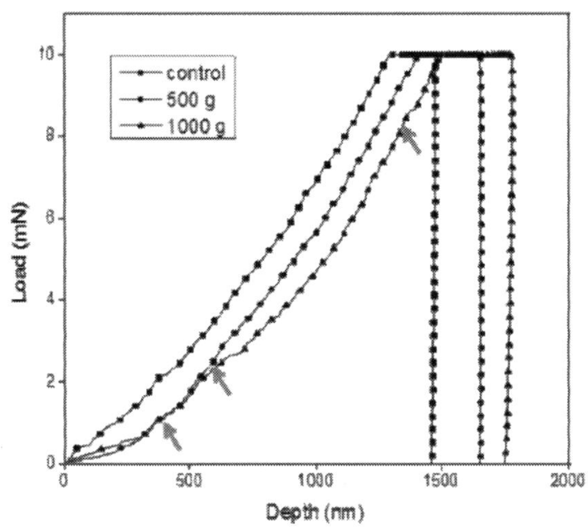

Figure 2. *P-h* profile of SAC 0307 solder on PCB with different blast source

Figure 3 shows the relationship between hardness and plastic work of solder joint on PCB with different weight of blast source (control, 500 g and 1000 g). It is observed that the hardness of SAC 0307 decrease while the plastic work increased when the weight of blast source increased. The migration of interstitial atoms and lattice vacancies along the gradient of a grain boundary cause deformation when presence of tension or compression pressure in reversed directions. These lattice defects tend to move in whichever direction will relieve the imbalance of pressure, and cause creep deformation due to under pressure [10]. The brittleness of a material may be gain from the value of hardness. With an increase in the weight of blast source, the hardness was decreased. This hardness is in line with the result from the indentation depth indicates that the solder material is becoming soft after exposure to blast source. Softening behavior in the metal usually occur due to changes in the microstructure. However, the accurate measure of hardness should be avoid fracture events that may lead to an indentation size effect in order to estimate a yield stress [11].

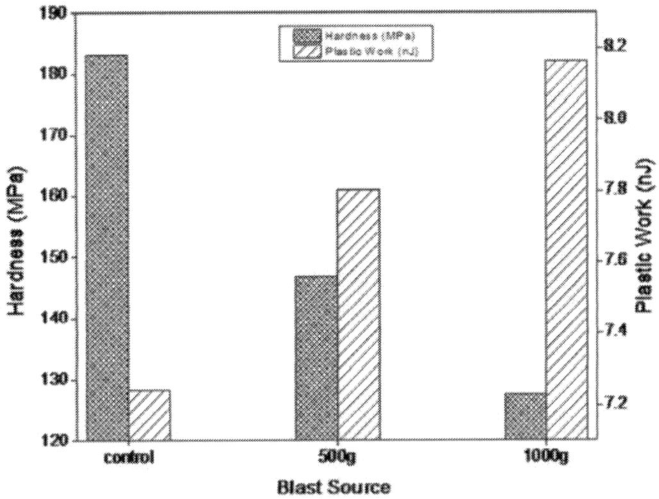

Figure 3. Relationship of Hardness and Plastic Work of SAC0307 with different blast source

The reduced modulus and elastic work relationship is shown in figure 4. The reduced modulus (atomic binding) of samples decreased when the weight of blast source increased. The elastic work for before blast and after 500 g decreased but when the blast source increased to 1000 g the elastic work increased. Cheng et al. [12] stated that the elastic work of Sn-0.3Ag-0.7Cu (SAC 0307) shows higher bulk compliance and higher plastic energy dissipation compare to SAC 405 or SAC 305. Regardless, the consequent of a lower Ag content will rising the primary b-Sn phase (large b-Sn grains) and degradation of thermal mechanical fatigue. Thus, Sn-0.3Ag-0.7Cu solder have lower tensile strength and elastic modulus compare to higher of Ag content solder [13]. Exposure to blast resulted microstructure change as shown in the micromechanical result using nanoindentation test.

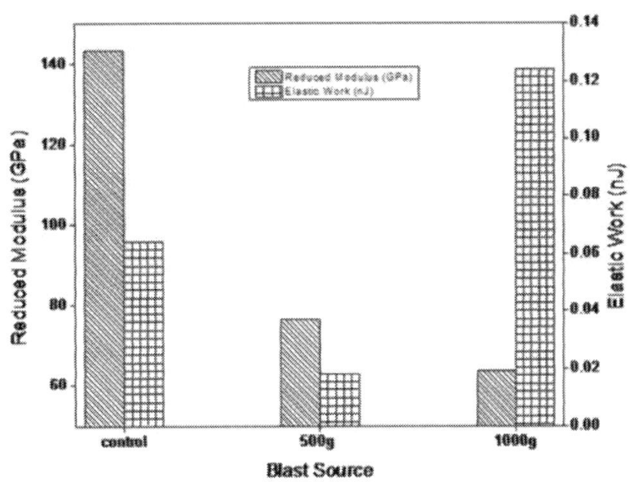

Figure 4. Relationship of Reduced Modulus and Elastic Work of SAC 0307 with different blast source

4. Conclusion

The micromechanical properties of SAC 0307 solder on printed circuit board with different blast source has successfully investigated using nanoindentation technique. The hardness and reduced modulus values decreased with the increment of blast source, while the plastic and elastic work increased after exposure to high blast source. As subjected to blast, the solder become softer. Blast exposure resulted the changes in the microstructure of the solder materials.

Acknowledgement

The authors are gratefully acknowledged the financial support from FRGS/1/2015/SG06/UPNM/03/1, Universiti Kebangsaan Malaysia (UKM) and Redring Solder (M) Sdn. Bhd. for research materials and collaboration work.

References

[1] Y. Lai, P. Yang, C. Yeh, Experimental studies of board-level reliability of chip-scale packages subjected to JEDEC drop test condition, Microelectron. Reliab. 46 (2006) 645–650.

[2] A. Fortier, M. G. Pecht, A perspective of the IPC report on lead-free electronics in military/aerospace applications,Microelectron. Reliab. 69 (2017) 66-70.

[3] L. M. Lee, A. A. Mohamad, Interfacial reaction of Sn-Ag-Cu lead-free solder alloy on Cu: A review, Advances in Mater. Sci. Eng. (2013), ID123697, 11 pages.

[4] B. Medgyes, D. Rigler, B. Illés, G. Harsányi, L. Gál, Investigating of Electrochemical Migration on Low- Ag Lead-Free Solder Alloys, 18th International Symposium for Design and Technology of Electronic Packaging (2012) 147-150.

[5] X. Deng, N. Chawla, K.K. Chawla, M. Koopman, Deformation behavior of (Cu, Ag)-Sn intermetallics by nanoindentation,Acta Materialia, 52(14) (2004) 4291-4303.

[6] M. A. Bakar,A.Jalar, A. R. Daud, R. Ismail, N. A. C. Lah, N. S Ibrahim, Nanoindentation approach on investigating micromechanical properties of joining from green solder materials. Sains Malaysiana45(8) (2016) 1275–1279.

[7] M. A. Bakar, A. Jalar, R. Ismail, A. R. Daud, Nanoindentation of Sn3.0Ag0.5Cu/ENIG Solder Joint after High Temperature Storage, Materials Science Forum 857(3)(2016) 40–43.

[8] C. A. Schuh, Nanoindentation Studies of Material, Materials Today 9(5)(2006) 32-40.

[9] B. Yang, H. Vehoff, Grain size effect on the mechanical properties of Nanonickel examined by nanoindentation, Mat. Sci. Eng. A-Struct 400-401(2005) 467-470.

[10] H.Ma, J. C. Suhling, A review of mechanical properties of lead-free solders for electronic packaging, J. Mater. Sci. 44 (2009)1141–1158.

[11] R.R.Chromik, R. P.Vinci, S. L. Allen, M. R Notis, Measuring the Mechanical Properties of Pb-Free Solder and Sn-Based Intermetallics by Nanoindentation. JOM 6 (2003) 66-69.

[12] F. Cheng, F. Gao, J. Zhang, W. Jin, X. Xiao, Tensile properties and wettability of SAC0307 and SAC105 low Ag lead-free solder alloys. J. Mater. Sci.46(2011) 3424–3429.

[13] Y. Tang, S. M. Luo, W. F. Huang, Y. C. Pan, G. Y. Li, Effects of Mn nanoparticles on tensile properties of low-Ag Sn-0.3Ag-0.7Cu-xMn solder alloys and joints. J. Alloy. Compd. 719(2017) 365–375.